volume I

MECHANISM DESIGN
Analysis and Synthesis
second edition

ARTHUR G. ERDMAN
Professor
of Mechanical Engineering
University of Minnesota

GEORGE N. SANDOR
Research Professor
of Mechanical Engineering
University of Florida

PRENTICE HALL, Englewood Cliffs, New Jersey 07632

Library of Congress Cataloging-in-Publication Data

ERDMAN, ARTHUR G.
  Mechanism design: analysis and synthesis/Arthur G. Erdman,
George N. Sandor. — 2nd ed.
    p.   cm.
  Includes bibliographical references.
  ISBN 0-13-569872-3 (v. 1)
    1. Machinery—Design. I. Sandor, George N. II. Title.
TJ230.E67 1990
621.8' 15—dc20
                              89-23097
                              CIP

Editorial/production supervision
  and interior design: Rob DeGeorge
Cover design: Wanda Lubelska
Manufacturing buyer: Denise Duggan
Color insert design and layout: Rosemarie Paccione

PRENTICE-HALL INTERNATIONAL (UK) LIMITED, *London*
PRENTICE-HALL OF AUSTRALIA PTY. LIMITED, *Sydney*
PRENTICE-HALL CANADA INC., *Toronto*
PRENTICE-HALL HISPANOAMERICANA, S.A., *Mexico*
PRENTICE-HALL OF INDIA PRIVATE LIMITED, *New Delhi*
PRENTICE-HALL OF JAPAN, INC., *Tokyo*
SIMON & SCHUSTER ASIA PTE. LTD., *Singapore*
EDITORA PRENTICE-HALL DO BRASIL, LTDA., *Rio de Janeiro*

Art Erdman
dedicates this work
to his daughters
Kristy *and* **Kari**
and son **Aaron**

George Sandor
dedicates this work
to his wife
Magdi

Contents

8 INTRODUCTION TO KINEMATIC SYNTHESIS: GRAPHICAL AND LINEAR ANALYTICAL METHODS   *480*

Contents

Preface to the First Edition

This two-volume work, consisting of Volume 1, *Mechanism Design: Analysis and Synthesis,* and Volume 2, *Advanced Mechanism Design: Analysis and Synthesis,* was developed over a 15-year period chiefly from the teaching, research, and consulting practice of the authors, with contributions from their working associates and with adaptations of published papers. The authors represent a combination of over 30 years of teaching experience in mechanism design and collectively have rendered consulting services to over 35 companies in design and analysis of mechanical systems.

The work represents the culmination of research toward a general method of kinematic, dynamic, and kineto-elastodynamic analysis and synthesis, starting with the dissertation of Dr. Sandor under the direction of Dr. Freudenstein at Columbia University, and continuing through a succession of well over 100 publications. The authors' purpose was to present texts that are timely, computer-oriented, and teachable, with numerous worked-out examples and end-of-chapter problems.

The topics covered in these two textbooks were selected with the objectives of providing the student, on one hand, with sufficient theoretical background to understand contemporary mechanism design techniques and, on the other hand, of developing skills for applying these theories in practice. Further objectives were for the books to serve as a reference for the practicing designer and as a source-work for the researcher. To this end, the treatment features a computer-aided approach to mechanism design (CAD).

Useful and informative graphically based techniques are combined with computer-assisted methods, including applications of interactive graphics, which provide the student and the practitioner with powerful mechanism design tools. In this manner, the authors attempted to make all contemporary kinematic analysis and synthesis readily available for the student as well as for the busy practicing designer, without the need for going through the large number of pertinent papers and articles and digesting their contents.

Many actual design examples and case studies from industry are included in the books. These illustrate the usefulness of the complex-number method, as well as other techniques of linkage analysis and synthesis. In addition, there are numerous end-of-chapter problems throughout both volumes: over 250 multipart problems in each volume, representing a mix of SI and English units.

The authors assumed only a basic knowledge of mathematics and mechanics on the part of the student. Thus Volume 1 in its entirety can serve as a first-level text for a comprehensive one- or two-semester undergraduate course (sequence) in Kinematic Analysis and Synthesis of Mechanisms. For example, a one-semester, self-contained course of the subject can be fashioned by omitting Chapters 6 and 7 (cams and gears). Volume 2 contains material for a one- or two-semester graduate course. Selected chapters can be used for specialized one-quarter or one-semester courses. For example, Chapter 5, Dynamics of Mechanisms — Advanced Concepts, with the use of parts from Chapters 2 and 3, provides material for a course that covers kinetostatics, time response, vibration, balancing, and kineto-elastodynamics of linkage mechanisms, including rigid-rotor balancing.

The foregoing are a few examples of how the books can be used. However, due to the self-contained character of most of the chapters, the instructor may use other chapters or their combinations for specialized purposes. Copious reference lists at the end of each book serve as helpful sources for further study and research by interested readers. Each volume is a separate entity, usable without reference to the other, because Chapters 1 and 8 of Volume 1 are repeated as Chapters 1 and 2 of Volume 2.

The contents of each volume may be briefly described as follows. In Volume 1, the first chapter, Introduction to Kinematics and Mechanisms, is a general overview of the fundamentals of mechanism design. Chapter 2, Mechanism Design Philosophy, covers design methodology and serves as a guide for selecting the particular chapter(s) of these books to deal with specific tasks and problems arising in the design of mechanisms or in their actual operation. Chapter 3, Displacement and Velocity Analysis, discusses both graphical and analytical methods for finding absolute and relative velocities, joint forces, and mechanical advantage; it contains all the necessary information for the development of a complex-number-based computer program for the analysis of four-bar linkages adaptable to various types of computers. Chapter 4, Acceleration Analysis, deals with graphical and analytical methods for determining acceleration differences, relative accelerations and Coriolis accelerations; it explains velocity equivalence of planar mechanisms, illustrating the concept with examples. Chapter 5 introduces dynamic and kinetostatic analyses with various methods and emphasizes freebody diagrams of mechanism links. Chapter 6 presents design methods for both simple cam-and-follower systems as well as for cam-modulated linkages. Chapter 7 acquaints the student with involute gears and gear trains, including the velocity ratio, as well as force and power-flow analysis of planetary gear trains. The closing chapter of Volume 1, Chapter 8, is an introduction to dimensional synthesis of planar mechanisms using both graphical and closed-form linear analytical methods based on a "standard-form" complex-number approach. It treats the synthesis of single and multiloop mechanisms as function, path, and motion generators, with first- and higher-order approximations.

Volume 2 starts with the same introductory chapter, Chapter 1, as the first volume. Chapter 2 of Volume 2 is the same as Chapter 8 of Volume 1. Chapter 3 extends planar analytical kinematic synthesis to greater than three-condition precision, accomplished by way of closed-form nonlinear methods, including Burmester theory, and describes a computer package, "LINCAGES," to take care of the computational burden. Cycloidal-crank and geared linkages are also included. Chapter 4 presents a new computer-oriented, complex-number approach to planar path-curvature theory with new explicit forms of the Euler-Savary Equation (ESE) and describes all varieties of Bobillier's Construction (BC), demonstrating the equivalence of the ESE and BC methods. Chapter 5 is a comprehensive treatment of the dynamics of mechanisms. It covers matrix methods, the Lagrangian approach, free and damped vibrations, vibration isolation, rigid-rotor balancing, and linkage balancing for shaking-forces and shaking-moments, all with reference to computer programs. Also covered is an introduction to kineto-elastodynamics (KED), the study of high-speed mechanisms in which the customary rigid-link assumption must be relaxed to account for stresses and strains in elastic links due to inertial forces. Rigid-body kinematics and dynamics are combined with elastic finite-element techniques to help solve this complex problem. The final chapter of Volume 2, Chapter 6, covers displacement, velocity, and acceleration analysis of three-dimensional spatial mechanisms, including robot manipulators, using matrix methods. It contains an easily teachable, visualizable treatment of Euler-angle rotations. The chapter, and with it the book, closes with an introduction to some of the tools and their applications of spatial kinematic synthesis, illustrated by examples.

In view of the ABET accreditation requirements for increasing the design content of the mechanical engineering curriculum, these books provide an excellent vehicle for studying mechanisms from the design perspective. These books also fit in with the emphasis in engineering curricula placed on CAD/CAM and computer-aided engineering (CAE). Many computer programs are either included in the texts as flow charts with example input-output listings or are available through the authors.

The complex-number approach in this book is used as the basis for interactive computer programs that utilize graphical output and CRT display terminals. The designer, without the need for studying the underlying theory, can interface with the computer on a graphics screen and explore literally thousands of possible alternatives in search of an optimal solution to a design problem. Thus, while the burden of computation is delegated to the computer, the designer remains in the "loop" at each stage where decisions based on human judgment need to be made.

The authors wish to express their appreciation to the many colleagues and students, too many to name individually, who have made valuable contributions during the development of this work by way of critiques, suggestions, working out and/or checking of examples, and providing first drafts for some of the sections. Among the latter are

Dr. Ashok Midha (KED section), Dianne Rekow (Balancing section), Dr. Robert Williams (Spatial Mechanisms), and Dr. Donald R. Riley, who taught from the preliminary versions of the texts and offered numerous suggestions for improvements. Others making significant contributions are John Gustafson, Lee Hunt, Tom Carlson, Ray Giese, Bill Dahlof, Tom Chase, Sern Hong Wang, Dr. Sanjay G. Dhandi, Dr. Patrick Starr, Dr. William Carson, Dr. Charles F. Reinholtz, Dr. Manuel Hernandez, Martin Di Girolamo, Xirong Zhuang, and others.

Acknowledgment is also due to the Mechanical Systems Program, Civil and Mechanical Engineering Division, National Science Foundation, for sponsoring Research Grant No. MEA-8025812 at the University of Florida, under which parts of the curvature chapter were conceived and which led to the publication of several journal articles. Sources of illustrations and case studies are acknowledged in the text and in captions. Other sponsors are acknowledged in many of the authors' journal papers (listed among the references), from which material was adapted for this work.

The authors and their collaborators continue to develop new material toward possible inclusion in future editions. To this end, they will appreciate comments and suggestions from the readers and users of these texts.

Arthur G. Erdman
George N. Sandor

Preface to the Second Edition

The authors have received much valuable feedback from the instructors, students, and engineers using the books: *Mechanism Design: Analysis and Synthesis, Volume 1*, and *Advanced Mechanism Design: Analysis and Synthesis, Volume 2*, since the publication of these works in 1984. This feedback came from over a hundred institutions in the United States and abroad, including the authors' own universities, where the first volume, the second volume, or both of these books were adopted. A set of questionnaires distributed to these readers brought in still more helpful input toward improving the texts to make them more up-to-date and useful to students, teachers, and practitioners. The authors themselves, with the aid of their coworkers, undergraduate, graduate, and postgraduate students, developed much new material and many new applications in mechanism design. These advances in addition to the publisher's encouragement prompted the authors to undertake the revision of the first volume. The result is *Mechanism Design: Analysis and Synthesis, Volume I, Second Edition*.

In some chapters major reorganization has taken place. Important equations and concepts are highlighted and graphical methods are separated from analytical techniques. Also, computer software is now provided with the text.

In Chapter 1, Section 1.9 was added: Mechanism Design Example: Variable Speed Transmission. Also, fifteen new multipart problems were added raising the total to 43 problems.

The new title of Chapter 2 is: Computer-Aided Mechanism Design Philosophy. In this chapter, Section 2.1, Introduction, has been thoroughly revised. The new title of Section 2.2 is: The Seven Stages of Computer-Aided Engineering Design, in which descriptions of available computer aids at each of the stages as well as a guideline for incorporating safety considerations in design were added. Furthermore, a complicated design flowchart which was purely descriptive was deleted. Section 2.3, How The Seven Stages Relate to This Text, has been condensed and thereby made more readable. Section 2.4 was replaced by a History of Computer-Aided Mechanism Design. It covers the period from the 1950s through the 1980s and provides a glance into the future. Section 2.5, Design Categories and Mechanism Parameters, and Section 2.6, Troubleshooting Guide, are essentially unchanged. A new appendix to Chapter 2 enumerates Contemporary Theories and Artificial Intelligence Implementations of Design Methodology and features a comprehensive bibliography.

Chapter 3, Section 3.1, which discusses position analysis of linkages, has been revised and extensively augmented in the areas of Grashof criteria, alternative linkage-circuits, branching, transmission, and deviation angles. Section 3.2, Displacement Analysis: Graphical Method, is new and contains detailed explanations and examples of geometric methods. Section 3.3, Displacement Analysis: Analytical Method, is revised based on plane vectors in complex-number form — the algebra for which is reviewed in the appendix to this chapter. Section 3.4, Concept of Relative Motion, is essentially unchanged. Section 3.5, Graphical Velocity Analysis, includes now only graphical methods while Section 3.6, Analytical Velocity Analysis, is presented by way of two fully worked-out examples using complex numbers. Instant centers and their use in velocity analysis are fully treated in sections 3.7 and 3.8. Mechanical Advantage is the topic of Section 3.9; the subject is expanded for easier reading and utilization. Section 3.10 carries the title: Analytical Methods for Velocity and Mechanical Advantage Determination. It includes a new subsection on the limits of motion of four-bar linkages. The chapter winds up with Section 3.11, Computer Program for the Kinematic Analysis of Four-Bar Linkages, and terminates with an appendix, in which complex-number vector algebra is reviewed. The end-of-chapter problem set has been increased from 62 to 77 multi-part problems.

In Chapter 4, Acceleration Analysis, Section 4.1, Introduction, Section 4.2, Acceleration Difference, and Section 4.3, Relative Acceleration, are unchanged except for standardizing the notation. Section 4.4, Coriolis Acceleration, has been augmented with several fully worked-out in-text examples. The title of Section 4.5 was revised to read Mechanisms with Curved Slots and Higher-Pair Connections; this section was expanded and completely rewritten with step-by-step, worked-out in-text examples. The end-of-chapter problem set has been expanded from 43 to 57 multi-part problems.

In Chapter 5, Introduction to Dynamics of Mechanisms, Sections 5.1 through 5.4, Kinetostatic Analysis, were expanded to cover both graphical and analytical approaches. Section 5.5, A Design Example, is practically unchanged. Section 5.6, The Matrix Method, was augmented with a comprehensive example, worked out in detail by both the matrix and superposition methods. The two methods are discussed and implemented by a computer program flowchart in Section 5.7. The end-of-chapter problem set has been updated with the addition of numerous subproblems.

In Chapter 6, Cam Design, the first three sections are largely unchanged. Section 6.4, Displacement Diagrams: Graphical Development, is new and is followed by Section 6.5, Displacement Diagrams: Analytical Development. Section 6.6, Advanced Cam Profile Techniques, is substantially expanded and illustrated with example computer outputs. The essentially unchanged Graphical Cam Profile Synthesis is now Section 6.7. Section 6.8, Analytical Cam Profile Synthesis, has been augmented by a more complete explanation and reference to the computer disk containing the CAMSYN program, an updated version of the CAMSYNG program described in the First Edition of this book. Section 6.9, Cam Synthesis for Remote Follower and Section 6.10, Cam-Modulated Linkages, are unchanged. The end-of-chapter problem set has been expanded from 29 to 36 multi-part problems.

Chapter 7, Gears and Gear Trains, has been augmented by a detailed discussion of conjugacy and conjugate profiles, and by the presentation of a new in-text example and a new end-of-chapter problem. The chapter is otherwise unchanged.

In Chapter 8, which discusses synthesis, Section 8.3 has been rewritten and is now called Type Synthesis. The presentation of precision-point spacing has been relocated to precede the three-position synthesis of function generators. Software packages, such as LINCAGES and KINSYN, are referred to wherever applicable and a new, fully worked out in-text example is included demonstrating the correlation of graphical and complex-number analytical methods in three-precision-point motion generation synthesis. The end-of-chapter problem set is augmented with a new multi-part problem of rack-and-gear-mechanism synthesis for path generation with prescribed timing.

Several color pages were added to the Second Edition, illustrating mechanisms referred to in the text or copies of computer-generated mechanism design methods. Contributions of artwork were made by Dr. Sridhar Kota, University of Michigan (who has contributed new four-bar symmetric coupler curve design charts, a geared five-bar coupler curve design chart, and dwell mechanism sample output from MINN-DWELL) and others credited in the captions of the photographs. Also new in this edition is an IBM AT/PS2 disk which supplements chapters 3, 4, 6, and 8. Readers will be able to design four-bar linkages for three design positions and then analyze either the synthesized mechanism or one created by keyboard input for position, velocity, or acceleration type information. Also, a full cam design option will illustrate the concepts in Chapter 6. Major contributions for software are acknowledged from Chris Huber, Ralph Peterson, Mike Lucas, and the Productivity Center of the University of Minnesota.

In closing this preface, the authors wish to express their appreciation to the many undergraduate and graduate students whose work has been utilized for in-text examples and end-of-chapter problems. Among these from the University of Minnesota, particular mention is due to Dr. Tom Chase and John Titus, who also contributed by suggesting and drafting improvements in the text itself. Dr. Don Riley, also of the University of Minnesota, was the leader in the development of the software associated with this book. Dr. Albert C. Esterline of the University of Minnesota Computer Science Department contributed the draft for the Appendix to Chapter 2 on the history of computer aids and artificial intelligence in the area of mechanism design. Dr. Harold Johnson of Georgia Tech made numerous valuable suggestions, among them the substantial expansion of curved-slot mechanisms in the Acceleration Analysis chapter.

Other contributors to this edition include Jon Thoreson of 3M, Elizabeth Logan of the University of Minnesota, Greg Vetter from Truth, SPX, and Gary Bistran for photography.

Dr. Suren Dwivedi of West Virginia University and other users of the First Edition, too numerous to list here by name, helped considerably by sending the authors their suggestions. Mr. Doug Humphrey, Senior Engineering Editor at Prentice Hall, and his staff including Rob DeGeorge, Production Editor, deserve special acknowledgment for their thorough professional work in bringing this Second Edition into existence. And last, but by no means least, the authors are deeply grateful to their respective wives, Magdi B. Sandor and Mary Jo Erdman, for their patience and forbearance in spending endless days and nights as "book-widows" during the years of preparation of this edition.

Art Erdman
George Sandor

chapter one

Introduction to Kinematics and Mechanisms

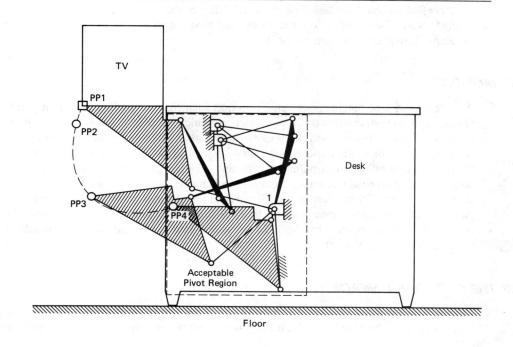

1.1 INTRODUCTION

Engineering is based on the fundamental sciences of *mathematics, physics,* and *chemistry.* In most cases, engineering involves the analysis of the conversion of energy from some source to one or more outputs, using one or more of the basic principles of these sciences. *Solid mechanics* is one of the branches of physics which, among others, contains three major subbranches: *kinematics,* which deals with the study of relative motion; *statics,* which is the study of forces and moments, apart from motion; and *kinetics,* which deals with the action of forces on bodies. The combination of kinematics and kinetics is referred to as *dynamics.* This text describes the appropriate mathematics, kinematics, and dynamics required to accomplish mechanism design.

A *mechanism* is a mechanical device that has the purpose of transferring motion and/or force from a source to an output. A *linkage* consists of links (or bars) (see Table 1.1), generally considered rigid, which are connected by joints (see Table 1.2), such as pins (or revolutes), or prismatic joints, to form open or closed chains (or loops). Such *kinematic chains,* with at least one link fixed, become (1) *mechanisms* if at least two other links retain mobility, or (2) *structures* if no mobility remains. In other words, a mechanism permits relative motion between its "rigid" links; a structure does not. Since linkages make simple mechanisms and can be designed to perform complex tasks, such as nonlinear motion and force transmission, they will receive much attention in this book. Some of the linkage design techniques presented here are the result of a resurgence in the theory of mechanisms based on the availability of the computer. Many of the design methods were discovered before the 1960s, but long, cumbersome calculations discouraged any further development at that time.

1.2 MOTION

A large majority of mechanisms exhibit motion such that all the links move in parallel planes. This text emphasizes this type of motion, which is called *two-dimensional, plane,* or *planar* motion. Planar rigid-body motion consists of *rotation* about axes perpendicular to the plane of motion and *translation* —where all points in the body move along parallel straight or planar curvilinear paths and all lines embedded in the body remain parallel to their original orientation. *Spatial* mechanisms, introduced in Chap. 6 of Vol. 2, allow movement in three dimensions. Combinations of rotation around up to three nonparallel axes and translations in up to three different directions are possible depending on the constraints imposed by the joints between links (spherical, helical, cylindrical, etc.; see Table 6.1, Vol. 2).

1.3 THE FOUR-BAR LINKAGE

Mechanisms are used in a great variety of machines and devices. The simplest closed-loop linkage is the four-bar, which has three moving links (plus one fixed link)* and

---

* A *linkage* with one link fixed is a *mechanism.*

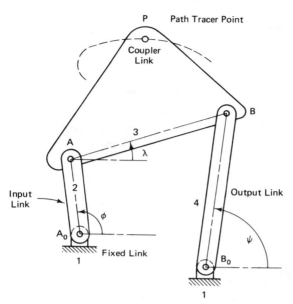

Figure 1.1

four "revolute," "pivoted" or "pin" joints (see Fig. 1.1). The link that is connected to the power source or prime mover is called the *input* link (A_0A). The *output* link connects the moving pivot B to ground pivot B_0. The *coupler* or *floating* link connects the two moving pivots, A and B, thereby "coupling" the input to the output link.

Figure 1.2 shows three applications where the four-bar has been used to accomplish different tasks. The *level luffing crane* of Fig. 1.2a is a special type of four-bar that generates approximate *straight-line motion* of the *path tracer* point (point P). Cranes of this type can be rated at 50 tons capacity and typically have an approximate straight-line travel of the coupler tracer point about 9 m long.

Figure 1.2b is a drive linkage for a lawn sprinkler, which is adjustable to obtain different ranges of oscillation of the sprinkler head. This *adjustable linkage* can be varied in its function by changing the length and angle of the output link by way of the clamping screw. Figure 1.2c shows a four-bar automobile hood linkage design. The linkage controls the relative motion between the hood and the car frame.

The three applications shown in Fig. 1.2 are quite different and in fact represent three different *tasks* by which all mechanisms may be classified by application: path generation, function generation, and motion generation (or rigid-body guidance). In *path generation* (Fig. 1.2a), we are concerned with the path of a tracer point. A *function generator* (Fig. 1.2b) is a linkage in which the relative motion (or forces) between links (generally) connected to ground is of interest. In *motion generation* (Fig. 1.2c), the entire motion of the coupler link is of concern. These tasks are discussed in greater depth in Chaps. 2 and 8.

The four-bar has some special configurations created by making one or more links infinite in length. The slider-crank (or crank and slider) mechanism of Fig. 1.3 is a four-bar chain with a slider replacing an infinitely long output link. The internal combustion

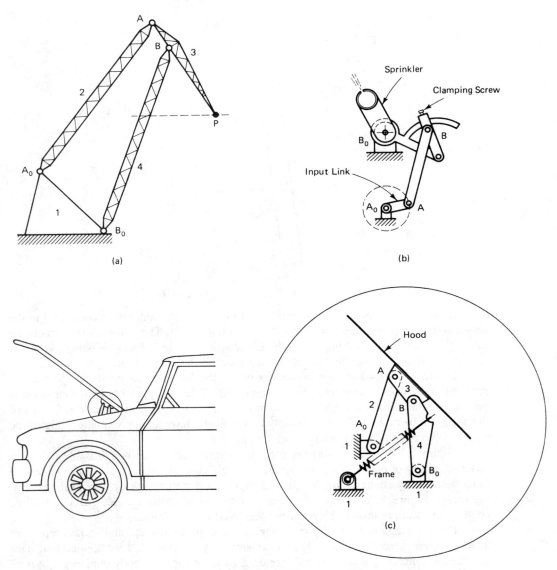

(a)

(b)

(c)

Figure 1.2

engine is built around this mechanism—the crank is link 2, the connecting rod is the coupler (link 3), and the piston is the slider (link 4).

Other forms of four-link mechanisms exist in which a slider is guided on a moving link rather than on the fixed link. These are called *inversions* of the slider-crank, produced when another link (the crank, coupler, or slider) is the fixed link. Section 3.1 shows some applications of the inversions of the slider-crank.

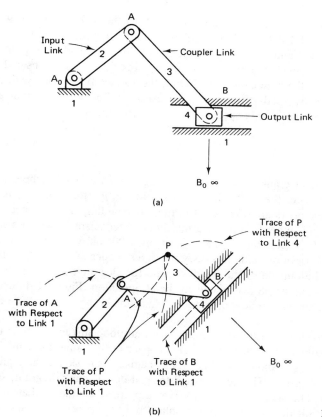

Figure 1.3

1.4 THE SCIENCE OF RELATIVE MOTION

All motion observed in nature is relative motion; that is, the motion of the observed body relative to the observer. For example, the seated passenger on a bus is moving relative to the waiting observer at the bus stop, but is at rest relative to another seated passenger. Conversely, the passenger moving along the aisle of the bus is in motion relative to the seated passenger as well as relative to the waiting observer at the bus stop.

The study of motion, kinematics, has been referred to as the science of relative motion. Design and analysis of machinery and mechanisms relies on the designer's ability to visualize relative motion of machinery components. One major objective of this chapter is to familiarize the reader with motion generated by a variety of linkage mechanisms and thus prepare for topics in both analysis and synthesis based on this fundamental understanding. Figure 1.3b shows a slider-crank linkage with a triangular coupler link ABP. Each point of the coupler link traces a different path, called *coupler curves*, with respect to ground (link 1). Point A traces out a circular arc centered at A_0, point B travels in a straight line, and point P traces out a more complex curve. All these coupler curves are part of the *absolute motion** of link 3. Suppose that the path of point *P* with respect to link 4 instead of link 1 is desired. This relative motion may be found by envisioning oneself sitting on link 4 and observing the motion of link 3, in particular point *P* of link 3. In other words, we invert the mechanism, fixing link 4 (the slider) instead of link 1, and move the rest of the mechanism (including the former fixed link) with respect to link 4. Here the relative path of point *P* with respect to link 4 is a circular arc centered at *B*. Thus absolute motion is a special case of relative motion.

1.5 KINEMATIC DIAGRAMS

Although the four-bar and slider-crank are very useful linkages and are found in thousands of applications, we will see later that these linkages have limited performance capabilities. Linkages with more members are often used in more demanding circumstances.

Figure 1.4 shows a typical application of a multiloop mechanism in which a mechanical linkage is required. A casement window must open 90° outward from the sill and be at sufficient distance from one side to satisfy the egress codes and from the other side to provide access to the outside of the window pane for cleaning. Also, the force required to drive the linkage must be reasonable for hand operation. Figure 1.4a and b show one of the popular casement window operator mechanisms in the 90° and 30° positions, respectively.

It is often difficult to visualize the movement of a multiloop linkage such as that shown in Fig. 1.4, especially when other components appear in the same diagram. The first step in the motion analysis of more complicated mechanisms is to sketch the equivalent *kinematic* or *skeleton diagram*. This requires a "stripped-down" stick diagram, such as that shown in Fig. 1.5. The skeleton diagram serves a purpose similar to that of the electrical schematic or circuit diagram in that it displays only the essential skeleton of the mechanism, which, however, embodies the key dimensions that affect its motion. The kinematic diagram takes one of two forms: a sketch (proportional but not exactly to scale), and the scaled kinematic diagram (usually used for further analysis: position, displacement, velocity, acceleration, force and torque transmission, etc.). For convenient reference, the links are numbered (starting with ground link as number 1), while the joints are lettered. The input and output links are also labeled. Table 1.1 shows typical skeleton diagrams of planar links. One purpose of the skeleton diagram is to provide a kinematic schematic of the relative motions in the mechanisms. For example, a pin joint depicts relative rotation, a slider depicts relative straight-line translation, and so on.

Figure 1.5 shows the kinematic diagram (sketch) for the casement window linkage. Notice that there are six links, five pin joints, one slider joint, and one roller in this sketch. Note also that one loop of the mechanism contains a slider-crank linkage (1,5,4,6). Connected to the slider crank is a bar and a roller (2,3), which provides the input for opening and closing the window. The kinematic diagram simplifies the mechanism for visual inspection and, if drawn to scale, provides the means for further analysis.

---

* In mechanism analysis it is convenient to define one of the links as the fixed frame of reference. All motion with respect to this link is then termed absolute motion.

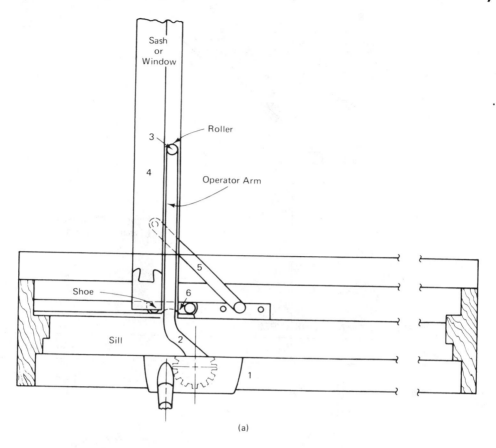

(a)

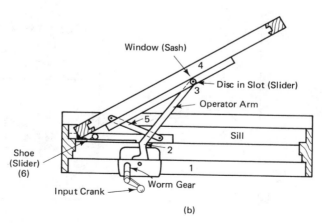

(b)

Figure 1.4 *(Courtesy of Truth Division, SPX Corp.)*

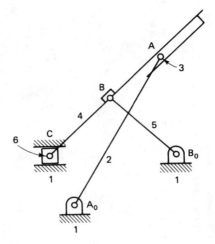

Figure 1.5 Equivalent *kinematic diagram* (sketch) of Fig. 1.4.

TABLE 1.1 PLANAR LINK TYPES

Link types	Typical form	Skeleton diagram(s)
Binary		
Ternary		
Quaternary		

Another application where a multiloop mechanism has been suggested is a proposed variable-stroke engine [126]* (Fig. 1.6). This linkage varies the piston stroke in response to power requirements. The operation of the stroke linkage is shown in Fig. 1.7.

For each piston, the lower end of a control link is adjusted along an arc prescribed by the control yoke shown. The top of the control link is connected to the main link which, in turn, connects to a component that plays the role of a conventional connecting rod. In essence, the result is an engine with variable crank-throw.

When control-yoke divergence from vertical is slight [Fig. 1.7a] the main link is restricted in its movement, and the resulting piston stroke is small. As the control nut moves inward on its screw, the angle between the control yoke and "the axis of the cylinder" is increased. This causes the main link to move in a broader arc, bringing about a longer stroke. The angle between the control yoke and the cylinder axis varies between 0 and 70°; the resulting stroke varies from 1 in. to 4.25 in. "The linkage is designed so that the compression ratio stays approximately the same, regardless of piston stroke."

The equivalent kinematic diagram of this adjustable mechanism is shown in Fig. 1.8. Notice that there are nine links, nine pins, and two sliders in this sketch.

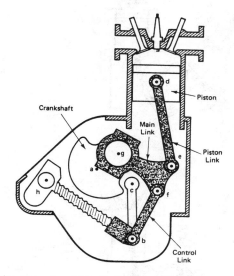

Figure 1.6   Section view of variable displacement engine showing crankshaft, main link, piston link, and stroke control link. Stroke is varied by moving the lower end of the control link.

* Numbers in square brackets pertain to References starting on page 616.

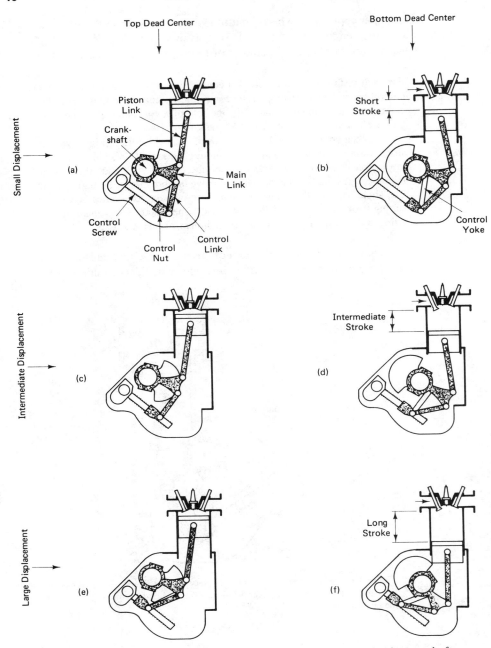

Figure 1.7 Variable-displacement linkage; stroke is varied by moving lower end of control link.

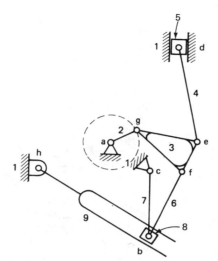

Figure 1.8 Slider (link 8) is fixed once Control Screw 9 is adjusted.

1.6 SIX-BAR CHAINS

If a four-bar linkage does not provide the type of performance required by a particular application, one of two single-degree-of-freedom six-bar linkage types (with seven revolute joints) is usually considered next: the *Watt chain* or the *Stephenson chain* (see Sec. 1.7 and Figs. 1.9 to 1.13). These classifications depend on the placement of the ternary[†] links (members with three revolute joints; see Table 1.1). In the Watt chain, the ternary links are adjacent; in the Stephenson chain, the ternary links are separated by binary links (links with only two revolute joints). Several applications where six-bar chains have been employed will help us become familiar with these linkages.

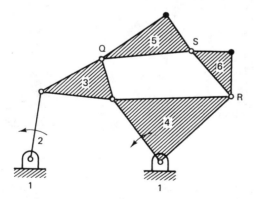

Figure 1.9 Watt I six-bar linkage.

[†] Notice in Figs. 1.9 to 1.13 that some of the triangular-shaped links are truly ternary, while others are shown as triangular to indicate possible path tracer points on floating links.

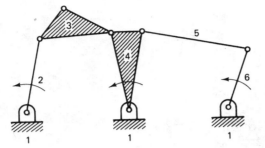

Figure 1.10   Watt II six-bar linkage.

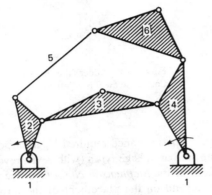

Figure 1.11   Stephenson I six-bar linkage.

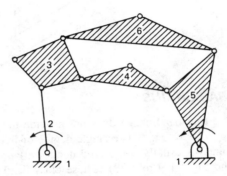

Figure 1.12   Stephenson II six-bar linkage.

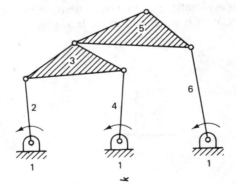

Figure 1.13   Stephenson III six-bar linkage.

Example 1.1 [48]

In the manufacture of cassette tape cartridges, it is sometimes necessary to thread a leader tape contained in the assembled cassette onto a device which winds blank magnetic tape into the cassette. A mechanical linkage is sought to thread the leader tape.

Figure 1.14 shows the position of the cassette, the leader tape, and the device through which the tape must be threaded at the time it is desired that the guiding linkage begin operation. The dashed line is the final configuration of the threaded leader tape.

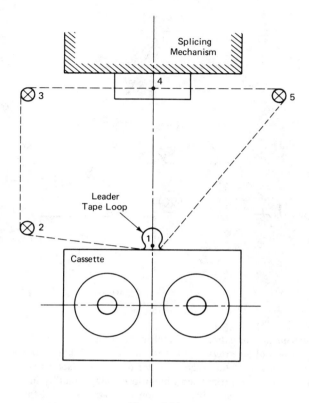

Figure 1.14

The tape unwinds from both sides of the cassette and so forms a loop as it is pulled out. The numbers 1 to 5 indicate the successive positions the tape must pass. The crossed circles at positions 2, 3, and 5 (Fig. 1.14) are posts that hold the tape loop. These posts are initially below the tape deck and each comes up to hold the tape once the tape loop has passed over its position. The tape loop must be guided clear of the posts at positions 2, 3, and 5 in the proper direction of travel.

The following were considered necessary requirements for the linkage:

1. 360° input crank rotation.
2. Input rotation must be timed to the positions of the path point (tape loop threading guide) to allow posts 2, 3, and 5 to be brought up at the correct time.
3. Angular orientation of the coupler link containing the path point must be specified at each prescribed position.
4. The Stephenson III chain was chosen for this example. The computer-aided techniques of Chap. 3 of Vol. 2 were used to produce the final design shown in Fig. 1.15.

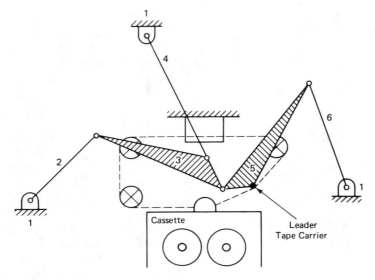

Figure 1.15

Example 1.2 [128]

Mechanisms are extremely useful in the design of biomechanical devices. For example, in the design of an external prosthesis for a through-knee amputee, it is desirable to duplicate the movement of the relative center of rotation (see Chap. 3) between the thigh (femur) and the leg bones (tibia and fibula) to maintain stability in walking. Figures 1.16 and 1.17 show a Stephenson I six-bar motion generator designed for this purpose. The zero-degree flexion (fully extended) position is shown in Fig. 1.16a together with the trajectory of the instant center of rotation of link 1, the artificial leg, with respect to the femur (link 6). The 90° flexion (bent knee) position is shown in Fig. 1.17, and the kinematic diagram (sketch) of this linkage is shown in Fig. 1.16b.

Example 1.3 [83]

A feeding mechanism (see Fig. 1.18, not to scale) is required to transfer cylindrical parts one-by-one from a hopper to a chute for further machining. A Watt II mechanism was chosen for this task. The timing link (6) provides rotation to the cupped platform (whose rotation is a prescribed function of input-link rotation) which transfers one cylinder from the hopper to the chute, while the prescribed path of the output coupler (point P) positions the cylinder on the platform and then pushes the cylinder into the chute.

Example 1.4 [130]

Another example of a dual-task requirement for a linkage is simultaneous path and function generation (Fig. 1.19). The specified function is $\psi = \phi^2$, both shown on linear scales, for $1 \leq \phi \leq 3$, while the required path is an approximate straight line. Figure 1.19 shows the

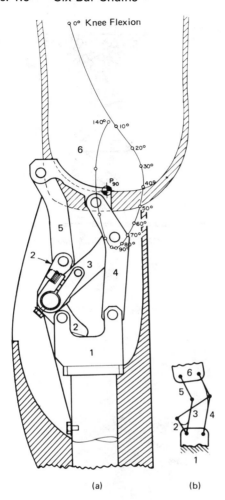

(a)   (b)

Figure 1.16 Six-bar linkage prosthetic knee mechanism. (*Biomechanics Laboratory, University of California, Berkeley.*)

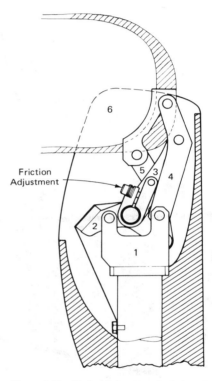

Figure 1.17 Six-bar linkage prosthetic knee mechanism. (*Biomechanics Laboratory, University of California, Berkeley.*)

first prescribed position of a Stephenson III linkage synthesized using the techniques described in Chap. 3 of Vol. 2. The other four prescribed positions are shown in Fig. 1.20.

The previous four examples consist of links with pin (or revolute) connections. If one or more of the links in one of the six-bars in Figures 1.9 to 1.13 is changed to a slider, different six-link mechanisms are obtained. Numerous possible six-link mechanisms exist with combinations of links, pins, and sliders. (See the appendix to Chap. 8 for a sample case study.)

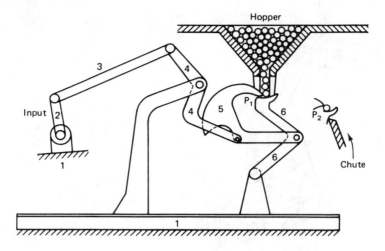

Figure 1.18

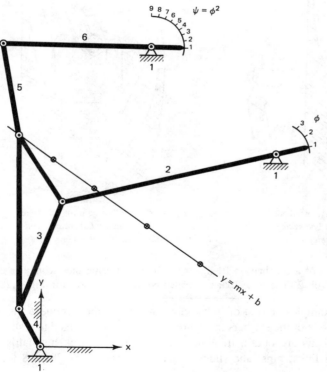

Figure 1.19

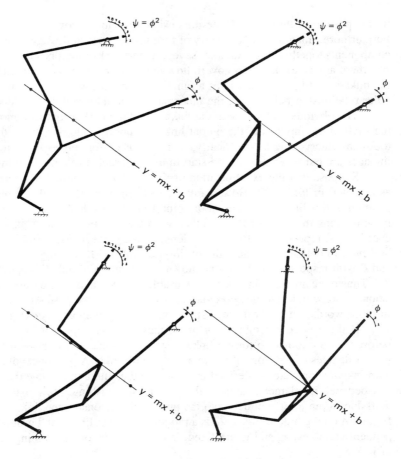

Figure 1.20

1.7 DEGREES OF FREEDOM

The next step in the kinematic analysis of mechanisms, following the drawing of the schematic, is to determine the number of *degrees of freedom* of the mechanism. By degrees of freedom we mean the number of independent inputs required to determine the position of all links of the mechanism with respect to ground. There are hundreds of thousands of different linkage types that one could invent. Envision a bag containing a large variety of linkage components from Tables 1.1 and 1.2; binary, ternary, quaternary, and so on, links; pin joints, slider joints; cams and cam followers; gears, chains, sprockets, belts, pulleys, and so on. (Spherical and helical as well as other connections that allow three-dimensional relative motion are not included here, as only planar motion in parallel planes is discussed in this portion of the book. Three-dimensional motion is covered in Chap. 6 of Vol. 2.) Furthermore, imagine the possibility of forming all sorts of linkage types by putting these components together. For example, several binary links might be connected by pin joints. Are there any rules that help govern how these mechanisms are formed? For instance, is the linkage in Fig. 1.21 usable as a function generator, where we wish to specify the angular relationship between ϕ, the independent variable, and ψ, the dependent variable?

The obvious problem with the linkage of Fig. 1.21 is that, if a motor is attached to the shaft of the input link, the output link may not respond directly — there appear to be too many intervening links. Clearly, there is a need for some rule of mobility by which linkages are put together. We can start to develop such a rule by examining a single link.

Suppose that the exact position of rigid link K is required in coordinate system XY as depicted in Fig. 1.22. How many independent variables will completely specify the position of this link? The location of point A can be reached, say, from the origin by first moving along the X axis by x_A and then y_A in the direction of the Y axis. Thus, these two coordinates, representing two translations, locate point A. More information is required, however, to define completely the position of link K. If the angle of the line of points A and B with respect to the X axis is known, the position of link K is specified in the plane XY. Thus there are three independent variables: x_A, y_A, and θ (two translatioñs and one rotation, or three independent coordinates) associated with the position of a link in the plane. In other words, an unconstrained rigid link in the plane has *three degrees of freedom.*

If there is an assembly of n links, they possess a total of $3n$ degrees of freedom before they are joined to form a linkage system. Connections between links result in the loss of degrees of freedom of the total system of links. A pin (or revolute) joint is called a *lower-pair* connector — defined in some previous literature as one that has surface contact between its elements such as the pin and the bushing. How many degrees of freedom does a pin joint subtract from the previously unconstrained links which it connects? If point A on the link in Fig. 1.22 is a pin joint between link K and ground, then two independent variables, x_A and y_A are fixed, leaving θ as the one remaining degree of freedom of link K.

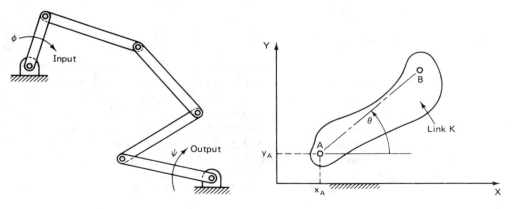

Figure 1.21   **Figure 1.22**

In an assembly of links such as that in Fig. 1.21, each pin connection will remove two degrees of freedom of relative motion between successive links. This observation suggests an equation that will determine the degrees of freedom of an n-link chain connected by f_1 pin joints, with ground (the fixed link) considered as one of the links:

$$\text{degrees of freedom} = F = 3(n - 1) - 2f_1 \qquad (1.1)$$

Equation (1.1) is known as *Gruebler's equation*. The number of mobile links is $(n - 1)$. The pin joint permits *one* degree of relative freedom between two links — thus the notation f_1. This equation is one of the most popular mobility equations used in practice. For other versions, see Ref. 76, and Chap. 8 of this book.

Most mechanism tasks require a single input to be transmitted to a single output. Therefore, single-degree-of-freedom mechanisms, those that have *constrained motion,* are the types used most frequently. For example, it is easy to see intuitively that the four-bar of Fig. 1.1 is a single-degree-of-freedom linkage. An intuitive degree-of-freedom analysis may proceed as follows. Once the independent variable ϕ is specified, the position of point A is known with respect to A_0 and B_0; since the lengths of the coupler base AB and the output link B_0B are known, B_0AB is a triangle with no further mobility (zero degrees of freedom) and the position of the rest of the linkage is determined.*

Using Gruebler's equation to determine the number of degrees of freedom of the linkage in Fig. 1.1, we have

$$n = 4, \qquad f_1 = 4$$
$$F = 3(4 - 1) - 2(4) = +1$$

The $+1$ indicates a single degree of freedom for the linkage. As a further demonstration of the use of Gruebler's equation, refer to the Watt I six-bar in Fig. 1.9:

$$n = 6, \qquad f_1 = 7$$
$$F = 3(6 - 1) - 2(7) = +1$$

Intuitively, one can be satisfied that this linkage has a single degree of freedom as predicted by the equation. Once assembled, links 1 to 4 form a four-bar linkage which has already been demonstrated to have a single degree of freedom. Observe that links 4, 3, 5, and 6 form a second four-bar linkage with the position of links 3 and 4 already determined. Since the positions of points Q and R are determined, QSR forms a "rigid" triangle and the position of the entire mechanism is specified.

---

* Actually, there are two possible *circuits* (sometimes called *branches*) for the rest of the four-bar (ABB_0), mirror images about the diagonal B_0A in which the links AB and BB_0 could be assembled. However, the linkage cannot move from one circuit to the other without disassembly. Thus the number of degrees of freedom of a linkage is independent of the fact that the mechanism may have several different circuits. Therefore, one may formulate the intuitive definition of degrees of freedom thus: When, after specifying n coordinates (x, y, and/or θ) of link positions, the possible positions of the remaining links are finite, the number of degrees of freedom is n. The concept of branching is addressed in Chap. 3 (see especially Fig. 3.14).

Determine the degrees of freedom of the trench hoe of Fig. 1.23. This linkage system has an element that has not been included in the degree-of-freedom discussion up to this point—the slider (hydraulic cylinder in this case). Let us therefore determine how many degrees of freedom of relative motion a sliding connection subtracts between adjacent links: in other words, how many relative constraints a slider imposes. In Fig. 1.3(a), the slider (link 4) is constrained with respect to ground (link 1) against moving in the vertical direction as well as being constrained from rotating in the plane. Thus the slider joint allows movement only along the slide and subtracts two degrees of freedom of relative motion: one rotation and one translation. Equation (1.1) may now be expanded in scope so that f_1 *equals the sum of the number of pin joints plus the number of slider joints*—since they both allow only one degree of relative motion.

The trench hoe has 12 links (consider the cab as the ground link), 12 pin joints, and three slider joints (the piston-cylinder combinations). If you counted only 11 pin connections, look more carefully at point Q in the figure. Three links are connected by the same pin constraint. There are two pin joints at Q, one connecting links 9 and 10,

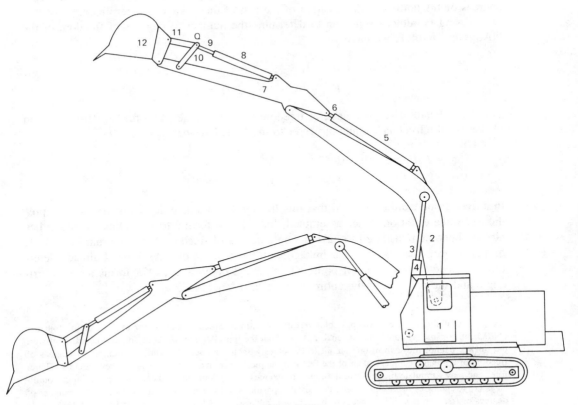

Figure 1.23

the other connecting links 10 and 11. In general the number of pin joints at a common connection is

$$f_1 = m - 1 \tag{1.2}$$

where m is the number of links joined by a single revolute joint.

The number of degrees of freedom of the trench hoe is therefore

$$F = 3(12 - 1) - 2(15) = +3$$

Thus the trench hoe linkage requires three input coordinates to determine the position of all its links relative to the cab. These are supplied by the three hydraulic cylinders that are attached along the boom.

Are there other types of joints besides pins and sliders that can be used to connect mechanism members in plane motion? If so, do they all subtract two degrees of freedom? Five other types of planar joints are shown in Table 1.2. Whereas pin and slider joints (lower pairs) allow only one degree of freedom of relative motion, the *higher-pair* joints (joints that have been defined in the literature as joints that have either point or line contact only) may permit a higher number (two or three) degrees of freedom of relative motion. Each has a lower-pair equivalent, consisting of as many lower pairs as the number of degrees of freedom of relative motion allowed by the higher-pair joint.

Rolling contact with no sliding allows only one degree of freedom of relative motion, due to the absence of sliding, which leaves only the relative rotation θ (see Table 1.2). The pure rolling joint can therefore be included as an f_1-type joint. The lower-pair equivalent for instantaneous velocity equivalence* is simply a pin joint at the relative instant center (see Chap. 3), which is the contact point between the two links for rolling contact with no sliding. Thus this essentially higher-pair joint allows only one degree of freedom because of the additional constraint against sliding.[†]

The *roll-slide contact* constrains only one degree of freedom (relative motion in the y direction in Table 1.2). First, let us consider the lower-pair combination for instantaneous velocity equivalence, which is a slider and pin joint combination. This allows two degrees of freedom ($n = 3$, $f_1 = 2$) of relative motion. The degrees of freedom of the roll-slide joint can be verified by a modified *Gruebler's equation extended to include roll-slide joints:*

$$F = 3(n - 1) - 2f_1 - 1f_2 \tag{1.3}$$

where f_2 is the number of roll-slide contact joints (those that permit *two* degrees of relative motion across the joint).

---

* Average velocity is a measure of displacement in the interval in which it occurs, $\Delta s/\Delta t$. The limiting value of this as $\Delta t \rightarrow 0$, namely ds/dt, is the instantaneous velocity at a point in time. Instantaneous velocity equivalence means that, if a higher-pair joint in a mechanism is replaced by its lower-pair equivalent, the instantaneous velocity of the relative motion allowed between the two original links of the higher pair will remain the same but the relative acceleration will, in general, be different.

[†] This is a force-closed joint: it requires a force, such as gravity, to prevent separation.

TABLE 1.2 PLANAR KINEMATIC PAIRS—LINK JOINTS

Joint name	Diagram	Lower pair with equivalent instantaneous velocity
Pin (revolute)	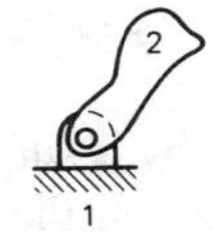 $n = 2$ $f_1 = 1$ $F = +1$ $F = $ degrees of freedom	
Slider (prismatic)	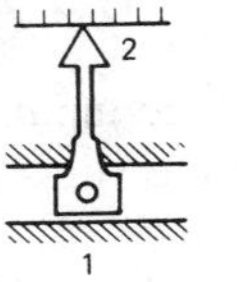 $n = 2$ $f_1 = 1$ $F = +1$	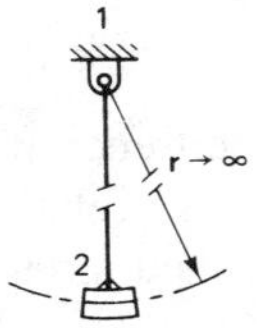 $r \to \infty$ $n = 2$ $f_1 = 1$ $F = +1$
Rolling contact (no sliding)	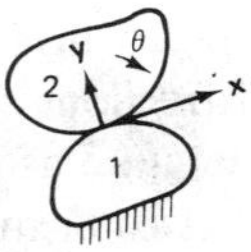 $n = 2$ $F = +1$	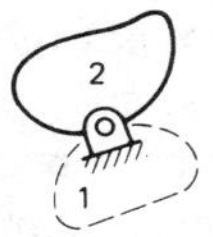 $n = 2$ $f_1 = 1$ $F = +1$
Roll-slide contact	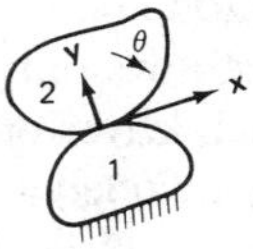 $n = 2$ $f_1 = 0$ $f_2 = 1$ $F = +2$	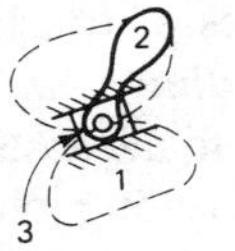 or  $n = 3$ $f_1 = 2$ $f_2 = 0$ $F = +2$
Gear contact (includes roll-slide contact between gear-teeth and rolling contact between pitch circles)*	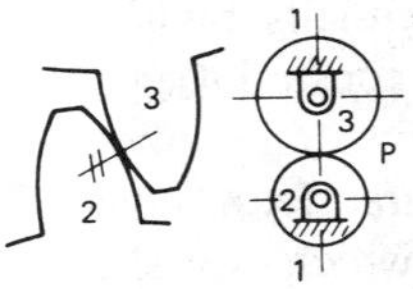 $n = 3$ $f_1 = 2$ $f_2 = 1$ $F = 1$	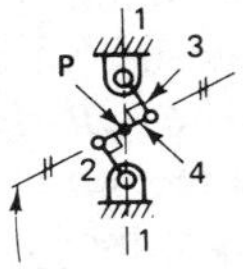 P $n = 4$ $f_1 = 4$ $f_2 = 0$ $F = +1$ Parallel to common normal of contacting tooth flanks; P, pitch point
Spring	 $n = 2$ $f_1 = 0$ $f_2 = 0$ $F = +3$	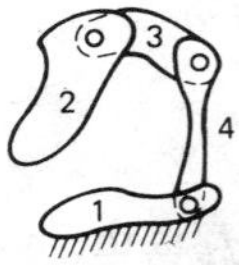 $n = 4$ $f_1 = 3$ $f_2 = 0$ $F = +3$
Belt and pulley (no sliding) or chain and sprocket	$F = +1$ 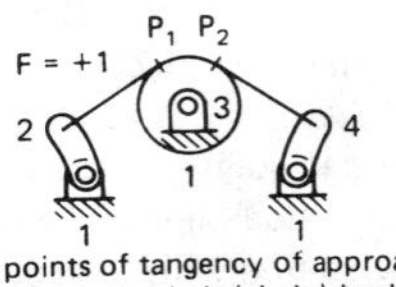 P_1 P_2 P_1, P_2: points of tangency of approaching and receding belt (chain) leads	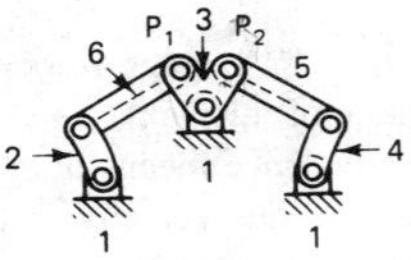 P_1 P_2 $n = 6$ $f_1 = 7$ $f_2 = 0$ $F = +1$

*See also Fig. 1.27.

Using Eq. (1.3) for the higher-pair model itself:

$$F = 3(2 - 1) - 1 = +2$$

For the gear set shown in Table 1.2, in which link 1 is fixed, the gear bearings are pin joints and the *gear tooth contact* is roll-slide. Therefore, $f_1 = 2$ and $f_2 = 1$, so that

$$F = 3(3 - 1) - 2(2) - 1 = +1$$

The lower-pair linkage for instantaneous velocity equivalence is a four-bar with fixed pivots located at the center of the gears, and moving pivots at centers of curvature of the tooth profiles in contact. The coupler goes through the pitch point P along the line of action of the gear mesh, perpendicular to the common tangent of the contacting tooth flank surfaces and also to the two links connected to ground. Thus the lower-pair model of a gear set predicts the same number of degrees of freedom:

$$F = 3(4 - 1) - 2(4) = +1$$

A *spring connection* (Table 1.2) produces a mutual force between the two links it connects, but it does not kinematically constrain the relative motion between the two links (assuming that the spring is within its range of extension and compression). Two binary links and three pin joints form the instantaneous-velocity-equivalent lower-pair model to the spring connection, allowing the same number of degrees of freedom of relative motion between the links connected by the spring. Thus for the equivalent lower-pair linkage, the number of degrees of freedom is

$$F = 3(4 - 1) - 2(3) = +3$$

The *belt and pulley* or *chain and sprocket* (see Table 1.2), where the belt or chain is maintained tight, are also possible planar connections. A Watt II linkage is the instantaneous-velocity-equivalent lower-pair connection to the belt and pulley (no sliding allowed). Using Eq. (1.3) for the equivalent six-bar linkage,

$$F = 3(6 - 1) - 2(7) = +1$$

Example 1.5

Determine the degrees of freedom of the mechanism shown in Fig. 1.24.

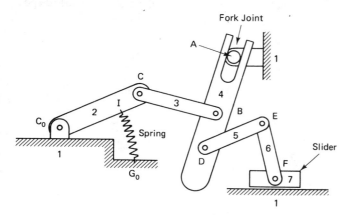

Figure 1.24

Solution   There are seven links, seven lower pairs, one roll-slide contact, and one spring connection. From Eq. (1.3),

$$F = 3(7 - 1) - 2(7) - 1(1) - 0(1) = +3$$

Let us check this by way of the velocity-equivalent lower-pair connections shown in Fig. 1.25. The spring has been replaced by two binary links, and the *fork joint* (or *pin-in slot joint* — rolling contact with sliding) has been replaced by a pin and slider. Therefore,

$$F = 3(10 - 1) - 2(12) = +3$$

This answer can be verified by intuition. If both the pin and slider joint at point *A* are fixed (taking away two degrees of freedom), link 4 is fixed in the plane. The slider-crank (*DEF*) is still free to move, however. Since the slider-crank has a single degree of freedom, the entire mechanism has a total of three degrees of freedom.

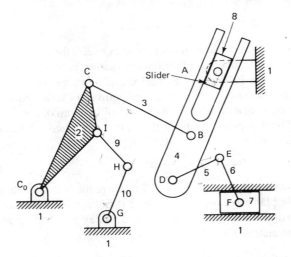

Figure 1.25

Before leaving the subject of degrees of freedom, it should be pointed out that there are linkages whose computed degrees of freedom may be zero (indicating a structure) or negative (indicating an indeterminate structure). They can, nevertheless, move due to special linkage proportions. For example, for the five-bar of Fig. 1.26,

$$F = 3(5 - 1) - 2(6) = 0,$$

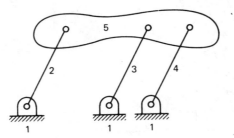

Figure 1.26

but because of the parallelogram configuration, the linkage can still move. This is called an *overconstrained* linkage, in which the third grounded link provides a *redundant* constraint. If there are manufacturing errors in link lengths or pivot locations, this linkage will jam. Figure 1.27 shows another overconstrained example. Here we see two nip rollers in pure rolling contact. Gruebler's equation yields $F = 3(3 - 1) - 2(3) = 0$. This simple mechanism does move, owing simply to the fact that the sum of the radii of the rollers equals the distance between the ground pivots for all positions of the rollers.

There are also instances when Gruebler's formula yields a seemingly excessive number of freedoms. This may involve a "passive or redundant degree of freedom" (see also Example 6.1 of Vol. 2) and does not alter the constraint between the output versus input motions of a mechanism. Take, for example, the cam-and-follower mechanism of Fig. 1.28.

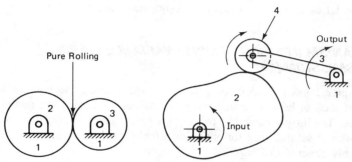

| Figure 1.27 | Figure 1.28 |

Here the rotation of the follower roller 4 does not affect the output oscillation of the follower arm 3. Even if roller 4 were welded to arm 3, the motion of arm 3 would remain unchanged. Checking this condition by Gruebler's equation, regarding the cam and roller contact as a roll-slide,

$$F = 3(4 - 1) - 2(3) - 1(1) = +2$$

Now, if we weld roller 4 to arm 3,

$$F = 3(3 - 1) - 2(2) - 1(1) = +1$$

Also, if slipping (the redundant freedom between cam and roller) is prevented by friction—in other words, their contact becomes rolling only:

$$F = 3(4 - 1) - 2(4) = +1$$

The reader can verify that the casement window mechanism in Figs. 1.4 and 1.5 contains a similar passive degree of freedom.

1.8 ANALYSIS VERSUS SYNTHESIS

The process of drawing kinematic diagrams and determining degrees of freedom of more complex mechanisms are the first steps in both the kinematic analysis and synthesis process. In *kinematic analysis,* a particular given mechanism is investigated based on the mechanism geometry plus possibly other known characteristics (such as input angular velocity, angular acceleration, etc.). *Kinematic synthesis,* on the other hand, is the process of designing a mechanism to accomplish a desired task. Here, both the type (*type synthesis*) as well as the dimensions (*dimensional synthesis*) of the new mechanism can be part of kinematic synthesis (see Chaps. 2 and 8 in this book and Chap. 3 of Vol. 2).

The fundamentals described in this chapter are most important in the initial stages of either analysis or synthesis. The ability to visualize relative motion, to reason why a mechanism is designed the way it is, and the ability to improve on a particular design are marks of a successful kinematician. Although some of this ability comes in the form of innate creativity, much of it is a learned skill that improves with practice. Chapter 2 will help put mechanism design into perspective: the structure or methodology of design is described, including the place of kinematic analysis and synthesis. Before that, however, let us look at a mechanism design case study.

1.9 MECHANISM DESIGN EXAMPLE: VARIABLE SPEED TRANSMISSION [27]

Chapter 1 has provided some tools for approaching mechanism design. Let us take a brief look at how an actual problem has been solved using methods developed in this book. The insight gained by this sample case study may help motivate learning techniques as well as set the stage for obtaining a feel for the mechanism design process (more thoroughly covered in Chap. 2).

The example chosen is that of the redesign of a control mechanism for a V-belt variable speed transmission similar to that in Fig. 1.29. The old design (Fig. 1.30) made use of the inertia of cam-shaped flyweights subject to the centrifugal force in a rotating two-piece sheave to change the axial position of the moving sheave half of the driving V-belt sheave of the transmission. The inertia force will change the axial position of the sheave half and, thus, alter the distance of the belt from the centerline of the sheave. The new concept uses slider-crank mechanisms (Figs. 1.31-1.36) for a much improved design.

The relative angular speed between input and output shafts connected by V-belts are inversely proportional to the radii between the centerline of the drive and output shafts and the belt. As these radii change (as depicted in Figs. 1.31-1.36), variable speed is obtained. Figure 1.31 (taken from a patent application) illustrates the initial difficulty of interpreting technical drawings of machinery. Kinematic diagrams are most useful to discern links from structural members. Note that member 84 is connected to input link 76, the two forming a weighted bell crank with the weight at 84, and 82 is a weight on the coupler 78. Without knowledge of kinematic chains, one would be hard put to interpret these figures.

A check on degrees of freedom of the axially moving sheave half with respect to the axially fixed sheave half will yield +1 by Gruebler's equation for each of the three slider-crank mechanisms. All movement occurs in a plane through the axis of the shaft. In this rotating plane, the axially fixed link is made up of the spider 66, the shaft 42,

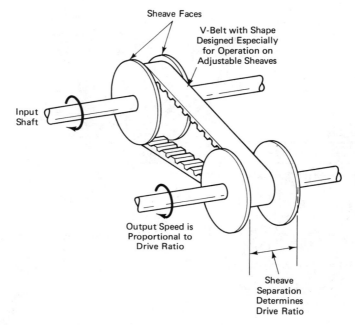

Figure 1.29   Variable-speed V-belt drive.

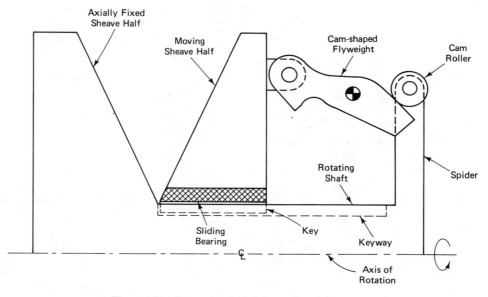

Figure 1.30   Schematic of original cam flyweight system.

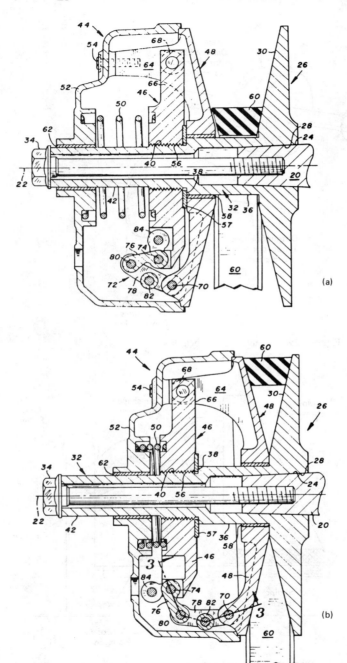

(a)

(b)

Figure 1.31 Two positions of the linkage controlled v-belt transmission taken from a patent application. The centerline -22, the axially fixed rotating disk (or spider) 66, the axially fixed shave half 26, the flyweights 84 and 82, belt 60, the links 76 and 78, and axially movable sheave half 48 are of primary interest here. (Courtesy of *Yamaha Motor Corporation, USA*)

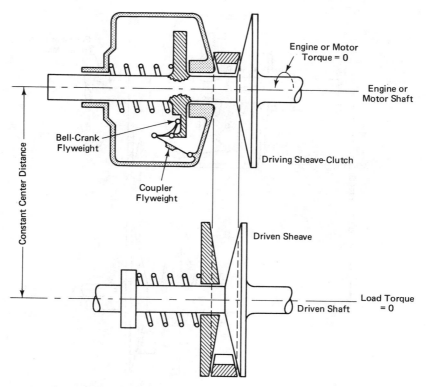

Figure 1.32 Engine is at idle speed, driving sheave is open, does not squeeze the v-belt, vehicle is at a standstill. (Note, the final design requires no spring on the driver.)

and the axially fixed sheave half 30 (Fig. 1.31). The movable links are the bell crank 76, the coupler 78, and the axially movable sheave half 48. The joints are revolutes 70, 74, and 80, and the slider joint between the axially movable sheave half bushing 58 and the axially fixed shaft 32. Therefore, $F = 3(4 - 1) - 2(4) = +1$. We are now ready for further review of the mechanism design process.

Rubber-belt-type variable-pitch drives represent a low-cost and smooth-running option for implementing a continuously-variable transmission. Combining the rubber belt drive with the automatic stepless shift mechanism described here produces an economical and reliable power train. The function of the flyweight mechanism on the driving sheave of the continuously-variable V-belt-type transmission is to produce a prescribed belt force as a function of the axial position of the sheave when the sheave is rotated at a constant angular velocity. A typical plot of force versus axial position is provided in Fig. 1.37.

A simplifed schematic of the original cam flyweight system is shown in Fig. 1.30, including the shaft, spider, and movable sheave. The cam itself consists of a plate cam

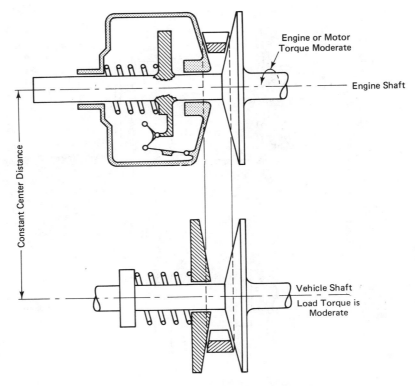

Figure 1.33 Engine speed is increased, driving sheave is partially closed by centrifugal force of bellcrank and coupler flyweights, squeezes *v*-belt. Vehicle at half speed, transmission is in midrange.

that is pinned to the movable sheave half near to its outer periphery. The cam roller is pinned to the spider. The axial force that squeezes the belt is exerted by the centrifugal force of the flyweight. The driven sheave is spring-loaded to maintain the correct belt tension and sense the load torque. The force varies as a function of the position of the movable sheave due to variation in the position of the center of gravity of the cam and the varying pressure angle (see Chap. 6).

The operational life of the original cam system was limited due to wear at the cam surface. The wear was aggravated by the vibration inherent in attaching the driving sheave directly to the internal combustion engine and the relatively high pressure angles required to minimize the size of the flyweight mechanism.

Type Synthesis of the Improved Variable-Sheave Drive

An improved design was obtained by application of type synthesis, analysis, and computer-aided dimensional synthesis. The methodologies involved at each stage are only briefly referred to here.

The improved variable-sheave-drive design was initiated by itemizing several possible design alternatives, that is, type synthesis, for an improved driving sheave-clutch.

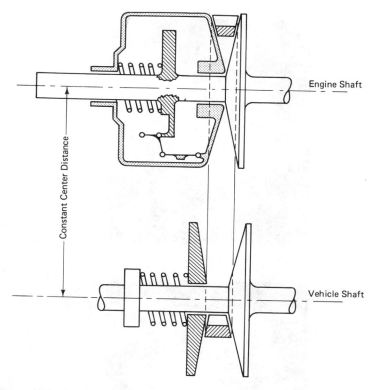

Figure 1.34 Engine at full speed, flyweights close driving sheave, forcing v-belt to run on high radius. Since v-belt length is constant, driven sheave is forced open. Transmission is at highest drive ratio.

Figure 1.35 Photograph of spider (66 in Fig. 1.31) containing the input ground pivots for the slider-crank control mechanisms.

(a)

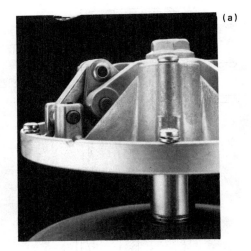

(b)

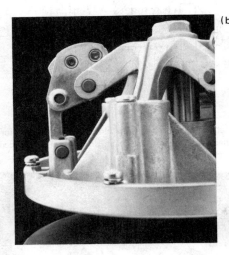

Figure 1.36 Photographs of the (a) fully open and (b) fully closed position of the moving sheave half.

Specifically, improved cam systems, four-bar linkages, six-bar linkages, and hybrid mechanisms were considered. The merits and drawbacks of each are summarized in Table 1.3.

The existing cam system was attractive in that it enabled precise control of the axial belt force over the total stepless shift range of the driving sheave-clutch. Therefore, improving the pressure angle of the cam system (enabling transmittal of a higher percentage of the cam force into moving the sheave) was considered as a redesign option. However, improving the pressure angle would require increasing the size of the cam, and manufacturing the precision cam surface is costly. Furthermore, the cam contact was considered undesirable for application due to the severe vibration associated with the internal combustion engine power plant.

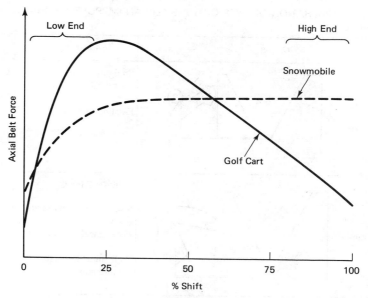

Figure 1.37   Typical axial belt force versus moving-sheave position profile. (*Courtesy of Yamaha Corporation, USA*)

Two four-link chain design options were considered: a slider-crank (Figure 1.38) and a double-slider (Table 1.3). Both options included at least one slide to use the inherent sliding action of the axially moving sheave half. The slider-crank was considered to be a more attractive design option than the double-slider because of the higher durability of its pin jointed crank, rather than a slide. In addition, the crank design appeared to offer more flexibility for the axial belt force versus position profile. However, dimensioning the slider-crank to obtain an arbitrary axial belt force profile appeared difficult.

Using one of the five possible six-bar chains (one of which appears in Table 1.3) appeared attractive in that more complex axial belt force profiles could be expected, increasing the chances for obtaining a desired arbitrary input profile. In addition, the slide inherent to the moving sheave half could still be used to advantage as one of the seven necessary joints of a six-bar. However, the added costs, added weight, and reduced reliability associated with adding two additional links to each of three positioning mechanisms posed serious drawbacks. Using still more complex linkages, such as eight- or ten-link chains, would aggravate this problem further.

Finally, a hybrid linkage with a spring-loaded crank abutting a limit stop on the spider was considered as a way to use the simplicity of a slider-crank, while increasing the control over the axial belt force profile. Nevertheless, the cost and potential vibrational problems associated with introducing spring-loaded links were considerable potential drawbacks.

The simple slider-crank appeared to constitute the preferred design alternative, as it offered the minimum number of components and the durability associated with simple

TABLE 1.3 ENUMERATION OF TYPE OPTIONS FOR AN IMPROVED VARIABLE SHEAVE-CLUTCH FOR V-BELT DRIVE

Option	Advantages	Drawbacks
Cam	Straightforward control of axial belt force Existing technology	Costly to manufacture Improving pressure angle increases size Poorly suited to harsh vibrational environment
Four-Link slider-crank	Simplest topology Robustness of pin joints Economical manufacturing Well suited to harsh vibrations	Difficult to match desired axial belt force
Four-Link double-slider	Simple topology Robustness of lower-pair joints Economical manufacturing	Very difficult to match desired axial belt force Potential friction problems at second slider
Six-Link chain	High control of axial belt force Robustness of pin joints	Increase in expense and weight, decrease in reliability due to addition of extra links
Hybrid Linkage	Added control of axial belt force	Potential vibration problems Increased cost

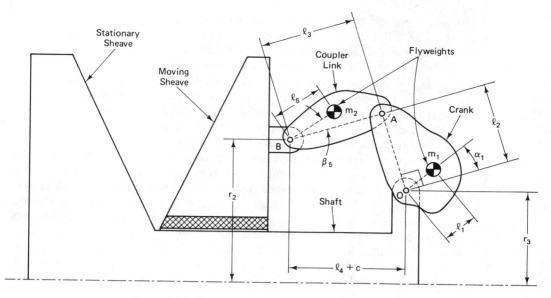

Figure 1.38   Schematic of slider-crank system showing design variables.

pin joints. However, this design option would be feasible only if the system could be designed to generate an axial belt force profile close to that of Fig. 1.37. A suitable mechanism was found using the procedures described in the following.

Comparison of the Slider-Crank and Cam-Flyweight Systems

The slider-crank system demonstrates several performance and manufacturing advantages over the cam system. Specifically, the slider-crank demonstrates improved wear properties, reduced size and mass, lower manufacturing costs, smoother shift action, and enhanced adjustability.

The slider-crank uses single-degree-of-freedom pin joints exclusively to complement the basic sliding action of the movable sheave half. The pin joints have been found to be very robust, leading to nearly indefinite life of the slider-crank mechanism. The overall mass and space requirements of the slider-crank system are less than that of the cam system. Reduction in mass is largely attributable to improved transmission-angle (Chapter 3) characteristics of the slider-crank. The improved force-transmission attributes of the slider-crank also make it possible to reduce the size envelope of the mechanism, giving a notable reduction in the radial space requirements. Furthermore, the improved transmission angle produces lower bearing loads on the pin joints.

The manufacturing costs of the slider-crank are reduced due to elimination of manufacturing the precision cam surfaces of the plate cams. The cam required manufacture from special heat-treated steel to provide sufficient surface strength. In contrast, the coupler and crank of the slider-crank can be manufactured economically from die cast aluminum. Standard dowel pins provide effective and economical revolute joints for the system. Furthermore, the engagement control spring, required to tension the cam system, was found to be totally unnecessary for the slider-crank system.

The slider-crank system has been found to produce a smoother start-up and shift action than the cam system. The improvement is again attributable to elimination of the cam contact in the noisy environment created by the internal combustion engine. The use of slider-cranks also results in quieter operation of the drive train. Other design advantages found in the new design are described in further detail in [27].

Members of the mechanisms community are constantly faced with the decision of whether to use cams or linkages to produce a desired motion or force in a machine. Cams offer the advantage of enabling continuous control over the output parameter, while linkages offer potential benefits in durability and manufacturing. The foregoing design of the variable-sheave-clutch V-belt drive provides a case study where substantial improvements were obtained by replacing a cam system with a simple slider-crank system. Specifically, the slider-crank resulted in nearly indefinite life, reduced size and weight, lower manufacturing costs, independent control of the axial belt force at high and low transmission ratios, and smoother stepless shift action. Furthermore, the axial force versus slider dis-

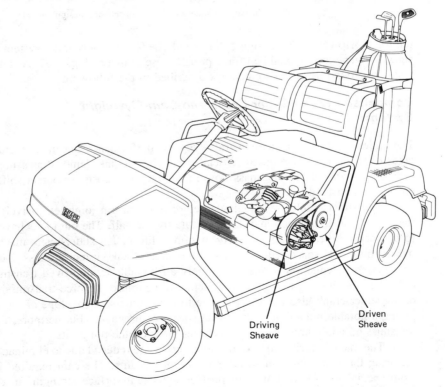

Driving
Sheave

Driven
Sheave

Figure 1.39 Actual installation of the improved variable-sheave V-belt drive.

placement profile can be controlled by the proper location of the flyweights, namely the choices α_1, β_5, ℓ_1, and ℓ_5 in Figure 1.38 (see Sections 5.3 and 5.4).

However, finding the simple slider-crank to be capable of replacing the cam function was challenging. The final mechanism design was found by writing a dedicated computer-aided design program (using the methods in Chap. 5) that provided extensive information feedback on the drive characteristics at various speeds. An iterative analysis scheme was facilitated by use of a dual-mode question and answer, direct command computer-to-user interface. The redesign was successfully implemented by using the program to survey literally hundreds of linkage design possibilities in a matter of a few days. The resulting design is now being marketed successfully as a golf cart transmission (Fig. 1.39).

PROBLEMS

1.1. As described in the chapter, all mechanisms fall into the categories of motion generation (rigid-body guidance), path generation, or function generation (including input–output force specification). Find and sketch an example of each task type (different from those presented in this book). Identify the type of linkage (four-bar, slider-crank, etc.), its task, and why this type of linkage was used for this task.

1.2. A linkage used for a drum foot pedal is shown in Fig. P1.1. Identify the linkage type. Why is this linkage used for this task? Can you design another simple mechanism for this task?

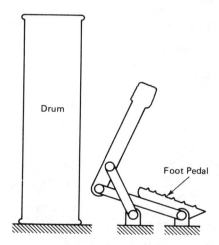

Drum

Foot Pedal

Figure P1.1

1.3. Figure P1.2 shows a surgical tool used for shearing. A spring (two spring steel leafs) between the two handles keeps the cutting surfaces apart and allows the shears to be used easily in one hand. Disregarding the spring connection and considering the straight handle on the right as the grounded link,

(a) What task does this mechanism perform?

(b) Sketch the kinematic diagram of this mechanism.

(c) What mechanism is this?

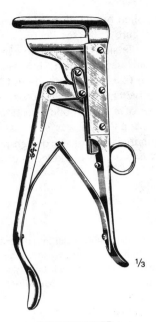

SCHUMACHER
FB 911
Sternum shears
210 mm, 8¼"   **Figure P1.2**   (Courtesy of AESCULAP)

1.4. A dump truck mechanism is depicted in Fig. P1.3.
  (a) What type of six-bar mechanism is this?
  (b) What task does it satisfy?

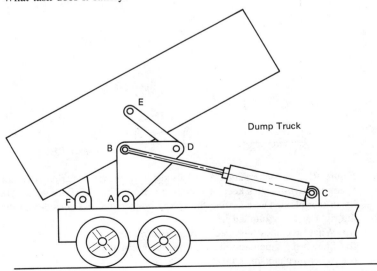

Figure P1.3

1.5. A mechanism for controlling the opening of an awning window is shown in Fig. P1.4. The vertical member is fixed to the frame of the house. The link with the label on it carries the window. The mechanism causes the window to move straight out from the building (clearing the metal lip around the window) before rotating the window out in a counterclockwise direction.

(a) What task does this mechanism satisfy?

(b) Calculate the degrees of freedom of this mechanism.

(c) Which type of six-bar is used here?

Vertical Member

Figure P1.4 Courtesy of Truth Division, SPX Corp.

1.6. Figure P1.5 shows a pair of locking toggle pliers. Identify the type of linkage (four-bar, slider-crank, etc.), its task, and why this type of linkage was used for this task. Notice that there is an adjusting screw on the mechanism. What is its function? Why is it located where it is?

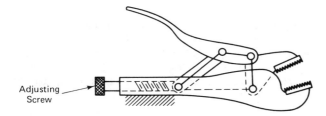

Adjusting Screw

Figure P1.5

1.7. A desolventizer (see Fig. P1.6) receives a fluid, pulpy food material (e.g., "spent" soybean flakes) and passes it on to the successive trays by gravity. These trays are heated by passing steam through them. The flakes are completely dried up by the time they come out of the last tray. The material is forced through an opening on the bottom of each plate to the next tray. The task of helping the material through the opening is accomplished by a "sweeping arm" attached to a central rotating shaft that runs vertically through the desolventizer. A control is needed for the gate opening to correspond with the rise in the level of the food material; that is, the gate opening should increase as the level of the material rises. A linkage* is used to perform this task. In order to sense any increase in the level of the material,

---

* Suggested by P. Auw, S. Royle, and F. Kwong [49].

a paddle is rigidly connected to the input link of the linkage while the gate on the bottom of the plate is attached to the output link.

What is the task of this linkage (motion, path, or function generation)? Why was this linkage chosen for this task?

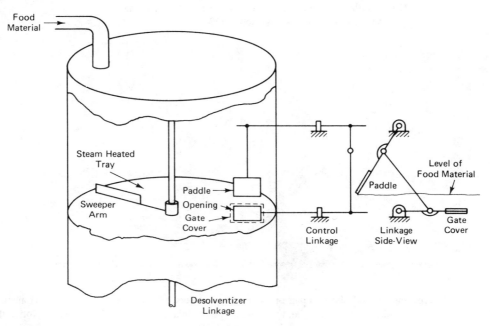

Figure P1.6

1.8. Figure P1.7 shows a proposed speed-control device[†] that could be mounted on an automobile engine and would serve a twofold purpose:

  (a) To function as a constant-speed governor for cold mornings so that the engine will race until the choke is reset. This speed control would enable the engine speed to be regulated, thus maintaining a preset idle speed. The idle speed would be selected on the dash-mounted speed-control lever.

  (b) To function as an automatic cruise control for freeway driving. The desired cruising speed could be selected by moving the indicator lever on the dashboard to the desired speed.

     (1) Sketch the kinematic diagram (unscaled) of the portion of this linkage that moves in planar motion.

     (2) Sketch the lower-pair equivalent linkage.

1.9. Those who are participating in the two-wheel revolution are aware that a derailer mechanism helps to change speeds on a 10-speed bicycle. A 10-speed, as the name implies, has 10 gear ratios that may be altered while the bike is in operation. The rear wheel has a five-sprocket cluster and the crank has two sprockets. The gear ratio is altered by applying a side thrust on the drive chain, the thrust causing the chain to "derail" onto the adjacent sprocket.

[†] Suggested by G. Anderson, R. Beer, and W. Gullifer [49].

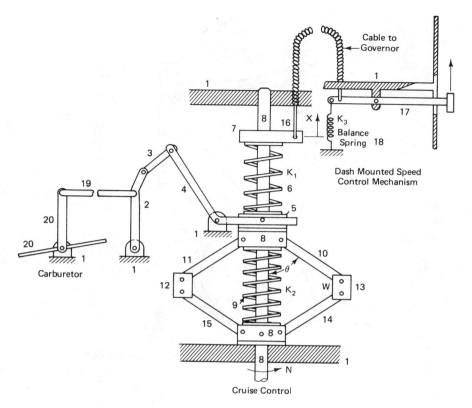

Figure P1.7

The operation of the rear derailer is adequate. However, the front derailer is less efficient due to the larger step necessary in transferring from one sprocket to the other.

This calls for a design that would enhance the life of the chain–sprocket system by reducing the side thrust on it and by enabling more teeth to be in contact with the chain during the initial stages of the transfer. These objectives were accomplished* by lifting the chain off one sprocket, moving it along the path as shown in Fig. P1.8, and then setting it down onto the adjacent sprocket.

(a) Draw the scaled kinematic diagram of this mechanism.

(b) Is this a motion-, path-, or function-generator linkage?

1.10. As steam enters into a steam trap, it is condensed and allowed to flow out of the trap in liquid form. The linkage in Fig. P1.9 has been suggested[†] to be a feedback control valve for the steam trap. The float senses the level of the condensate while the linkage adjusts the exit valve.

(a) Draw the unscaled kinematic diagram for this linkage.

(b) Is this a function, path, or motion generator?

(c) Can you design another simple linkage for this task?

* By G. Fichtinger and R. Westby [56, 66].

† By M. L. Pierce, student, University of Minnesota.

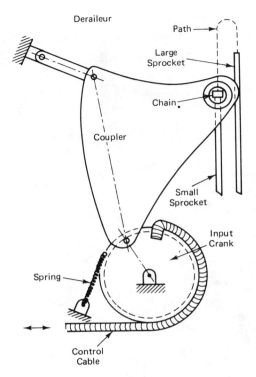

Figure P1.8

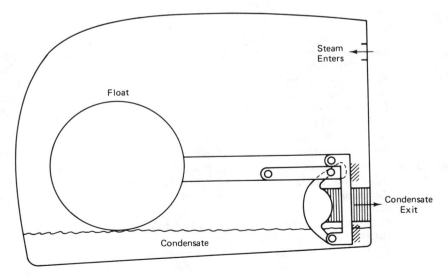

Figure P1.9

1.11. A typical automotive suspension system is shown in Fig. P1.10. A cross-sectional schematic is shown in Fig. P1.11.
   (a) What type of linkage is this (motion, path or function generator)?
   (b) Why is a linkage used in this application?
   (c) If the dimensions of the linkage were changed, what would be the effect on the vehicle?

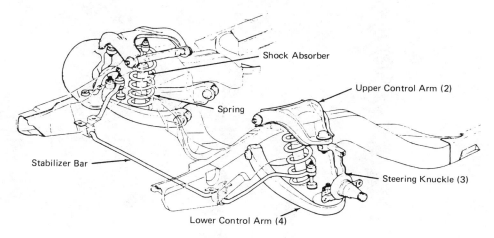

Figure P1.10

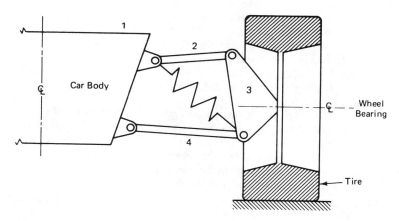

Figure P1.11

1.12. Frequently in the control of fluid flow, a valve is needed that will regulate flow proportional
to its mechanical input. Unfortunately, very few valves possess this characteristic. Gate
valves, needle valves, ball valves, and butterfly valves, to name a few, all have nonlinear
flow versus mechanical input characteristics. A valve with linear characteristics would do
much to simplify the proportional control of fluid flow.

The linkage* in Fig. P1.12 appears to be one means of providing a simple, durable, and inexpensive device to transform a linear mechanical control signal into the nonlinear valve positions which will produce a flow proportional to the control signal. The butterfly valve is connected to the short link on the right. The input link is on the left.

(a) What type of linkage is this?

(b) Which task does this linkage perform (function, path, or motion generation)?

Figure P1.12

1.13. In converting x-ray film from the raw material to a finished product, a multiloop mechanism was designed to transport the film from the sheeting operation, to the stenciling operation, to a conveyor belt.

The linkage shown in Fig. P1.13 must pick up the film from beneath the stenciling and sheeting devices with a vertical or nearly vertical motion to prevent sliding between the film and mechanism. The mechanism follows a horizontal path (with no appreciable rotation) slightly above the stenciling and sheeting devices while transporting the film from pickup to delivery.

Although the double-parallelogram-based linkage in Fig. P1.13 accomplished the task adequately, a simpler linkage[†] (shown in Fig. P1.14) was synthesized using the techniques of Chap. 3 of Vol. 2.

(a) Draw the unscaled kinematic diagrams of both linkages.

(b) Verify the degrees of freedom of both mechanisms.

(c) What type of six-bar is shown in Fig. P1.14?

1.14. Figures P1.15 and P1.16 show two gripper mechanisms suggested for use in industrial robots [29]. For each gripper,

(a) Determine the task performed.

(b) Find the number of degrees of freedom.

(c) Can you find any four-bar or six-bar chains?

---

* Designed by B. Loeber, B. Scherer, J. Runyon, and M. Zafarullah using the technique described in Sec. 8.16 (see Ref. 49).

[†] Designed by D. Bruzek, J. Love, and J. Riggs [49].

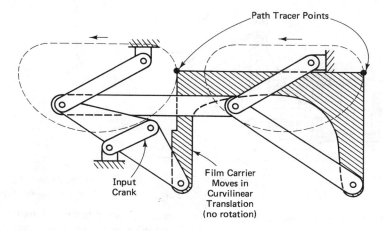

Figure P1.13 Eight-bar transport mechanism.

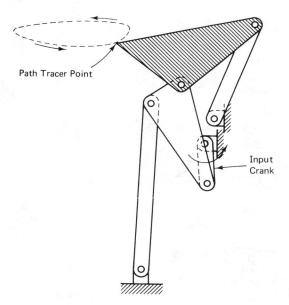

Figure P1.14 Six-bar transport mechanism.

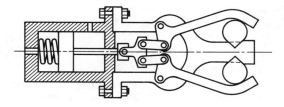

Figure P1.15 Spring-loaded gripper of linkage type with double fingers.

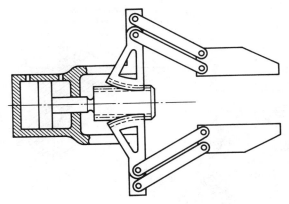

Figure P1.16 Gripper of dual gear-and-rack type actuated by pneumatic source.

1.15. In order to emboss characters onto credit or other data cards, a multiloop linkage has been designed [10] which exhibits high mechanical advantage (force out/force in; see Chap. 3). Separately timed punch and die surfaces are required so that the card is not displaced during the embossing process. (The desired motions are shown in Fig. P1.17.) The embossing linkage (see Figs. P1.18 to P1.20) makes use of an interposer arrangement wherein two os-

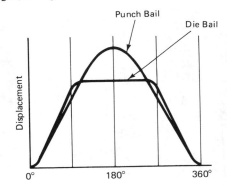

Figure P1.17

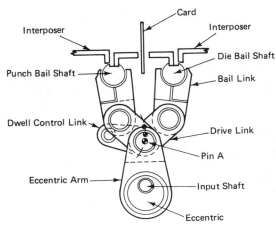

Figure P1.18 90° position. (*Courtesy of Data Card Corporation.*)

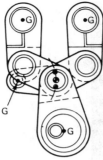
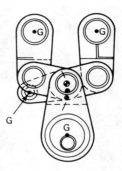

Figure P1.19 180° position (*G*, grounded pivot).

Figure P1.20 270° position (*G*, grounded pivot).

cillating bail shafts drive respective punches and dies, provided that the interposers are inserted in the keyways on top of the shafts.

(a) Draw the scaled kinematic diagram of this linkage.

(b) Determine the degrees of freedom of this linkage by both intuition and Gruebler's equation.

1.16. An idea came to mind to design and build a mechanism inside a box that, once turned on, would send a finger out of the box, turn itself off, and return back into the box [10, 13]. Two different types of linkages were designed for this task. The linkage shown in Figs. P1.21 to P1.23 was created by D. Harvey while the mechanism in Figs. P1.24 to P1.26 was invented by T. Bjorklund. (Note that the external switch and the internal limit switch are in parallel, so that the latter keeps the motor running until the finger has been withdrawn into the box.)

(a) Draw the kinematic diagrams of these linkages.

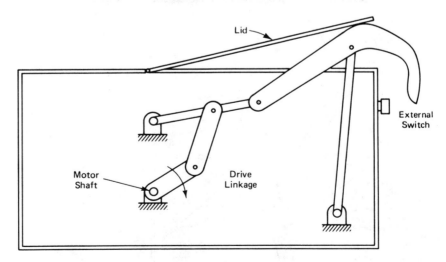

Figure P1.21

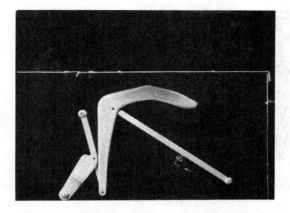

Figure P1.22

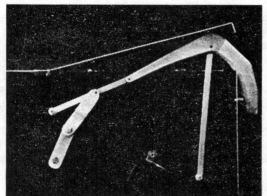

Figure P1.23

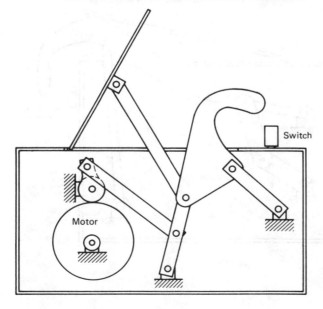

Switch

Motor

Figure P1.24

Figure P1.25   **Figure P1.26**

(b) Show (by intuition and Gruebler's equation) that both these mechanisms have a single degree of freedom. (Disregard the lid in Fig. P1.21.)

(c) In Fig. P1.21 what type of six-bar linkage is this? What is its task?

1.17. A six-bar lift mechanism for a tractor is shown in two positions (solid and dashed lines) in Fig. P1.27. What type of six-bar is this? What task does it satisfy?

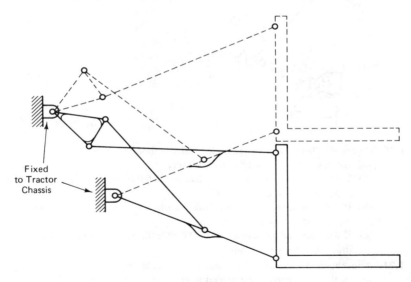

Fixed
to Tractor
Chassis

Figure P1.27

1.18. An agitator linkage for a washing machine is shown in Fig. P1.28 (ground pivots are identified by the letter G).

(a) What type of six-bar is this?

(b) What task does this linkage fulfill (motion, path, or function generation)?

(c) Why use a six-bar linkage in this application?

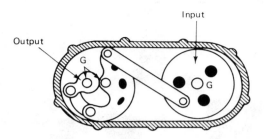

Figure P1.28   *G* signifies a ground pivot.

1.19. An automobile hood linkage is shown in Fig. P1.29. Notice the difference from the linkage in Fig. 1.2c.
   (a) Neglecting the spring, what type of six-bar is this linkage?
   (b) Draw the instantaneous-velocity-equivalent lower-pair diagram of this linkage (including the spring).

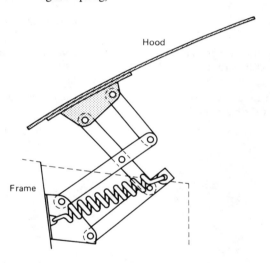

Figure P1.29

1.20. Figure P1.30 shows a surgical tool called a thoracic retractor which is used to pull and hold soft tissues out of the way during surgery.
   (a) What are the total degrees of freedom of this mechanism if the left curved member is considered as a ground link?
   (b) If we consider the curved link on the left side to be a ground link, which six-bar mechanism results?
   (c) If we consider the screw and the short link, which makes a T with the screw, the ground link, which six-bar results?

1.21. A mechanism was desired to automatically fold, in thirds, letters exiting from a laser printer so that they are ready to be placed in envelopes. Figure P1.31 shows part of the final mechanism* which is driven from a cam shaft. Figures a, b, and c show positions where the sheet of paper has just been fed onto the top of the mechanism, the right third halfway through the

---

* Designed by Ann Guttisberg, Chris Anton, and Chris Lentsch [3].

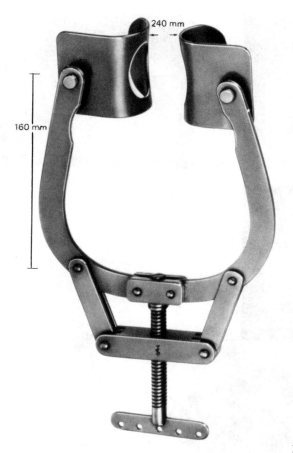

240 mm

160 mm

Figure P1.30   (Courtesy of AESCULAP)

fold, and the full-fold position, respectively. A similar mechanism folds the left side driven by the same cam. Considering the right side only (note that there is also another follower on the right side which drives another function of this mechanism, which can be disregarded here):

(a) What task does this mechanism perform?

(b) Draw an unscaled kinematic diagram of this mechanism.

(c) Which type of six-bar is part of this mechanism?

(d) Why is a six-bar used rather than just a four-bar?

1.22. Figure P1.32 shows a cutaway view of a Zero-Max variable-speed drive [41, 103]. This drive yields stepless variable speed by changing the arc through which four one-way clutches drive the output shaft when they move back and forth successively. Figure P1.33 shows one of these linkages, which is referred to as a "single lamination." The drive has sets of equally spaced out-of-phase linkages which use three common fixed shafts, A_0, C_0, and D_0. The rotation of the input A_0A causes the output link DD_0 to oscillate, thus rotating the output shaft D_0 in one direction (due to the one-way clutch assembly). The position of pivot B_0 is adjusted by rotating the speed control arm about C_0 to change the output speed of the drive. As B_0

(a)

(b)

(c)

Figure P1.31

Figure P1.32   *(Courtesy of Zero-Max Inc.)*

approaches the line BD, the output speed decreases since B_0, the center of curvature of the trajectory of B will approach point D, causing link 6 to become nearly stationary.

(a) What type of six-bar is this (with B_0 considered fixed)? What is its task?

(b) If link C_0B_0 is considered mobile, how many degrees of freedom does the linkage have? (Use Gruebler's equation.)

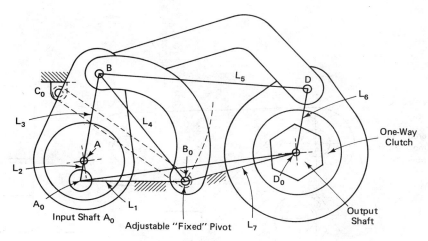

Figure P1.33

1.23. Based on the concept that mechanisms can be things of beauty besides having functional value, a mechanism clock was conceived.* Mechanisms would manipulate small cubes with numbers on them in such a way as to indicate the time. It was determined that three sets of cubes would be used to read minutes and hours (two for minutes). The cubes would be turned over 90° to use four sides of each for numbers—thus five cubes for the 0–9 set and three cubes each for the 0–5 and 1–12 set. Rather than work against nature, it was decided to remove the bottom cube and allow gravity to settle other cubes into their place, while the cube was placed on top of the stack. The devices had to be reasonable in size, be reliable, and be manufacturable. The motion was separated into three steps, as shown in Fig. P1.34. The problem is (1) how to remove the bottom cube from the stack, (2) how to rotate the cube, and (3) how to transport the cube to the top of the stack. Figures P1.35 and P1.36 show the final design for the mechanism clock.

(a) Draw the unscaled kinematic diagram for (1) each of three steps separately; (2) the entire mechanism.

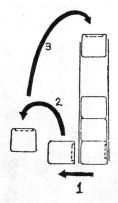

Figure P1.34

* By Jim Turner [161].

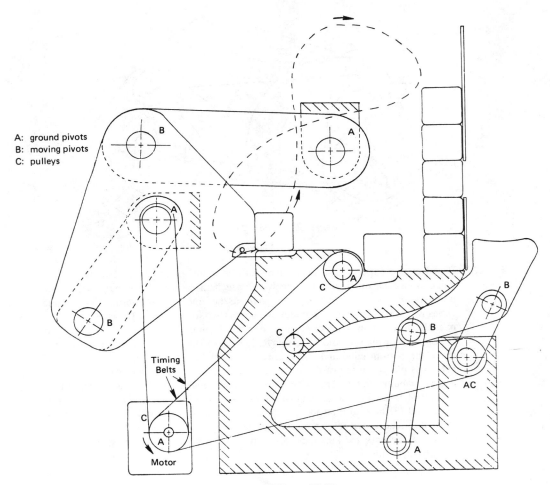

A: ground pivots
B: moving pivots
C: pulleys

Timing Belts

Motor

Figure P1.35

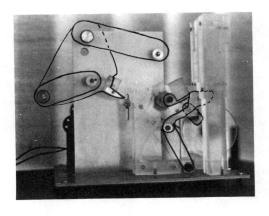

Figure P1.36

(b) Determine which task (motion, path, or function generation) is accomplished in each step.

(c) Determine the degrees of freedom of the entire linkage.

(d) Design your own mechanism clock and show it in a conceptual diagram.

1.24. Multiloop mechanisms have numerous applications in assembly line operations. For example, in a soap-wrapping process, where a piece of thin cardboard must be fed between rollers which initiate the wrapping operation, a seven-link mechanism[†] is employed such as that shown schematically in Fig. P1.37.

    The motion of the suction cups is prescribed in order to pick up one card from a gravitation feeder (the suction cups mounted on the coupler approach and depart from the card in the vertical direction) and insert the card between the rollers (the card is fed in a horizontal direction). The input timing is prescribed in such a fashion that the cups pick up the card during a dwell period (a pause in the motion) and also in a way that the card is fed into the rollers at approximately the same speed as the tangential velocity of the rollers.

(a) Sketch the instantaneous-velocity-equivalent lower-pair diagram of this mechanism.

(b) Determine the degrees of freedom of this linkage as shown in Fig. P1.37 and verify your answer by determining the degrees of freedom of the lower-pair equivalent linkage.

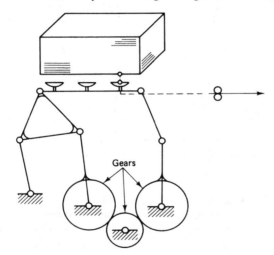

Gears

Figure P1.37

1.25. The linkage in Fig. P1.38 has been suggested[*] for stamping packages automatically at the end of an assembly line. The ink pad is located at the initial linkage position while the packages will travel along an assembly line and stop at the final position in order to be stamped. It is desirable that the linkage have straight-line motion toward the box so that the stamp imprint will not be smudged. A solenoid will drive the input link through its range of motion.

(a) What type of six-bar linkage is this?

(b) Draw this linkage in at least four other positions in order to (1) determine the range of rotation of the input link; (2) check if the linkage indeed does hit the ink pad in a straight-line motion approaching and receding from the box.

[†] This application was brought to the authors' attention by D. Tesar of the University of Texas, Austin.

[*] By J. Sylind (synthesized by the techniques presented in Chap. 3 of Vol. 2).

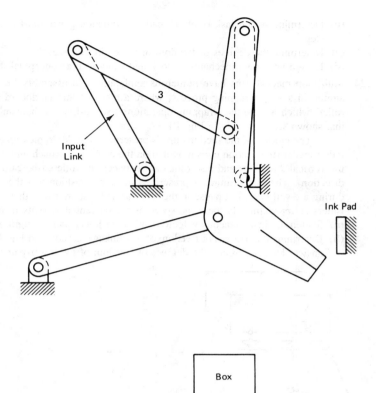

Figure P1.38

(c) Is this type of linkage a good choice for this task? Why?

(d) If the input link and link 3 are changed in length and orientation and the input pivot location moved, what are the consequences on the performance of the entire linkage?

1.26. Figure P1.39 shows a schematic diagram of a card feeder mechanism in its initial configuration. Cards are placed in the mailing list "file feed" by the machine operator. The file feed then intermittently feeds cards into the lower hopper. The cards must be joggled to align them against the hopper back plate so that they will feed out of the hopper properly when they reach the hopper feedroll. A cam causes the joggler movement.

If there should be a jam or misfeed in the hopper, the operator must pull on the joggler handle to pivot it open so that he can remove the cards from the hopper. To get at the joggler handle, the operator must first open an outer cover. (The purpose of this outer cover is to reduce noise levels from the machine.)

The linkage in Fig. P1.40 has been suggested[†] to avoid the inconvenience for the operator to have to open the outer cover as well as to open the joggler.

[†] By R. E. Baker of IBM, Rochester, MN.

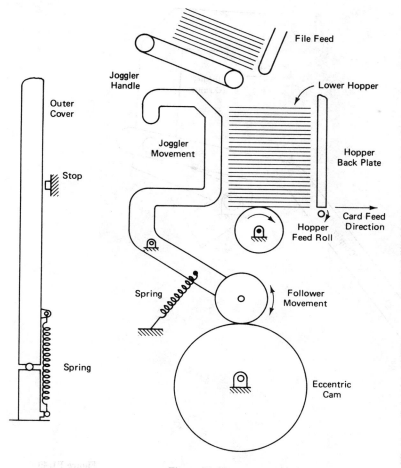

Figure P1.39

(a) Draw the kinematic diagram of the linkage in Fig. P1.39.
(b) Add the change suggested in Fig. P1.40 to Fig. P1.39 and draw the new kinematic diagram.
(c) Determine the degrees of freedom of both mechanisms [parts (a) and (b)].
(d) Determine by graphical construction the total rotation of the joggler if the cover is rotated 90° counterclockwise.

1.27. Double-boom cranes and excavation devices are commonly used in the building construction industry. Their popularity is due primarily to their versatility, mobility, and high load-lifting capacity. This type of equipment is typically actuated by means of hydraulic cylinders. Figure P1.41 shows a typical knuckle boom crane [149].
(a) Draw the unscaled kinematic diagram for this mechanism.
(b) Determine the degrees of freedom for this linkage.

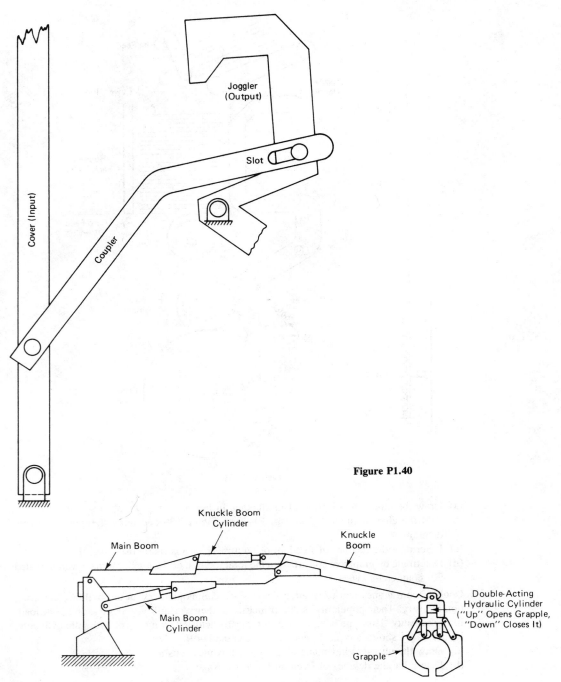

Figure P1.40

Figure P1.41

1.28. (a) Draw the unscaled kinematic diagram of the linkage in Fig. P1.42.
   (b) Determine the degrees of freedom of both the original linkage and a lower-pair equivalent kinematic diagram.

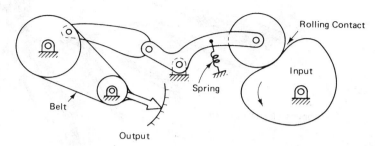

Figure P1.42

1.29. (a) Draw the unscaled lower-pair equivalent kinematic diagram of the linkage in Fig. P1.43.
   (b) Determine the degrees of freedom of both the original linkage and the lower-pair equivalent.**

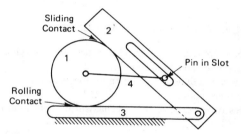

Figure P1.43

1.30. Figure P1.44 shows a surgical tool used for dilating (enlarging) valves. The surgeon squeezes the handle against the spring return, causing a tube running down the center of the long, slender cylinder to extend, thereby opening the mechanism at the end of the cylinder. Disregarding the spring return,
   (a) Draw an unscaled kinematic diagram of the entire tool.
   (b) Determine the degrees of freedom of this device.

1.31. Figure P1.45 shows a linkage-driven pump (U.S. Patent #3927605)* which "is described for pumping fluids at an elevated pressure, wherein the fluid is monitored and activates a transducer member, which in turn adjusts a linkage means for converting reciprocating pumping motion to a transverse oscillatory motion, whereby the pumping stroke is controlled to regulate fluid pressure."

   In layman's language, the input shaft (13), driven at constant speed, drives through the mechanism to deliver required fluid pressure at orifice 36. The stroke (displacement) of piston 26 must be able to change from maximum displacement to zero depending on the pres-

---

** As will be shown later, equivalence of higher and lower pair mechanisms prevails for degrees of freedom, displacements and velocities, but not for accelerations or higher order motion derivatives.

* Courtesy of Graco Inc., Minneapolis, MN.

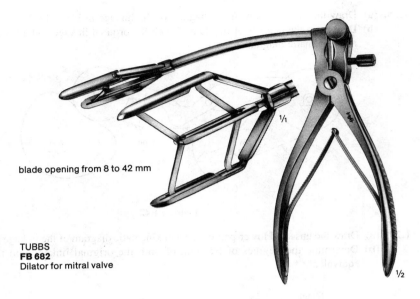

blade opening from 8 to 42 mm

TUBBS
FB 682
Dilator for mitral valve

Figure P1.44   (Courtesy of AESCULAP)

sure at orifice 36. This pressure causes piston 32 to rotate link 24 (which pivots at pin 28) to find a new position against balancing spring 30.

(a) Draw an unscaled kinematic diagram of the entire mechanism.

(b) What type of joint is there between link 25 and 26?

(c) Determine the degrees of freedom of this mechanism by both intuition and Gruebler's equation.

1.32. (a) Draw the unscaled kinematic diagram of the linkage in Fig. P1.46.

(b) Determine the degrees of freedom of this linkage.*

1.33. An end loader is to be designed for attachment onto crawler-type tractors.[†] The linkage must have two degrees of freedom: one allowing the system to lift the bucket and the other to allow the bucket to be tipped while the first is held fixed. Assume that all joints will be those allowing only one relative degree of freedom.

(a) What is the minimum number of binary links that your design must contain?

(b) What is the minimum number of links that would produce a linkage with $F = 2$? [*Hint*: Consider part (a).]

(c) Assuming that a linkage with the fewest number of links in part (b) will not work, what is the fewest number of links in the next more complicated linkage that will work?

1.34. Could one build a 10-link linkage in Prob. 1.33 with the desired degrees of freedom?

If the chosen linkage has nine links ($n = 9$):

(a) How many pairs (joints) will have to be purchased or designed?

(b) What is the maximum number of joint elements that can occur on one link?

(c) How many links can have the maximum number of joint elements?

---

* As will be shown later, equivalence of higher and lower pair mechanisms prevails for degrees of freedom, displacements, and velocities, but not for accelerations or higher order motion derivatives.

[†] This question is used by courtesy of Wm. Carson, University of Missouri, Columbia.

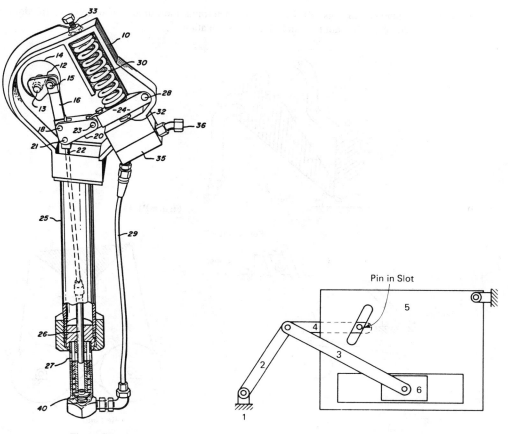

Figure P1.45   Figure P1.46

1.35. The linkage in Fig. P1.47 was designed as a function generator ($y = -x + 2$, for $1 \leq x \leq 4$) using the methods described in Chap. 8. Verify that this geared mechanism has a single degree of freedom.

Figure P1.47

1.36. For the linkages in Figs. P1.48 to P1.57, determine the number of degrees of freedom of the mechanisms by intuition and Gruebler's equation.

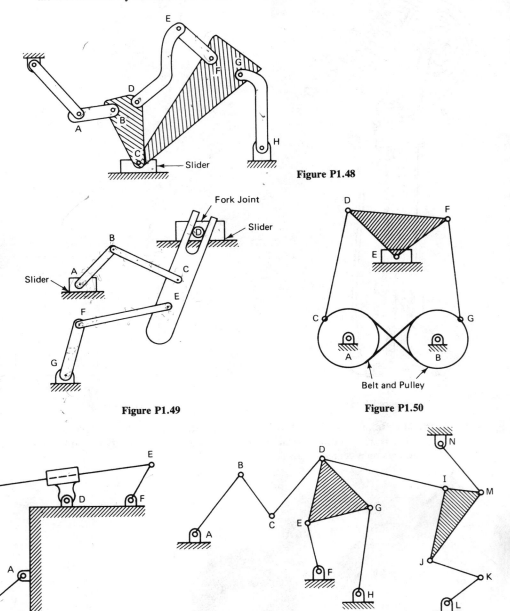

Figure P1.48

Figure P1.49

Figure P1.50

Figure P1.51

Figure P1.52

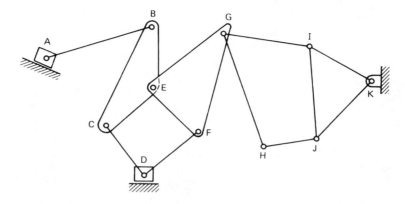

Figure P1.53

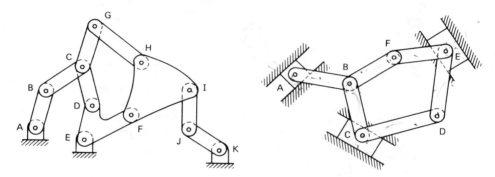

Figure P1.54

Figure P1.55

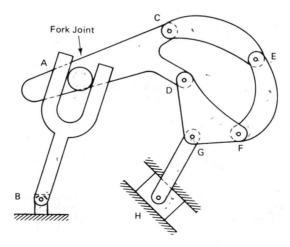

Figure P1.56

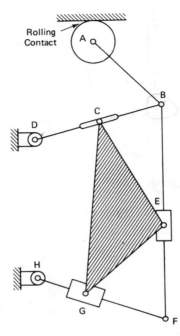

Figure P1.57

1.37. A metering pump [84] was designed such that a movable pivot controls the stroke of the slider (see Fig. P1.58). The pivot is adjustable to any position through a 90° arc about

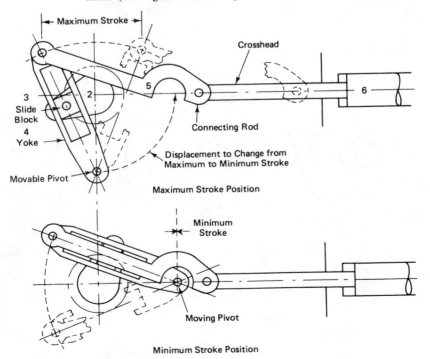

Figure P1.58

the center of the crank. When the crankshaft-to-movable-pivot line is perpendicular to the crosshead motion, the stroke is maximum. When it is in line with the crosshead motion, the stroke is minimum. Draw the unscaled kinematic diagram of this linkage. Determine the degrees of freedom of the linkage.

1.38. What type of six-bar is shown in Fig. P1.51?

1.39. Figures P1.59(a) and (b) are taken from U.S. Patent #3853289, "Trailing Edge Flap and Actuating Mechanism Therefore."* Both the extended and retracted positions are shown. (Note that the trailing flap segment 52 is movably connected to the main segment 34 through track 54 which guides two rollers connected to the trailing segments (the details of which are not shown.)

(a) Draw an unscaled kinematic diagram of this mechanism.

(b) Point out familiar four- and six-bar chains.

(c) How many ternary links are there?

(d) Find the number of degrees of freedom of this mechanism by both intuition and Gruebler's equation.

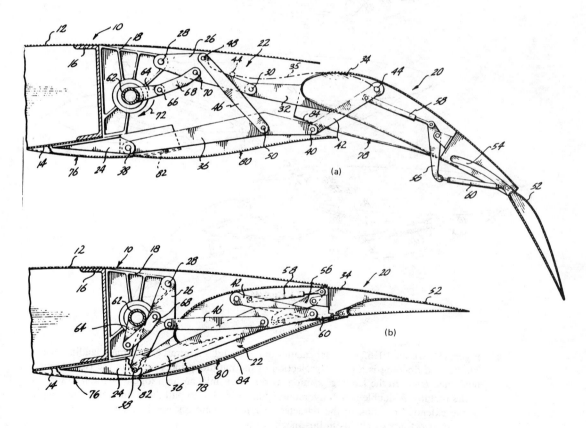

Figure P1.59

* Inventors, C. H. Nevermann and Ellis J. Roscow; Assignee, The Boeing Company, December 10, 1974.

1.40. Figures P1.60(a), (b), and (c) show the cruise, take-off, and landing positions of a short take-off and landing wing design of U.S. Patent #3874617.* The spoiler, *S*, and the forward and aft flaps, *FF* and *AF*, are deployed by the linkage system.

(a) Draw an unscaled kinematic diagram of this mechanism.

(b) Point out familiar four- and six-bar chains.

(c) How many ternary links are there?

(d) Find the number of degrees of freedom of this mechanism by both intuition and Gruebler's equation.

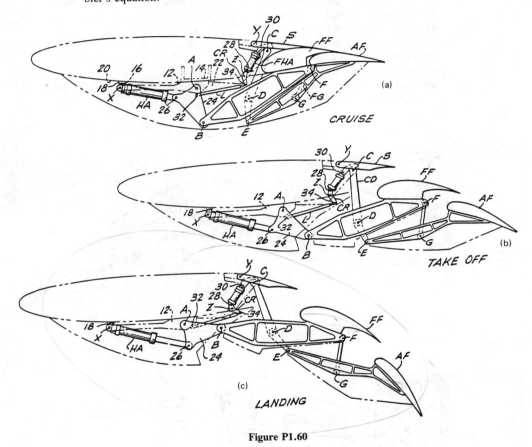

Figure P1.60

1.41. Figures P1.61 and P1.62 show schematics of the Douglas Aircraft MD80 main landing gear.[†] The second drawing is a planar projection of the multiloop mechanism in the two extreme positions. *Hint*: In the landing position, two sets of links are in a toggle position to form a stiff structure. A double action hydraulic cylinder is used to pull these links from the toggle in the extended position to the retracted position, where one set is again in toggle. Also look for two four-bar chains in this mechanism.

* Inventor, Robert E. Johnson; Assignee, McDonnell Douglas Corporation, April 1, 1975.

[†] Courtesy of Douglas Aircraft Company, Long Beach, California.

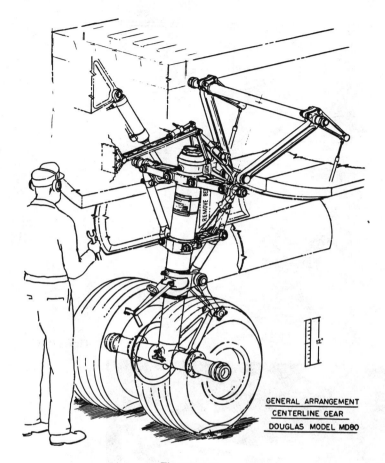

Figure P1.61

(a) Draw an unscaled kinematic diagram of the mechanism.

(b) Determine the number of degrees of freedom of the mechanism.

(c) Can you find any embedded six-bars? If so, where are they and what types are they?

1.42. As shown in Figures P1.63 and P1.64, a new mechanism was designed* to turn the pages of music books automatically. To the upper right corner of each right-hand page a magnetic strip is attached. The linkage (possibly actuated by a foot switch) will generate a path, such that the path tracer point (also magnetized) will turn the page.

(a) What task does this mechanism satisfy?

(b) Prepare an unscaled kinematic diagram of this mechanism.

(c) Verify the degrees of freedom by both Gruebler's equation and by intuition.

(d) What type of six-bar linkage is this?

(e) Can you suggest a different six-bar or a four-bar mechanism for this task?

* By Brad Wilke and Steve Toperzer.

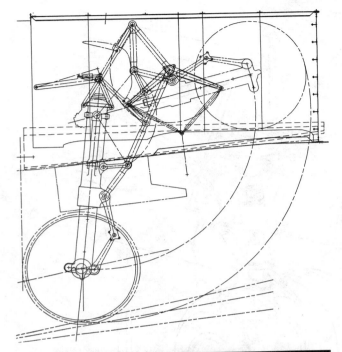

Figure P1.62

Figure P1.63

Figure P1.64

1.43. For the mechanisms in Figures P1.65–P1.71, answer the following questions:
  (a) What task does this mechanism satisfy?
  (b) Verify the degrees of freedom by both Gruebler's equation and by intuition.
  (c) Can you find any four-link or six-link chains? (For the latter, what type of six-bar did you find?)

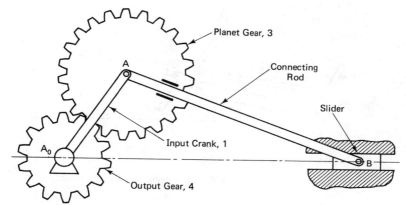

Figure P1.65 Gear-slider mechanism.

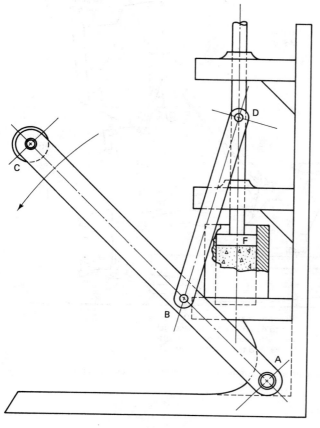

Figure P1.66 Mechanical press mechanism.

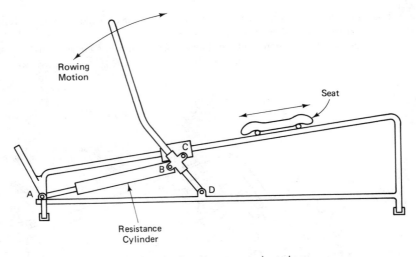

Figure P1.67   Rowing type exercise trainer.

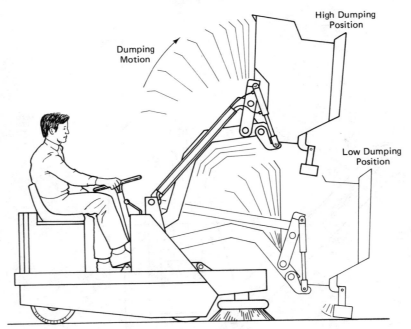

Raises hydraulically to 5 ft. — rolls
out to dump into containers, pickup
beds. Also dumps floor level, off
docks, into trash compactors.

Figure P1.68   Dumping mechanism for floor sweeper. (*Courtesy of Tennant Inc.*)

Figure P1.69 Bidirectional skylight. Rotating the gears clockwise will open the skylight straight up and then tilt it as shown. Counterclockwise rotation of the gears will tilt the skylight in the opposite direction.

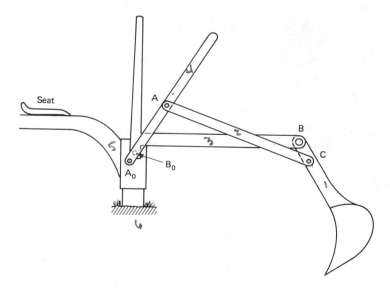

Figure P1.70 Playground shovel mechanism.

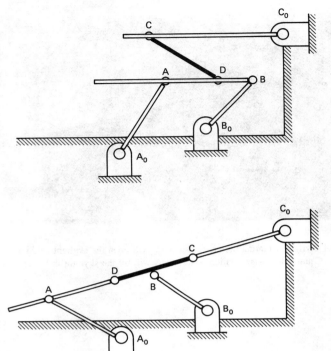

Figure P1.71   The step to ramp mechanism, for the handicapped, was designed by James Wandzel, Jeffrey Kivi and Beky Feist.

chapter two

Computer-Aided Mechanism Design Philosophy

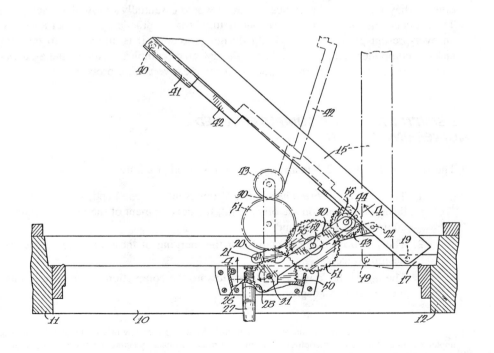

2.1 INTRODUCTION

What is design? Can creativity be taught? How do we begin the design process? Where does kinematic analysis and synthesis fit into engineering design? Can engineering design be rationalized and systematized? Where does the computer fit into the design process? Can a design methodology or philosophy be formulated, practiced, and taught to student engineers? Can scientific background and innate human intuition be augmented by a design discipline to enhance creative engineering performance?

These and related questions can be answered in the affirmative [18, 34, 79, 122, 141]. Computer-aided design discipline is attaining a degree of maturity and importance that warrants its discussion in a text on kinematics.

The complex process of creative engineering design is subject to infinite variations [16, 33, 91, 92, 131, 137, 165]. One purpose here is to present a general guideline, in the form of an uncomplicated flowchart, which is readily kept in mind by designer and student alike, and can thus serve as an aid of broad applicability in both practice and education. Another purpose is to show what kind of computer software is appropriate in the successive stages of the design process.

The Seven Stages of Engineering Design [142] (Fig. 2.1) were evolved some years ago, partly on the basis of published works and partly on the basis of experience in the practice and teaching of engineering design. Although different successful designers may use different terminology or have slight variations in the sequence (depending on which domain they may have experience in), most would essentially follow these seven steps.* The sequence quickly becomes "second nature" for the practicing designer and serves as an ever-present guideline in tackling design problems. It is applicable to the simplest tasks in component design as well as to the design of complex systems, and assures complete coverage of the significant phases in the creative design process.

2.2 THE SEVEN STAGES OF COMPUTER-AIDED ENGINEERING DESIGN

The flowchart in Fig. 2.1 is arranged in a Y-shaped structure.

1. The two *upper branches* of the Y represent, on one hand, the evolution of the design task, and on the other hand, the development of the available, applicable engineering background.
2. The *junction* of the Y stands for the merging of these branches: generation of design concepts.
3. The *leg* of the Y is the guideline toward the completion of the design, based on the selected concept.

* The reader is referred to the Appendix to this chapter for a review of contemporary theories and AI implementation of design methodology. Here modern research topics spanning the design process are related to the Seven Stages of Engineering Design.

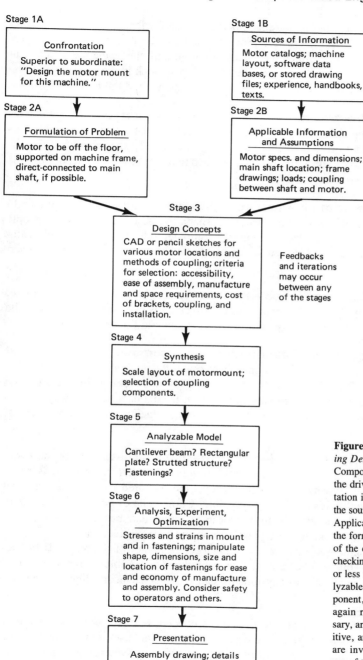

Stage 1A

Confrontation

Superior to subordinate:
"Design the motor mount
for this machine."

Stage 1B

Sources of Information

Motor catalogs; machine
layout, software data
bases, or stored drawing
files; experience, handbooks,
texts.

Stage 2A

Formulation of Problem

Motor to be off the floor,
supported on machine frame,
direct-connected to main
shaft, if possible.

Stage 2B

Applicable Information
and Assumptions

Motor specs. and dimensions;
main shaft location; frame
drawings; loads; coupling
between shaft and motor.

Stage 3

Design Concepts

CAD or pencil sketches for
various motor locations and
methods of coupling; criteria
for selection: accessibility,
ease of assembly, manufacture
and space requirements, cost
of brackets, coupling, and
installation.

Feedbacks
and iterations
may occur
between any
of the stages

Stage 4

Synthesis

Scale layout of motormount;
selection of coupling
components.

Stage 5

Analyzable Model

Cantilever beam? Rectangular
plate? Strutted structure?
Fastenings?

Stage 6

Analysis, Experiment,
Optimization

Stresses and strains in mount
and in fastenings; manipulate
shape, dimensions, size and
location of fastenings for ease
and economy of manufacture
and assembly. Consider safety
to operators and others.

Stage 7

Presentation

Assembly drawing; details
of mount and coupling
elements; parts list.

Figure 2.1 *The Seven Stages of Engineering Design* as applied to component design. Component involved: mounting bracket for the drive-motor on a machine. The confrontation is simply a superior's instruction, and the sources of information are clearly at hand. Applicable information is easily looked up; the formulation of the problem and selection of the design concept may involve repeated checking with the superior. Synthesis is more or less routine for this example, but the analyzable model, even of such a simple component, calls for some thought. Analysis is again routine, experiment is hardly necessary, and optimization, if any, is purely intuitive, and is justified only if large quantities are involved. The presentation may make use of the graphics capability of a CAD system including solid modeling and/or color shading for display of maximum stress.

The flowchart implies, but is not encumbered by, the feedbacks and iterations that are essential and inevitable in the creative process. As it stands, it is one possible representation of the design process, which has been applied in both the academic area and in professional practice. Referring now only to the titles of each block in the diagram and disregarding the rest of the text in each block for the moment, we find the following stages.

Stage 1A: Confrontation.   The confrontation is not a mere problem statement, but rather the actual encounter of the engineer with a need to take action. It usually lacks sufficient information and often demands more background and experience than the engineer possesses at the time. Furthermore, the real need may not be obvious from this first encounter with an undesirable situation.

Stage 1B: Sources of information.   The sources of information available to the engineer encompass all human knowledge. Perhaps the best source is other people in related fields. Information data bases are useful computer aids at this stage. Parts catalogs and design information may be readily available on the computer as part of expert systems.

Stage 2A: Formulation of problem.   Since confrontation is often so indefinite, the engineer must clarify the problem that is to be solved: It is necessary to ferret out the real need, and define it in concrete, quantitative terms suitable for engineering action.

Stage 2B: Preparation of information and assumptions.   From the vast variety of sources of information the designer must select the applicable areas, including theoretical and empirical knowledge and, where information is lacking, fill the gap with sound engineering assumptions. Retrieving information from a data base by way of keywords entered in a computer terminal or by a hierarchy of menus, multiple screens, or icons can be helpful here.

Stage 3: Generation and selection of design concepts.   Here the background developed by the foregoing preparation is brought to bear on the problem as it was just formulated, and all conceivable design concepts are prepared in schematic skeletal form, drawing on related fields as much as possible. Compendia of designs and standard component banks, stored in graphic form, and/or in relational data bases, are useful here. Computer graphics is most useful for trying out concepts for preselection.

It should be remembered that creativity is largely a matter of diligence. If the designer lists all the ideas that can be generated or assimilated, workable design alternatives are bound to develop, and the most promising can be selected in the light of requirements and constraints.

Stage 4: Synthesis.   The selected design concept is a skeleton. We must give it substance: fill in the blanks with concrete parameters with the use of systematic design methods guided by intuition. Compatibility with interfacing systems is essential. In some areas, such as kinematic synthesis, advanced analytical, graphical, and combined, computer-aided methods have become available [6, 23, 25, 26, 46, 50, 57-62, 93, 97-99,

152, 166]. Synthesis algorithms, program libraries for linkage synthesis, spring and dash-pot designs, electrical and electronic circuit synthesis, analog and digital control system synthesis (and more) are all available from the software literature. However, intuition, guided by experience, is the traditional approach.

Stage 5: Analyzable model.   Even the simplest physical system or component is usually too complex for direct analysis. It must be represented by a model amenable to analytic or empirical evaluation. In abstracting such a model, the engineer must strive to represent as many of the significant characteristics of the real system as possible, commensurate with the available time, methods, and means of analysis or experimental techniques. Typical models are simplified physical versions, mechanical electrical analogs, models based on nondimensional equivalence, mathematical models, free-body diagrams, and kinematic skeletal diagrams. Computer models, like solid or wire frame models of objects on graphics terminals, both two- and three-dimensional; mathematical models of mechanisms [20-22, 59, 60, 93, 147, 169-171] flow patterns and conformal mappings for potential fluid flow and for conductive heat transfer are examples of useful computer aids.

Stage 6: Experiment, analysis, optimization.   Here the objective is to determine and improve the expected performance of the proposed design.

1. Design-oriented experiment, either on a physical model or on its analog, must take the place of analysis where the latter is not feasible. Computer aids in experimentation include direct data acquisition, real-time data processing while the experiment is in progress, graphic representation and computer analysis of experimental results.

2. Analysis or test of the representative model aims to establish the adequacy and responses of the physical system under the entire range of operating conditions. As a computer aid, software exists for kinematic and dynamic analysis of mechanisms and structures, linear and nonlinear analysis of control systems, finite element analysis for stress and strain of complex geometries, and for dynamic responses in physical systems.

3. In optimizing a system or a component, the engineer must decide three questions in advance: (a) With respect to what criterion or weighted combination of criteria should the designer optimize? (b) What system parameters can be manipulated? (c) What are the bounds on these parameters; and what constraints is the system subject to? After these decisions are made, various computer aids can be used for the actual numerical work. These include linear and nonlinear programming, curve fitting, and classical extrema-seeking methods of the first, second, and higher order, to mention a few.

Although systematic optimization techniques have been and are being worked out (such as linear, nonlinear, and dynamic programming; digital computational and heuristic methods in kinematic synthesis) [65, 67-70, 168, and the appendix of this chap.], this stage is largely dependent on the engineer's intuition and judgment. The amount of optimizing effort should be commensurate with the importance of the function or the system component and/or the quantity involved.

Experiment, analysis, and optimization form a closed-loop stage in the design process. The loop itself may be iterative, and the results may give rise to feedbacks and iterations involving any or all of the previous stages, including a possible switch to another design concept.

Stage 7: Presentation.   No design can be considered complete until it has been presented to (and accepted by) two groups of people:

1. Those who will use it, and
2. Those who will make it.

The engineer's presentation must therefore be understandable to the prospective user, and contain all the necessary details to allow manufacture and construction by the builder. Computer preparation of documentation, such as drawings, renderings in perspective form, performance data and charts obtained from computer models of the physical system, financial charts and graphs, instant display of effects of changes, are useful computer aids at this final stage. Technological advancements allow merging of design results (drawings, charts, etc.) with word processing in "cut-and-paste" type operations often performed right on the monitor screen. This permits easy flow from the design-documentation stages without need for recreating information. (Refer to the color inserts in this book for examples of types of computer output that is very useful in the presentation stage.)

Iterations.   Clearly, creative design is not a one-way, single-pass effort. It is often necessary to retrace one's steps: feedbacks and iterations may occur at any stage. If, at the analysis stage, undesirable responses are discovered and resynthesizing cannot correct these, perhaps a new concept is in order. If no suitable concept can be generated, perhaps the problem should be redefined. The designer should not be distressed by these setbacks, but rather regard them as opportunities to create superior designs on a better-informed basis. Well-thought-out data structures allow one to save numerous potential designs in computer memory which may be called up later.

One aspect of iteration, partial or even complete redesign, may result from safety analysis. It is incumbent upon the designer to consider the safety of the user or operator and the public, not only in the normal use of the product, but also in foreseeable misuses or even abuses. To discharge this responsibility, the designer must

1. Seek to identify each possible hazard;
2. Change the design to eliminate the hazard;
3. If 2 is not feasible, guard the hazard;
4. Warn against the hazard by instruction and placard.

In order to satisfy the safety aspect of the design, the designer should become familiar with federal, state, local and industry standards and guidelines applicable to the design.

2.3 HOW THE SEVEN STAGES RELATE TO THIS TEXT

The purpose of this book and Vol. 2 is to serve both as an educational tool and as a resource book for the mechanism designer. Here are a few ways in which these books can be utilized in following the seven stages of engineering design.

1. *Confrontation and sources of information.* The mechanism nomenclature and typical examples of Chap. 1 can be drawn on in this stage.

2. *Formulation of the problem.* See Chaps. 1, 2 and 8 (Vol. 1) for tasks of synthesis, and Table 2.3 for a troubleshooting guide.

3. *Generation of design concepts.* The many examples throughout this book, as in Chap. 1, as well as the type synthesis of Chap. 8, are helpful at this stage.

4. *Synthesis.* The chief sources here are Chapters 8 (Vol. 1) and 3 (Vol. 2) on kinematic synthesis, as well as portions of Chap. 6 (Vol. 1) on cams and Chap. 7 (Vol. 1) on gear trains. Additional synthesis information is offered in Chap. 4 (Vol. 2) on path curvature, and Chap. 6 (Vol. 2) on spatial mechanisms.

5. *Analyzable model.* Chapters 3 through 7 (Vol. 1) and Chaps. 4, 5, and 6 (Vol. 2) all hold valuable information on this subject.

6. *Analysis, experiment, and optimization.*
   a. For analysis, see Chaps. 3 to 7 (Vol. 1) and Chaps. 4, 5, and 6 (Vol. 2).
   b. For experiment, see Chap. 5 (Vol. 1) and Chap. 5 (Vol. 2).
   c. For optimization, see Chap. 3 (Vol. 2).

7. *Presentation.* The simplified kinematic diagrams of Chap. 1 (Vol. 1) and simulation using computer graphics (Chap. 8, Vol. 1 and Chap. 3, Vol. 2) will often help describe the merits of a proposed mechanism.

2.4 HISTORY OF COMPUTER-AIDED MECHANISM DESIGN

Many of the basics of mechanism analysis and synthesis presented in this book and Vol. 2 were known over one hundred years ago. Many of these techniques, which tend to be graphical in nature, can be made more useful to the mechanism designer by having the computer carry out the repetitive portions of the constructions with much greater precision than is possible manually. The designer can then concentrate on the more creative aspects of the design process, which occur in stages 4, 5 and 6, namely synthesis, abstracting the analyzable model, and experimenting with various designs interactively on the computer. Thus, although the drudgery is delegated to the computer, the designer's innate creativity remains in the "loop."

Application of the computer to mechanism problems has had a relatively short history. The evolution started from mainframe analysis codes and has progressed to user-friendly design methods on the desktop or laptop personal computer. Table 2.1 shows a historical perspective on the first thirty years of computers applied to mechanisms [59], and the following paragraphs summarize the events decade by decade.

1950s

The 1950s saw the first introduction and availability of the digital computers in industry and engineering programs at universities. Some 36 programs are referenced [75], most originating in universities. Several programs were developed by Al Hall et al. at Purdue, C. W. McLarnan's group at Ohio State, J. E. Shigley et al. at Michigan, F. Freudenstein's group at Columbia, and J. Denavit and R. Hartenberg at Northwestern. Freudenstein reviewed the computer programs developed for mechanism design prior to 1961 [75]. In 1951 Kemler and Howe introduced "perhaps the earliest published reference on computer applications in mechanism design; illustrates calculations of displacements, velocities, and accelerations in quick-return mechanisms" [94].

One of the early contributions which used the computer for linkage synthesis was that of Freudenstein and Sandor [78], who adapted the graphical-based techniques suggested by Burmester in 1876 and reformulated these for computer solution. The resulting complex synthesis equations were solved in batch mode on an IBM 650. This work formed the technical basis for the KINSYN and LINCAGES codes which emerged in the 1970s.

1960s

The computer became more available to university researchers in the early to mid 1960s. Many researchers began to utilize the power of the computer for solving equations which were too tedious by either graphical, slide rule, or electromechanical desk-calculator techniques. The mid to late 1960s saw synthesis problems being solved in the batch mode on the computer by either precision-point or optimization type techniques. The area of dynamic rigid-body mechanism analysis and linkage balancing began to emerge based on the power of the digital computer. Although there was some initial success with analog and hybrid (combined analog and digital) computers in solving differential equations of motion, numerical methods for integration, such as Runge-Kutta, caused the analog devices to be phased out.

1970s

The early 1970s saw a spurt in applications on the computer. Codes such as IMP, developed by P. Sheth and J. Uicker at the University of Wisconsin, and DRAM and ADAMS, developed at the University of Michigan by D. Smith, N. Orlandea, and M. Chace, had early roots in this decade. Computing slowly switched from strictly batch to interactive, which was a significant step in making techniques more useful to designers. Also, computer graphics applied to mechanism design received its christening in the early 1970s by Kaufman. KINSYN I was a custom-built program at M.I.T. and should be recognized as the major milestone in kinematic design. The digital computer alone took us halfway toward useful computer-aided design of mechanisms. Computer graph-

TABLE 2.1 HISTORY OF COMPUTER-AIDED MECHANISM DESIGN

	1951	1960	1970	1980	1990
COMPUTER GRAPHICS (1)	●Light Pen Concept		●In House Systems (Industry) ●"Turn Key" Systems	●Raster Graphics ●Significant "Turn Key" Use	●Solid Modeling
FINITE ELEMENTS (2)		●In House Programs (Industry)	●Start of Commercial Programs ●Serious Commercial Use	●Pre and Post Processors	●CAD* Integration
MECHANISM ANALYSIS (3)	●In House Programs (Univ.)	●Cam Design and Dynamic	●Linkage Balancing, Significant Industrial Use ●General Packages Start	●Numerous Commercial Packages	
MECHANISM SYNTHESIS (3)		●First Four Bar Synthesis ●Multiloop Linkages	●Interactive Graphics	●Serious Industrial Use ●Micro Computer Programs	
MECHANISM OPTIMIZATION		●First Activity	●Increased Major Activity	●Rebirth of Interest	
HIGH-SPEED MECHANISMS		●Partially Flexible Linkages	●Use of Finite Element Theory ●Effect of Clearances	●More Serious Industrial Interest	
TYPE SYNTHESIS (4)			●Enumeration by Computer ●Increased Activity	●Use of Expert Systems	
CAD* INTEGRATION				●Identification of Need	●Significant Progress ●Computer-Aided Eng.

●Start of Activity

Sources of information and colleagues who assisted in generating this table are: (1) S. H. (Chase) Chasen Lockheed, Georgia Company and D. R. Riley University of Minnesota; (2) J. Glaser CDC Corp; (3) F. Freudenstein, Columbia Univ. and A. H. Soni, University of Cincinnati; and (4) A. H. Soni.

*Computer Aided Design

81

ics for input, output as well as to enhance interaction in design-decision making was the second required ingredient. By the mid to late 1970s, several other software packages for synthesis and analysis became available.

1980s

The 1980s exhibited a burst in activity in mechanisms for several reasons. Microcomputers became generally available, and several different groups began to develop and market software on micros like LINCAGES, IMP, KADAM 2 (Williams) and MCADA (Orlandea). Refer to the color inserts in this book for examples of some of this software. The desktop and laptop computers of the 1980s replaced the main frames of yesterday. The area of robotics also played a role in raising interest of colleagues in related fields toward the importance of kinematics in this "hot topic." Many researchers have used kinematic theories to investigate different aspects of robotics, such as three-dimensional animation, work space prediction, interference calculations, dynamic response, etc.

The 1980s have also seen the beginning of integration of mechanism analysis, synthesis, and dynamics with other computer-aided design areas, such as drafting, finite elements and simulation.

The 1990s and On

Integration of the computer into mechanism design looks very exciting. The mechanism designer has available an impressive set of tools at his or her disposal for optimal analysis and design of mechanisms. Several specific areas will see increased activity. These include (1) use of solid modelers for the display and analysis of 2-D and 3-D mechanisms, (2) integration of mechanism analysis and synthesis software into other phases of computer-aided design and manufacture, (3) many more custom applications to specific needs of industry, (4) more computer-assisted analysis and design for machine elements: gears, cams, indexers, etc., (5) better techniques for analysis and simulation of more complex problems, including clearances, deflections of links, friction, damping, etc., (6) the development of computer-aided type synthesis techniques for designers, useful in both stages 3 and 4 of Fig. 2.1, which include expert systems and artificial intelligence techniques, (7) the use of sophisticated graphical interfaces resulting in very user-friendly software, (8) increased development of mechanism design software on micro-, desktop, and laptop computers, and (9) use of the super computer that permits large-scale design optimization, parallel processing, and simulation.

2.5 DESIGN CATEGORIES AND MECHANISM PARAMETERS

The purpose of this section is to discuss the full range of possible design parameters (variables) that may be designated in a particular problem solution. The discussion here is applicable to mechanisms in general even though the parameters commonly needed to be specified are illustrated in Fig. 2.2 with the planar four-bar linkage. The four-bar of this figure is also used throughout this section to describe the design categories.

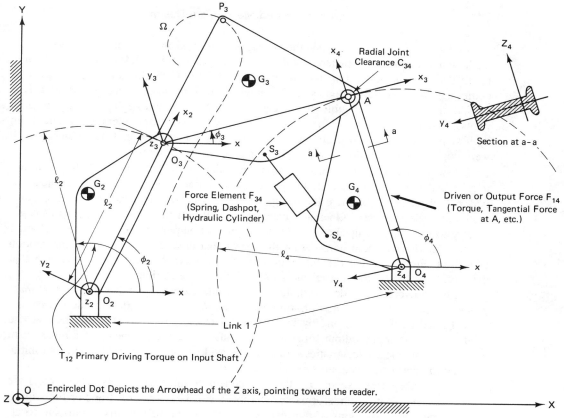

Figure 2.2

General Nomenclature for Planar Mechanism

i: link number (link 1 is the frame)

X, Y, Z: fixed or inertial coordinate axes attached to the ground

x_i, y_i, z_i: coordinate axes attached to link i

ϕ_i: angular position of link i measured from the direction of the X axis

Ω: path traced by point P_i in the plane of the fixed link

Design Parameters

A_i: set of cross-sectional properties of link i at each section a-a (including area, area moments of inertia)

C_{ij}:   set of joint clearances between links i and j

E_i: material properties of link i (modulus of elasticity, Poisson's ratio, shear modulus, etc.)

F_{ij}:   set of parameters describing a force-producing element; also referred to as force F_{ij}, attached between links and directed from i to j

G_i: mass center coordinates $(x, y, z)_i$ of link i

J_i: mass moment and products of inertia of link i

L_i: set of geometric parameters for link i (link length l_i, orientations of slider joints, cam profile, gear ratios, etc.)

M_i: mass of link i

O_i: global coordinates X, Y, Z of link i attachment points to the ground

P_i: local coordinates $(x, y, z)_i$ of a point on link i whose path is of interest

S_i: local coordinates $(x, y, z)_i$ of attachment point to link i of a force-producing element

T_{ij}:   similar to F_{ij} but describes the primary driving torque associated with the input shaft and link

ϕ_{io}:   initial value of ϕ_i

Mechanism analysis and synthesis can be classified into two major groups: kinematic and dynamic, as depicted in Table 2.2. Categories K1 to K4 are kinematic while D1 to D5 are considered dynamic. In categories K1 to D4 the links are treated as rigid bodies and in D5 as elastic bodies. In all categories with the exception of D4, synthesis is carried on assuming negligible clearances and in D4 with non-negligible clearances in joints. As shown in Table 2.2, a distinct group of design parameters are available in each step, and in general each successive step has some additional parameters which were not available and do not affect the objectives in prior steps.

Not all categories are pertinent to a particular mechanism problem. If one is not required, it is simply bypassed. Frequently, consideration of the kinematic objectives may be sufficient to complete the design of slowly operated mechanisms such as automotive throttle linkages. Kinematic categories deal with the effects of mechanism geometry on the relationships between (1) the input and output motions (position, velocity, and acceleration) without considering the forces that created the *assumed* input motion $\phi_2 = f(t)$ (Fig. 2.2), and (2) the input–output and internal joint forces, assuming that the forces created by the inertial mass of the links are negligible.

Dynamic categories deal with the effects of mechanism geometry and inertial mass properties on the (1) input motion–time response, $\phi_2 = f(t)$, created by the torque and force T_{ij} and F_{ij} (Fig. 2.2) and by friction forces in the joints; and (2) the shaking and joint forces, and input–output force transmissions. Since the geometry of a mechanism affects its dynamic characteristics, kinematic considerations are almost always an initial and integral part of a dynamic synthesis problem.

In all categories except D5, the mechanism components (links, cams, etc.) are treated as rigid, that is, undeformable, members. For heavily loaded and/or high-speed mechanisms, the deflections of the members may become sufficiently significant to affect adversely the attainment of the kinematic and dynamic objectives; thus the members must be considered as elastic bodies.

As shown in Table 2.2, in general, each successive category helps determine additional design parameters. Therefore, an approach that breaks a design synthesis problem into smaller steps is to start with, say, a kinematic synthesis for a limited number of design objectives of either function generation, rigid-body guidance (motion generation), or path generation, and then determine geometric parameters to satisfy these objectives. Then, one may step up to a dynamic synthesis, say, by taking forces into account and determining additional unknown parameters to satisfy further design objectives, such as limits on dynamic loads. One must realize, however, that with the design parameters from previous steps fixed, it may be impossible to meet the design objectives of a subsequent step. If this occurs, or additional objectives emerge, an iteration in the design process is necessary, resulting in the familiar trek "back to the drawing board." This could involve returning to a previous step to select an alternate set of parameters which satisfy the objectives in that step as well as in subsequent steps, or the iteration could involve selecting a new type of mechanism or a revision in the objectives. The probability of achieving better designs generally improves with each such iteration.

TABLE 2.2 UNKNOWN DESIGN PARAMETERS TO BE DETERMINED BY SYNTHESIS AND ANALYSIS

	Design Category	L_i Dimensions	ϕ_{i0} Initial positions	O_i Joint locations	P_i Tracer point positions	G_i Mass-center locations	M_i Link masses	J_i Link mass-moments of inertia	S_i Locations of force elements	F_{ij} Type and magnitude of force elements	T_{ij} Required input torque	C_{ij} Joint clearances	A_i Cross-sectional properties	E_i Material properties
K1	Function generation	×	×											
K2	Rigid-body guidance	×	×	×	×									
K3	Path generation	×	×	×	×									
K4	Static force analysis	×	×	×		×*	×*		×	×				
D1	Balancing	×				×	×	×	×					
D2	Dynamic force analysis	×		×	×	×	×	×	×	×				
D3	Motion–time response analysis	×		×	×	×	×	×	×	×	×			
D4	Effects of clearances	×		×	×	×	×	×	×	×	×	×		
D5	Elastic body dynamics	×		×	×	×	×	×	×	×	×		×	×

*Only component weights are significant, not inertial properties.

Table 2.2* serves a very useful function in the iterative design process by showing which parameters affect the attainment of each design objective. For example, if balancing is a problem, relocating the centers of mass G_i can be done without affecting the attainment of the kinematic objectives, with the possible exception of K4. However, changing link geometry L_i will in general affect attainment of the kinematic as well as balancing objectives.

Design Categories and the Parameters of Fig. 2.2

The kinematic "K" and the dynamic "D" categories can be illustrated with the four-bar linkage shown in Fig. 2.2. Alternate key titles or words associated with a category are given in parentheses.

K1 *Function generation* (coordination of input and output position, velocity, and/or acceleration)

Objective: The output angle ϕ_4 is to change in a prescribed manner with respect to input position ϕ_2; that is, the function $\phi_4 = f(\phi_2)$ is to be generated.

Examples: (a) Automotive throttle linkage; (b) automotive valve lifter mechanism transforms camshaft rotations to desired valve positions.

K2 *Rigid-body guidance* (motion generation)

Objective: A body, link 3, not directly jointed to the ground, must pass through specified positions and orientations.

Examples: (a) The bucket of a front loader; (b) power tailgate of a truck used as a lifter platform; (c) single-piece overhead garage door.

K3 *Path generation* (coupler curve generation; position, velocity, and/or acceleration at points along a point path).

Objective: Some point on a link, P_3, is to trace a desired path, Ω, on another link (usually the fixed link).

Examples: (a) D-shaped curve of a movie film transport mechanism; (b) Watt's linkage producing straight-line motion of his steam-engine piston, replaced by the piston rod guide (crosshead) on later steam engines.

K4 *Static forces* (transmission angle, mechanical advantage)

Objectives: (1) Attainment of desired magnitudes of the driven output force F_{14} at specified corresponding input positions ϕ_2; the driving torque T_{12} is to be a specified value; and/or (2) the forces transmitted through the joints must be held below the load capacity of the bearings.

Examples: (a) Scissors jack; (b) designing the hydraulic cylinder of a front loader to create a desired lifting capacity.

* See Sect. 5.5 for a case study that uses this table.

D1 *Balancing* (inertial shaking force and/or moment)

Objective: The net cyclic forces and moments due to the inertia of the moving links, which are transmitted to the foundation and which cause vibrations, are to be reduced.

Example: Locating and sizing counterweights for a reciprocating engine or compressor.

D2 *Dynamic forces* (inertia forces, dynamics of machines, kinetostatic analysis)

Objectives: The objectives are the same as in K4, finding static forces, but with the addition that: (1) the inertial forces are considered significant, and (2) the input link angular velocity, $\dot{\phi}_2$, and acceleration, $\ddot{\phi}_2$, are also assumed to be known or specified.

Example: To determine (for example) the required driving torque, horsepower, and resulting joint forces of a punch press.

D3 *Motion–time response* (input-torque balancing, force-system synthesis)

Objective: Determine an applied force system for a mechanism, including inter-link torques and forces T_{ij} and F_{ij}, which result in a desired input motion–time relationship, $\phi_2 = f(t)$, and/or specific output forces, F_{14}, during a specified period of time.

Examples: (a) Flywheel or spring design to obtain more nearly constant punch press drive shaft speed; (b) spring-absorber design for an electric circuit breaker, such that closure time and rebound are minimized and that the desired contact force is exerted.

D4 *Effects of clearances* (and tolerances)

Objectives: To limit (1) the inaccuracies in attainment of the kinematic objectives, (2) the joint-force increases due to impact, and (3) associated vibrations and noise caused by clearances in joints.

Example: Misregister in offset printing presses due to gear backlash and bearing clearances.

D5 *Elastic-body dynamics* (flexible link, kinetoelastodynamics)

Objective: Reduce the problems of link deformations, buckling, and "miss" of kinematic objectives by limiting the stresses and elastic deflections within components to acceptable levels.

Example: In the design of high-speed mechanisms, there is a trade-off between reducing inertial forces by decreasing the mass of components on one hand, and on the other hand, the increased deformability that results, causing distortion of the desired motion path. For instance, the "hand" or gripper of a high-speed automation module may miss its target.

TABLE 2.3 TROUBLESHOOTING GUIDE

Problem symptom	Possible mechanism cause	Design category	Chapter reference
Missing a desired position	1. Incorrect geometric parameters due to (a) design or (b) machining error(s)	K1, K2, K3, K4	1, 6, 7, and 8 (Vol. 1) and 3, 4, and 6 (Vol. 2)
	2. Incorrect motion–time response	D3	5 (Vol. 2)
	3. Excessive clearances	D4	—
	4. Excessive component deflections	D5	5 (Vol. 2)
Transmissibility (inadequate force transducer: i.e., unable to lift an object)	1. Poor transmission angle or mechanical advantage	K4	3 (Vol. 1)
	2. Excessive friction	K4, D2	5 (Vol. 1) and 5 (Vol. 2)
	3. Excessive inertial forces	D2	5 (Vol. 1) and 5 (Vol. 2)
	4. Incorrect attached force elements	D3	5 (Vol. 1) and 5 (Vol. 2)
Shaking the supporting base	1. Mechanism out of balance	D1	5 (Vol. 2)
	2. Varying drive speed and torque	D3	5 (Vol. 1) and 5 (Vol. 2)
	3. Excessive accelerations/decelerations, impact of links against stops	D3	5 (Vol. 2)
	4. Excessive clearances	D4	—
	5. Excessive elastic deflections of components	D5	5 (Vol. 2)

Breaking at the connections

1. Mechanism cannot execute the required movement	Phase 1: K1, K2, K3	1, 3, 6, 7, and 8 (Vol. 1) and 3, 4, and 6 (Vol. 2)
2. Excessive static forces	K4	3 and 5 (Vol. 1)
3. Excessive dynamic forces	D2	5 (Vol. 1) and 5 (Vol. 2)
4. Excessive clearances and impact forces	D4	5 (Vol. 1) and 5 (Vol. 2)

Bending of links

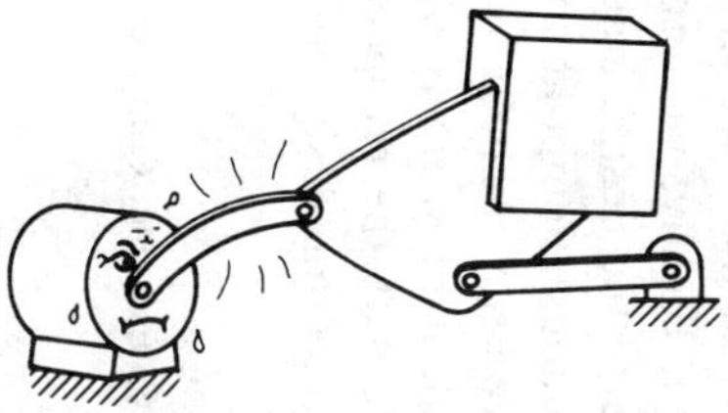

1. See "Breaking"		
2. Excessive component vibrations and deflections	D5	5 (Vol. 2)

Noise

1. Impact against rigid stops	D3	5 (Vol. 2)
2. Excessive clearances	D4	—
3. Resonance	D5	5 (Vol. 2)

2.6 TROUBLESHOOTING GUIDE: SYMPTOMS, CAUSES, AND SOURCES OF ASSISTANCE

A practicing designer may not recognize a problem by one of the technical names listed in the preceding section. Unfortunately, many of the dynamic malfunctions in Table 2.3 are only dealt with after the trouble occurs either in a prototype device or in the field. An engineer may be assigned to solve the mechanism problem based on symptoms of failure. Table 2.3 provides a means of translating the physically *perceived symptoms* of "miss," "break," "bend," "shake," "transmit," and "noise" to possible causes, design categories, and sources of assistance. Section 5.5 provides a design case study in which the strategies of this table were used in an industrial situation.

APPENDIX: CONTEMPORARY THEORIES AND AI* IMPLEMENTATIONS OF DESIGN METHODOLOGY[1]

In this appendix we discuss some of the more important theories of design, especially as they have been motivated by attempts to automate the design process. The automation we have in mind goes beyond computer-aided design, since it calls upon the software to make decisions. However, in the 1980s complete automation of the design process was still well in the future; a practical goal was a system with narrow coverage or an interactive system with which a human designer collaborates.

Computer automation has motivated research in design theory because automation requires explicit representation of design knowledge and design procedures. We are interested in design theories with scope approximately that of Fig. 2.1; footnotes are used to relate the discussion to this figure. Bibliographical references given in this Appendix cite the literature that has been chosen for its importance or clarity. We first look at a design methodology that prescribes explicitly and thoroughly how a human designer should proceed; it was one of the most extensively developed positions on design methodology up to the end of the 1980s. We then turn to *artificial intelligence* (AI), a discipline that has grown out of computer science and that has developed techniques for explicit action where only implicit suggestions have existed. The discussion covers not only expert systems, which have already found commercial application, but also other AI and even data-base techniques. The goals of the research include developing a descriptive and more thorough theory of design. Finally, we turn to formal theories of design, which have arisen largely in response to the design automation goal. There is a summary of what is probably the most ambitious attempt at a mathematically rigorous theory of design. Other formalisms that have been or could be applied to design are discussed; these are applications of abstract mathematics, many of which have previously been used in computer science. We conclude with the prospect of design theory as a general discipline, covering all types of engineering design and perhaps more.

---

*Artifical Intelligence.

[1]Part of contribution by Albert C. Esterline, Computer Science Dept., University of Minnesota [63].

One school of design theorists, centered in Germany, advocates *systematic design.* These theorists, rather than describing how design *is* carried out, prescribe how design *should be* carried out if we are to be guaranteed of finding a promising solution to a design problem when one exists. The model, as presented by Pahl and Beitz [AI1],[2] consists of the following steps:[3]

1. Clarify the task.
2. Establish the function structure.
3. Choose the physical processes and define forms for subfunctions.
4. Determine the embodiment.
5. Perform detail design.
6. Finalize and review the entire design.

Step 1 is self explanatory.

Steps 2 and 3 together constitute *conceptual design. Function* here refers to the general input-output relationship, independent of any particular physical realization, of a system whose purpose is to perform a task. The *function structure* defines the structural breakdown of the overall function into subfunctions. The inputs and outputs of these subfunctions are linked; energy, material, and signals are seen as crossing subfunction boundaries. Five generally valid subfunctions, which convert energy, material, and signals in various ways, are *change, vary, connect, channel,* and *store*; task-specific subfunctions are also recognized. A *physical process,* such as friction or elasticity, is based on physical effects. A *solution principle* is developed by assigning a physical process to each subfunction, and then defining form, or geometry, for the resulting solution.

Step 3 is intended to produce several alternative solution principles for a given problem.

Step 4 starts from a solution principle, determines the layout and forms, and develops a technical product or system satisfying the technical and economic constraints.

In step 5, the arrangement, form, dimensions, and surface properties of all the individual parts are finally laid down, the materials are specified, the technical and economic feasibility is rechecked, and production documents are prepared.

Failure at any step is accommodated by redoing a previous step. It is emphasized that all steps must be followed to guarantee that no promising solution is overlooked. Furthermore, at any step, we must search for a solution in a systematic way. This may be done analytically by combining concepts, but may be as straightforward as looking up previous solutions in a catalogue.

Computer automations of the design process have attempted to formulate theories of design that use *descriptive*, rather than the *prescriptive* theory of Pahl and Beitz [AI1]. AI research in this field has explicitly formalized many issues that were ignored as long

---

[2]References in this Appendix are found at the end of this section.

[3]These steps cover the entire scope of Fig. 2.1.

as design was a uniquely human endeavor. For example, one AI approach is to formulate and encode the expertise of expert designers. Expertise is competence gained from years of quality experience. An AI program that incorporates expertise to achieve expert-level performance is called an *expert system* [AI2]. The process of formulating expertise in a form usable in an expert system is called *knowledge engineering*. Frequently, but by no means always, expertise is represented by *production rules* of the form *IF* ⟨*condition*⟩ *THEN* ⟨*action*⟩, where the action might simply consist of asserting that a certain state of affairs holds. Expertise is domain-specific: a designer who is expert at designing in one field or domain may have little ability in another domain. It follows that expert systems have narrow coverage. Expertise is typically qualitative, not quantitative knowledge.

To get the advantages of numerical methods, an expert system can be integrated with numerical software [AI3]. The core of most expertise is *heuristics*, or rules of thumb, that express empirical associations. A heuristic program may not be guaranteed to find a solution when one exists and, when it does find a solution, the solution may be suboptimal. However, such a heuristic program usually finds a near-optimal solution. The advantage of heuristics can be seen if we think of design as a search through a space of candidate solutions for a solution that meets the constraints and performance goals stated in the problem.[3]

A brute-force search, which checks one candidate after another in some predetermined order, is very expensive. Heuristic search [AI4], in contrast, proceeds intelligently by eliminating entire regions of the search space: an experienced human designer may solve a design problem heuristically in a matter of hours, while a computer running a nonheuristic program might take years to solve the same problem. One view of expert systems is that they implement theories of the cognitive processes of experienced designers, and thus exhibit performance comparable to the performance of experts.

The most successful design expert systems up to the end of the 1980s have been those that perform routine design [AI5]. In routine design, both the knowledge and the problem-solving that are needed are known in advance.[4] Although the choices are simple at each decision point, the combinations of design requirements are too numerous to be looked up in a catalogue.

An expert system capable of creating novel designs requires a larger repertoire of AI techniques that are not domain specific. Edison [AI6], for example, creates novel mechanical designs using knowledge of naive physics, qualitative reasoning, planning and invention: heuristics applied to abstract device models.

Planning [AI7] finds a sequence of operations to move from an initial state to a goal state; in this context, planning lays out the steps in a design project. Naive physics [AI8] formalizes common-sense knowledge of the everyday physical world, on which every engineer tacitly relies. Qualitative physics [AI9] represents continuous properties by discrete systems of symbols that refer to states relevant to the kind of reasoning per-

---

[3]Search in this sense relates most directly to stage 3 of Fig. 2.1.

[4]Relates to stages 4-6 in Fig. 2.1.

formed. A qualitative simulation of a physical system typically describes the evolution of that system in terms of changes of state and equilibrium.[5]

All approaches we have considered rely in part on catalogues of models or previous solutions. A large catalogue should be implemented as a data base.[6] A data-base management system (DBMS) efficiently accesses large amounts of data kept on secondary storage. Just as importantly, a DBMS imposes a uniform model on the data. It often requires considerable analysis to formulate such a model; once formulated, the model enforces a disciplined use of the data. To this extent, data base design can be viewed as theory formation. Design data bases generally require information to be more structured than was convenient with most commercially available DBMSs in the 1980s, which were generally intended for business applications. However, object-oriented DBMSs [AI10], which allow structured objects to be represented in a natural way, are the subject of research. Indeed, more general object-oriented approaches have attracted the attention of the automated design community. Objects supported by object-oriented languages [AI11] not only represent the properties and parts of the real-world entities they model, but also encode procedures to operate on such representations. AI researchers frequently use frames [AI12, AI13], which are object-like and store procedures along with (passive) data. All these object-oriented approaches enforce disciplined use of the data as well as procedures that operate on the data.

For design, a catalogue lookup alone is usually not sufficient: a catalogue usually does not contain exactly the solution to a given problem specification. However, catalogues are still useful, especially if we consider most design to be redesign, as does the Dominic system [AI14]. Dominic takes as input a set of parameters describing constraints of a design, a set of performance goals, and an initial design procedure. It identifies the weakness in the initial design, proposes a change in a single variable, and then evaluates the effect of this change; if the effect is positive, the change is retained. This evaluate-and-redesign cycle continues until an acceptable design is achieved.[7]

Case-based reasoning [AI15] goes a step further, and considers not only previous solutions, but also previous methods or plans for finding a solution. This approach finds a previous *case* that solved a problem similar to the problem at hand. If the previous solution in fact satisfies the constraints and performance requirements of the current problem, it is accepted, and the design work is finished. Otherwise, the design plan followed in the previous case in question is modified as required. In addition to following past successes, case-based reasoning can refer to previous cases to avoid past errors. In a real sense, case-based systems are capable of learning.

On a more abstract level, Yoshikawa's General Design Theory [AI16, AI17] attempts to determine the requirements of an automated design system by establishing a mathematically rigorous model of the design process. Yoshikawa specifically concentrates on rigorous definitions of the input (specifications) to and output (design solutions)

[5]Relate to stages 5 and 6 in Fig. 2.1.

[6]Data bases relate to stages 1B, 2B, and 3 of Fig. 2.1.

[7]Relates to stages 4-6 in Fig. 2.1 and to the loops through these stages.

from the design process and on definitions of the knowledge and concepts on which the design process operates.[8] General Design Theory is presented as a deductive theory: definitions are stated and axioms are laid down; then theorems are deduced. The goodness of the axioms is to be judged from the number of useful theorems deducible from them. Yoshikawa maintains that a neccessary condition for design is to form concepts of nonexistent things from knowledge about existing things.

Yoshikawa's General Design Theory attempts to introduce mathematical rigor into a broad view of the design process. There are several formal models of design with narrower coverage.[9] These models typically draw from areas of abstract mathematics and often apply methods pioneered in computer science for reasoning about programs and their design.

The final formalism we mention is formal logic, which has been used in foundation studies in mathematics, in reasoning about the meanings and correctness of computer programs, and as a tool in philosophical analysis. Its use in design theory is in its infancy. Fitzhorn, for example, uses logic to implement his *graph grammar* for physical solids. Philosophers have investigated the logic of practical reasoning [AI22], which is concerned not with what is true, but with what satisfies requirements. Despite its obvious relevance to design, this logic is yet to be applied to design theory. Although it is not clear how far mathematically rigorous models can be extended to the various aspects of design, it is clear that such models will find increasing use in formulating, testing, and implementing theories of design. The prospect arises that design theory may sink its own roots into the mathematical substratum, absorb the relevant methodologies from AI, and branch out to all design domains. In that case, design theory would become a general discipline in its own right.

REFERENCES FOR CONTEMPORARY DESIGN THEORY AND DESIGN-RELATED ARTIFICIAL INTELLIGENCE

AI1 Pahl, G. and Beitz, W., *Engineering Design,* ed. by Ken Wallace, Bath, U.K.: The Pitman Press, 1984 (original German edition: Pahl, G. and Beitz, W., *Konstrukstionslehre,* Berlin, Heidelberg: Springer-Verlag, 1987).

AI2 Waterman, Donald A., *A Guide to Expert Systems,* Reading, MA: Addison-Wesley, 1986.

AI3 Kowalik, J. S. (ed.), *Coupling Symbolic and Numerical Computing in Expert Systems,* papers from a workshop, Bellevue WA, Aug. 27-29, 1985, Amsterdam: North-Holland, 1986.

AI4 Korf, R. E., "Search: A Survey of Recent Results," in Shrobe, H. E. (ed.), *Exploring Artificial Intelligence,* San Mateo, CA: Morgan Kaufmann, 1988, pp. 197-237.

AI5 Brown, D. C., "Capturing Mechanical Design Knowledge," *ASME CIE,* Boston, 1985, pp. 121-129.

AI6 Dyer, M. G., Flowers, M., and Hodges, J., "Edison: An Engineering Design Invention System Operating Naively," in Sriram, D. and Adey, R. (eds.), *Proc. First International Con-*

[8]The details of stages 4-6 of Fig. 2.1 tend to be suppressed in this account.

[9]Generally, such models relate to some or all of the stages in Fig. 2.1 with the possible exception of stage 6.

ference on Applications of Artificial Intelligence in Engineering Problems, Southampton Univ., U.K., April 1986, New York: Springer-Verlag, 1986, pp. 327-342.

AI7 Georgeff, M. P., "Reasoning about Plans and Actions," in Shrobe, H. E. (ed.), *Exploring Artificial Intelligence,* San Mateo, CA: Morgan Kaufmann, 1988, pp. 173-196.

AI8 Hayes, P., "The Second Naive Physics Manifesto," in Bachman, R. and Levesque, H. (eds.), *Readings in Knowledge Representation,* San Mateo, CA: Morgan Kaufmann, 1985, pp. 468-485.

AI9 Forbus, K. D., "Qualitative Physics: Past, Present, and Future," in Shrobe, H. E. (ed.), *Exploring Artificial Intelligence,* San Mateo, CA: Morgan Kaufmann, 1988, pp. 239-296.

AI10 Spooner, D. L., Milicia, M. A., and Faatz, D. B., "Modeling Mechanical CAD Data with Data Abstractions and Object-Oriented Techniques," *Proceedings: International Conference on Data Engineering,* Washington, D.C.: IEEE Computer Society Press, 1986, pp. 416-424.

AI11 Stroustrup, B., "What is Object-Oriented Programming?," *IEEE Software,* May 1988, pp. 10-20.

AI12 Minsky, M., "A Framework for Representing Knowledge," in Winston, P. (ed.), *The Psychology of Computer Vision,* New York: McGraw-Hill, 1975, pp. 211-277.

AI13 Bobrow, D. G., and Winograd, T., "An Overview of KRL, a Knowledge Representation Language," *Cognitive Science* 1, 1977, pp. 3-46.

AI14 Howe, A., Cohen, P., Dixon, J., and Simmons, M., "Dominic: A Domain-Independent Program for Mechanical Engineering," in Sriram, D. and Adey, R. (eds.), *Proc. First International Conference on Applications of Artificial Intelligence in Engineering Problems,* Southampton Univ., U.K., April 1986, New York: Springer, 1986, pp. 289-299.

AI15 Kolodner, J. (ed.), *Proc. Workshop on Case-Based Reasoning,* Clearwater, FL, May 10-13, 1988, San Mateo, CA: Morgan Kaufmann, 1988.

AI16 Yoshikawa, H., "General Design Theory and a CAD System," in Sata, T. and Warman, E. (eds.), *Man-Machine Communication in CAD/CAM,* Amsterdam: North-Holland, 1981, pp. 35-53.

AI17 Yoshikawa, H., "General Design Theory and Artificial Intelligence," in Bernold, T. (ed.), *Artificial Intelligence in Manufacturing: Key to Integration?* Amsterdam: Elsevier, 1987, pp. 35-61.

AI18 Freiling, M. J., Rehfuss, S., Alexander, J. H., Messick, S. L., and Shulman, S. J., "The Ontological Structure of a Troubleshooting System for Electronic Instruments," in Sriram, D. and Adey, R. (eds.), *Proc. First International Conference on Applications of Artificial Intelligence in Engineering Problems,* Southampton Univ., U.K., April 1986, New York: Springer, 1986, pp. 609-620.

AI19 Nagl, M., "A Tutorial and Bibliographical Survey on Graph Grammars," in Claus, V., Ehrig, H., and Rozenberg, G. (eds.), *Graph Grammars and their Applications in Biology and Computer Science,* New York: Springer-Verlag.

AI20 Fitzhorn, Patrick, *Linguistic Solid Modeling Using Graph Grammars,* Ph.D. dissertation, Fort Collins, Colo.: Dept. of Mechanical Engineering, Colorado State University, 1985.

AI21 Stiny, G., *Pictorial and Formal Aspects of Shape and Shape Grammars,* Basel and Stuttgart: Birkhaeuser, 1975.

AI22 Kenny, A. J., "Practical Inference," *Analysis* 26, no. 3, Jan. 1966, pp. 65-75.

chapter three
Displacement and Velocity Analysis

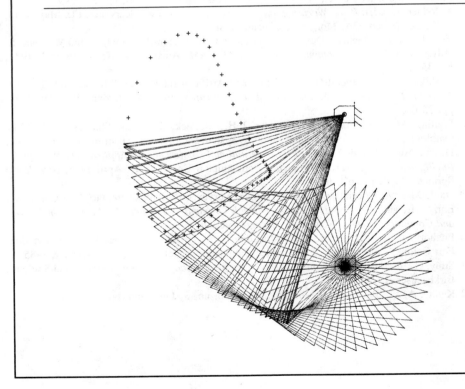

3.1 DISPLACEMENT ANALYSIS: USEFUL INDICES FOR POSITION ANALYSIS OF LINKAGES

One of the simplest and most useful mechanisms is the four-bar linkage. Most of the development in this and the following chapters concentrates on the four-bar, but the procedures are also applicable to more complex linkages.

Chapter 1 categorized three tasks for which mechanisms (in particular the four-bar) are used: path, motion, and function generation. Also, through Gruebler's equation it was found that the four-bar linkage has a single degree of freedom. Are there more distinguishing features that are useful to know about four-bar linkages? Indeed there are! These features include the Grashof criteria, the concept of inversion, dead-center position (change point condition),* circuits, and transmission angle.

The four-bar may take the form of a so-called crank-rocker, double-rocker or a double-crank (drag-link) linkage, depending on the *range of motion* of the two links connected to the ground link. In Figs. 3.1 to 3.4 four different possibilities are illustrated. The input crank of the *crank-rocker* type (Fig. 3.1) can rotate continuously through 360° while the output link just "rocks" (or oscillates). Both the input and output links of the *double-crank* or *drag-link* (Fig. 3.2) make complete revolutions, while the *double-rocker* has limited rotations of both the input and output links (Fig. 3.3). In the *parallelogram linkage* (Fig. 3.4), where the length of the input link equals the output link and the lengths of the coupler and ground link are also equal, both the input and output link may rotate entirely around or switch into a crossed configuration called an *antiparallelogram linkage*. One might guess that a particular four-bar would become one of these types, depending on some relationship involving its link lengths. The *Grashof criteria* provide this relationship. Grashof's law states that *the sum of the shortest and longest links of a planar four-bar linkage cannot be greater than the sum of the remaining two links if there is to be continuous relative rotation between two links.* If we identify the

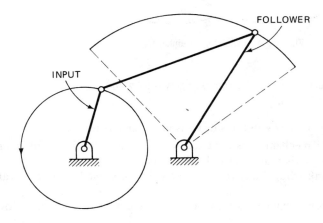

Figure 3.1 Crank and rocker.

* See Fig. 3.12.

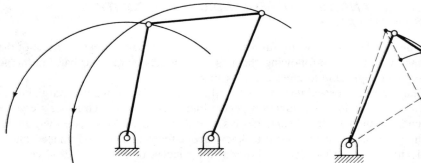

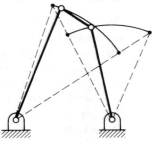

Figure 3.2   Drag-link or double-crank.   **Figure 3.3**   Double-rocker.

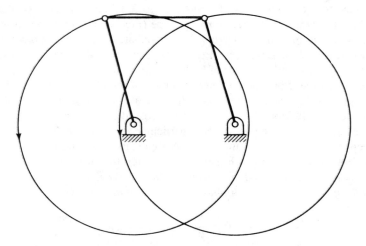

Figure 3.4   Parallelogram linkage.

longest link as l, the shortest as s, and the remaining two as p and q, the following relationships [86] are valid (Fig. 3.5):

1. If $l + s < p + q$, four possibilities or Grashof mechanisms exist.
   a. A crank-rocker mechanism is obtained when the shortest link is the crank,
      the frame being either adjacent link (Fig. 3.6a).
   b. A double-crank (drag-link) results when the shortest link is the frame
      (Fig. 3.6b).
   c. A rocker-crank mechanism is formed when the shortest link is the follower
      (Fig. 3.6c).
   d. A double-rocker mechanism results when the link opposite the shortest is
      the frame (Fig. 3.6d).

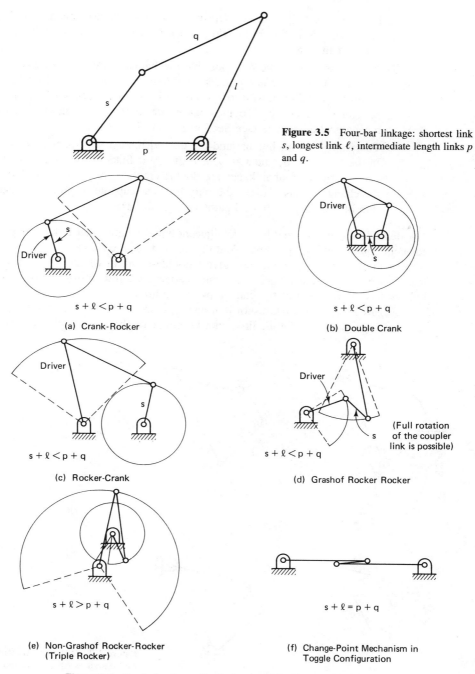

Figure 3.5 Four-bar linkage: shortest link s, longest link ℓ, intermediate length links p and q.

$s + \ell < p + q$

(a) Crank-Rocker

$s + \ell < p + q$

(b) Double Crank

$s + \ell < p + q$

(c) Rocker-Crank

$s + \ell < p + q$

(Full rotation of the coupler link is possible)

(d) Grashof Rocker Rocker

$s + \ell > p + q$

(e) Non-Grashof Rocker-Rocker (Triple Rocker)

$s + \ell = p + q$

(f) Change-Point Mechanism in Toggle Configuration

Figure 3.6 Grashof and non-Grashof types of four-bar mechanisms [24].

2. If $l + s > p + q$, four non-Grashof triple-rocker mechanisms result depending on which link is the frame (Fig. 3.6e). Continuous relative motion is not possible for this case.

3. If $l + s = p + q$, the four possible mechanisms are those of case 1, but they all suffer from the *change point condition:* the centerlines of all links become collinear, creating also a toggle condition (which occurs when the input and coupler are lined up). Toggles can be desirable, for example, to obtain high mechanical advantage (see Section 3.9 and Fig. 3.6f).

4. The parallelogram linkage and the deltoid linkage are special cases of item 3. In the first, $l = q$, and $s = p$, and the short links are separated by a long link (Fig. 3.4). All four linkages are double-cranks if they are controllable through the change points. This is the only four-bar capable of producing parallel motion of the coupler, but all paths are circular arcs.

The deltoid is a linkage that has two adjacent equal-length short links connected to two adjacent equal-length longer links. With a long side as the frame, a crank-rocker is possible; a short side as the frame may give a double-crank in which the short rotating link makes two turns to the longer link's one (called a Galloway mechanism). Again, this linkage has the problem of the change-point condition.

Figure 3.7a–d are the Grashof four bars of case 1, where $l + s < p + q$. The same Grashof configuration of four links can be either of the cases under 1, depending on

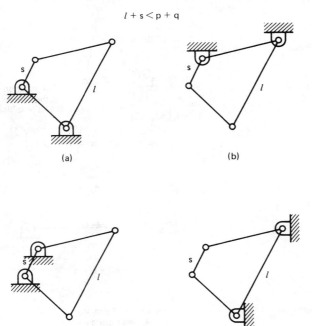

(a)

(b)

(c)

(d)

Figure 3.7 Different mechanisms formed by inversions of the four-bar linkage of Fig. 3.5.

which link is specified as the frame (or ground). *Kinematic inversion* is the process of fixing different links of a chain to create different mechanisms. Note that *the relative motion between links of a mechanism does not change in different inversions*. This property will be used to good advantage several times in this book. Other linkages also have kinematic inversions. For example, inversions of the slider-crank mechanisms are used for different purposes (Figs. 3.8 to 3.11). In Fig. 3.8, link 1 is fixed. We recognize this linkage; it is used, for example, in the internal combustion engine, wherein the input force is the gas pressure on the piston (link 4). When link 2 is fixed (Fig. 3.9), the linkage becomes the type used, for example, in the Gnome aircraft engine. Here, the crank shaft is held stationary (secured to the aircraft frame), while the connecting rod, crank case (integral with the cylinders), and cylinders (link 1) rotate. The propeller is attached to the crank case. This inversion has also been used for a quick-return mechanism in machine tools.

Figure 3.10 shows the inversion, where link 3, the connecting rod, is fixed, and the piston and cylinder are interchanged; it is used in marine engines and in toy steam engines. The fourth inversion may be recognized as a hand pump (Fig. 3.11).

Notice in Fig. 3.12a and b that the four bar has two alternate configurations for a given position of the input (driver). These are called *geometric inversions*. All four bars have geometric inversions. One cannot move from the first to the second geometric inversion without traveling through the dead center position (change point) (Fig. 3.12c). But both ranges of motion in a and b can be reached without taking the four bar apart if the change point can be negotiated. One can make use of a "dead center" position as is done with a rear seat linkage (Fig. 3.13). This is a sketch of a four-bar linkage that guides the motion of both the seat back and the cover plate pan of a 1986 Ford Mustang. When

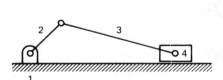

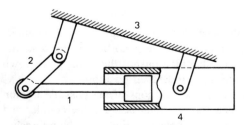

Figure 3.8 Slider-crank.

Figure 3.10 Inversion of the slider-crank: oscillating slide.

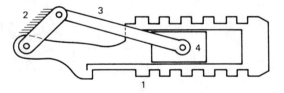

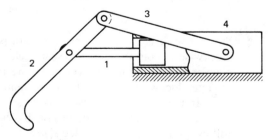

Figure 3.9 Inversion of the slider-crank: rotating slide.

Figure 3.11 Inversion of the slider-crank: stationary slide.

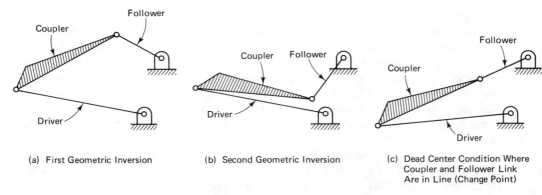

(a) First Geometric Inversion   (b) Second Geometric Inversion   (c) Dead Center Condition Where Coupler and Follower Link Are in Line (Change Point)

Figure 3.12 Two geometric inversions of a four-bar linkage.

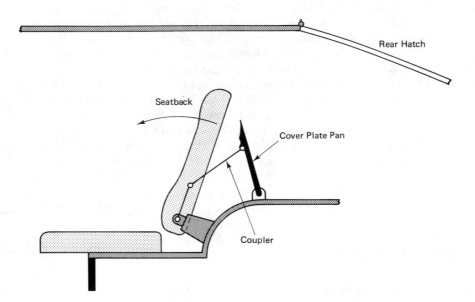

Figure 3.13 Sketch of 1986 Ford Mustang rear seat linkage moves through the dead center position in the seat down position.

the seat back is to be lowered, one grasps the seat (input link) and rotates it counterclockwise. About the time the seat back is parallel to the lower portion of the seat, the coupler and output link (cover plate) are in line. One then pushes the cover plate down (counterclockwise) through the dead center into a stable latched position.

In the case of all Grashof four bars, there are two sections of the possible motion that can only be obtained by physically disconnecting the joint between the coupler and follower links. These are called *separate circuits* as illustrated in Fig. 3.14. Non-Grashof mechanisms only have a single circuit (Fig. 3.15) containing both geometric inversions.

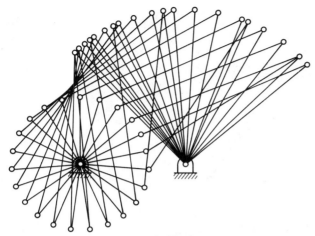

(a) First Circuit

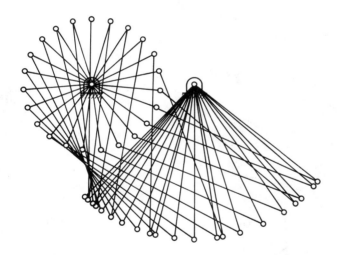

(b) Second Circuit

Figure 3.14 Two circuits of a crank-rocker mechanism.

Crank-rocker and double-crank mechanisms never reach a dead center position; so the two geometric inversions always fall on the two different circuits (Fig. 3.16). Conversely, each circuit is composed of the same geometric inversion. Rocker-cranks or double-rockers (Fig. 3.17) have two dead center positions on both circuits (four different configurations). Each circuit has a distinct range of motion of the driver.

Besides knowing the extent of rotation of the input and output links, it would be useful to have a measure of how well a mechanism might run before actually building it. Hartenberg and Denavit [86] mention that "*run* is a term that more formally means the effectiveness with which motion is imparted to the output link; it implies smooth opera-

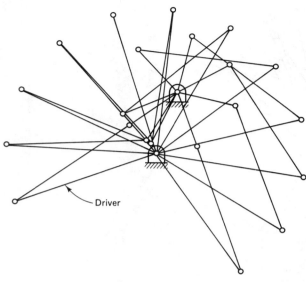

(a) Geometric Inversion 1

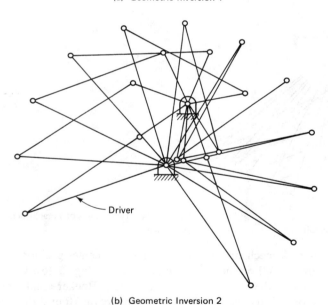

(b) Geometric Inversion 2

Figure 3.15 A non-Grashof triple-rocker can reach a dead center; both geometric inversions must fall on the only possible circuit. The two geometric inversions are shown individually for clarity. Both can be reached without disconnecting link members by pushing the follower down in a and up in b.

tion, in which a maximum force component is available to produce a torque or a force, whatever the case might be, in an output member." The resulting output force or torque is not only a function of the geometry of the linkage but is generally the result of dynamic or inertia forces (see Chap. 5) which are often several times as large as the static forces and act in quite different directions. For the analysis of low-speed operation or for an

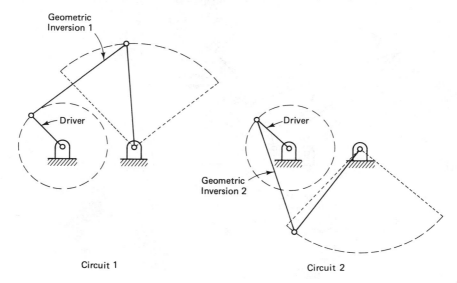

Figure 3.16 If the four-bar cannot reach a dead center position, each geometric inversion will fall on a distinct circuit.

easily obtainable index of how any mechanism might run at moderate speeds, the concept of the transmission angle is extremely useful.

Alt [1] defines the *transmission angle* as the smaller (acute) angle between the direction of the velocity-difference vector (see Sec. 3.5) of the floating link and the direction of the absolute velocity of the output link, both taken at the point of connection. He describes the transmission angle as a measure of the aptness of motion transmittal from the floating link (not the input link of the mechanism) to the output link, but recognizes in a later publication [2] that this kinematically determined transmission angle does not reflect the action of gravity or dynamic forces.

The transmission angle γ is illustrated in the four-bar linkage of Figs. 3.18 and 3.19. The velocity-difference vector, denoted as V_{BA} (velocity of point B relative to point A) is perpendicular to the floating link (link 3 in this case), while the absolute velocity of the output is perpendicular to link 4.

Another approach was suggested by Bloch [14] involving the *deviation angle δ*, which is the smallest angle between the direction of the static force F_{34}, transmitted through the floating link, and the absolute velocity of the output link, V_B, at the point of connection. Figures 3.18 and 3.19 also show the deviation angle. The direction of the static force of the floating link is along the line of its pin joints, since the link is a two-force member (due to the absence of any other force on the link and the assumption of frictionless pin joints at its ends). The *pressure angle* used in cam and cam-follower systems (Chap. 6) is equivalent to the deviation angle. The authors prefer to use the deviation angle δ rather than transmission angle γ, because it is quicker to find the absolute velocity and the static force.

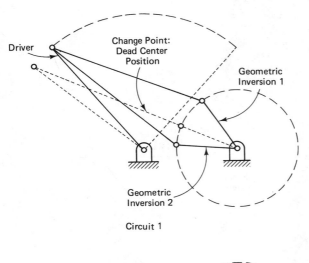

Circuit 1

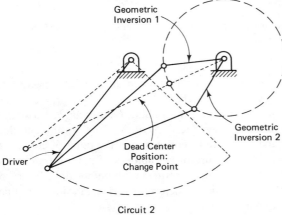

Circuit 2

Figure 3.17 If the four-bar can reach a dead center position, both geometric inversions occur on each circuit.

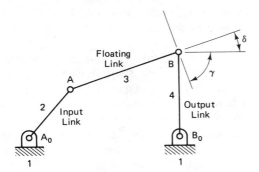

Figure 3.18 Transmission and deviation angles, γ and δ, of a four-bar linkage.

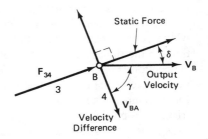

Figure 3.19 Transmission and deviation angles.

Notice that in this case $\gamma + \delta = 90°$. This relation is true whenever the coupler link has just the two opposite joint forces acting on it. The optimum transmission angle is 90° while the optimum deviation angle is 0°. During the motion of a mechanism, these angles will of course change in value. A transmission angle of 0° occurs at a change-point position, at which point the output link being in line with the coupler, will not move regardless of how large a force is applied to the input link. In fact, due to friction in the pin joints, the general rule of thumb is to reject mechanisms with transmission angles of less than 30°. This limiting value will, of course, depend somewhat on the specific application for the linkage. See Sec. 3.3 for determination of maximum and minimum transmission-angles.

Example 3.1

Find the transmission and deviation angles for the mechanisms in Fig. 3.20 and 3.22.

Solution   In the slider-crank mechanism, the velocity of the output is along the slide and the static force $\mathbf{F}_{34}$ is along link 3, which is a two-force member. Figure 3.21 shows the resulting transmission and deviation angles for the slider-crank linkage.

The six-bar linkage (Fig. 3.22) with input on link 2 and output on link 6 will create two locations of concern for transmission and deviation angles. Four-bar $A_0 ABB_0$ will bind up if ABB_0 are in one line forming a dead center position, regardless of how good or bad the situation is at point D (provided there are no forces acting in the dyad of links 5 and 6). The same statement is true at point D: there could be a 90° transmission angle at B, but if CDD_0 are lined up, the mechanism will not move. Figure 3.23 shows the set of transmission and deviation angles for this case.

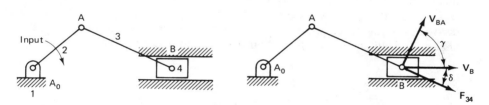

Figure 3.20 Slider-crank mechanism.

Figure 3.21 Transmission and deviation angles of the slider-crank mechanism of Fig. 3.20.

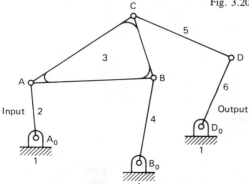

Figure 3.22 Stevenson six-bar mechanism.

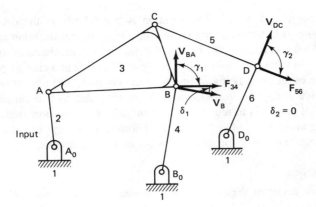

Figure 3.23 Transmission and deviation angles for six-bar with input on link 2. Note that the deviation angle at D is zero in this position.

It should be pointed out that if we reverse the input and output links for the six-bar in Fig. 3.22, then the analysis becomes more difficult because link 3 is no longer a two force member. A method has been suggested [167] that creates a *virtual equivalent linkage* for this mechanism. The four-bar A_0ACBB_0 is replaced by a *virtual link* C_0C that is kinematically equivalent for position and velocity. Later in this chapter, we will discuss instant centers and find that the extended plane of link 3 has a unique point C_0 (the instant center between links 1 and 3) that has momentarily zero velocity with respect to ground. C_0 is found (as we will see) at the intersection of the extensions of links 2 and 4. The instantaneous velocity of point C is in the same direction (and has the same magnitude) for the original six-bar as it would be for the virtual equivalent four-bar linkage D_0DCC_0. Therefore, the transmission and deviation angles shown in Fig. 3.24 are derived. This case exemplifies the complexity of finding the transmission and deviation angles for multiloop mechanisms. You may want to reread this subsection after you have

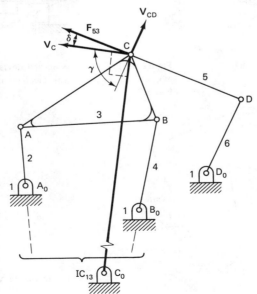

Figure 3.24 Transmission and deviation angle of the six-bar mechanism of Fig. 3.22 with input and output link reversed.

studied instant centers. Reference [167] suggests a method for dealing with multiloop mechanisms. The key is to look for locations in the mechanism where there is a possible dead center position. In addition, look for possible virtual equivalent linkages to reduce the complexity.

Matrix-based definitions have also been developed which measure the ability of a linkage to transmit motion. The value of a determinant (which contains derivatives of motion variables with respect to an input motion variable for a given linkage geometry and called the *Jacobian*) is a measure of the ease of movability of the linkage in a particular position.

3.2 DISPLACEMENT ANALYSIS: GRAPHICAL METHOD

A single-degree-of-freedom mechanism such as the four-bar can be analyzed graphically for relative displacements without great effort. Although the accuracy depends on one's care in construction and the scale of the drawing, acceptable precision can usually be obtained. A quick method for generating a number of positions of a mechanism (or full animation) is illustrated in Fig. 3.25. The only required drawing instruments are a scale (straightedge), compass, and one overlay (drafting paper or velum, or tissue paper).

In Fig. 3.25a, a crank-rocker four-bar with input crank A_0A is to be analyzed for displacements of the path tracer point P (and perhaps the relative angles of the coupler link AB and output link B_0B with respect to the input crank. The coupler link is reproduced on the overlay of Fig. 3.25b. Since points A and B of the coupler are constrained to move along the circular arcs drawn with a compass through A around A_0 and through B around B_0, one only has to move overlay b over the four-bar a, being careful to keep points A and B on their respective arcs and mark each successive location of points A, B, and P (by pressing the point of compass through to a or by placing the overlay under a). Figure 3.25c shows the result of this construction for a portion of the cycle of motion.

Although this method is fairly quick, it is quite cumbersome for a great deal of analysis and, of course, not very accurate. However, more complex mechanisms can also be analyzed this way by constraining joints to move on their respective paths.

Somewhat more accurate results are achieved if, instead of using an overlay, the graphical analysis proceeds with the use of compasses and drafting triangles. Figures 3.26, 3.27, and 3.28 and the accompanying texts exemplify this method. However, precision still suffers from limitations of drafting accuracy, flat intersections of arcs and lines, and intersections that are too far off the paper. These difficulties and the general availability of computers are strong motivations for using analytical methods, especially because even so-called computer-graphics approaches require analytically developed software.

Graphical Displacement Analysis of the Four-bar Mechanism.   As shown in Figure 3.26, we wish to construct the jth position of the four-bar path-generator mechanism A_0AB_0BP. The first or starting position is marked with the subscript 1. To construct the jth position, proceed as follows.

1. Draw arcs about A_0 with radius A_0A_1 and about B_0 with radius B_0B_1. These are the paths of joints A and B, respectively.

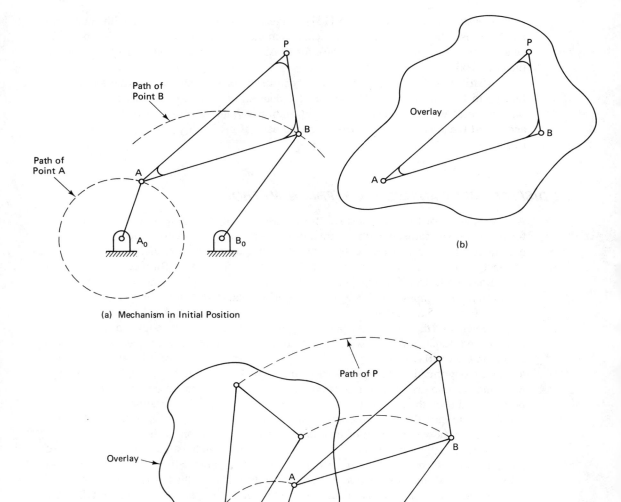

Figure 3.25 The four-bar of Figure 3.1 is analyzed graphically for the path of point P using the overlay (b) placed on circular arc path constraints of points A and B.

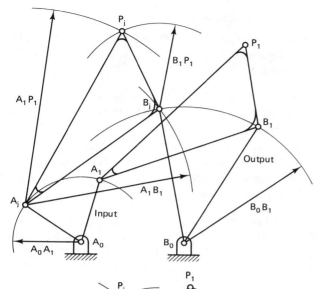

Figure 3.26 Graphical displacement analysis of the four-bar mechanism.

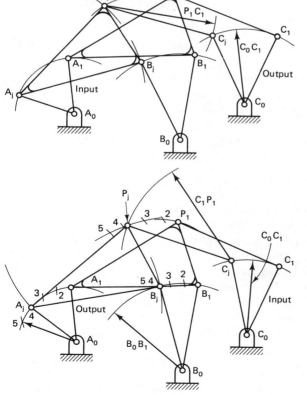

Figure 3.27 Graphical displacement analysis of the Stevenson III six-bar mechanism with $A_0 A$ as the input and $C_0 C$ as the output.

Figure 3.28 Graphical displacement analysis of the Stevenson III six-bar mechanism with link $C_0 C$ as the input and $A_0 A$ or $B_0 B$ as the output.

2. Draw the jth position of the input link, A_0A_j.

3. With radius A_1B_1 draw an arc about A_j to intersect the path of B. This is B_j.*

4. Construct the point P_j by intersecting the arc of radius B_1P_1 centered at B_j with the arc of radius A_1P_1 centered at A_j.

This completes the construction. Also note that, if the four-bar linkage is used as a function-generator, in which case point P of the coupler would be of no concern, then the construction would be complete after step 3 above.

Graphical Displacement Analysis of the Stevenson III Six-bar Mechanism. This mechanism, with revolutes $A_0AB_0BPCC_0$, is shown in Fig. 3.27. Its starting position is identified by subscript 1 for all moving joints. To construct the jth position when A_0A is the input and C_0C is the output, proceed as follows.

1. Construct the jth position of the four-bar linkage, $A_0A_jB_0B_jP_j$, as before.

2. Construct the jth position of the output dyad by drawing an arc centered on C_0 with radius C_0C_1 and intersect it with an arc centered on P_j with radius P_1C_1. This locates the joint C in its jth position, C_j, and the construction is complete.

Figure 3.28 shows the construction when C_0C is the input and A_0A or B_0B is the output. The procedure is as follows.

1. Draw the jth position of the input link and swing an arc centered on C_j with radius C_1P_1.

2. Now, using the method of Figure 3.26, construct four or five successive positions of the point P in the vicinity of the above-drawn C_1P_1 arc, thus constructing a portion of the path of point P. P_j is at the intersection of the C_1P_1 arc and the path of point P.

This completes the graphical construction.

3.3 DISPLACEMENT ANALYSIS: ANALYTICAL METHOD[†]

The development of displacement and velocity analyses which follow assume familiarity of the reader with complex numbers. The appendix to this chapter, which covers the fundamentals of complex numbers should, therefore, be reviewed at this time.

In Chap. 1, a method for determining the number of degrees of freedom of a mechanism was discussed. If a mechanism has one degree of freedom (like the four-bar), prescribing one position parameter, like the angle of the input link, will completely specify the position of the rest of the mechanism when one of the two possible geometric inversions is selected. Let us develop an analytical expression relating the absolute angular positions of the links of a four-bar linkages. This will be much more useful than a graphi-

---

* Since two circles intersect at two points, there will be two solutions. The second solution, not shown in Fig. 3.26, constitutes the other "geometric inversion" of this four-bar mechanism.

[†] This section has been revised with the help of Dr. Tom Chase, University of Minnesota.

cal analysis procedure when analyzing a number of positions and/or a number of different mechanisms, because the expressions developed below are easily programmed for automatic computation.

In Fig. 3.29a, let the independent variable parameter be the angle θ_2 of the input link (link $\mathbf{r}_2$) with respect to a fixed x axis. Since we are analyzing this linkage, the lengths of the links, $\mathbf{r}_1$, $\mathbf{r}_2$, $\mathbf{r}_3$, $\mathbf{r}_4$, $\mathbf{r}_5$ and $\mathbf{r}_6$, plus the angular position of the ground link (θ_1) and the angle α (which represents the fixed angle between the two sides of the coupler triangle, $\mathbf{r}_3$ and $\mathbf{r}_5$) are the known constant parameters. Unknown, however, are the dependent variable parameters, angles θ_3 and θ_4, as well as the x and y coordinates of the path tracer point, P. The following simple procedure will determine these dependent parameters in terms of the independent variable. Steps 1 through (and including) 3 will be performed only once for the initial position of the four-bar mechanism to determine the proper geometric inversion. When analyzing the same mechanism for other positions

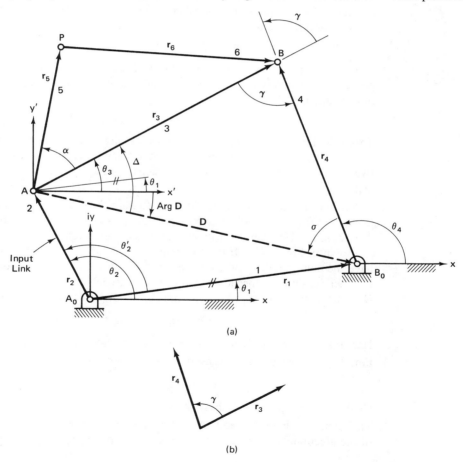

(a)

(b)

Figure 3.29 Position variables of a four-bar mechanism.

of the input crank ($\mathbf{r}_2$), we assumed that the linkage does not move through its dead center position to another geometric inversion. Thus, we must know the coordinates of the four-bar pivots in the initial position of the mechanism.

1. Expressing the angle of vectors representing links uniformly in the range $-180° < \theta \leq 180°$ is convenient. Therefore, if in an arithmetic calculation with vectors in polar form the resulting argument is greater than $180°$, subtract $360°$; if it is smaller than or equal to $-180°$, add $360°$. In other words, apply Eq. (3.1).

$$\text{If } \theta_i > 180°, \quad \text{then } \theta_i \leftarrow \theta_i - 360°$$
$$\text{If } \theta_i \leq -180°, \quad \text{then } \theta_i \leftarrow \theta_i + 360° \qquad (3.1)$$

To determine the angle of vectors expressed in cartesian form, one can simply subtract the coordinates of the tail end of the link vector from its arrow end coordinates. For example, the cartesian complex form of the link vector between points A and B would be

$$\mathbf{r}_3 = (x_B - x_A) + i(y_B - y_A) \qquad (3.2)$$

The argument of $\mathbf{r}_3$ with respect to the x' (real) axis (which axis moves with point A but remains parallel to the fixed x axis throughout the mechanisms movement) is

$$\theta_3 = \text{ATAN2}(y_B - y_A, x_B - x_A) \qquad (3.3)$$

This FORTRAN arctangent function yields angles expressed in the interval $-\pi° < \theta_i \leq \pi°$. Note that angles measured counterclockwise from the positive x axis are positive. The remaining steps assume all angles are converted to degrees.

2. Determine the geometric inversions by finding the value of the transmission angle, γ, (Fig. 3.29a) between vectors $\mathbf{r}_3$ to $\mathbf{r}_4$. Note that this angle is measured from $\mathbf{r}_3$ to $\mathbf{r}_4$. This can be visualized by placing the vectors $\mathbf{r}_3$ and $\mathbf{r}_4$ tail-to-tail as shown in Fig. 3.29b. If the trigonometric sine of the angle γ is positive, we have the configuration shown in Fig. 3.29a; if it is negative, the four bar is in its other possible geometric inversion. Point B would be reflected about the line D in this configuration. This determination is implemented by Eq. (3.4).

$$\gamma = \text{Arg } \mathbf{r}_4 - \text{Arg } \mathbf{r}_3 \qquad (3.4)$$

Here Eq. (3.1) may have to be applied to express γ in the $-180°$ to $180°$ interval.

3. Find the coupler angle α, measured from $\mathbf{r}_3$ to $\mathbf{r}_5$, as follows.

$$\alpha = \text{Arg } \mathbf{r}_5 - \text{Arg } \mathbf{r}_3 \qquad (3.5)$$

Here again, Eq. (3.1) may have to be applied.

4. The geometric inversion and coupler triangle orientation are now fixed by way of the algebraic signs of the angles γ and α respectively (the sense and magni-

tude of γ and α are now known). For any new position of $\mathbf{r}_2$ find the new vector $\mathbf{D}$ by Eq. (3.6):

$$\mathbf{D} = \mathbf{r}_1 - \mathbf{r}_2 \qquad (3.6)$$

5. Calculate the new value of the angle Δ, which is measured from $\mathbf{D}$ to $\mathbf{r}_3$, see Fig. 3.29a. This angle has the same sense and the same algebraic sign (signum) as γ.

$$\Delta = (\text{signum } \gamma)\left|\cos^{-1}[(r_3^2 + D^2 - r_4^2)/(2r_3D)]\right| \qquad (3.7)$$

6. Calculate the new value of the angle σ, which is measured from vector $\mathbf{r}_4$ to vector $(-\mathbf{D})$, and has therefore the same sense and the same algebraic sign (signum) as γ.

$$\sigma = (\text{signum } \gamma)\left|\cos^{-1}[(r_4^2 + D^2 - r_3^2)/(2r_4D)]\right| \qquad (3.8)$$

With the new values of the angles Δ and σ known, Figure 3.29a shows that now

$$\theta_3 = \text{Arg } \mathbf{D} + \Delta, \quad \text{and} \quad \theta_4 = \text{Arg } \mathbf{D} + (180° - \sigma) \qquad (3.9)$$

Again, Eq. (3.1) may need to be applied in order to express θ_3 and θ_4 in the $-180°$ to $180°$ interval.

Next, calculate the new value of the vector $\mathbf{r}_5$ as follows:

$$\mathbf{r}_5 \leftarrow r_5 \exp[i(\theta_3 + \alpha)] \qquad (3.10)$$

where α is taken from Eq. (3.5).

The new position vector of the path tracer point is

$$\mathbf{P} = \mathbf{r}_2 + \mathbf{r}_5 \qquad (3.11)$$

where, for the two vectors on the right side, their new values are to be used.

The foregoing method of analysis is readily programmed for automatic computation according to the flowchart of Fig. 3.30. The flowchart leaves the output statements to the programmer. Typical output options would be the following:

For function generation: θ_2 and θ_4
For path generation: θ_2 and $\mathbf{P}$
For motion generation: θ_2, θ_3 and $\mathbf{P}$

These should be printed out just before every iterative incremental rotation of the input link $\mathbf{r}_2$.

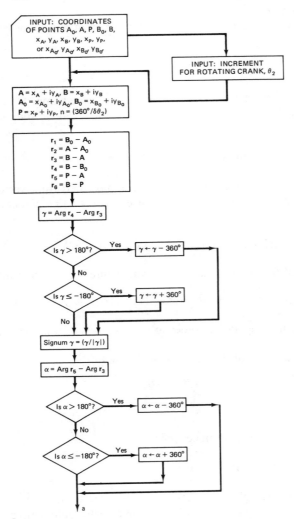

Figure 3.30 Flowchart of computer program for the displacement analysis of a four-bar mechanism (see Fig. 3.29).

*Analytic Method of Finding Limits of Rotation of Four-bar Input Link**

From Figs. 3.31, 3.32, and the Grashof criteria, the following conditions are observed for complete crank rotation.

$$r_1 + r_2 \leq r_3 + r_4 \tag{3.12}$$

and because the magnitude of the length difference of the first two exceeds that of the second two,

$$r_1 - r_2 \geq r_3 - r_4 \tag{3.13}$$

* Also see Sec. 3.10: Limits of motion of the four-bar linkage.

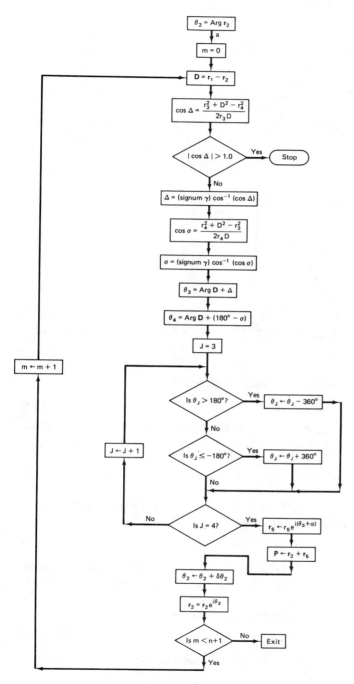

Figure 3.30

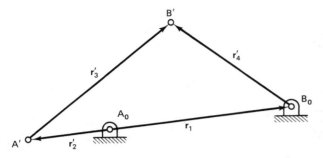

Figure 3.31 Crank rotation unlimited in the outward direction because $r_1 + r_2 \leq r_3 + r_4$ [condition (3.12)].

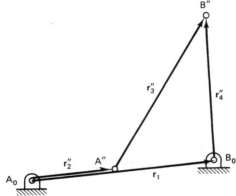

Figure 3.32 Crank rotation unlimited in the inward direction because $r_1 - r_2 \geq r_3 - r_4$ [condition 3.13].

Now, if condition (3.12) does not hold, in other words, if

$$r_1 + r_2 > r_3 + r_4 \tag{3.14}$$

but if condition (3.13) does hold, the rotation of link r_2 is limited in the outward direction only (Fig. 3.33). The range of rotation for this case can be calculated thus:

$$\cos(\theta_2')''' = \cos(\arg \mathbf{r}_2''' - \arg \mathbf{r}_1) = \frac{r_1^2 + r_2^2 - (r_4 + r_3)^2}{2r_1r_2} \tag{3.15}$$

Equation (3.15) is satisfied by both $(\theta_2')'''$ and $-(\theta_2')'''$, values that represent the limits of the range of rotation of input link $\mathbf{r}_2$. The range is thus

$$\Delta(\theta_2')''' = |2(\theta_2')'''| \tag{3.16}$$

Note that the transmission angle $|\gamma|$ is 180° for both outward limit positions of the driving link. Also observe that the linkage can "branch" (change geometric conversions) in these positions.

If condition (3.12) does hold but condition (3.13) does not hold, that is,

$$r_1 - r_2 < r_3 - r_4 \tag{3.17}$$

the rotation of input link $\mathbf{r}_2$ is limited in the inward direction only (Fig. 3.34). Here

$$\cos(\theta_2')^v = \cos(\arg \mathbf{r}_2^v - \arg \mathbf{r}_1) = \frac{r_1^2 + r_2^2 - (r_3 - r_4)^2}{2r_1r_2} \tag{3.18}$$

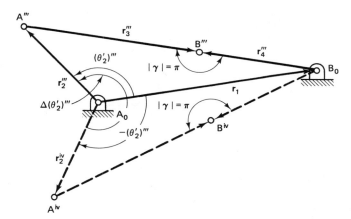

Figure 3.33 Condition (3.12) does not hold, but condition (3.13) does hold; therefore, input link ($\mathbf{r}_2$) rotation is limited in the outward direction.

which is satisfied by both $(\theta_2')^v$ and $-(\theta_2')^v$, yielding the range of input rotation

$$\Delta(\theta_2')^v = 2\pi - 2|(\theta_2')^v| \qquad (3.19)$$

Note that the transmission angle γ for each inward limit position is 0°. Also note that the mechanism can change geometric conversions in these positions.

When neither condition (3.12) nor (3.13) holds, input link rotation is limited in both the outward and inward directions, and the transmission angle varies all the way from 0 to 180°. However, when input link rotation is unlimited in one or both directions, the transmission angle γ will reach an extreme value during the "unlimited" rotation of the driving link. It is easy to show by inverting the mechanism that the extreme value or values of the transmission angle γ short of 0° or 180° are reached as the driving link passes through collinearity with the fixed link, that is, when $\theta_2' = 0$ or 180° (Fig. 3.35a and b). This condition can be used for finding the extreme value(s) of the transmission angle as follows.

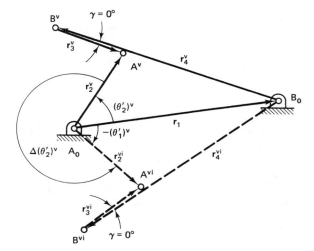

Figure 3.34 Condition (3.12) holds, but condition (3.13) does not hold; input link ($\mathbf{r}_2$) rotation is limited in the inward direction.

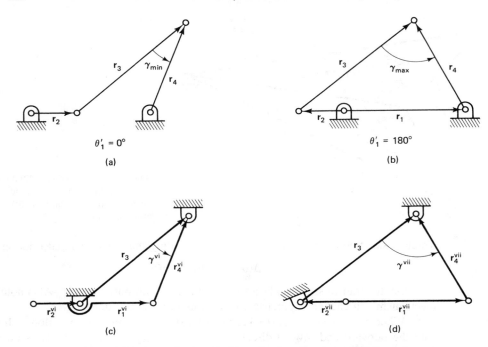

Figure 3.35　Determination of minimum and maximum transmission angles.

We start by declaring the former coupler to be the fixed link, and the former follower to be the driver. Now the former transmission angle γ is the new "input" angle. To illustrate this, we take the configuration of Fig. 3.35 (c) and (d), letting $\mathbf{r}_3$ be fixed and $-\mathbf{r}_4$ be the driver. For determining the extreme values of the transmission angle, we use the cosine law:

$$\cos \gamma_{\min} = \frac{r_4^2 + r_3^2 - (r_1 - r_2)^2}{2r_4r_3} \quad \text{and} \quad \cos \gamma_{\max} = \frac{r_4^2 + r_3^2 - (r_2 + r_1)^2}{2r_4r_3}$$

$$\sin \gamma_{\min} = |\sqrt{1 - \cos^2\gamma_{\min}}| \qquad \qquad \sin \gamma_{\max} = |\sqrt{1 - \cos^2\gamma_{\max}}|$$

$$\gamma_{\min} = \arg(\cos \gamma_{\min} + i \sin \gamma_{\min}) \qquad \gamma_{\max} = \arg(\cos \gamma_{\max} + i \sin \gamma_{\max})$$

3.4 CONCEPT OF RELATIVE MOTION

Notice that in Fig. 3.29 the position angles of all link vectors were measured from the x axis of a fixed reference frame. Hence they were absolute angles. Unless otherwise stated, all coordinate systems in the text will be fixed or pinned to a link so as to move parallel to a fixed inertial frame of reference. For example, in Fig. 3.29, reference frame $x'y'$ remains parallel to xy but its origin is pinned to point A which moves in a circular arc about A_0. The position angle of the coupler link is expressed in terms of coordinate system $x'y'$.

The following discussion of the difference of motion between points of the same link and relative motion between different links will help in solving position, velocity, and acceleration problems.

Table 3.1 shows the four possible cases* that are applicable when examining the motion of various points in a mechanism. The 2×2 matrix in this table represents combinations of the same or different points on the same or different links. Each case is worthy of comment with respect to the complexity of a motion analysis (Fig. 3.36):

Case 1: Same point-same link.   For example, the motion of point Q on link 2 with respect to itself. This is a trivial analysis. There is no motion of Q relative to itself.

Case 2: Different points-same link.   Case 2 is called a "difference" [86] motion. Examples are the motion between points Q and P on link 2 or the motion between points R and S on link 3.

Case 3: Same point-different links (momentarily coincident points).
For instance, the motion of R on link 2 with respect to point R on link 3, or the motion

TABLE 3.1 THE FOUR CASES OF REFERRED MOTION IN LINKAGE MECHANISMS

	Same point	Different point
Same link	*Case 1* Trivial	*Case 2* Difference motion
Different links	*Case 3* Relative motion	*Case 4* Manageable through a series of case 2 and case 3 steps

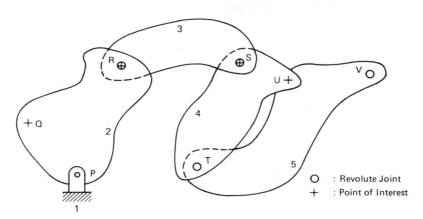

Figure 3.36

* This concept was formulated on the basis of discussions with J. Uicker, University of Wisconsin.

of point U on link 4 with respect to the momentarily coincident point U on link 5. Case 3 motion is called "relative motion." In some instances the analysis is trivial, as with point R of link 2 with respect to point R on link 3—that is, when the point happens to be a revolute joint joining the two links. In other instances, as with point U, which is not a joint, the analysis can be quite complex—requiring knowledge of the instantaneous paths of the point of interest as a point of each link with respect to the fixed frame of reference.

Linkage motion analysis often includes both case 2 and case 3 analyses.

Case 4: Different points-different links.   For example, the motion of point V on link 5 with respect to points P, Q, R, or S on different links. In most cases not enough information is known to perform a single-step case 4 analysis. Usually several intermediate steps of case 2 and/or case 3 analyses need to be performed (dictated by the physical constraints of a mechanism) in place of a single-step case 4 analysis.

These four cases of *referred* motion become more and more important as the analyses become more complex (e.g., acceleration analysis) and it becomes more difficult to keep track of the *relative* motion components. The understanding of which of the four cases of motion are involved in a particular case is fundamental in the kinematics of linkages. Most errors in kinematic analysis of mechanisms result from the misinterpretation of relative motion.

3.5 VELOCITY ANALYSIS: GRAPHICAL METHOD

The concept of velocity can be developed by beginning with just two elements of a mechanism: a ground link (link 1) and an input link (link 2) which is pinned directly to ground at A_0 (see Figs. 3.37 to 3.38). The radius vector $\mathbf{R}_A$ of an arbitrary point A on link 2 is located instantaneously at angular position θ_A with respect to the x axis of an

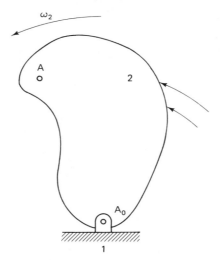

Figure 3.37 Link 2 pivoted to ground link 1 at A_0 and rotating CCW at angular velocity ω_2.

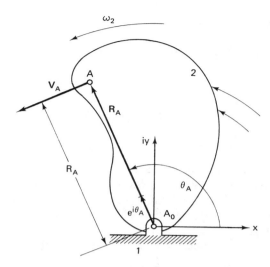

Figure 3.38   Absolute velocity $\mathbf{V}_A$ of point A on rotating link 2.

absolute reference system fixed to link 1. Notice that the axes in Fig. 3.38 represent a complex plane x, iy. Complex numbers* are utilized in many instances in this text due to their ready applicability to planar mechanism analysis and synthesis.

Link 2 is in the process of changing position with respect to link 1. The time rate of change of angular position of link 2 with respect to the fixed x axis is called the *angular velocity* (ω_2) of link 2,

$$\omega_2 \equiv \frac{d\theta_A}{dt} \tag{3.20}$$

positive counterclockwise. The position of point A with respect to A_0 (case 2 analysis — see Table 3.1) may be defined mathematically in either polar or Cartesian form. Two scalar quantities, the length R_A and the angle θ_A with respect to the x axis, define the vector locating point A.

Polar form:

$$\mathbf{R}_A = R_A e^{i\theta_A} \tag{3.21}$$

Using Euler's equation we obtain the Cartesian form:

$$\mathbf{R}_A = R_A \cos \theta_A + iR_A \sin \theta_A \tag{3.22}$$

The *absolute linear velocity* of a point is the time rate of change of position vector of that point with respect to the ground reference frame

$$\mathbf{V}_A = \frac{d\mathbf{R}_A}{dt} \tag{3.23}$$

---

* See the appendix to this chapter for a review of complex numbers.

or polar form

$$\mathbf{V}_A = iR_A \frac{d\theta_A}{dt} e^{i\theta_A} = R_A \omega_2 i e^{i\theta_A} = i\mathbf{R}_A \omega_2 \tag{3.23a}$$

or Cartesian form

$$\mathbf{V}_A = R_A(-\sin\theta_A)\frac{d\theta_A}{dt} + iR_A(\cos\theta_A)\frac{d\theta_A}{dt} = R_A\omega_2(-\sin\theta_A + i\cos\theta_A)$$

$$\tag{3.23b}$$

The polar form of $\mathbf{V}_A$ is very informative. The scalar value of the velocity is the radius R_A times the angular velocity ω_2, while the direction is 90° from the unit position vector $e^{i\theta_A}$ in the same sense as ω_2. The notation followed when using complex numbers is that *counterclockwise rotations are positive;* thus ω_2 is positive here and the absolute velocity $\mathbf{V}_A$ is to the left in Fig. 3.38.

Notice that the magnitude of the linear velocity and angular velocity are related.

$$|\mathbf{V}_A| = V_A = R_A|\omega_2| \tag{3.24}$$

Also, the direction of the velocity vector is always perpendicular to the position vector originating at the point of reference since, if a component of the velocity were to be along the position vector, the link would deform, which contradicts the rigid link assumption.

Suppose that link 2 contained another point of interest, point B (see Fig. 3.39). The absolute velocity of point B would be

Polar form:

$$\mathbf{V}_B = R_B\omega_2 i e^{i\theta_B} = \mathbf{R}_B i\omega_2 \tag{3.25}$$

Cartesian form:

$$\mathbf{V}_B = R_B\omega_2(-\sin\theta_B + i\cos\theta_B) \tag{3.26}$$

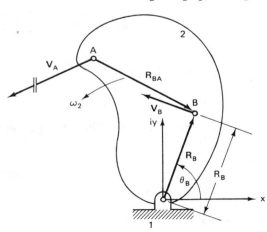

Figure 3.39 Absolute velocities $\mathbf{V}_A$ and $\mathbf{V}_B$ of points A and B on rotating link 2.

Thus far, we have discussed only absolute velocities of a point on a link pinned to ground. What is the difference between the absolute velocity at point B and the absolute velocity at point A (still case 2 motion); that is, if you were to sit on link 2 at point A, keep looking in the fixed x direction and out of the corner of your eye watch point B, what would be the apparent velocity of point B with respect to you at A? You would observe the *velocity difference,* which is the difference of two absolute velocities of two points *on the same link.* The velocity difference is in this case, $\mathbf{V}_{BA}$, where the second subscript is the point of reference and the first subscript is the point of interest. It follows that

$$\mathbf{V}_{BA} = \mathbf{V}_B - \mathbf{V}_A = \mathbf{R}_B i\omega_2 - \mathbf{R}_A i\omega_2 = (\overrightarrow{AB})i\omega_2 = \mathbf{R}_{BA} i\omega_2$$

from which

$$\mathbf{V}_B = \mathbf{V}_A + \mathbf{V}_{BA} = \mathbf{R}_B i\omega_2 = \mathbf{R}_A i\omega_2 + (\overrightarrow{AB})i\omega_2 \qquad (3.27)$$

A vector solution to this equation appears in Fig. 3.40. Notice that the absolute velocities have only one subscript. The absent second subscript is understood to be ground. In Fig. 3.40, point O_V is an arbitrary origin for drawing the velocity diagram. All absolute velocities are laid out starting at this origin so that $\mathbf{V}_B$ [from the left side of Eq. (3.27)] is drawn to a convenient scale, parallel to $\mathbf{V}_B$ in Fig. 3.39. Then the right side of the equation is drawn in by starting at O_V and drawing $\mathbf{V}_A$ to the same scale. The vector difference between $\mathbf{V}_B$ and $\mathbf{V}_A$ is $\mathbf{V}_{BA}$ closing the vector polygon. Note that $\mathbf{V}_{BA}$ is perpendicular to $\mathbf{R}_{BA}$ (as it should be since link 2 is rigid) and that

$$\mathbf{V}_{BA} = -\mathbf{V}_{AB} \quad \text{or} \quad \mathbf{R}_{BA} i\omega_2 = -\mathbf{R}_{AB} i\omega_2 \qquad (3.28)$$

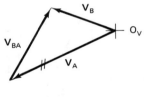

$$\mathbf{V}_B = \mathbf{V}_A + \mathbf{V}_{BA}$$

Figure 3.40 Finding the velocity difference $\mathbf{V}_{BA}$ of point B with respect to point A by means of the vector triangle. Both A and B are points of rotating link 2 from Fig. 3.34.

and

$$\mathbf{V}_{BA} = i\omega_2 \mathbf{R}_{BA} \qquad (3.29)$$

Solving Eq. (3.29) for ω_2, we obtain

$$\omega_2 = \frac{\mathbf{V}_{BA}}{i\,\mathbf{R}_{BA}} \quad \text{or} \quad |\omega_2| = \frac{V_{BA}}{R_{BA}} \qquad (3.30)$$

Notice that ω_2 is a positive or negative real according as $\mathbf{V}_{BA}$ and $i\mathbf{R}_{BA}$, which are collinear vectors, point in the same or opposite direction.

Example 3.2

Link J in Fig. 3.41 is moving with respect to ground. Points P and Q of link J are locations of known absolute velocities $\mathbf{V}_P$ and $\mathbf{V}_Q$. Find the angular velocity ω_J of this link with respect to ground, or, which is the same thing, with respect to the fixedly oriented $xPiy$ system, attached to link J at P and moving along with it, while remaining parallel with the fixed $x_0 Oiy_0$ system.

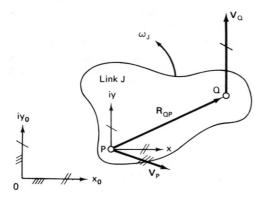

Figure 3.41 Motion of link J known only by way of known velocity vectors $\mathbf{V}_P$ and $\mathbf{V}_Q$ of its points P and Q.

Solution   Using Eq. (3.27) yields

$$\mathbf{V}_Q = \mathbf{V}_P + \mathbf{V}_{QP}$$

Figure 3.42 shows the graphical solution of this equation. Finally, from Eq. (3.30),

$$\omega_J = \frac{\mathbf{V}_{QP}}{i\mathbf{R}_{QP}} \quad \text{or} \quad |\omega_J| = \frac{V_{QP}}{R_{QP}}$$

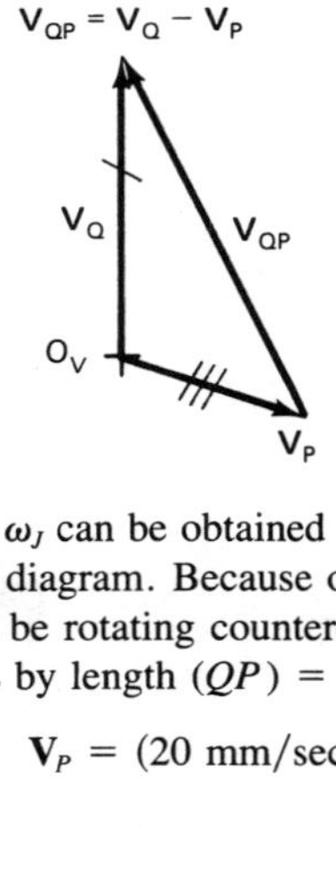

Figure 3.42   Graphical construction to find velocity difference of point Q with respect to point P of link J of Fig. 3.38 in preparation for calculating the unknown angular velocity ω_J.

Notice that the sense of ω_J can be obtained by observing the direction of the velocity difference $\mathbf{V}_{QP}$ on the velocity diagram. Because of the direction of the velocity of point Q with respect to P, link J must be rotating counterclockwise. The magnitude ω_J is obtained by dividing the magnitude V_{QP} by length $(QP) = R_{QP}$. For example, let

$$\mathbf{V}_P = (20 \text{ mm/sec})e^{i(-18.89°)}$$

and

$$\mathbf{V}_Q = i(30 \text{ mm/sec})$$

From the vector triangle of Fig. 3.42, by graphical construction, we find that

$$\mathbf{V}_{QP} = (40 \text{ mm/sec})e^{i(117°)}$$

Checking this by substituting in Eq. (3.27), we have

$$\mathbf{V}_{QP} = \mathbf{V}_Q - \mathbf{V}_P = i30 - 18.92 + i6.48$$

$$\mathbf{V}_{QP} = -18.92 + i(36.48) = (41.10 \text{ mm/sec})e^{i(117°)}$$

If $\mathbf{R}_{QP} = (42 \text{ mm})e^{i(27°)}$, then from Eq. (3.30),

$$\omega_J = \frac{(41.10)e^{i(117°)}}{e^{i(90°)}(42)e^{i(27°)}} = 0.98 \text{ rad/sec}$$

counterclockwise, because $\omega_J > 0$.

Thus the example above demonstrates that the complex-number notation will determine the angular velocity with the correct algebraic sign to indicate whether it is clockwise (cw) or counterclockwise (ccw), without rules of thumb or visual inspection of the geometry involved. This approach is therefore well adapted for automatic digital computation where there is no opportunity for visual verification.

It is now possible to tackle the velocity analysis of a linkage made up of several links. Graphical solution of complex number equations (such as Figures 3.40 and 3.42) will be emphasized initially due to the inherent visual feedback offered. Analytical solutions for the same examples are included later. The slider-crank linkage in Fig. 3.43 is a good mechanism to begin with. The objective is to determine the velocity of point B on the slider (link 4), given the input angular velocity ω_2. Since this mechanism is being analyzed for velocity considerations, the displacements should already be known (i.e., the positions of point A, point B, θ_2, and θ_3 are given).

Step 1.   Find the absolute velocity of point A on link 2 (case 2 analysis):

$$\mathbf{V}_A = \mathbf{R}_A i \omega_2 \overleftarrow{} \qquad (\perp A_0 A) \tag{3.31}$$

Step 2.   Find the absolute velocity of point A of link 3 (case 3 analysis). This is a trivial step since a pin joint connects link 2 and link 3 at A, and $\mathbf{V}_{(A3)} = \mathbf{V}_{(A2)}$.

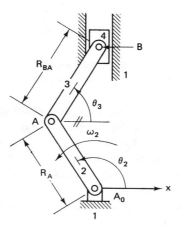

Figure 3.43   The motion of this slider-crank mechanism is known by way of given input angular velocity ω_2. The velocity of slider 4 is to be found.

Step 3. Find the velocity of point B_3 by using point A_3 (both points on link 3) and Eq. (3.27) (case 2 analysis):

$$\mathbf{V}_B = \mathbf{V}_A + \mathbf{V}_{BA} \tag{3.32}$$

Recall that the vector equation (3.32) is equivalent to two independent scalar equations: the summation of the x components and the summation of the y components. Also (using polar notation this time), each velocity vector has two scalar unknowns: its magnitude and its direction. A useful accounting scheme that keeps track of knowns and unknowns in a vector equation is to place a D under the vector if the direction is known (accompanied with an arrow showing the approximate direction) and an M if the magnitude is known. After this is done for each vector in the equation, at most two scalars can be left as unknowns, and the unknowns can be found either graphically or analytically. In this case, Eq. (3.32) becomes

$$
\begin{array}{ccc}
\mathbf{V}_B & = \mathbf{V}_A & + \mathbf{V}_{BA} \\
\updownarrow & \leftarrow & \leftrightarrow \\
D & D & D \\
 & M &
\end{array} \tag{3.33}
$$

Both components of $\mathbf{V}_A$ are known (the magnitude is $V_A = R_A \omega_2$). The direction of $\mathbf{V}_B$ is vertical since the slider is constrained to move in the vertical slot. Also, the direction of $\mathbf{V}_{BA}$ is known to be perpendicular to link AB. With just two unknowns remaining, Eq. (3.33) can be solved graphically as in Figure 3.44 by choosing an appropriate scale for $\mathbf{V}_A$.

$$\mathbf{V}_B = \mathbf{V}_A + \mathbf{V}_{BA}$$

Figure 3.44 Vector construction to find slider velocity for the mechanism of Fig. 3.40.

Step 4. Find the velocity of point B on link 4 (case 3 analysis). Again, this is a trivial step. Thus, the velocity of the slider is found by simply measuring the length of V_B in Figure 3.44.

Notice that this example was formally broken into four steps, two of which were trivial. There is no need to write down the trivial steps once one becomes accustomed to thinking about each individual step. Again, a warning: The more complicated the analysis and the problem, the more critical the need to keep Table 3.1 in mind when working with graphical (or analytical) methods. (See Sec. 3.6 for analytical solution to this problem.)

Example 3.3

The four-bar linkage shown in Fig. 3.45 is driven by a motor connected to link 2 at 600 rpm clockwise. Determine the linear velocities of points A and B and the angular velocities of links 3 and 4 in the position shown in the figure.

Solution

Step 1.   Calculate $\mathbf{V}_A$, as part of link 2.

To obtain ω_2 in radians per second, we use the relationship

$$\omega_2\left(\frac{rad}{sec}\right) = \left[N\left(\frac{rev}{min}\right)\right]\left[2\pi\left(\frac{rad}{rev}\right)\right]\left[\frac{1}{60}\left(\frac{min}{sec}\right)\right]$$

and give it the algebraic sign $+$ for ccw and $-$ for cw rotation. Thus

$$\omega_2 = \frac{-(600)(2\pi)}{60} = -62.8 \text{ rad/sec}$$

$$\mathbf{V}_A = i(\overrightarrow{A_0 A})(\omega_2) = i(2.5)e^{i(118.72°)}(-62.8) = (157)e^{i(28.72°)} \text{ cm/sec} \nearrow$$

Step 2.   Find $\mathbf{V}_B$.

Using Eq. (3.27),

$$\mathbf{V}_B = \mathbf{V}_A + \mathbf{V}_{BA}$$
$$\leftrightsquigarrow \qquad \nearrow \qquad \updownarrow$$
$$\text{D} \qquad \text{D} \qquad \text{D}$$
$$\text{M}$$

Measuring from the velocity diagram (Fig. 3.46),

$$\mathbf{V}_B = (150)e^{i(-17.7°)} \text{ cm/sec} \rightarrow, \qquad \mathbf{V}_{BA} = (118)e^{i(-87.99°)} \text{ cm/sec} \downarrow$$

Step 3.   Calculate ω_3 and ω_4.

From Figs. 3.45 and 3.46, the magnitudes of the angular velocities in radians per second are found by the graphical (geometric) method as follows.

1. Measure the lengths of the linear velocities according to the velocity scale;
2. divide these by their respective radius vectors.

For example, from Fig. 3.46, $\mathbf{V}_A = 157$ cm/sec, $\mathbf{V}_B = 147$ cm/sec and $\mathbf{V}_{BA} = 120$ cm/sec.

From Figure 3.45, $Z_2 = 2.5$ cm, $Z_3 = 5.5$ cm and $Z_4 = 5.0$ cm. Therefore,

$$|\omega_2| = \frac{157 \text{ cm/sec}}{2.5 \text{ cm}} = 62.8 \text{ rad/sec},$$

$$|\omega_4| = \frac{147 \text{ cm/sec}}{5.0 \text{ cm}} = 29.4 \text{ rad/sec} \quad \text{and}$$

$$|\omega_3| = \frac{120 \text{ cm/sec}}{5.5 \text{ cm}} = 21.8 \text{ rad/sec}.$$

The sense for ω_3 and ω_4 may be verified by observing the velocity directions in Fig. 3.46 and envisioning $\mathbf{V}_{BA}$ located at point B as part of link 3 and $\mathbf{V}_B$ located at point B at the end of link 4. It is clear that both links have a clockwise rotation sense and, therefore, ω_3 and ω_4 are negative. The same conclusion will be arrived at using complex numbers Eq. (3.30):

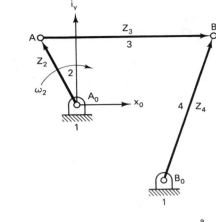

Figure 3.45 Four-bar mechanism with given input angular velocity and link vectors:
$$\mathbf{Z}_2 = \overrightarrow{A_0A} = 2.5 \text{ cm } e^{i(118.72°)} = -1.20 + i2.19$$
$$\mathbf{Z}_3 = \overrightarrow{AB} = 5.5 \text{ cm } e^{i(2.01°)} = 5.50 + i0.20$$
$$\mathbf{Z}_4 = \overrightarrow{B_0B} = 5.0 \text{ cm } e^{i(72.30°)} = 1.52 = i4.76$$

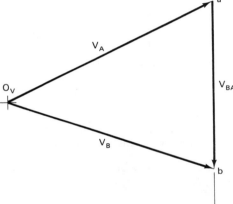

Figure 3.46 Velocity-vector triangle for joints A and B of the mechanism of Fig. 3.45. Note that $\mathbf{V}_B$ is perpendicular to $\mathbf{Z}_4$, $\mathbf{V}_A$ is perpendicular to $\mathbf{Z}_2$, and $\mathbf{V}_{BA}$ is perpendicular to $\mathbf{Z}_3$.

$$\omega_3 = \frac{\mathbf{V}_{BA}}{i\mathbf{Z}_3} = \frac{(120)e^{i(-87.99°)}}{i(5.5)e^{i(2.01°)}} = -21.8 \text{ rad/sec (cw)}$$

$$\omega_4 = \frac{\mathbf{V}_B}{i\mathbf{Z}_4} = \frac{(147)e^{i(-17.70°)}}{i(5)e^{i(72.30°)}} = -29.4 \text{ rad/sec (cw)}$$

Example 3.4

Figure 3.47 shows the same four-bar mechanism as that of Example 3.3 (Fig. 3.45) with the addition of point P on the coupler link. Input velocity is the same as in Example 3.3. Calculate $\mathbf{V}_P$.

Solution   Using the velocity-difference equation between P and A,

$$\mathbf{V}_P = \mathbf{V}_A + \mathbf{V}_{PA}$$
$$\qquad\quad \nearrow \qquad \searrow$$
$$\qquad\quad \text{D} \qquad \text{D}$$
$$\qquad\qquad\quad \text{M}$$

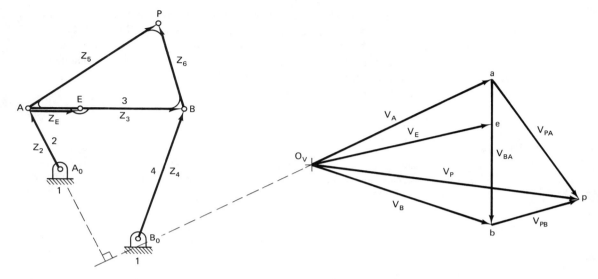

Figure 3.47 The four-bar mechanism of Fig. 3.45 with added coupler point P.

Figure 3.48 Velocity-vector diagram for the mechanism of Fig. 3.47.

There are not enough knowns to solve this equation, but we have not made use of all pertinent information. The velocity-difference equation between P and B can be expressed as

$$\mathbf{V}_P = \mathbf{V}_B + \mathbf{V}_{PB}$$

$$\qquad \overset{\rightarrow}{} \quad \overset{\leftrightarrow}{}$$

$$\qquad D \quad D$$

$$\qquad M$$

This equation also contains three unknowns and cannot be solved by itself, but the two equations can be solved simultaneously.

$$\mathbf{V}_A + \mathbf{V}_{PA} = \mathbf{V}_B + \mathbf{V}_{PB}$$

$$\overset{\nearrow}{} \quad \overset{\searrow}{} \quad \overset{\rightarrow}{} \quad \overset{\swarrow}{}$$

$$D \quad D \quad D \quad D$$

$$M \qquad \quad M$$

There are now two scalar equations and two scalar unknowns (the magnitudes of $\mathbf{V}_{PB}$ and $\mathbf{V}_{PA}$). Figure 3.48 shows that the intersection of the directions of the velocity difference vectors $\mathbf{V}_{PB}$ and $\mathbf{V}_{PA}$ yields point p, and thus

$$V_P = 210 \text{ cm/sec} \longrightarrow$$

It should be noted that in the vector diagram of Fig. 3.48, triangle abp is similar to coupler triangle ABP (Fig. 3.47), because $ab \perp AB$, $bp \perp BP$, and $ap \perp AP$.

Triangle abp is called the *velocity image* of link ABP. Also, the velocity image (triangle in this case) is rotated 90° from the original link in the direction of the angular velocity of that link. The reason is that all velocity difference components are related to the original link vectors by $\mathbf{V} = i\omega\mathbf{r}$. The relationship between a rigid link with three or more points of interest and the corresponding velocity diagram yields a very useful shortcut analysis procedure. Once velocities of two points on a link are calculated, the velocity difference of any other point can be obtained by similar triangles. For example, the absolute velocity of point E (Figs. 3.47 and 3.48) is

$$V_E = 144 \text{ cm/sec} \nearrow$$

Example 3.5

Figure 3.49 shows a six-bar linkage which is actually a four-bar connected to an inverted slider-crank mechanism (see Fig. 3.10). With $\omega_2 = -186$ rpm cw, find $\mathbf{V}_D$, $\mathbf{V}_{(F5)}$ (velocity of F as a point of link 5), and ω_5.

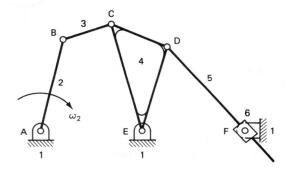

Figure 3.49 Six-bar mechanism. Link 2 is input, ω_2 is given, $\mathbf{V}_D$, $\mathbf{V}_{F_5}$, and ω_5 are to be determined.

Solution   Figure 3.50 shows the graphical solution of this problem based on the successive solutions of the following equations:

$$\mathbf{V}_C = \mathbf{V}_B + \mathbf{V}_{CB}$$
$$\quad \nearrow \quad\ \searrow \quad\ \searrow$$
$$\quad\ \text{D} \quad\ \ \text{D} \quad\ \ \text{D}$$
$$\quad\quad\quad\ \ \text{M}$$

$$\mathbf{V}_D = \mathbf{V}_C + \mathbf{V}_{DC}$$
$$\quad \searrow \quad\ \nearrow \quad\ \updownarrow$$
$$\quad\ \text{D} \quad\ \ \text{D} \quad\ \ \text{D}$$
$$\quad\quad\quad\ \ \text{M}$$

$$\mathbf{V}_{(F5)} = \mathbf{V}_D + \mathbf{V}_{(F5)D}$$
$$\quad \searrow \quad\quad \searrow \quad\ \nearrow$$
$$\quad\ \text{D} \quad\quad\ \text{D} \quad\ \ \text{D}$$
$$\quad\quad\quad\quad\ \ \text{M}$$

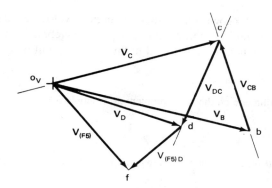

Figure 3.50   Velocity-vector diagram for the six-bar mechanism of Fig. 3.49.

From the velocity polygon:

$$\mathbf{V}_D = (52)e^{i(\arg \overrightarrow{ED} \,-\, 90°)} \text{ cm/sec} \longrightarrow$$

$$\omega_5 = \mathbf{V}_{(F5)D}/i\overrightarrow{DF} = -6.0 \text{ rad/sec (cw)}$$

$$V_{(F5)} = 45 \text{ cm/sec} \searrow$$

The complex-number analytical solution is left to the reader as an exercise (see Exer. 3.2).

3.6 VELOCITY ANALYSIS: ANALYTICAL METHOD

The velocity analysis method described in the last section resulted in a fairly quick graphical solution. When more precision or repeated analyses (of a large number of positions of the same mechanism or of several different mechanisms) are required, the equivalent analytical method should be used, or a generalized analysis package employed.

In order to illustrate this method, which lends itself to computer implementation, let us again look at the slider crank example of Fig. 3.43.

To solve this example analytically by complex numbers, we substitute in Eq. (3.27).

$$V_B e^{i(\pi/2)} = i\omega_2 \mathbf{R}_A + i\omega_3 \mathbf{R}_{BA} \tag{3.34}$$

Here the unknowns are the two reals V_B and ω_3. We therefore separate the real and imaginary parts of Eq. (3.34)

$$V_{Bx} = 0 = -\omega_2 R_{Ay} - \omega_3 R_{BAy}$$

$$V_{By} = \omega_2 R_{Ax} + \omega_3 R_{BAx}$$

from which

$$\omega_3 = -\omega_2 \frac{R_{Ay}}{R_{BAy}}$$

$$V_{By} = \omega_2 \left(R_{Ax} - \frac{R_{BAx} R_{Ay}}{R_{BAy}} \right) \quad \text{and} \quad \mathbf{V}_{BA} = -i\omega_2 \frac{R_{Ay}}{R_{BAy}} \mathbf{R}_{BA}$$

In these computations, all known reals, such as angular velocities and real and imaginary parts of vectors, should be entered with their proper algebraic signs. Then, if the unknown V_{By} turns out to be negative, as it will in this case, $\mathbf{V}_B$ will point downward (see Exer. 3.1).

Example 3.6

Use complex-number arithmetic on the same problem as Example 3.3.

Solution

$$\mathbf{V}_B = i\omega_2\mathbf{Z}_2 + i\omega_3\mathbf{Z}_3 = i\omega_4\mathbf{Z}_4 \tag{3.35}$$

where ω_2 is known, and ω_3 and ω_4 are unknown. Separating real and imaginary parts, we have

$$V_{Bx} = -\omega_2 Z_{2y} - \omega_3 Z_{3y} = -\omega_4 Z_{4y} \tag{3.36}$$

$$V_{By} = \omega_2 Z_{2x} + \omega_3 Z_{3x} = \omega_4 Z_{4x} \tag{3.37}$$

or

$$\begin{bmatrix} -Z_{3y} & Z_{4y} \\ Z_{3x} & -Z_{4x} \end{bmatrix} \begin{bmatrix} \omega_3 \\ \omega_4 \end{bmatrix} = \begin{bmatrix} Z_{2y}\omega_2 \\ -Z_{2x}\omega_2 \end{bmatrix} \tag{3.38}$$

from which

$$\omega_3 = \frac{\begin{vmatrix} \omega_2 Z_{2y} & Z_{4y} \\ -\omega_2 Z_{2x} & -Z_{4x} \end{vmatrix}}{\begin{vmatrix} -Z_{3y} & Z_{4y} \\ Z_{3x} & -Z_{4x} \end{vmatrix}} = \frac{-\omega_2 Z_{2y} Z_{4x} + \omega_2 Z_{2x} Z_{4y}}{Z_{3y} Z_{4x} - Z_{3x} Z_{4y}}$$

$$= \frac{(62.8)(2.19)(1.52) - (62.8)(-1.2)(4.76)}{(0.19)(1.52) - (5.5)(4.76)} = -21.93 \text{ rad/sec}$$

$$\omega_4 = \frac{\begin{vmatrix} -Z_{3y} & Z_{2y}\omega_2 \\ Z_{3x} & -Z_{2x}\omega_2 \end{vmatrix}}{\begin{vmatrix} -Z_{3y} & Z_{4y} \\ Z_{3x} & -Z_{4x} \end{vmatrix}} = \frac{\omega_2 Z_{2x} Z_{3y} - \omega_2 Z_{2y} Z_{3x}}{Z_{3y} Z_{4x} - Z_{3x} Z_{4y}}$$

$$= \frac{(-62.8)(-1.2)(0.19) + (62.8)(2.19)(5.5)}{(0.19)(1.52) - (5.5)(4.76)} = -29.82 \text{ rad/sec}$$

$$\left.\begin{array}{l} V_{Bx} = (29.82)(4.76) = 141.94 \\ V_{By} = (-29.82)(1.52) = -45.32 \end{array}\right\} \quad \mathbf{V}_B = (149.00)e^{i(-17.71°)}$$

$$\mathbf{V}_{BA} = i\omega_3\mathbf{Z}_3 = i(-21.16)(5.5)e^{i(2.01°)} = (116.38)e^{i(-87.99°)}$$

The slight differences in numerical values between the results of the two approaches is due to the inaccuracies in graphical construction in Fig. 3.46 and/or rounding-off errors in computation.

Example 3.7

Solve Example 3.4 analytically

Solution   We will need to determine the arguments of $\mathbf{Z}_5$ and $\mathbf{Z}_6$, say β_5 and β_6.

$$Z_5 e^{i\beta_5} = Z_3 e^{i(2.01°)} + Z_6 e^{i\beta_6} \qquad (3.39)$$

The complex conjugate of this equation also holds (see the appendix to this chapter).

$$Z_5 e^{-i\beta_5} = Z_3 e^{-i(2.01°)} + Z_6 e^{-i\beta_6} \qquad (3.40)$$

Multiplying Eqs. (3.39) and (3.40) together eliminates β_5:

$$Z_5^2 = Z_3^2 + Z_3 Z_6 e^{i(2.01°)} e^{-i\beta_6} + Z_3 Z_6 e^{-i(2.01°)} e^{i\beta_6} + Z_6^2 \qquad (3.41)$$

from which

$$e^{i(\beta_6 - 2.01°)} + e^{i(2.01° - \beta_6)} = \frac{Z_5^2 - Z_3^2 - Z_6^2}{Z_3 Z_6} \qquad (3.42)$$

$$= 2\cos(\beta_6 - 2.01°)$$

Therefore,

$$\beta_6 - 2.01° = \cos^{-1}\left(\frac{Z_5^2 - Z_3^2 - Z_6^2}{2 Z_3 Z_6}\right) = \cos^{-1}\left[\frac{(5.6)^2 - (5.5)^2 - (3.3)^2}{2(5.5)(3.3)}\right] = \pm 105.63°$$

so $\beta_6 = 2.01 \pm 105.63$, signifying possible geometric inversion. From the drawing it is clear that the $+$ sign is to be used, and $\beta_6 = 107.64°$.

Similarly, we can eliminate β_6 from Eqs. (3.39) and (3.40).

$$Z_5 e^{i\beta_5} - Z_3 e^{i(2.01°)} = Z_6 e^{i\beta_6}$$

$$Z_5 e^{-i\beta_5} - Z_3 e^{-i(2.01°)} = Z_6 e^{-i\beta_6}$$

Multiplying, we obtain

$$Z_5^2 - Z_3 Z_5 e^{i(\beta_5 - 2.01°)} - Z_3 Z_5 e^{-i(\beta_5 - 2.01°)} + Z_3^2 = Z_6^2$$

$$e^{i(\beta_5 - 2.01°)} + e^{-i(\beta_5 - 2.01°)} = \frac{Z_6^2 - Z_5^2 - Z_3^2}{-Z_3 Z_5} = 2\cos(\beta_5 - 2.01°)$$

$$\beta_5 - 2.01° = \cos^{-1}\left[\frac{(3.3)^2 - (5.6)^2 - (5.5)^2}{-2(5.5)(5.6)}\right] = \pm 34.58°$$

with the $+$ sign: $\beta_5 = 36.59°$. Thus $\mathbf{Z}_5 = (5.6)e^{i(36.59°)}$ and $\mathbf{Z}_6 = (3.3)e^{i(107.64°)}$.

$$\mathbf{V}_P = \mathbf{V}_A + \mathbf{V}_{PA} = (157)e^{i(28.72°)} + i\omega_3 \mathbf{Z}_5 = (157)e^{i(28.72°)} + i(-21.93)(5.6)e^{i(37.02°)}$$

$$\mathbf{V}_P = (157)e^{i(28.72°)} + (122.8)e^{i(-52.98°)}$$

$$V_{Px} = 137.69 + 73.97 = 211.66$$

$$V_{Py} = 75.44 - 98.09 = -22.65$$

$$\mathbf{V}_P = (212.87)e^{i(-6.11°)} \text{ cm/sec}$$

If E is one-third of the way toward B from A,

$$\mathbf{Z}_E = \frac{5.5}{3}e^{i(2.01°)} \quad \text{and} \quad \mathbf{V}_E = \mathbf{V}_A + \mathbf{V}_{EA} = (157)e^{i(28.72°)} + i\omega_3\frac{5.5}{3}e^{i(2.01°)}$$

$$\mathbf{V}_E = (157)e^{i(28.72°)} + i(-21.19)\frac{5.5}{3}e^{i(2.01°)} = (157)e^{i(28.72°)} + (38.85)e^{i(-87.99°)}$$

$$\mathbf{V}_{Ex} = 137.69 + 1.36 = 139.05$$

$$\mathbf{V}_{Ey} = 75.44 - 38.83 = 36.61$$

$$\mathbf{V}_E = (143.79)e^{i(14.75°)} \text{ cm/sec}$$

3.7 INSTANT CENTERS

The relative velocity (velocity-polygon) method of performing a velocity analysis of a mechanism is just one of several methods available. One drawback of the relative velocity method is the number of steps required in analyzing a complex linkage such as the one in Fig. 3.49. The instantaneous center or instant center method is a very useful technique which is often quicker in complex linkage analysis.

An *instantaneous center* or *instant center* or *centro* is a point at which there is no relative velocity between two links of a mechanism at that instant. The two-link system of Fig. 3.51 consists of a link 2 and ground (link 1) which are connected together at A_0 by a pin (or revolute) joint. The point at which links 1 and 2 have no relative velocity is obviously point A_0. In fact, for all positions in the motion of link 2, the instant center (1,2) is located at A_0.

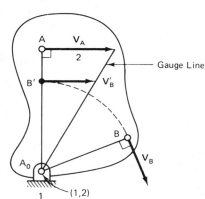

Figure 3.51 Pivot A_0 is the *instant center* for all positions of moving link 2 with respect to link 1. Graphical construction of the velocity of point B of link 2 with the aid of the instant center.

Notice that with the absolute velocity of one point, say A, on link 2 known, with the aid of the instant center of link 2 with respect to ground, a simple construction yields the absolute velocity of any other point, such as B. A *gauge line* is drawn from A_0 through the tip of the velocity vector $\mathbf{V}_A$. There is a linear relationship between the magnitude of the velocity and the distance from the instant center (1,2) (since $V = |R\omega|$). A circular arc centered at A_0 through B locates B' on line $A_0 A$. $\mathbf{V}_B'$ is drawn parallel to $\mathbf{V}_A$ extending to the gauge line. Since B' and B are equidistant from A_0, $\mathbf{V}_B'$ has the same magnitude as the velocity of B. Furthermore, $\mathbf{V}_B \perp A_0 B$, and thus the velocity of B is determined.

In the four-bar linkage of Fig. 3.52, we can identify several instant centers: instant center (1,2) is located at A_0 and center (1,4) is located at B_0. Also, center (2,3) and center (3,4) are located at A and B, respectively. Note that as the linkage moves, these last two pin joints remain instant centers, but their positions do not remain fixed with respect to ground.

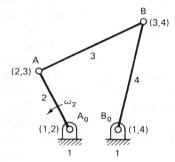

Figure 3.52 Obvious instant centers in a linked mechanism are at the pin joints between links.

Are there instant centers between links 1 and 3 and links 2 and 4? Let us examine this question by means of Fig. 3.53. This figure shows two finitely separated positions of the coupler link of the linkage of Fig. 3.52. The two positions of pins A and B are represented by A_1, A_2 and B_1, B_2. If the perpendicular bisector of $B_1 B_2$ is drawn, any point along this line could serve as a ground pivot which, if connected to B, would serve as the center of rotation of B from B_1 to B_2. If the perpendicular bisector of $A_1 A_2$ is also drawn, one will observe the intersection of these bisectors at P_{12} (*pole* 1,2). This is the point which, if connected rigidly to the coupler link AB, can serve as the fixed pivot about which link AB can be rotated from position 1 to 2. The motion of the coupler link between positions would not (in general) be duplicated, but the two end positions would be exact. Now, if these two positions are brought closer and closer together, line $A_1 A_2$

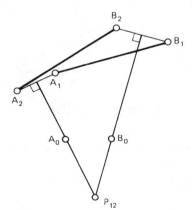

Figure 3.53 The *pole* 12 (P_{12}) about which the coupler could rotate from position 1 to 2.

will approach the direction of $\mathbf{V}_A$, B_1B_2 that of $\mathbf{V}_B$, and P_{12} would approach the instant center between links 1 and 3, (1,3). Also, the perpendicular bisectors would approach links 2 and 4, respectively. Thus the instant center $I_{1,3}$ will be at the intersection of the extension of links 2 and 4. This construction is valid when the two links of interest (1 and 3) are connected by binary links (2 and 4). But how can we find the locations of less obvious instant centers?

Kennedy's Theorem. Kennedy's theorem deals with the three instant centers between three links of a system of rigid members. In Fig. 3.54 there are three links; ground (link 1) and two other links (2 and 3) which are pinned to ground.* The instant centers (1,2) and (1,3) are located at the respective pin joints, but where is instant center (2,3)? For instance, is point P the instant center? Think of P first as being part of link 2 (P_2) and then as part of link 3 (P_3). P_2 has an absolute velocity which is perpendicular to $\overline{O_2P_2}$ while P_3 has an absolute velocity which is perpendicular to $\overline{O_3P_3}$. Now, if P were the instant center I_{23}, then $\mathbf{V}_{P2P3} = -\mathbf{V}_{P3P2}$ would be zero. Notice that point P could not be the instant center (2,3) because the directions of $\mathbf{V}_{(P2)}$ and $\mathbf{V}_{(P3)}$ are not the same. Where, at least, would the directions of the two absolute velocity vectors be the same? At any point along the line of instant centers (1,2) and (1,3) (say at point Q) the absolute velocities of that point, either as a point on link 2 or of link 3, would be perpendicular to the line between these centers. Depending on the angular velocities ω_2 and ω_3, the magnitudes will be equal somewhere along the line of centers or its extensions. This brings us to *Kennedy's theorem of three centers: The three instantaneous centers of three bodies moving relative to one another must lie along a straight line* (see Fig. 3.55 for illustration).

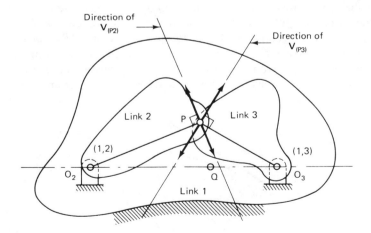

Figure 3.54   Three links (1, 2, and 3) moving with respect to each other. Where is the instant center of links 2 and 3?

* Link 3 passes behind link 2.

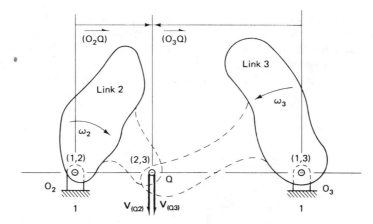

Figure 3.55 Proof of Kennedy's theorem that all three instant centers of three links moving with respect to one another must lie on a straight line.

Before returning to the four-bar linkage of Fig. 3.52 with this theorem, let us develop this theory further. Suppose that the instant center (2,3) were located at point Q, as shown in Fig. 3.55; then

$$\mathbf{V}_{(Q2)} = i\omega_2(\overrightarrow{O_2Q}) = i\omega_2(\overrightarrow{1,2 - 2,3}) \tag{3.43}$$

$$\mathbf{V}_{(Q3)} = i\omega_3(\overrightarrow{O_3Q}) = i\omega_3(\overrightarrow{1,3 - 2,3}) \tag{3.44}$$

but

$$\mathbf{V}_{(Q2)} = \mathbf{V}_{(Q3)} \tag{3.45}$$

so the angular velocity ratio is

$$\frac{\omega_2}{\omega_3} = \frac{(\overrightarrow{O_3Q})}{(\overrightarrow{O_2Q})} = \frac{(\overrightarrow{1,3 - 2,3})}{(\overrightarrow{1,2 - 2,3})} \tag{3.46}$$

which is a positive or negative real. The absolute value of Eq. (3.46) is

$$\boxed{\left|\frac{\omega_2}{\omega_3}\right| = \frac{(\overline{1,3 - 2,3})}{(\overline{1,2 - 2,3})}} \tag{3.47}$$

If the relative instant center (2,3) lies *between* the absolute instant centers (1,2 and 1,3), the angular velocity ratio is *negative* (i.e., the two links rotate in opposite directions). If the relative instant center lies *outside* the other two, the angular velocity ratio is *positive* (i.e., links rotate in the same direction). Eq. (3.46) relates an *angular velocity ratio to a ratio of distances between instant centers*. This fundamental concept is the basis for the "formula method" of the next section.

Another way of looking at instant centers involves replacing a linkage *instantaneously* with a pair of gears. Figure 3.56 shows the gear pair that produces instantaneously the same angular velocity ratio as that of links 2 and 3 in Fig. 3.55. The pitch radius vector of gear 2 is $(1,2 - 2,3)$ and $(1,3 - 2,3)$ is the pitch radius vector of gear 3. A positive angular velocity ratio case is depicted in Fig. 3.57.

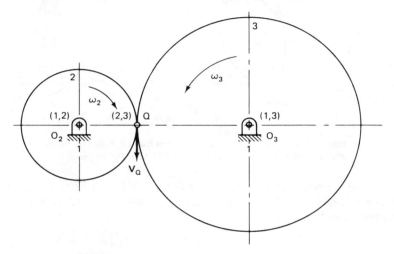

Figure 3.56 Instant centers of a gear pair with negative velocity ratio. This gear pair generates the same angular velocity ratio (ω_2/ω_3) as links 2 and 3 in Fig. 3.55.

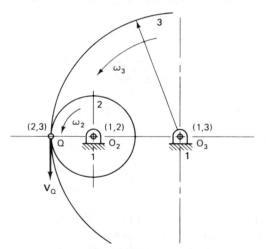

Figure 3.57 Instant centers of a gear pair with positive velocity ratio.

Returning to the four-bar linkage of Fig. 3.52, we now are able to find the locations of instant centers $(2,4)$ and $(1,3)$. According to Kennedy's theorem, instant center $(1,3)$ lies on a line drawn through the common instant centers between links 1, 3 and another link. If that other link is link 2, instant centers $(2,3)$, $(1,2)$, and $(1,3)$ lie on a straight line; so a line is drawn between centers (2,3) and (1,2) (see Fig. 3.58). If links 1, 3, and 4 are utilized, centers (3,4), (1,4), and (1,3) lie on a straight line. A line through (3,4), (1,4) intersects the other line, locating center (1,3). Instant center (2,4) is located by drawing lines through centers (2,3), (3,4) and through centers (1,2), (1,4). Compare this procedure with the intuitive method described by going to the limit (infinitesimal displacement) in connection with Fig. 3.53. Observe that pole P_{12} in Fig. 3.53 approximates the location of the instant center (1,3) in Fig. 3.58.

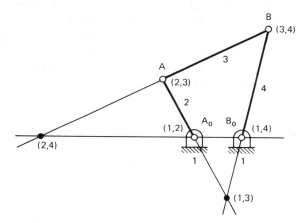

Figure 3.58 Construction of instant centers between opposite links of a four-bar mechanism.

The instant center between two links that are not connected directly together can be thought of as an instantaneous pin joint between those links. For example, we could pin link 3 to ground (link 1) at instant center 1,3 and obtain the same instantaneous relative velocity between those links as the four-bar mechanism has provided. The concept of lower-pair equivalency introduced in Table 1.2 is based on this velocity-equivalence concept.

There is a short graphical method (based on the principle of duality in graph theory) that helps keep track of the instant centers already obtained and those which are obtainable. Although this method may not be warranted for a simple linkage such as a four-bar, it is extremely useful in more complex mechanisms. A circle is drawn and divided by a number of tic marks equal to the number of links in the mechanism. For the four-bar of Fig. 3.58, four tic marks are labeled to represent the four links (see Fig. 3.59). Once an instant center is obtained, a line is drawn between the two numbered tics. The tics represent the links, and the line represents a known instant center between two links.

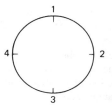

Figure 3.59 Successive steps in keeping track of instant center constructions in Fig. 3.58.

Thus one line represents one instant center. For example, once centers (1,2), (2,3), (3,4), and (1,4) are obtained in Fig. 3.58, solid lines connect 1, 2; 2, 3; 3, 4; and 1, 4 in Fig. 3.60. To obtain additional centers using Kennedy's theorem, we need to find intersections of lines containing appropriate instant centers in Fig. 3.58. The graphical accounting procedure of Figs. 3.59, 3.60 and 3.61 reminds us not only of which centers have not yet been obtained, but which are immediately available and which intersections are appropriate. If we wish to find (1,3), a dashed line is drawn between 1 and 3 (Fig. 3.60). Since there are two sets of two solid lines connecting 1 and 3, center (1,3) is attainable. The diagram (and Kennedy's theorem) says that a line through (1,4) and (4,3) together with a line through (1,2) and (2,3) will intersect at (1,3). Once this is completed as in Fig. 3.58, a solid line replaces the dashed line between 1 and 3 (see Fig. 3.61). The dashed line in Fig. 3.61 indicates that center (2,4) is attainable by lines through (1,4), (1,2) and (3,4), (2,3), as demonstrated in Fig. 3.58.

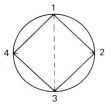

Figure 3.60 Successive steps in keeping track of instant center constructions in Fig. 3.58.

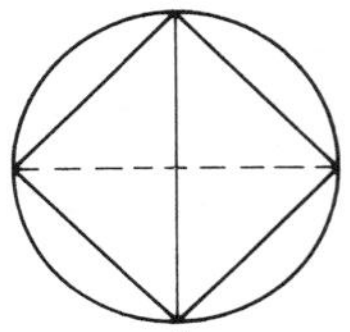

Figure 3.61 Successive steps in keeping track of instant center constructions in Fig. 3.58.

Before going to a more complex mechanism it is worthwhile noting that the number of instant centers N of a mechanism with n links is

$$N = \frac{n(n - 1)}{2} \tag{3.48}$$

This is intuitively true, because each of the n links has an instant center common with each one of the other $(n - 1)$ links, but center (j,k) is the same as (k,j), hence the product $n(n - 1)$ must be divided by 2.

Example 3.8

Determine the locations of all the instant centers for the six-bar mechanism in Fig. 3.62 in preparation for a velocity analysis of this mechanism.

Solution   The number of instant centers is, by Eq. (3.48),

$$N = \frac{6(6 - 1)}{2} = 15$$

The instant centers, which are obtained by inspection, are shown in the circular diagram (Fig. 3.62). Notice that the instant center between the slider (link 6) and the slide (link 5) is off at infinity in the direction perpendicular to the slide. The circular diagram (Fig. 3.62) indicates that several instant centers are now available (this is not the general rule with more

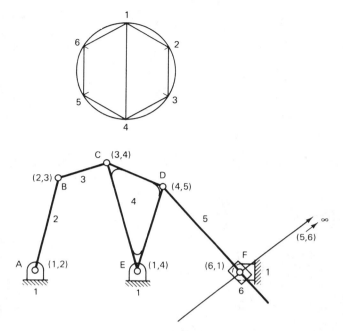

Figure 3.62 Obvious instant centers in a six-bar mechanism.

complex mechanisms; usually one at a time is attainable, thus enabling the determination of the next). Centers (1,3), (1,5), (2,4), and (4,6) can be determined as described above. Notice particularly the construction of center (4,6). Figure 3.63 dictates that (4,6) lies along lines connecting (1,6), (1,4) and (5,6), (4,5). In the second pair, (5,6) is at ∞ in the direction perpendicular to link 5. How can a line be drawn through (5,6) and (4,5)? Simply draw a line parallel to (1,6)–(5,6) through (4,5) as shown in Fig. 3.64. This can be justified if you picture yourself sitting at (4,5) and looking at (5,6) off in infinity in the direction of (1,6)–(5,6). It is clear that your line of sight would be parallel to (1,6)–(5,6).

Figure 3.65 indicates the stage of the analysis where (4,6), (1,5), (2,4), and (1,3) have been obtained. Instant center (2,5) is one possible next choice. The intersection of lines through (1,2), (1,5) and (2,4), (4,5) yields this center, which will be used in Ex. 3.9 for velocity analysis.

The rest of the instant centers — (3,5), (2,6), (3,6) — are obtained following the same procedure (see Fig. 3.64). Notice that these last three centers have more than two intersecting lines which locate them. This provides a redundant verification indicating the accuracy

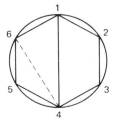

Figure 3.63 Construction of all instant centers for the six-link mechanism of Fig. 3.62. Figures 3.63 and 3.65 show several stages of the graphical "bookkeeping" during the construction.

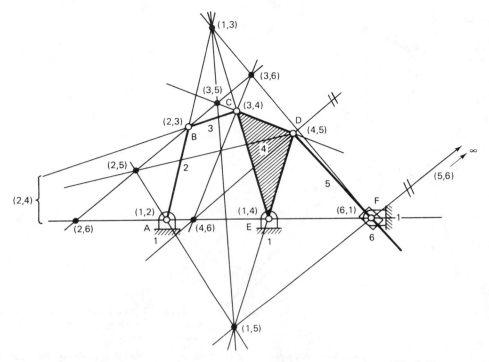

Figure 3.64 Construction of all instant centers for the six-link mechanism of Fig. 3.62. Figures 3.63 and 3.65 show several stages of the graphical "bookkeeping" during the construction.

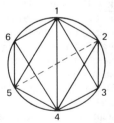

Figure 3.65 Construction of all instant centers for the six-link mechanism of Fig. 3.62. Figures 3.63 and 3.65 show several stages of the graphical "bookkeeping" during the construction.

of the construction. If more than two lines intersect at the same point, everything is fine. A slight discrepancy indicates a slight inaccuracy in the drawing. If the intersections are not close, a major mistake has been made and a complete review of previous steps is warranted.

3.8 VELOCITY ANALYSIS USING INSTANT CENTERS*

How can instant centers be used in velocity analysis? There are actually two methods: the *formula method* and the *graphical method* based on instant centers.

* As will be shown later, that while instant-center analysis is applicable to displacement and velocity analyses, it is not suitable for accelerations or higher-order motion derivatives.

Let us look first at the *formula* method. We will return to the four-bar of Figs. 3.45 and 3.47. Recall that $\omega_2 = -600$ rpm (cw) in this example. Determine ω_4, ω_3, V_P, and V_B. Eq. (3.46) is a fundamental formula relating angular velocity ratios and instant center locations. Any combination of subscript numbers in the correct sequence may replace those in Eq. (3.46).* Here

$$\frac{\omega_4}{\omega_2} = \frac{\overrightarrow{(1,2 - 2,4)}}{\overrightarrow{(1,4 - 2,4)}} = \frac{\ell_1}{\ell_2}$$

Figure 3.66 shows that instant center (2,4) lies outside (1,2) and (1,4), so that the angular velocity ratio ω_4/ω_2 is positive. This is also seen from the fact that ℓ_1 and ℓ_2 are collinear vectors of equal sense. Checking this by inspection, as link 2 rotates clockwise, so will link 4. From Fig. 3.66,

$$\frac{\omega_4}{\omega_2} = \frac{\ell_1}{\ell_2} = \frac{2.8 \text{ cm}}{5.9 \text{ cm}} = 0.47$$

Thus the instantaneous angular velocity of link 4 in the position shown is

$$\omega_4 = (0.47)(-600) = -282 \text{ rpm}$$

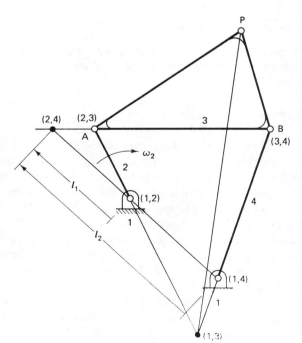

Figure 3.66 The angular velocity ratio ω_4/ω_2 is the quotient of the two collinear vectors ℓ_1/ℓ_2.

* One may either use Eq. (3.46) or Eq. (3.47) to solve these problems. The latter requires inspection for determination of cw or ccw rotation, but is usually preferred unless one is programming the solution.

or

$$\omega_4 = \frac{2\pi}{60}(-282) = -29.5 \text{ rad/sec (cw)}$$

Also from the Fig. 3.66,

$$\frac{\omega_3}{\omega_2} = \frac{\overrightarrow{(1,2-2,3)}}{\overrightarrow{(1,3-2,3)}} = \frac{2.1 \text{ cm}}{6.2 \text{ cm}} = 0.34$$

$$\omega_3 = -21 \text{ rad/sec (cw)}$$

Instant center (2,3) also lies outside of instant centers (1,2) and (1,3), yielding a positive angular velocity ratio between links 2 and 3.

The velocity $\mathbf{V}_B$ is now calculated (considering B as a point on link 4) see Fig. 3.45 for link dimensions.

$$\mathbf{V}_B = i\omega_4(\overrightarrow{1,4-3,4})^\dagger = 147.5 \text{ cm/sec} \searrow$$

Or substituting for ω_4, an expression for $\mathbf{V}_B$ is obtained.

$$\mathbf{V}_B = i\omega_2 \frac{\overrightarrow{(1,2-2,4)}}{\overrightarrow{(1,4-2,4)}}(\overrightarrow{1,4-3,4})^\dagger$$

Also, the vector $\mathbf{V}_P$ can be expressed in terms of ω_3,

$$\mathbf{V}_P = i\omega_3(\overrightarrow{1,3-P})^{**} = (204 \text{ cm/sec})e^{i(\arg(\overrightarrow{1,3-P})-90°)} \searrow$$

(perpendicualr to $(\overrightarrow{1,3-P})$, pointing to the right, because $\omega_3 < 0$). This equation is based on the fact that link 3 is rotating instantaneously about center (1,3). Substituting for ω_3, we obtain

$$\mathbf{V}_P = i\omega_2 \frac{\overrightarrow{(1,2-2,3)}}{\overrightarrow{(1,3-2,3)}}(\overrightarrow{1,3-P})^{**}$$

The same problem will now be solved by the *graphical method* based on instant centers and the fact that *the magnitude of velocity is proportional to the radius from the point of reference to point of interest*. The velocity of point B may be found as follows (Fig. 3.67). First the velocity of point A is drawn to scale on the figure. Then the instant center (2,4) is found at the intersection of the extensions of lines (1,2) (1,4) and (2,3) (3,4). Now, recalling that instant center (2,4) is a point at which link 2 and link 4 (link 4 contains point B) have no relative velocity or, stated in other words, link 2 and link 4 have the same absolute velocity at (2,4). Therefore, instant center (2,4) is the unique point at which we may transfer a known absolute velocity from link 2 to link 4.

A line is drawn connecting the three key instant centers* (1,2), (1,4), and (2,4). An arc of radius A_0A is drawn from A to this line locating A'. After the magnitude of $\mathbf{V}_A$ is redrawn to scale perpendicular to A_0A', a gauge line is constructed from A_0 through the tip of $\mathbf{V}_A'$. The velocity of instant center (2,4) as part of link 2, $\mathbf{V}_{2,4}$, is now drawn

---

* Instant centers (1,2), (1,3), and (2,3) could also be used as described in the next part of this development because point B is also a part of link 3.

† Take link dimensions from Figure 3.45.

** In Figure 3.45, knowing that link 4 is 5 cm, get $(1,3\text{-}P)$ by proportion.

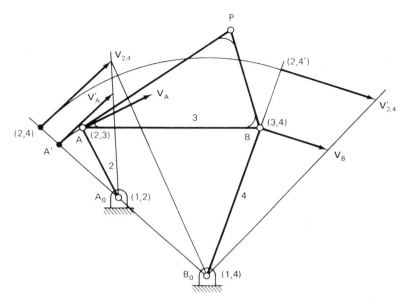

Figure 3.67 Construction of $\mathbf{V}_B$ by way of the instant center and gage line method when $\mathbf{V}_A$ is known.

between (2,4) and the gauge line. The transfer from link 2 to link 4 is now accomplished (using the property of instant center (2,4) described in the previous paragraph) by constructing an arc, centered at B_0 [center (1,4)], of radius $[B_0 - (2,4)]$, from (2,4) until it intersects a line through (1,4) and (3,4). At this intersection [point (2,4)′] the magnitude of $\mathbf{V}_{2,4}$ is laid out to find $\mathbf{V}'_{2,4}$. After a gauge line from (1,4) is constructed, the velocity of B is found ($V_B = 149$ cm/sec).

The velocity of point P is found using the same strategy except that centers (1,2), (1,3), and (2,3) are now of interest since a known velocity of link 2 is being transferred to a point on link 3. Starting with the known velocity of point A, the construction procedure is shorter since point A is the instant center (2,3) (see Fig. 3.68). A gauge line from (1,3) through the tip of $\mathbf{V}_A$ is drawn. After an arc of radius $[(1,3) - P]$ is constructed, P' is located. The velocity $\mathbf{V}'_P$ is then transferred to point P, yielding $V_P = 210$ cm/sec↘. Notice that $\mathbf{V}_B$ may also be obtained by transferring $\mathbf{V}_P$ to $\mathbf{V}''_P$ and down to $\mathbf{V}_B$ since B is also a part of link 3.

The foregoing procedure can be programmed for digital computation or performed on a hand calculator by the use of complex numbers, as demonstrated earlier in this chapter. However, instant centers 1,3 and 2,4 still need to be determined analytically. For this, the method will now be illustrated for instant center 1,3 as follows.

Referring to Fig. 3.58, we see that $[\overrightarrow{A_0(1,3)}] = \mathbf{I}_{1,3} = \lambda_2 \mathbf{Z}_2$, where λ_2 is an unknown real. Also $\mathbf{I}_{1,3} = \mathbf{Z}_2 + \mathbf{Z}_3 - \mathbf{Z}_4 + \lambda_4 \mathbf{Z}_4$, where λ_4 is also an unknown real. We equate the two expressions for $\mathbf{I}_{1,3}$.

$$\mathbf{Z}_2 \lambda_2 - \mathbf{Z}_4 \lambda_4 = \mathbf{Z}_2 + \mathbf{Z}_3 - \mathbf{Z}_4 \qquad (3.49)$$

Since all $\mathbf{Z}_k$ are known, Eq. (3.49) can be solved for λ_2 and λ_4 (see Exer. 3.3).

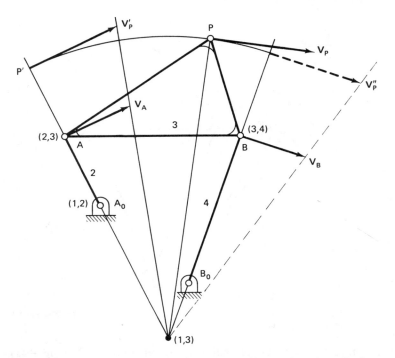

Figure 3.68 Gage-line construction of the velocity of a coupler point using instant centers.

Example 3.9

For the six-bar of Fig. 3.49, find the absolute velocity of point D and point F (as part of link 5) given ω_2.

Method 1: The formula method. Since points F and D are both part of link 5, instant centers (1,2), (1,5), and (2,5) will be used.

$$\frac{\omega_5}{\omega_2} = \frac{\overrightarrow{(1,2 - 2,5)}}{\overrightarrow{(1,5 - 2,5)}}$$

with $\omega_2 = -186$ rpm $= -19.5$ rad/sec. From Figure 3.64,

$$\omega_5 = -19.5 \text{ rad/sec}\left(\frac{1.5 \text{ cm}}{5.0 \text{ cm}}\right) = -5.9 \text{ rad/sec}$$

Therefore,

$$\mathbf{V}_D = i\omega_5(\overrightarrow{(1,5) - D})* = (51 \text{ cm/sec})e^{i(\arg\overrightarrow{(1,5 - D)}* - 90°)}$$

$$\mathbf{V}_{F5} = i\omega_5(\overrightarrow{(1,5) - F})* = (43 \text{ cm/sec})e^{i(\arg\overrightarrow{(1,5 - F)}* - 90°)}$$

Method 2: The graphical technique. Once the instant centers and other points of interest are found, *the linkage itself is not necessary*. In Fig. 3.69 instant centers (1,2), (1,5), and (2,5) are shown together with points B, D, and F. The known velocity $\mathbf{V}_B = 77$ cm/sec on link 2 is transferred by way of an arc about (1,2) to point B', which is along the line of

* Measure in Figure 3.64.

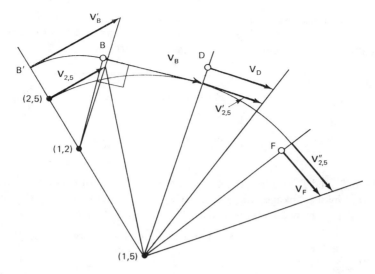

Figure 3.69   Gage-line construction of velocities in the six-link mechanism of Fig. 3.64 using instant centers.

centers $(1,5)$, $(1,2)$, and $(2,5)$. Construction of a gauge line from $(1,2)$ through the tip of $\mathbf{V}_B'$ will lead to $\mathbf{V}_{2,5}$. $\mathbf{V}_{2,5}$ is transferred to $\mathbf{V}_{2,5}'$ and $\mathbf{V}_{2,5}''$ along the arc of radius $(1,5) - (2,5)$, centered at $(1,5)$, until it intersects lines $[(1,5) - D]$ and $[(1,5) - F]$, respectively. $\mathbf{V}_D$ and $\mathbf{V}_F$ are then found by using gauge lines. They are

$$\mathbf{V}_D = 51 \text{ cm/sec} \searrow$$

$$\mathbf{V}_F = 45 \text{ cm/sec} \searrow$$

3.9 MECHANICAL ADVANTAGE

One of the major criteria of which a designer must be aware is the ability for a particular mechanism to transmit torque or force. Some mechanisms, such as a gear train, transmit a constant torque ratio between the input and output because there is a constant speed ratio between input and output (see Chap. 7). In a linkage, however, this is not the case. How might we determine a relationship between *force out* and *force in?* Two observations can be made without further analysis.

1. As hinted in the gear train mentioned above, the torque ratio is a function of the speed or angular velocity ratio between output and input links of the mechanism.
2. The torque ratio is a function of geometric parameters, which, in the case of a linkage, will generally change during the course of the mechanism motion.

If we assume that a mechanism is a conservative system (i.e., energy losses due to friction, heat, etc., are negligible compared to the total energy transmitted by the system), and if we assume that there are no effects of inertia forces, *power in* (P_{in}) is equal to *power out* (P_{out}) (see Fig. 3.70). Thus the *torque in* times the *angular velocity in* is equal to the *torque out* times the *angular velocity out:*

$$P_{in} = T_{in}\omega_{in} = T_{out}\omega_{out} = P_{out} \tag{3.50}$$

or,

$$P_{in} = F_{in}V_{in} = F_{out}V_{out} = P_{out} \tag{3.51}$$

where T_{in} and F_{in} are torque and force exerted *on* the linkage, and T_{out} and F_{out} are those exerted *by* the linkage; where V_{in} and V_{out} are the velocities of the points through which $\mathbf{F}_{in}$ and $\mathbf{F}_{out}$, respectively, act; and where, in vector form

$$\mathbf{V} \cdot \mathbf{F} = VF \cos(\arg \mathbf{F} - \arg \mathbf{V}) \tag{3.51a}$$

Also,

$$\mathbf{V} \cdot \mathbf{F} = V_x F_x + V_y F_y \tag{3.51b}$$

(For a proof, see Exer. 3.4.)

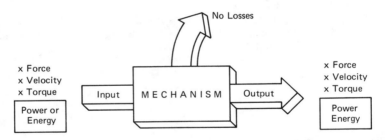

Figure 3.70 Power and energy are conserved through the mechanism. Force, velocity and torque are not.

Figure 3.70 reminds us that neither force, velocity, nor torque alone is constant through a linkage mechanism. The designers of many misguided perpetual motion machines disregarded this fact.

Notice that the units of torque times angular velocity and the scalar product of force and velocity both represent power. From Eq. (3.50),

$$\frac{T_{out}}{T_{in}} = \frac{\omega_{in}}{\omega_{out}} \tag{3.52}$$

By definition, *mechanical advantage* (M.A.) is the ratio of the magnitudes of *force out* over *force in*

$$\boxed{\text{M.A.} = \frac{F_{\text{out}}}{F_{\text{in}}}} \tag{3.53}$$

where $F = |\mathbf{F}|$.

Combining Eqs. (3.53) and (3.50) by noticing that torque is the product of force times a radius,

$$\text{M.A.} = \left(\frac{T_{\text{out}}}{r_{\text{out}}}\right)\left(\frac{r_{\text{in}}}{T_{\text{in}}}\right) = \left(\frac{r_{\text{in}}}{r_{\text{out}}}\right)\left(\frac{T_{\text{out}}}{T_{\text{in}}}\right) \tag{3.54}$$

and

$$\boxed{\text{M.A.} = \left(\frac{r_{\text{in}}}{r_{\text{out}}}\right)\left(\frac{\omega_{\text{in}}}{\omega_{\text{out}}}\right)} \tag{3.55}$$

Thus the mechanical advantage is a product of *two* factors: (1) a ratio of distances that depend on the placement of the input and output forces and (2) an angular velocity ratio. The first factor may not change in value as the mechanism moves, but the second one will change in most linkage mechanisms. Since the angular velocity ratio can be expressed entirely in terms of directed distances (based on the instant center development), the mechanical advantage can be expressed entirely in terms of ratios of distances (see Sec. 3.8).

Let us look at the four-bar mechanism in Fig. 3.71. If we neglect the weight of links 2, 3, and 4, what reading would you expect the scale to display as a result of the block weighing 10 lbf on link 2 of the mechanism? Using Eq. (3.55),

$$\frac{F_{\text{out}}}{F_{\text{in}}} = \text{M.A.} = \left(\frac{\omega_{\text{in}}}{\omega_{\text{out}}}\right)\left(\frac{r_{\text{in}}}{r_{\text{out}}}\right) \tag{3.56}$$

In this case link 2 is the input while link 4 is the output. From Eq. (3.55),

$$\text{M.A.} = \frac{F_{\text{out}}}{F_{\text{in}}} = \left(\frac{\omega_2}{\omega_4}\right)\left(\frac{r_{\text{in}}}{r_{\text{out}}}\right) = \frac{(1,4 - 2,4)}{(1,2 - 2,4)}\frac{(r_{\text{in}})}{(r_{\text{out}})} \tag{3.56a}$$

Note that the common instant center (2,4) is outside the others, making the angular velocity ratio positive. Measuring distances on Fig. 3.71* and solving for F_{out},

$$F_{\text{out}} = F_{\text{in}}(\text{M.A.}) = (10)\frac{(2)}{(.5)}\frac{(1.5)}{(.5)} = (10)(4)(3) = 120 \text{ lbf}$$

* 51 mm in Fig. 3.71 = 2 in.

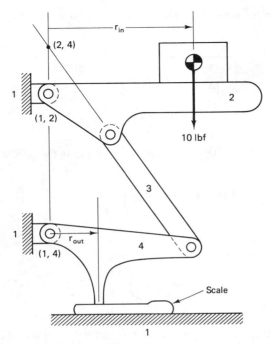

Figure 3.71 Determination of the scale reading based on a weight of 10 lbs.

where M.A. was $(4)(3) = 12$. The gain in mechanical advantage has contributions both from the radius ratio and the angular velocity ratio. Both are distances measured on the diagram.

This result can be verified by way of free-body diagrams of Fig. 3.72. Here too, the mechanical advantage is purely in terms of distances.

$$\text{M.A.} = \left(\frac{1.5}{0.3}\right)\left(\frac{1.2}{0.5}\right) = (5)(2.4) = 12$$

Expressions in the form of Eq. (3.56) are *powerful design tools* and can usually be verified by inspection. In many design situations, the mechanical advantage expression for a mechanism will allow the optimal redesign of that device for improved mechanical advantage. Practical considerations such as the maximum permitted size of the mechanism will usually limit the amount of change allowable from an original design (see Chap. 8 appendix).

For example, suppose that the four-bar linkage of Fig. 3.73 is being used as the driving mechanism of a manually operated pump. In the position shown, the handle is being pulled left with force $\mathbf{F}_{in}$. Meanwhile the pressure difference across the piston in the cylinder is resisting the movement by a force equal and opposite to $\mathbf{F}_{out}$. What is the mechanical advantage of this device in the position shown? (The piston rod is instantaneously perpendicular to link 4.)

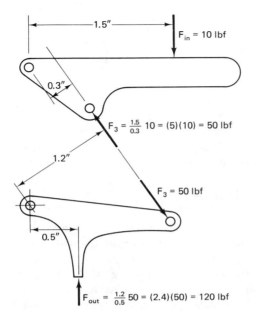

F_{in} = 10 lbf

1.5"

0.3"

$F_3 = \frac{1.5}{0.3}$ 10 = (5)(10) = 50 lbf

1.2"

F_3 = 50 lbf

0.5"

$F_{out} = \frac{1.2}{0.5}$ 50 = (2.4)(50) = 120 lbf

Figure 3.72 Determination of mechanical advantage by way of free-body diagrams.

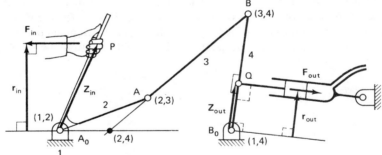

Figure 3.73 Mechanical advantage of this four-bar handpump drive mechanism increases as the toggle position (A_0, A, and B collinear) is approached.

If the input link is identified as link 2 and the output link as link 4, then according to Eq. (3.52),

$$\frac{T_4}{T_2} = \frac{\omega_2}{\omega_4}$$

Instant center (2,4) is found to lie between (1,2) and (1,4), so that

$$\frac{T_4}{T_2} = \frac{\omega_2}{\omega_4} = \frac{\overrightarrow{(1,4 - 2,4)}}{\overrightarrow{(1,2 - 2,4)}} \tag{3.57}$$

Using complex numbers, Eq. (3.57) will have the form

$$\frac{T_4}{T_2} = \frac{z_{out} F_{out} \sin(\arg \mathbf{F}_{out} - \arg \mathbf{z}_{out})}{z_{in} F_{in} \sin(\arg \mathbf{F}_{in} - \arg \mathbf{z}_{in})}$$

and

$$\text{M.A.} = \frac{F_{out}}{F_{in}} = \frac{z_{in}}{z_{out}} \left(\frac{\overrightarrow{(1,4 - 2,4)}}{\overrightarrow{(1,2 - 2,4)}} \right) \frac{\sin(\arg \mathbf{F}_{in} - \arg \mathbf{z}_{in})}{\sin(\arg \mathbf{F}_{out} - \arg \mathbf{z}_{out})} \qquad (3.58)$$

If we disregard the algebraic sign of the instant center distance ratio and write r instead of z for the arm lengths of the input and output forces, Eq. (3.58) becomes

$$\text{M.A.} = \left(\frac{r_{in}}{r_{out}} \right) \frac{(1,4 - 2,4)}{(1,2 - 2,4)} \qquad (3.59)$$

Notice the factors that make up the mechanical advantage. The mechanical advantage is greater if the ratio z_{in}/z_{out} is greater. This checks intuitively with Fig. 3.73. The second ratio can be checked intuitively also. Notice that as point P is moved to the left (link 2 rotated counterclockwise) to the position shown in Fig. 3.74, where the four-bar linkage is in its "toggle" position, centers $(1,2)$ and $(2,4)$ coincide. According to the expression [Eq. (3.58)], the mechanical advantage goes to infinity in this position. Since links 2 and 3 line up, (ideally) no force is required at P to resist an infinite force at Q. Of course, bending of link 4 would occur before an infinite force could be applied. The ratio of sines is a measure of the relative closeness to perpendicularity of each force to its arm vector. With these considerations, Eq. (3.58) is intuitively verified.

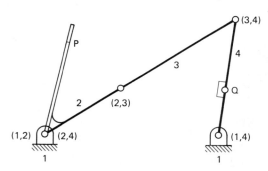

Figure 3.74 Toggle position of the pump-drive mechanism of Fig. 3.73: M.A. is theoretically infinite: a small force at P can overcome a very large force at Q.

If more mechanical advantage is required with $A_0 P$ in the position shown in Fig. 3.73, then Eq. (3.58) dictates the following possibilities:

1. Increase z_{in}.
2. Decrease z_{out}.
3. Move B_0 away from $(2,4)$ (keeping the rest of the linkage the same).
4. Move A_0 toward $(2,4)$ (keeping the rest of the linkage the same).
5. Move point A until links 2 and 3 line up.
6. Make $\mathbf{F}_{in}$ perpendicular to $\mathbf{z}_{in}$.

Example 3.10

Determine the mechanical advantage of the adjustable toggle pliers in Fig. 3.75. Why is this device designed this way?

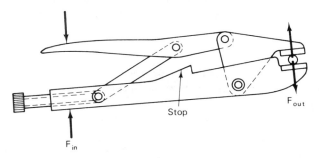

Figure 3.75   The stop prevents excessive overtravel (branching) beyond the toggle position of the upper handle and the coupler link.

Stop

F_{out}

F_{in}

Solution   Let us designate input link = 3, output link = 4, and ground link = 1 (see Fig. 3.76). Thus $T_3\omega_3 = T_4\omega_4$, since $\mathbf{F}_{out}$ is perpendicular to $\mathbf{r}_{out}$ and $\mathbf{F}_{in}$ is perpendicular to $\mathbf{r}_{in}$,

$$T_4 = [\mathbf{r}_{out} \times \mathbf{F}_{out}] = r_{out}F_{out}\sin(\arg \mathbf{F}_{out} - \arg \mathbf{r}_{out}) = r_{out}F_{out}\sin(-90°) = -1.9F_{out}$$

Similarly, $T_3 = [\mathbf{r}_{in} \times \mathbf{F}_{in}] = (5.1)F_{in}$. From Equation (3.52)

$$\frac{T_4}{T_3} = \frac{\omega_3}{\omega_4} = \frac{\overrightarrow{(3,4-1,4)}}{\overrightarrow{(3,4-1,3)}} = \frac{-1.6}{0.7} = -2.29^{\dagger}$$

from which

$$\text{M.A.} = \frac{F_{out}}{F_{in}} = \left|\frac{r_{in}}{r_{out}}\frac{\omega_3}{\omega_4}\right| = \left|\frac{5.1}{1.9}(-2.29)\right| = 6.15*$$

In the position shown, the pliers have only a 6.15:1 mechanical advantage. As we clamp down on the pliers, however, instant center (2,4) approaches center (1,2). When (1,2), (2,3), and (3,4) are close to being in a straight line, (2,4) is nearly coincident with (1,2), (1,3) approaches (3,4) and the mechanical advantage approaches infinity.

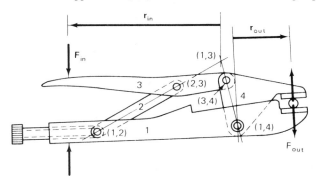

r_{in}

r_{out}

F_{in}

(1,3)

3   (2,3)

(3,4)   4

2

(1,2)   1

(1,4)

F_{out}

Figure 3.76   As (3,4), (2,3), and (1,2) approach collinearity, M.A. approaches infinity.

* Notice that we could have gone directly to this step as suggested in Equation (3.55).

† 15 mm in Fig. 3.76 = 1.9 units.

The screw adjustment should be set so that the maximum mechanical advantage occurs at the required distance between the jaws of the pliers. In fact, in some brands, there is a stop located in the "over center" position (just past the "dead center"), as shown in Fig. 3.74. This gives both a very high mechanical advantage and stable "grip" for the linkages, since it would take an ideally infinite force at the jaws to move the linkage back through its toggle position.

Example 3.11

Determine the mechanical advantage of the slider-crank mechanism shown in Fig. 3.77. The slider output link requires a different approach for the solution.

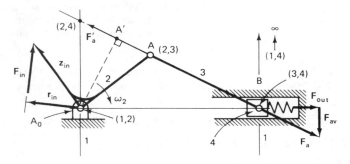

Figure 3.77 Geometric constructions toward finding the M.A. of a slider-crank mechanism.

Solution   If link 2 is considered the input link and link 4 the output link, the procedure described above must be modified (because link 4 is a slider so that $\omega_4 = 0$). We know that $(\text{power})_{\text{in}} = (\text{power})_{\text{out}}$. Thus, from Equations (3.50) and (3.51)

$$T_2\omega_2 = \mathbf{F}_{\text{out}} \cdot \mathbf{V}_B \tag{3.60}$$

where $T_2 = [\mathbf{z}_{\text{in}} \times \mathbf{F}_{\text{in}}]$. Notice that since the output link 4 is constrained to translate in the horizontal slot, any point considered as part of link 4 must have a velocity in the horizontal direction. Moreover, any point considered as part of link 4 has the same velocity, including the point of the extended <u>plane of link</u> 4 momentarily coincident with the center (2,4). This point has velocity $i\omega_2(\overrightarrow{1,2 - 2,4})$; therefore,

$$\mathbf{V}_B = i\omega_2(\overrightarrow{1,2 - 2,4}) \tag{3.61}$$

Combining (3.60) and (3.61), from (power in) = (power out), we get

$$[\mathbf{z}_{\text{in}} \times \mathbf{F}_{\text{in}}]\omega_2 = T_2\omega_2 = \mathbf{F}_a \cdot \mathbf{V}_B$$

$$\mathbf{F}_a \cdot (i\omega_2\overrightarrow{(1,2 - 2,4)}) = F_a|\omega_2\overrightarrow{(1,2 - 2,4)}|\cos(\arg \mathbf{F}_a - \arg \mathbf{V}_B)$$

$$\omega_2 z_{\text{in}} F_{\text{in}} \sin(\arg \mathbf{F}_{\text{in}} - \arg \mathbf{z}_{\text{in}}) = F_a|\omega_2\overrightarrow{(1,2 - 2,4)}|\cos(\arg \mathbf{F}_a - \arg \mathbf{V}_B)$$

from which, noting that

$$F_a \cos(\arg \mathbf{F}_a - \arg \mathbf{V}_B) = F_{\text{out}} \qquad \text{and} \qquad \frac{\omega_2}{|\omega_2|} = -1$$

we have

$$\text{M.A.} = \frac{F_{\text{out}}}{F_{\text{in}}} = -\frac{z_{\text{in}} \sin(\arg \mathbf{F}_{\text{in}} - \arg \mathbf{z}_{\text{in}})}{|\overrightarrow{(1,2 - 2,4)}|} = \frac{r_{\text{in}}}{|\overrightarrow{(1,2 - 2,4)}|} \qquad (3.62)$$

which is positive (as it should be) because $\sin(\arg \mathbf{F}_{\text{in}} - \arg \mathbf{z}_{\text{in}})$ is negative.

This expression makes sense: The longer the arm $\mathbf{r}_{\text{in}}$ on the input link, provided that its direction remains the same, the higher the mechanical advantage. Note also that, as the input link rotates cw, point A moves toward the toggle position and center $(2,4)$ moves toward $(1,2)$—increasing the mechanical advantage.

Another verification of Eq. (3.62) follows. Input torque $T_{\text{in}} = [\mathbf{z}_{\text{in}} \times \mathbf{F}_{\text{in}}]$ cw. T_{in} is resisted by the moment of a pin force F'_a at joint A. Since link 3 is a "two-pin" link, F'_a must act along link 3. Its resisting moment is $[((\overrightarrow{1,2}) - \overrightarrow{A'}) \times \mathbf{F}'_a]$ (ccw) where $((\overrightarrow{1,2}) - \overrightarrow{A'})$ is perpendicular to link 3. Link 3 transmits $\mathbf{F}'_a$ to slider 4 at $(3,4)$, where it is resisted by $\mathbf{F}_a$. The vertical component $\mathbf{F}_{av}$ of $\mathbf{F}_a$ is perpendicular to the motion of output link 4 and therefore does no work. Thus, the output force $\mathbf{F}_{\text{out}}$ is the horizontal component of $\mathbf{F}_a$, pointing to the right in Fig. 3.77. By similar triangles it is easy to show that

$$\frac{F_{\text{out}}}{F_a} = \frac{(1,2 - A')}{(1,2 - 2,4)} \qquad (3.63)$$

where, for instance,

$$(1,2) - A' = |\overrightarrow{(1,2) - A'}|$$

But

$$\frac{F_a}{F_{\text{in}}} = \frac{r_{\text{in}}}{(1,2 - A')} \qquad (3.64)$$

and by multiplication of these two ratios we find that

$$\frac{F_{\text{out}}}{F_{\text{in}}} = \frac{r_{\text{in}}}{(1,2 - 2,4)}$$

which agrees with Eq. (3.62).

Notice that link 3 could also be considered as the output link because point B is also a part of the connecting rod. In this case centers $(1,2)$, $(2,3)$, and $(1,3)$ would be used to obtain an expression for mechanical advantage that would be numerically equivalent to Eq. (3.62) (see Exer. 3.5).

For still another verification of Eq. (3.62), as well as for the verification of the graphical (geometric) construction to determine the mechanical advantage, we resort to free-body diagrams as follows. For the free-body diagram of link 2 (see Fig. 3.78), summing moments about A_0 yields

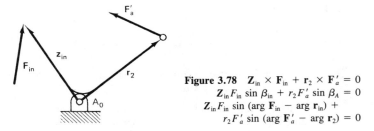

Figure 3.78 $\mathbf{Z}_{in} \times \mathbf{F}_{in} + \mathbf{r}_2 \times \mathbf{F}_a' = 0$
$Z_{in} F_{in} \sin \beta_{in} + r_2 F_a' \sin \beta_A = 0$
$Z_{in} F_{in} \sin (\arg \mathbf{F}_{in} - \arg \mathbf{r}_{in}) +$
$r_2 F_a' \sin (\arg \mathbf{F}_a' - \arg \mathbf{r}_2) = 0$

where we expressed β_{in} and β_A in terms of the arguments of the vectors involved.

Note that β_{in} is cw, and therefore negative, which checks with the fact that the input torque is cw. Also, $z_{in} \sin \beta_{in}$ is the perpendicular arm of $\mathbf{F}_{in}$ about A_0, and similarly, $z_2 \sin \beta_2$ is that of $\mathbf{F}_a'$ about A_0. Solving for F_a',

$$F_a' = -\frac{z_{in} F_{in} \sin \beta_{in}}{r_2 \sin \beta_2} > 0 \quad \text{because} \quad \frac{\sin \beta_{in}}{\sin \beta_2} < 0$$

Continuing in this manner, Eq. (3.62) is verified in still another way (see Exer. 3.6).

Example 3.12

A six-link function-generator linkage is shown in Fig. 3.79. (a) Find the location of all the instant centers for this mechanism; (b) if the velocity of point P is known to be 10 m/sec, find ω_3 and $\mathbf{V}_B$ by the instantaneous center method; (c) if force F_{in} is acting at P (see Fig. 3.79), find F_{out} by instant centers.

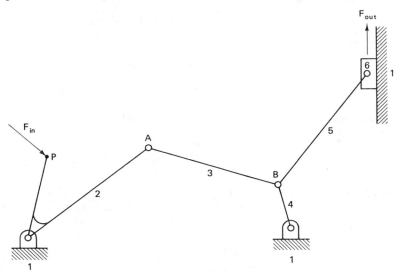

Figure 3.79   Six-link force transmission mechanism (see Exer. 3.9).

Solution

(a) Figure 3.80 shows the location of all the instant centers for this mechanism.

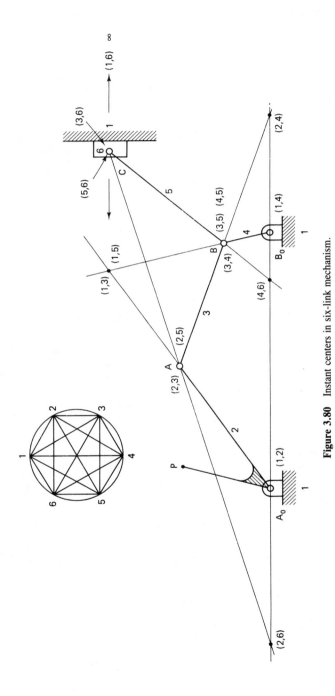

Figure 3.80 Instant centers in six-link mechanism.

(b) *Method 1:* Figure 3.81 shows how the instant center graphical technique can be used to solve for ω_3 and V_B. From this figure

$$V_A = 17.2 \text{ m/sec}$$

$$\omega_3 = \frac{V_A}{i(\overrightarrow{1,3 - 2,3})} = \frac{17.2}{0.054} = 318 \text{ rad/sec}$$

$$V_{2,4} = 44.7 \text{ m/sec}$$

$$V_B = (17.1 \text{ m/sec})i(-e^{i[\arg(\overrightarrow{B_0 B})]})$$

Method 2: Using instant center equations only (Fig. 3.80), since

$$\frac{\omega_3}{\omega_2} = \frac{(\overrightarrow{1,2 - 2,3})}{(\overrightarrow{1,3 - 2,3})} = -1.22$$

$$\omega_2 = \frac{V_P}{i(\overrightarrow{A_0 P})} = \frac{-V_P}{A_0 P} = -260.8 \text{ rad/sec (cw)}$$

then

$$\omega_3 = \frac{-V_P}{A_0 P}(-1.22) = 318.2 \text{ rad/sec (ccw)}$$

$$\omega_4 = \omega_2 \frac{(\overrightarrow{1,2 - 2,4})}{(\overrightarrow{1,4 - 2,4})} = (-260.8)(3.10) = -808.0 \text{ rad/sec (cw)}$$

Thus

$$V_B = i\omega_4(\overrightarrow{B_0 B}) = (17.3 \text{ m/sec})i(-e^{i \, \arg(\overrightarrow{B_0 B})})$$

(c) *Method 1:* Using instant centers (1,2), (1,5), and (2,5), (Fig. 3.81)

$$\frac{T_5}{T_2} = \frac{\omega_2}{\omega_5} = \frac{(\overrightarrow{1,5 - 2,5})}{(\overrightarrow{1,2 - 2,5})} = -\frac{0.054}{0.066} = -0.820$$

and (Figs. 3.79 and 3.80):

$$T_5 = F_{out}(0.054); \qquad T_2 = -F_{in}(0.031)$$

then

$$F_{out} = \frac{-F_{in}}{(0.054)}(0.031)(-0.820) = 0.471F_{in}$$

Method 2: Using instant centers (1,2), (1,6), and (2,6), from Eq. (3.62),

$$F_{out} = \left| \frac{T_2}{(1,2 - 6,2)} \right|$$

then

$$F_{out} = \left| \frac{-F_{in}(0.031)}{0.057} \right| = 0.54F_{in}$$

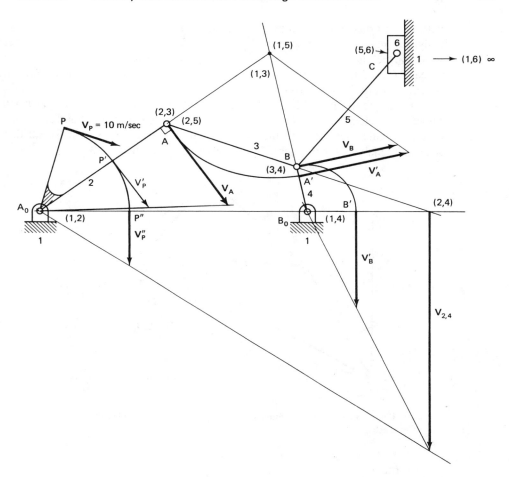

Figure 3.81   Velocity construction using instant centers and gage lines in six-link mechanism. Scale: 1 mm in Fig. 3.81 = 1.64 mm in the calculations of Example 3.12.

3.10 ANALYTICAL METHOD FOR VELOCITY AND MECHANICAL ADVANTAGE DETERMINATION

The procedures described above for velocity and mechanical advantage analysis are basically graphical solution procedures. Some analytical, complex-number equivalents have been presented. If only a finite number of positions of a linkage are to be analyzed, any of the graphical procedures are certainly warranted. If, however, a large number of positions and/or a large number of linkages need to be analyzed, the graphical procedures are too time consuming. With micro, desktop, or laptop computers or programmable hand calculators readily available, the analytical expressions based on complex-number representation are extremely valuable. After one becomes comfortable with the graphically based procedures (thereby gaining clear visualization of vector solutions), purely analytical methods can be used with greater confidence. When a question arises from the nonvisual techniques, a graphical spot check can verify the results. Also, computer graphics can be used to visually display the results of analytical methods (see Sec. 3.11 and the color inserts in this book).

Referring to Fig. 3.82, which shows a vector representation of a four-bar linkage, the angular velocities of links 3 and 4 may be determined as functions of the input link (link 2) angular velocity and the displacement position parameters. Recall that a displacement analysis for this same linkage was presented in Sec. 3.3.

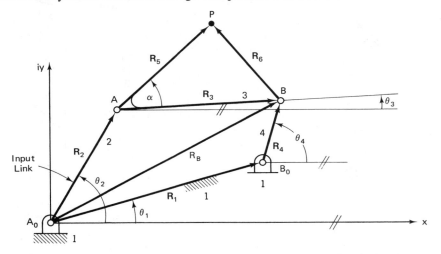

Figure 3.82   Vector notation for velocity analysis by complex numbers in four-bar mechanism.

The position vector from the base of the input link (A_0) to point B may be written via two routes:

$$\mathbf{R}_B = \mathbf{R}_2 + \mathbf{R}_3 \tag{3.65}$$

$$\mathbf{R}_B = \mathbf{R}_1 + \mathbf{R}_4 \tag{3.66}$$

The derivatives of these expressions will yield the velocity of point B. Using the polar form of these expressions for taking time derivatives leads to

$$\mathbf{R}_B = r_2 e^{i\theta_2} + r_3 e^{i\theta_3} = r_1 e^{i\theta_1} + r_4 e^{i\theta_4} \tag{3.67}$$

$$\mathbf{V}_B = r_2 \omega_2 i e^{i\theta_2} + r_3 \omega_3 i e^{i\theta_3} = r_4 \omega_4 i e^{i\theta_4} \tag{3.68}$$

Now using the Cartesian form of Eq. (3.68) in order to break it into real and imaginary parts, we obtain

$$\begin{aligned} r_2 \omega_2 \sin \theta_2 + r_3 \omega_3 \sin \theta_3 &= r_4 \omega_4 \sin \theta_4 \\ r_2 \omega_2 \cos \theta_2 + r_3 \omega_3 \cos \theta_3 &= r_4 \omega_4 \cos \theta_4 \end{aligned} \tag{3.69}$$

These two scalar equations contain only two scalar unknowns, ω_3 and ω_4. Multiplying the first equation by $\cos\theta_4$ and the second by $\sin\theta_4$, we have

$$r_2\omega_2\sin\theta_2\cos\theta_4 + r_3\omega_3\sin\theta_3\cos\theta_4 = r_4\omega_4\sin\theta_4\cos\theta_4$$

$$r_2\omega_2\cos\theta_2\sin\theta_4 + r_3\omega_3\cos\theta_3\sin\theta_4 = r_4\omega_4\sin\theta_4\cos\theta_4 \tag{3.70}$$

Subtracting the first from the second eliminates ω_4:

$$r_2\omega_2(\cos\theta_2\sin\theta_4 - \sin\theta_2\cos\theta_4) + r_3\omega_3(\cos\theta_3\sin\theta_4 - \sin\theta_3\cos\theta_4) = 0$$

or

$$r_2\omega_2\sin(\theta_4 - \theta_2) + r_3\omega_3\sin(\theta_4 - \theta_3) = 0 \tag{3.71}$$

Thus

$$\boxed{\omega_3 = -\frac{r_2}{r_3}\omega_2\frac{\sin(\theta_4 - \theta_2)}{\sin(\theta_4 - \theta_3)}} \tag{3.72}$$

Eliminating the terms containing ω_3 instead of ω_4 from Eq. (3.69) yields

$$\boxed{\omega_4 = \frac{r_2}{r_4}\omega_2\frac{\sin(\theta_3 - \theta_2)}{\sin(\theta_3 - \theta_4)}} \tag{3.73}$$

These two expressions are easily programmed for automatic computation. Refer to Sec. 3.11 for a computer program that includes these equations. Similar expressions may be derived for other linkages following the same procedure. There are several programs for finding angular velocity ratios in multiloop planar linkages. One such program is KADAM2*.

Notice that Eqs. (3.72) and (3.73) may also be utilized for mechanical advantage analysis where ω_3/ω_2 and ω_4/ω_2 are utilized in the M.A. expressions.

Example 3.13

Determine the analytical expression for the velocity of point P of Fig. 3.82.

Solution   The position and velocity of point P are expressed in vector form as

$$\mathbf{R}_P = r_2 e^{i\theta_2} + r_5 e^{i(\theta_3 + \alpha)} \tag{3.74}$$

The derivative of this expression yields

$$\mathbf{V}_P = r_2\omega_2 i e^{i\theta_2} + r_5\omega_3 i e^{i(\theta_3 + \alpha)} \tag{3.75}$$

* Available from the first author.

or in Cartesian form

Real part: $V_{P_x} = -r_2\omega_2 \sin \theta_2 - r_5\omega_3 \sin(\theta_3 + \alpha)$

Imaginary part: $V_{P_y} = r_2\omega_2 \cos \theta_2 + r_5\omega_3 \cos(\theta_3 + \alpha)$

(3.76)

where ω_3 is found by Eq. (3.72).

Correlation of Mechanical Advantage and Transmission Angle

In Sec. 3.1 we observed that the transmission angle is a means of determining the effectiveness with which motion is imparted to an output link of a particular mechanism. Mechanical advantage was defined as the instantaneous magnitude ratio of the output force to the input force of a particular mechanism (Sec. 3.9). Both of these indices of performance are static parameters (the effect of inertia is not included) that help us compare one linkage with another, or one position of a linkage with another position of the same linkage. Both of these indices can be expressed as a function of the geometry of the linkage. A comparison is warranted so as to avoid confusion of these terms.

In Fig. 3.83 the mechanical advantage may be expressed as [Eq. (3.56a)]

$$\text{M.A.} = \frac{F_{\text{out}}}{F_{\text{in}}} = \left(\frac{r_{\text{in}}}{r_{\text{out}}}\right)\left|\frac{\omega_2}{\omega_4}\right|$$

$$= \frac{r_{\text{in}}}{r_{\text{out}}}\frac{(1,4 - 2,4)}{(1,2 - 2,4)}$$

Constructing lines A_0A' and B_0B' perpendicular to the line containing $(2,4)$, A, and B, shows that by similar triangles

$$\frac{(2,4 - 1,4)}{(1,4 - B')} = \frac{(2,4 - 1,2)}{(1,2 - A')}$$

(3.77)

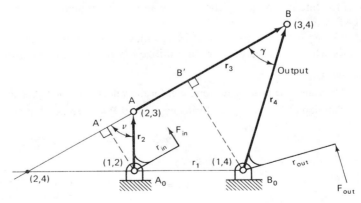

Figure 3.83 Correlation of M.A. with transmission angle γ and crank-coupler angle ν in a four-bar mechanism. See Eq. (3.79).

If γ and ν are the magnitudes of the smaller of the two angles made by the coupler (or its extension) with the output and input links, respectively, then $(1,4 - B') = r_4 \sin \gamma$, and $(1,2 - A') = r_2 \sin \nu$, and

$$\frac{(2,4 - 1,4)}{(2,4 - 1,2)} = \frac{r_4 \sin \gamma}{r_2 \sin \nu} \tag{3.78}$$

Thus

$$\text{M.A.} = \frac{r_{\text{in}}}{r_{\text{out}}} \left(\frac{r_4 \sin \gamma}{r_2 \sin \nu} \right) \tag{3.79}$$

Notice that the mechanical advantage becomes infinite when angle ν is either $0°$ or $180°$. This agrees with the analysis of Sec. 3.9 since in either of these cases instant center $(2,4)$ is coincident with center $(1,2)$. Notice also the relationship between the transmission angle γ and the mechanical advantage (disregarding the ratio $r_{\text{in}}/r_{\text{out}}$): If angle γ is $0°$ or $180°$, the mechanical advantage is zero. For other values of mechanical advantage, the transmission angle will vary. In Fig. 3.84 let $(2,4 - 1,4)$, $(2,4 - 1,2)$, r_2, and ν be fixed. The transmission angle γ and the length of the output link r_4 may have different values for a given value of M.A. as long as the product $(r_4 \sin \gamma)$ remains constant. The same amount of torque will be transferred to the output link in each of the cases shown in Fig. 3.84, but for the cases with smaller transmission angles, a larger component of the static force transmitted through the coupler will result in a larger bearing force at $(1,4)$ rather than in a usable force perpendicular to the output link. Thus a linkage with a good mechanical advantage may have an unacceptable transmission angle and a linkage with an excellent transmission angle in a particular position may not have a sufficient mechanical advantage. Since both the transmission angle and mechanical advantage vary with linkage position, either parameter can be critical to the designer in certain positions.

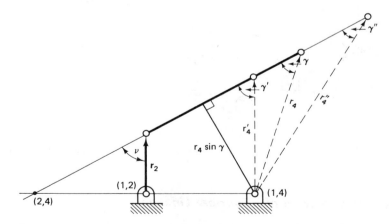

Figure 3.84 If $(2,4)$, $(1,2)$, $(1,4)$ and r_2 are unchanged, variation of transmission angle does not change the M.A.

Minimum Value of Mechanical Advantage

Section 3.9 showed that the mechanical advantage becomes infinitely large when the input link and the coupler are in line. Another useful design parameter would be the position in which a linkage attains the minimum value of mechanical advantage. *Freudenstein's theorem* [74] provides a method, expressible in terms of linkage geometry, for predicting the extreme value of angular velocity ratio ω_{out}/ω_{in}, which is the inverse of one of the factors in the mechanical advantage equation [Eq. (3.55)]. The therorem utilizes the two moving instant centers (2,4) and (1,3) shown in Fig. 3.85. The line between these centers is called the *collineation axis*. Freudenstein's theorem states: *At an extreme value of the velocity ratio in a four-bar linkage, the collineation axis is perpendicular to the connecting link AB*. In Fig. 3.85,

$$\left|\frac{\omega_4}{\omega_2}\right| = \frac{(2,4 - 1,2)}{(2,4 - 1,4)} = \frac{(2,4 - 1,2)}{(2,4 - 1,2) + (1,2 - 1,4)}$$

Since $(1,2 - 1,4)$ remains fixed as the linkage moves, the extreme values of the angular velocity ratio occur when the distance $(2,4 - 1,2)$ reaches an extreme value. These positions may occur when the instant center (2,4) is on either side of instant center (1,2). On the other hand, recall that the mechanical advantage is maximum when $(2,4 - 1,2)$ is minimum. During the motion of the linkage, instant center (2,4) moves along the line of centers (1,2) and (1,4). At an extreme value of the mechanical advantage, instant center (2,4) must come instantaneously to rest. This occurs when the velocity of (2,4), considered as part of link 3, is directed along *AB*. This will be true only when link 3 (extended to include (2,4)) is perpendicular to the collineation axis because center (1,3) is the instantaneous center of rotation of link 3. An inversion of this theorem is given by Shigley [148]: an extreme value of the velocity ratio ω_3/ω_2 of the four-bar linkage occurs when the collineation axis is perpendicular to the driven link (link 4).

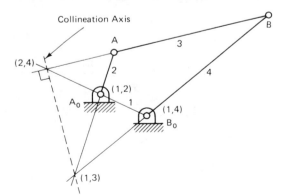

Figure 3.85 Freudenstein's theorem: At extreme of ω_4/ω_2 the collineation axis is perpendicular to coupler 3.

Limits of Motion of a Four-bar Linkage

Often it is desirable to determine the angular limits of motion of a four-bar linkage. For example, a Grashof rocker-rocker motion or path generator mechanism may be caused

TABLE 3.2 LIMITS OF MOTION OF A FOUR-BAR MECHANISM

Grashof type	Limits of Motion			
crank-rocker double-crank	crank may be rotated in a full circle			
rocker-crank double-rocker	$\theta_2' = \text{acos}\dfrac{r_2^2 + r_1^2 - \|(r_3 - r_4)\|^2}{2r_2r_1}$	first circuit	$\theta_{LO,1} = \theta_1 + \theta_2'$ $\theta_{HI,1} = \theta_1 + \theta_2''$	
	$\theta_2'' = \text{acos}\dfrac{r_2^2 + r_1^2 - \|(r_3 + r_4)\|^2}{2r_2r_1}$	second circuit	$\theta_{LO,2} = \theta_1 - \theta_2''$ $\theta_{HI,2} = \theta_1 - \theta_2'$	
triple-rocker	$r_2 + r_1 \geqq r_3 + r_4$ $\theta_2'' = \text{acos}\dfrac{r_2^2 + r_1^2 - \|(r_4 + r_3)\|^2}{2r_2r_1}$		$\theta_{LO} = \theta_1 - \theta_2''$ $\theta_{HI} = \theta_1 + \theta_2''$	
	$r_2 + r_1 < r_3 + r_4$ $\theta_2' = \text{acos}\dfrac{r_2^2 + r_1^2 - \|(r_4 - r_3)\|^2}{2r_2r_1}$		$\theta_{LO} = \theta_1 + \theta_2'$ $\theta_{HI} = \theta_1 - \theta_2'$	

to rock back and forth if a dyad is added to the four-bar as input. This "driving dyad" would have to form (together with the four-bar input link) a crank rocker, such that it would drive the original four-bar between its limits of motion. Table 3.2* presents the equations governing the extreme limits of angular motion of the input link for the Grashof four bars (Sec. 3.1). The equations are based on the linkage in Fig. 3.29.

3.11 COMPUTER PROGRAM FOR THE KINEMATIC ANALYSIS OF A FOUR-BAR LINKAGE**

Based on the previously derived formulas for position and velocity analysis of a four-bar linkage, an interactive program can easily be written. Typical program inputs are (see Table 3.3):

1. The lengths of the six sides of the linkage (see the figure in Table 3.3)
2. a. An indication of the direction of the coupler triangle, l_3, l_5, l_6
   b. An indication of the geometric inversion of the linkage.
3. The ground pivot specification [optional — if not specified then it is assumed that the ground link (l_1) is horizontal with the link 2 — link 1 pivot at location (0,0)]
4. The number of positions requested (equally divided into 360° to yield the increments of the input angle)
5. Specification of whether a constant angular velocity is specified or not

The linkage will remain on one side of the change point (Fig. 3.12c) formed by the output and coupler links. Some of the tabular program outputs are:

1. A table of angular velocities [using Eqs. (3.72) and (3.73)] of the coupler link and output link for given angular velocity of the input link.
2. A table of the coupler curve points [Eq. (3.74)], transmission angle [$0° \leq \gamma \leq 90°$, Eq. (3.4)], and a factor of the mechanical advantage (inverse angular velocity ratios between output and input links only) versus the input angle.

Linkage animation and graphs of angular position, velocity and acceleration are available as part of the LINCAGES[†] (Linkage INteractive Computer Analysis and Graphically Enhanced Synthesis) package [23, 46, 50, 57-62, 114, 124, 125, 152, 159]. The graphics output capability is illustrated by way of an example. The LINCAGES package is available through the first author. The synthesis portion is described in the

---

* Contributed by Dr. Tom Chase, Univ. of Minnesota

** The disc available with this book contains a version of this program.

[†] Available in CYBER, VAX, IBM AT/PS2, MACINTOSH, IRIS, SUN and other versions.

TABLE 3.3 DIRECTIONS FOR USE OF THE FOUR BAR ANALYSIS PACKAGE

Notation:

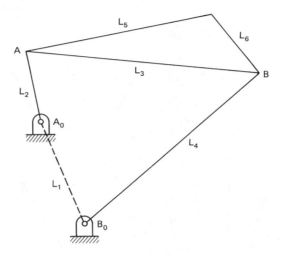

Program initialization inputs

Input a linkage by providing link lengths ($L_1 - L_6$) and the direction of the smaller rotations from L_3 to L_5 and from L_3 to L_4. Default values for these rotations are positive (counterclockwise).

Default values are always obtained by responding to a question with a return.

After the linkage is specified, the computer asks, "How many positions do you want to see?"

If the constant angular velocity of L_2 is not specified, the program will request values for theta 2, omega 2 and alpha 2 for each position the user wanted to see. Default value for theta 2 at position 1 is 5 degrees. Default value for omega 2 is 1 rad/sec. Default value for alpha 2 is 0.

appendix to Chap. 3 of Vol. 2. The color insert in this book shows typical outputs from analysis programs.*

Example 3.14

Analyze the motion generator (conveyor or linkage) synthesized in Ex. 8.3 (Fig. 8.64b).

Figure 3.86 shows an animation of the conveyor linkage through 15 positions. The only side of the coupler triangle shown is the link between moving pivots, but the path points are displayed as crosses. Figure 3.87 shows the coupler triangle in the same 15 positions, and Fig. 3.88 isolates the coupler curve (with letters indicating the timing of the input crank with 24° between letters).

Figure 3.89 is a plot of the transmission angle versus the input angle. Many other subroutine options are available in the LINCAGES package.

---

* General purpose planar software programs include KADAM2 and MCADA.

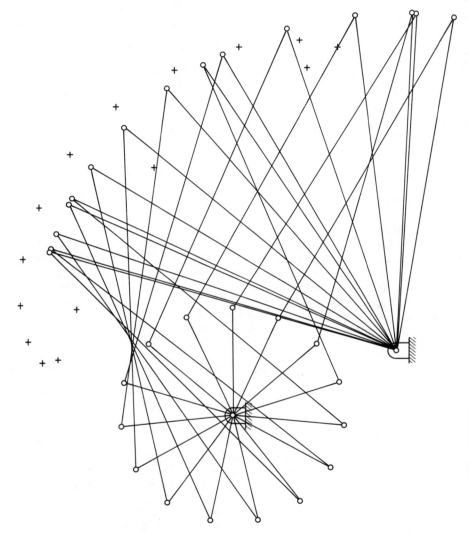

Figure 3.86 Successive positions of a four-bar mechanism 24 degrees of crank rotation apart. + signs are corresponding positions of a coupler point, plotted by the LINCAGES package.

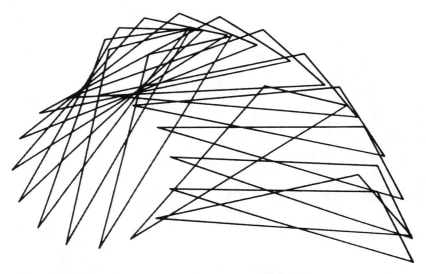

Figure 3.87 Successive positions of the coupler triangle for 24° crank positions of a four-bar mechanism.

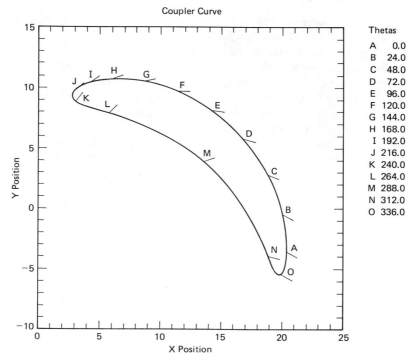

Figure 3.88 Hard copy of computer-plotted four-bar coupler curve, using the LINCAGES package. Timing and coupler angle line segments with respect to 24° crank rotations is shown by letters.

TRANSMISSION ANGLE VS THETA 1

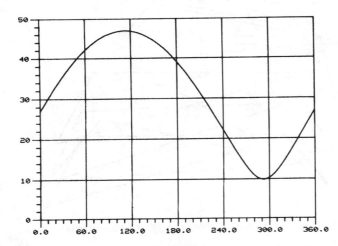

Figure 3.89 Hard copy of LINCAGES-plotted variation of transmission angle in a four-bar mechanism.

APPENDIX: REVIEW OF COMPLEX NUMBERS [86, 45]

Positive and negative numbers and zero are called *real numbers* and can be measured from a fixed point O on a straight line (Fig. 3.90). Any real number, positive or negative, has the property of designating a point by acting as its coordinate on the x axis. The set of all such numbers, each representing a single coordinate, forms one-dimensional space. To form sums of real numbers, we regard each to represent a directed length on the x axis. Positive numbers point to the right (see arrowheads in Fig. 3.90), negative numbers point to the left. A one-dimensional number composed of the sum of two others, a and b, each of which is itself a real number, is shown in Fig. 3.90. In symbolic form, $c = a + b$. The graphical procedure is to lay off a and then perform the operation indicated by the + sign. This means laying off b collinearly starting at the tip of a; the operation is that of graphical addition, and its result is the same as that of algebraic addition.

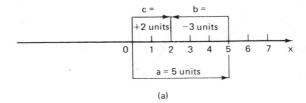

(a)

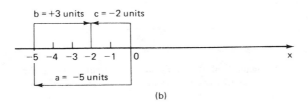

(b)

Figure 3.90 The one-dimensional number $c = a + b$. The elements a and b must be real numbers, and are laid off collinearly with the real axis x. (a) $c = a + b$, where $a = +5$, $b = -3$; (b) $c = a + b$, where $a = -5$, $b = +3$.

When an ordered pair of numbers, written (a,b), defines a third number $c = (a,b)$, c is called a *complex number*. To discuss complex numbers, it will be necessary first to develop the concept of the "imaginary" unit, $\sqrt{-1}$.

The Negative Sign as a 180° Turn

The coordinate OA in Fig. 3.91 is 3, and the coordinate OA' is -3. If we imagine OA to be pivoted about O, it could be turned counterclockwise until it coincided with OA'. Pivoting OA' about O in a counterclockwise turn of 180° would bring it into coincidence with OA. Thus the 180° turn has involved reversal of direction, as association with the negative sign or multiplication by (-1), as $(-1)OA = OA'$.

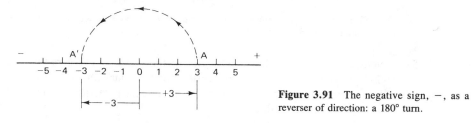

Figure 3.91 The negative sign, $-$, as a reverser of direction: a 180° turn.

The square root of a negative number such as $\sqrt{-x^2}$ can be written as $\sqrt{(-1)x^2} = \sqrt{(-1)}\,x$. This $\sqrt{-1}$ is then an operation to be performed on x. If the notation $i^2 = -1$ is introduced, it follows that $(i)(i) = i^2 = -1$, or two operations with i are the same as multiplying by -1. Now -1 has been seen to mean reversal of direction, and we may associate i^2 with -1; $i^2 = -1$ is thus an operator signifying a counterclockwise turn of 180°. Therefore, a single i indicates a turn of 90° in the *counterclockwise* direction. Conversely, $(-i)^2 = -1$ also, and we may interpret $-i$ to mean a 90° clockwise rotation. If we write $\sqrt{-1}\,x = ix$, the geometrical interpretation is that the number x will be turned through 90° counterclockwise. The symbol i was originated by Euler in 1777.

Imaginary Quantities. The number $\mathbf{c} = a + ib$ is called a *complex number* and is two-dimensional, since it involves both a and ib. Both a and b are real numbers; the number a is called the *real part* and ib the *imaginary part*.* A complex number of the form $a + i0$ is said to be a *real number,* whereas one of the form $0 + ib$ is called a *pure imaginary.* The validity of most of the theorems of arithmetic and algebra is unaffected by their transference to complex quantities—a very useful fact, since every algebraic equation in complex algebra has at least one root (Gauss).

The Complex Number. Suppose that $\mathbf{c}$ is represented as a distance r laid off from O as shown in Fig. 3.92. The double arrow indicates that r is not a directed line segment but rather an "absolute value," always positive. Geometrically, $\mathbf{c}$ is a radius vector: r times the length of a vector of unit length. We can describe the position of the end point P of the radius vector in two ways by the use of pairs of numbers. The *polar*

---

* While b may also be called the imaginary part, strictly speaking it is the coordinate on the imaginary axis. When multiplied by i it becomes the imaginary part ib.

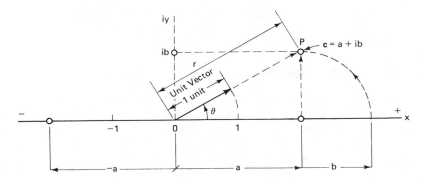

Figure 3.92 The complex number $\mathbf{c} = a + ib$ is two-dimensional, consisting of real and imaginary parts a and ib.

form includes the magnitude of radius r and the angle θ of the radius vector from the horizontal positive real axis. The *Cartesian* (rectangular) form specifies the real part of the radius vector, a, and the imaginary part, ib. The polar and Cartesian forms are related through Euler's equation:

$$\mathbf{c} = a + ib = r(\cos \theta + i \sin \theta) = re^{i\theta} \tag{3.85}$$

where

$$r = |\mathbf{c}| = |\sqrt{a^2 + b^2}|$$

$$\theta = \arg \mathbf{c} = \arg(a + ib)$$

Figure 3.92 is a picture of the complex number $\mathbf{c}$ shown in what is called the Gauss-Argand or *complex* plane. Any point in this plane represents a complex number. The absolute value of the complex number is r, also called the *modulus*; θ is called the *argument*, or angle, and is always measured from the positive real axis (Fig. 3.92) positive counterclockwise. When taking the argument of the complex number $a + ib$ (i.e., finding the direction of the radius vector representing that number measured from the positive real axis), it is not enough to take the arctangent of the ratio b/a. The reason for this is that the arctangent is multiple valued (as shown in Fig. 3.93):

$$\arctan \frac{b}{a} = \theta, \qquad \theta + \pi, \qquad \theta - \pi, \qquad \theta - 2\pi, \qquad \text{etc.} \tag{3.86}$$

Equation (3.86) give four values of the arctangent just between -2π and 2π, which include the direction diametrically opposite to the direction of $a + ib$.

In fact, since $b/a = -b/-a$, the arctangent of the ratio of imaginary to real parts of $a + ib$ would be the same as of $-(a + ib) = -a + i(-b)$, while the directions of their radius vectors differ by π. Therefore, it is important always to make a sketch of the radius vector of the complex number in order to see which of the multiple values of the arctangent gives the correct direction angle or "argument" of that vector.

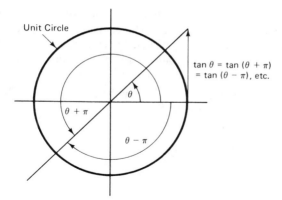

Figure 3.93

It is also useful for ease of visualization to define the direction angle θ in the interval

$$-\pi < \theta \leq \pi \qquad (3.87)$$

This will give the argument as defined in Fig. 3.94 for the four quadrants of the complex plane.

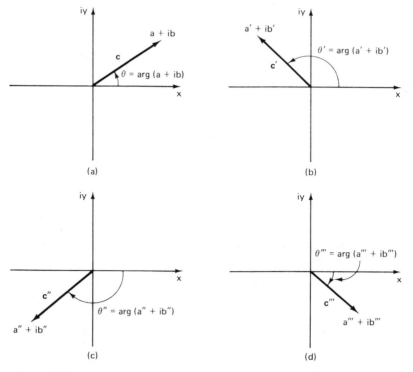

Figure 3.94 Arguments of complex numbers in the four quadrants of the complex plane.

Computer programmers have long realized this ambiguity of the arctangent and have therefore introduced the function ATAN2 (y, x) where y and x are signed reals, and which gives the argument of $x + iy$. Also, some hand calculators use this same function when the "rectangular to polar" transformation button is activated. Either will yield the angle in the interval given in Eq. (3.87). When using a hand calculator without such a transformation routine, the arctangent may be used in conjunction with a two-dimensional sketch of the radius vector to resolve the ambiguity.

$$\theta = \arg \mathbf{c} = \arg(a + ib)$$

Certain relations among complex numbers are shown in Fig. 3.95.

1. Two complex numbers can be equal only if their real and imaginary parts are separately equal.
2. Complex numbers add vectorially; their sum is found by adding their real parts to give the real part of their sum and adding their imaginary parts to give the imaginary part of their sum.
3. The difference of two complex numbers is found by taking the difference of their real parts to give the real part of their difference, and so on.

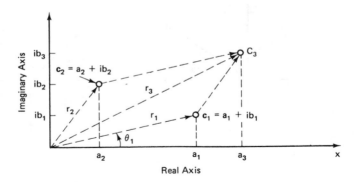

Figure 3.95 Addition of complex numbers $\mathbf{c}_1$ and $\mathbf{c}_2$ to yield $\mathbf{c}_3$.

Multiplication and division follow the rules of ordinary algebra with the additional relation $i^2 = -1$,

$$\begin{aligned}
(a + ib)(c + id) &= ac + (ib)c + a(id) + (ib)(id) \\
&= ac + i(bc + ad) + i^2(bd) \\
&= ac + i(bc + ad) + (-1)(bd) \\
&= (ac - bd) + i(bc + ad)
\end{aligned} \tag{3.88}$$

$$\frac{a + ib}{c + id} = \frac{(a + ib)(c - id)}{(c + id)(c - id)} = \frac{ac + (ib)c - a(id) - (ib)(id)}{c^2 - (id)^2}$$

$$= \frac{ac + i(bc - ad) - (-1)bd}{c^2 - i^2 d^2}$$

$$= \frac{ac + i(bc - ad) + bd}{c^2 + d^2} \qquad (3.89)$$

$$= \frac{ac + bd}{c^2 + d^2} + i\frac{bc - ad}{c^2 + d^2}$$

Multiplying and dividing in the manner described above illustrates the use of the Cartesian form. The polar form involves the use of exponentials and Euler's theorem.

Since

$$\mathbf{c} = r(\cos \theta + i \sin \theta) = re^{i\theta} \qquad (3.90)$$

we may write

$$\mathbf{c}_1 = r_1 e^{i\theta_1} \qquad \mathbf{c}_2 = r_2 e^{i\theta_2}$$

From this,

$$\mathbf{c}_3 = \mathbf{c}_1 \mathbf{c}_2 = (r_1 e^{i\theta_1})(r_2 e^{i\theta_2}) = r_1 r_2 e^{i(\theta_1 + \theta_2)}$$

$$= r_1 r_2 [\cos(\theta_1 + \theta_2) + i \sin(\theta_1 + \theta_2)] \qquad (3.91)$$

and

$$\frac{\mathbf{c}_1}{\mathbf{c}_2} = \frac{r_1 e^{i\theta_1}}{r_2 e^{i\theta_2}} = \frac{r_1}{r_2} e^{i(\theta_1 - \theta_2)}$$

$$= \frac{r_1}{r_2} [\cos(\theta_1 - \theta_2) + i \sin(\theta_1 - \theta_2)] \qquad (3.92)$$

The Gauss-Argand representation of the complex number makes use of the Cartesian coordinate plane xy and calls the complex number $\mathbf{Z}$, that is, $\mathbf{Z} = x + iy$. The number is thus displayed as the point whose coordinates are (x, y). Note that $\mathbf{Z}$ is the symbol for the complex number; it has no relation to the usual third Cartesian coordinate. The length of the radius vector is designated by r and is always positive. Figure 3.96 shows in summary many properties of the complex number in all quadrants of the complex, or $\mathbf{Z}$ plane.

Conjugate Complex Numbers. Two complex numbers are said to be complex conjugates of one another when they have the same real part and their imaginary parts are equal in magnitude but are opposite in sign. For example, the complex conjugate of $x + iy$ is $x - iy$. In general, we shall use a bar over the symbol of a complex number to denote its conjugate; thus, if $\mathbf{Z} = x + iy$, then $\overline{\mathbf{Z}} = x - iy$.

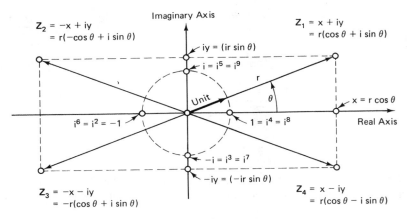

Figure 3.96   The complex plane (Gauss–Argand plane).

Points of the complex plane representing two conjugate complex numbers are symmetric with respect to the real axis (Fig. 3.97). Note that the magnitudes of two conjugate complex numbers are equal and that their angles are opposite in sense. Further, if $\mathbf{Z}$ is real, then $\overline{\mathbf{Z}} = \mathbf{Z}$; if pure imaginary, then $\overline{\mathbf{Z}} = -\mathbf{Z}$; also, $(\overline{\overline{\mathbf{Z}}}) = \mathbf{Z}$. Conjugate complex numbers are useful in a number of situations, such as in calculating the magnitude of a complex number,

$$|\mathbf{Z}|^2 = r^2 = x^2 + y^2 = (x + iy)(x - iy) = \mathbf{Z}\overline{\mathbf{Z}} \tag{3.93}$$

The square of the magnitude of a complex number is, therefore, its product with its own conjugate. The real and imaginary parts of a complex number may also be expressed in terms of the conjugate (Fig. 3.97).

$$\text{real part of } \mathbf{Z} = x = \frac{1}{2}[(x + iy) + (x - iy)] = \frac{\mathbf{Z} + \overline{\mathbf{Z}}}{2}$$

$$\text{imaginary part of } \mathbf{Z} = y = \frac{1}{2i}[(x + iy) - (x - iy)] = \frac{\mathbf{Z} - \overline{\mathbf{Z}}}{2i} \tag{3.94}$$

Even when a complex number is represented in terms of sums, products, ratios, and exponents of other numbers, its conjugate may be found by simply taking the conjugate of each of the numbers appearing in the expression. Consider, for example, the complex number

$$\mathbf{Z} = \frac{\mathbf{c}_1 e^{i\theta_1} + \mathbf{c}_2 e^{i\theta_2}}{\mathbf{d}_1 e^{i\alpha_1} + \mathbf{d}_2 e^{i\alpha_2}}$$

where $\mathbf{c}_1$, $\mathbf{c}_2$, $\mathbf{d}_1$, and $\mathbf{d}_2$ are themselves complex and the angles θ_1, θ_2, α_1, and α_2 are real. The conjugate of $\mathbf{Z}$ is

$$\overline{\mathbf{Z}} = \frac{\overline{\mathbf{c}}_1 e^{-i\theta_1} + \overline{\mathbf{c}}_2 e^{-i\theta_2}}{\overline{\mathbf{d}}_1 e^{-i\alpha_1} + \overline{\mathbf{d}}_2 e^{-i\alpha_2}}$$

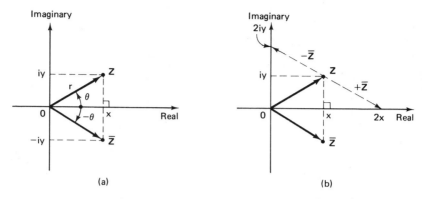

Figure 3.97 Conjugate complex numbers $\mathbf{Z} = x + iy$ and $\overline{\mathbf{Z}} = x - iy$.

The Angle of Turn. The angle of turn of a unit vector is defined by the operator $(\cos \theta + i \sin \theta)$. If we perform this operation again, turning through another angle θ, we shall have

$$(\cos \theta + i \sin \theta)(\cos \theta + i \sin \theta) = (\cos \theta + i \sin \theta)^2$$

However, the total angle of turn is 2θ, whence we may write the identity

$$(\cos \theta + i \sin \theta)^2 = \cos 2\theta + i \sin 2\theta$$

On extending this to n operations, we have

$$(\cos \theta + i \sin \theta)^n = \cos n\theta + i \sin n\theta \tag{3.95}$$

This last expression is known as De Moivre's theorem and is immediately obvious when considering it in terms of complex numbers.

PROBLEMS*

> Note: When needed dimensions are not given in the problem statement, choose one link to be of unit length and scale the drawing.

3.1. A four-bar linkage has the following dimensions: $A_0 B_0 = 2$ in., $A_0 A = 2 + \frac{1}{2}$ in., $AB = 1 + \frac{1}{2}$ in., $B_0 B = 2 + \frac{3}{4}$ in. (see Fig. P3.1).
 (a) Determine the type of four-bar by the Grashof criterion.
 (b) Determine the range of motion of this linkage if link 2 is the input.
 (c) Determine the range of motion of this linkage if link 4 is the input.

3.2. A four-bar linkage has the following dimensions: $A_0 B_0 = 2$ in., $A_0 A = 2 + \frac{1}{2}$ in., $AB = 1 + \frac{1}{2}$ in., $B_0 B = 3 + \frac{1}{4}$ in. (see Fig. P3.2).
 (a) Determine the type of four-bar by the Grashof criterion.
 (b) Determine the range of motion of this linkage if link 2 is the input.
 (c) Determine the range of motion of this linkage if link 4 is the input.

* Many of the new problems for the second edition were contributed by John Titus of the University of Minnesota.

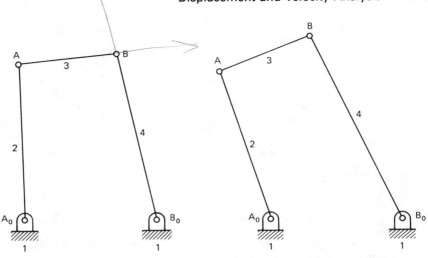

Figure P3.1   Figure P3.2

3.3. Determine the transmission and deviation angle(s) for the mechanisms in Figs. P3.3, P3.4, and P3.5 for the case where:
  (a) Link 2 is the input and link 4 is the output.
  (b) Link 4 is the input and link 2 is the output.

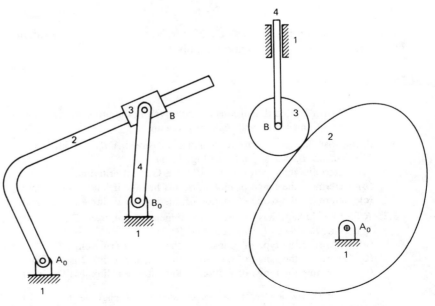

Figure P3.3   Figure P3.4

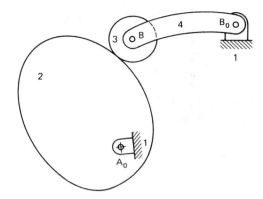

Figure P3.5

3.4. Determine the transmission and deviation angle(s) for the mechanisms in Fig. P3.6 for the case where:

(a) Link 2 is the input and link 5 is the output.

(b) Link 5 is the input and link 2 is the output.

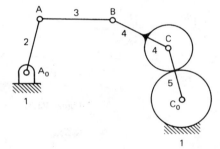

Figure P3.6

3.5. Determine the transmission and deviation angle(s) for the mechanisms in Figs. P3.7, P3.8, and P3.9 for the case where:

(a) Link 2 is the input and link 6 is the output.

(b) Link 6 is the input and link 2 is the output.

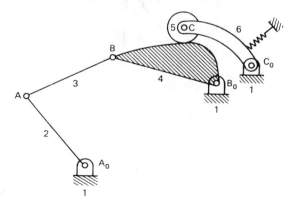

Figure P3.7

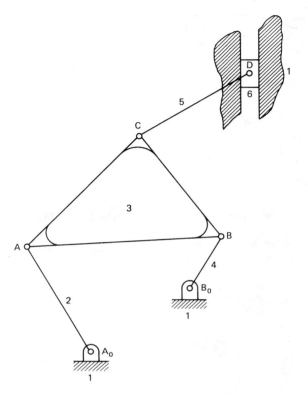

Figure P3.8

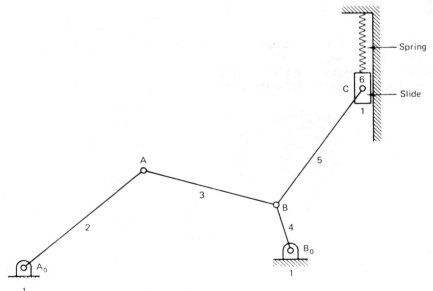

Figure P3.9

3.6. Determine the transmission and deviation angle(s) for the two quick-return shaper mechanisms of Figs. P3.10 and P3.11, where the input is link 2.

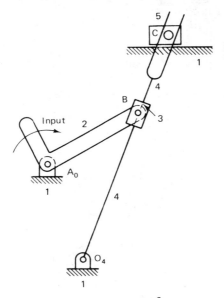

Figure P3.10

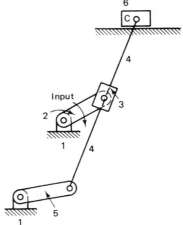

Figure P3.11

3.7. Find the transmission and deviation angle(s) for the convertible top linkage in the two positions shown in Fig. P3.12.

3.8. Determine the maximum and minimum transmission angles for the following linkage dimensions. The linkage topology is shown in Fig. P3.13. $A_0 B_0$ are along the x axis.

(a) $A_0 B_0 = 6''$, $A_0 A = 1''$, $AB = 2''$, $BB_0 = 6''$

(b) $A_0 B_0 = 6.93''$, $A_0 A = 4''$, $AB = 4''$, $BB_0 = 4''$

(c) $A_0 B_0 = 2''$, $A_0 A = 4''$, $AB = 3''$, $BB_0 = 3''$

(d) $A_0 B_0 = 3.5''$, $A_0 A = 4.5''$, $AB = 2''$, $BB_0 = 5''$

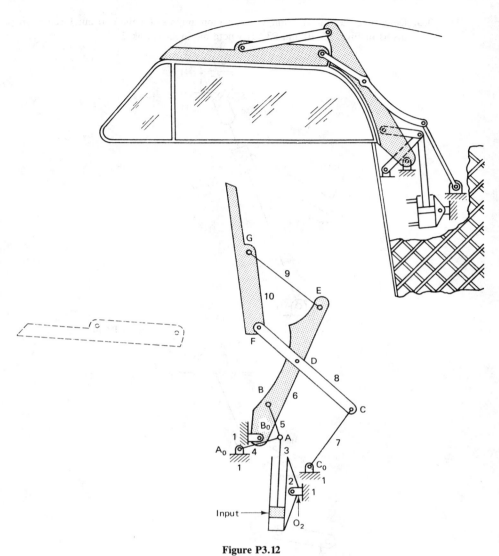

Figure P3.12

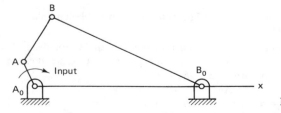

Figure P3.13

3.9. For the linkage shown in Fig. P3.14, find V_B, ω_3, and ω_4 if ω_2 is 1 rad/sec cw, $A_0A = AB = BB_0 = 4$ inches, the angle of A_0A is 60°, and AB is horizontal.
  (a) Use the relative-velocity method.
  (b) Comment upon the use of the instant-center method for this linkage position.

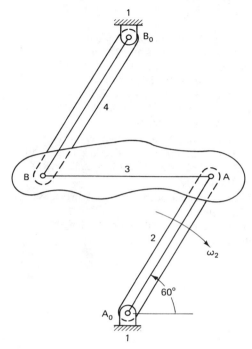

Figure P3.14

3.10. Determine the velocity of point E and the angular velocities ω_5 and ω_6 of the six-bar of Fig. P3.15 given $V_A = 40$ mm/sec.

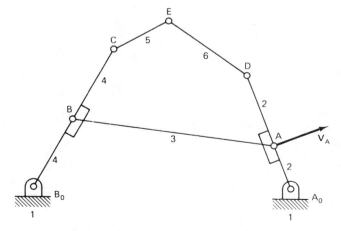

Figure P3.15

(a) Use the relative velocity method.

(b) Scale the figure to obtain numerical values of link lengths and directions, and use complex-number arithmetic.

3.11. Use the relative velocity method to find V_A, V_B, V_C, and ω_2, ω_3, ω_4 for each of the mechanisms shown in Figs. P3.16, P3.17 and P3.18 respectively. Compare the resulting linear and angular velocities.

(a) Linkage dimensions for Fig. P3.16 are $A_0A = 2''$, $AB = 2.6''$, $AC = 3.503''$, $BC = 3.983''$, $BB_0 = 2.5''$, $A_0B_0 = 6.5''$; A_0A is at $-15°$; $\omega_2 = 2$ rad/sec cw.

(b) Linkage dimensions for Fig. P3.17 are $A_0A = 3.065''$, $AB = 3.830''$, $AC = 3.368''$, $BC = 2.5''$, $BB_0 = 3.983''$, $A_0B_0 = 9.959''$; BB_0 is at $-50°$; $\omega_2 = 2$ rad/sec cw.

(c) Linkage dimensions for Fig. P3.18 are $A_0A = 3.503''$, $AB = 2.695''$, $AC = 2''$, $BC = 3.064''$, $BB_0 = 3.368''$, $A_0B_0 = 8.757''$; A_0A is at $+90°$; $\omega_3 = 2$ rad/sec cw. Note that the coupler angular velocity is given for this linkage.

3.12. Use the relative velocity method to find V_E and ω_3, ω_4, ω_5 and ω_6 for the Stephenson-I six-bar linkage of Fig. P3.19 if $\omega_2 = 2$ rad/sec cw. The link lengths are $A_0A = 2''$, $A_0D = 3.065''$, $AB = 2.6''$, $AC = 3.503''$, $BC = 3.983''$, $BB_0 = 2.5''$, $A_0B_0 = 6.5''$, $CE = 8.758''$, $DE = 3.368''$; A_0A is at $-10°$.

3.13. If point A of Fig. P3.20 has an instantaneous velocity of 9 m/second, find V_E, ω_4, and ω_7.

(a) Use the relative velocity method.

(b) Scale the figure to obtain numerical values of link lengths and directions, and use complex-number arithmetic.

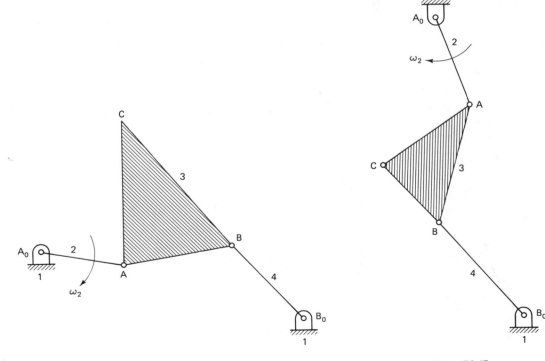

Figure P3.16   **Figure P3.17**

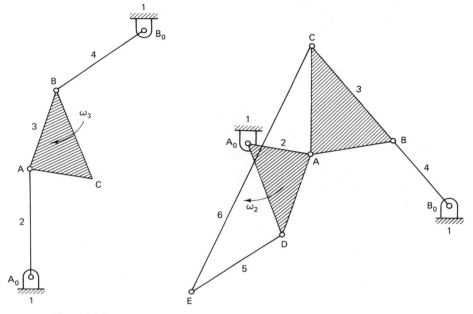

Figure P3.18   Figure P3.19

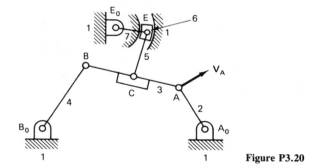

Figure P3.20

3.14. Find ω_6 and $\mathbf{V}_C$ (on 3) of the linkage in Fig. P3.21 if $\omega_2 = 1$ rad/sec cw.
   (a) Use the relative velocity method.
   (b) Use the instant center method.
   (c) Scale the figure to obtain numerical values of link lengths and directions, and use complex-number arithmetic.

3.15. Determine the angular velocity ratio ω_4/ω_2 of the linkage in Fig. P3.22.
   (a) Use the relative velocity method.
   (b) Use the instant center method.
   (c) Use complex-number arithmetic.

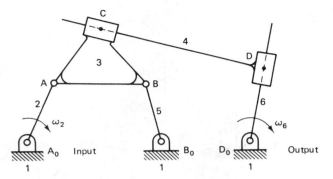

Figure P3.21

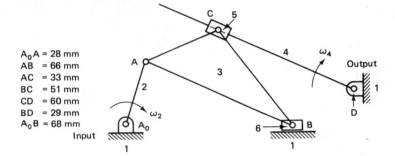

A_0A = 28 mm
AB = 66 mm
AC = 33 mm
BC = 51 mm
CD = 60 mm
BD = 29 mm
A_0B = 68 mm

Figure P3.22

3.16. Use the relative velocity method to find V_{B4}, V_{B5}, ω_3 and ω_5 for the two five-bar linkages shown below. In both cases $\omega_2 = 4$ rad/sec cw.

    (a) Linkage dimensions for Fig. P3.23 are $A_0A = 3''$, $AB = 2''$, $B_0B = 2.5''$ at this instant; A_0A and B_0B are at $+60°$; $V_{(B4)(B5)} = 9$ in/sec.

    (b) Linkage dimensions for Fig. P3.24 are $A_0A = 3''$, $AB = 2.5''$, $A_0B_0 = .5''$, $B_0B = 3.143''$ at this instant; A_0A is at $+60°$; $V_{(B4)(B5)} = 4$ in/sec.

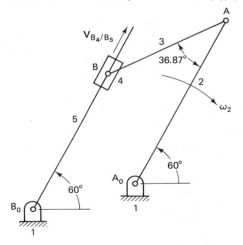

Figure P3.23

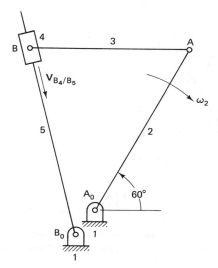

Figure P3.24

3.17. The scaled kinematic diagram of a match lighting mechanism is shown in Fig. P3.25. The match is placed at point C and lit by drawing across a striker when turning link 2. The link lengths and input angular velocity at this instant are $A_0A = 2.5''$, $A_0B = 4.33''$, $AB = 2.5''$; A_0A is at $-30°$; $\omega_2 = 4$ rad/sec ccw. Find the relative velocity between links 3 and the slider block (4) and ω_4.

(a) Use the relative velocity method.

(b) Use the instant center method.

(c) Use the complex-number arithmetic method.

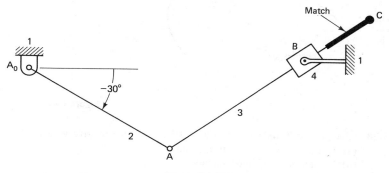

Figure P3.25

3.18. Find $\mathbf{V}_F$ and angular velocities ω_5 and ω_6 for the Watt I six-bar linkage shown in Fig. P3.26. The linkage dimensions are $A_0B_0 = 4''$, $A_0A = 5''$, $AB = 2''$, $B_0B = 5''$, $AC = 1''$, $CD = 2''$, $DE = 2.5''$, $DF = 1''$ at $45°$; A_0A is at $+126.87°$; $\omega_2 = 5$ rad/sec cw.

(a) Use the relative velocity method.

(b) Use the instant center method.

(c) Use the complex-number method.

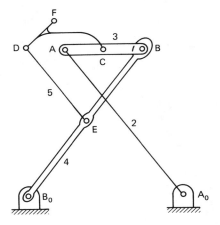

Figure P3.26

3.19. Find the velocity of point B on the follower of the cam mechanism in Fig. P3.27 given $\omega_2 = 10$ rad/sec ccw, using the relative velocity method. Assume slip-free rolling of link 3 on 2 and find the velocity of point B'.
  (a) Use the relative velocity method.
  (b) Use the instant center method.
  (c) Scale the figure to obtain numerical values of link lengths and directions, and use complex-number arithmetic.

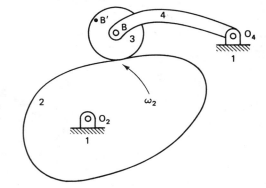

$O_2B = 1.25$ in.
$O_4B = 1.60$ in.
$O_2O_4 = 2.25$ in.

Figure P3.27

3.20. Given $\omega_2 = 10$ rad/sec cw in Fig. P3.28 find the relative velocity $\mathbf{V}_{(P3)(P2)}$.
  (a) Use the relative velocity method.
  (b) Use the instant center method.
  (c) Scale the figure to obtain numerical values of link lengths and directions, and use complex-number arithmetic.

3.21. Find $\mathbf{V}_C$ given $\mathbf{V}_A$ of the linkage in Fig. P3.29. (Assume true rolling of 4 on 1.)
  (a) Use the relative velocity method.
  (b) Use the instant center method.
  (c) Scale the figure to obtain numerical values of link lengths and directions, and use complex-number arithmetic.

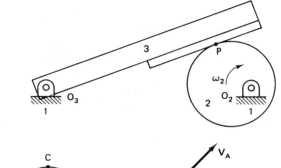

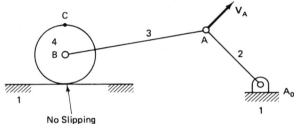

Figure P3.28

Figure P3.29

3.22. Given the angular velocity of the cam ($\omega_2 = 20$ rad/sec ccw), find the instantaneous vertical velocity of the follower (Fig. P3.30).
(a) Use the relative velocity method.
(b) Use the instant center method.
(c) Scale the figure to obtain numerical values of link lengths and directions, and use
complex-number arithmetic.

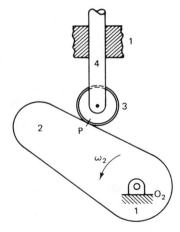

Figure P3.30

3.23. Find the angular velocity of link 7 and the linear velocity of point F on the slider of the
eight-bar linkage in Fig. P3.31; $\omega_2 = 1$ rad/sec ccw.
(a) Use the relative velocity method.
(b) Use the instant center method.
(c) Use complex-number arithmetic.

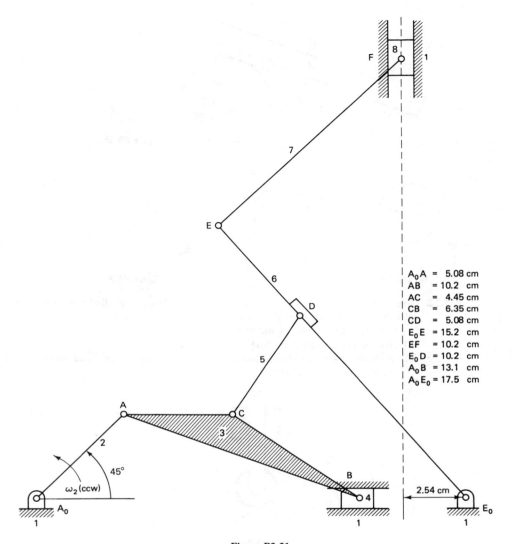

$A_0A = 5.08$ cm
$AB = 10.2$ cm
$AC = 4.45$ cm
$CB = 6.35$ cm
$CD = 5.08$ cm
$E_0E = 15.2$ cm
$EF = 10.2$ cm
$E_0D = 10.2$ cm
$A_0B = 13.1$ cm
$A_0E_0 = 17.5$ cm

Figure P3.31

3.24. A *polycentric* door is one in which the door is the coupler of a four-bar linkage. As it swings from the closed to open position, instant center 1,3 varies continuously, hence the name polycentric. Note that the coupler is the input link. Find $\mathbf{V}_C$ and angular velocities ω_2 and ω_4 for the door in the position shown in Fig. P3.32. The linkage inputs and dimensions are $A_0A = 24''$, $AB = 18''$, $AC = 12''$, $BB_0 = 18''$, $A_0B_0 = 12''$; AB is at $+60°$; $\omega_3 = 3$ rad/sec ccw (see Fig. P3.32).

(a) Use the relative velocity method.

(b) Use the instant center method.

(c) Use the complex-number method.

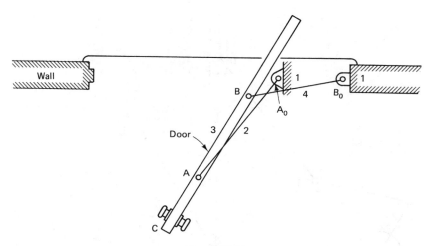

Figure P3.32

3.25. If the angular velocity of link 2 of Fig. P3.33 is 66.7 rad/sec cw, find the relative velocity of slider D with respect to slider B.
   (a) Use the relative velocity method.
   (b) Use the instant center method.
   (c) Scale the figure to obtain numerical values of link lengths and directions, and use complex-number arithmetic.

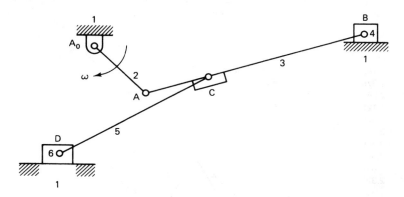

Figure P3.33

3.26. The angular velocity of crank 2 of Fig. P3.34 is $\omega_2 = 2000$ rpm ccw. Find ω_5.
   (a) Use the relative velocity method.
   (b) Use the instant center method.
   (c) Scale the figure to obtain numerical values of link lengths and directions, and use complex-number arithmetic.

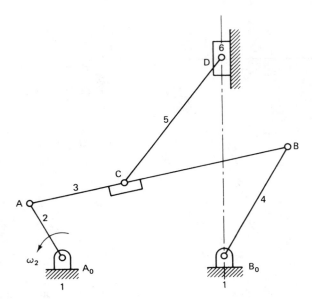

3.27. In the six-bar linkage shown in Fig. P3.35 find the velocity of points E and C and the angular velocity of link 6, given $\omega_2 = 10$ rad/sec cw.

   (a) Use the relative velocity method.

   (b) Use the instant center method.

   (c) Scale the figure to obtain numerical values of link lengths and directions, and use complex-number arithmetic.

Lengths (in.)

$A_0A = 1.188$
$AB\ = 1.375$
$AD\ = 2.750$
$B_0B = 1.125$
$CD\ = 2.625$
$C_0C = 0.938$
$DE\ = 0.750$
$A_0B_0 = 2.560$
$A_0C_0 = 3.040$
$B_0C_0 = 2.320$

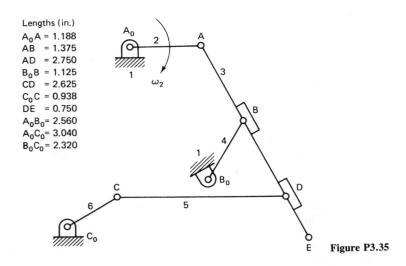

Figure P3.35

3.28. If the velocity of slider 6 is instantaneously 87.5 ft/sec upward (see Fig. P3.36), find the velocity of the center of mass of link 3 (V_{g3}) and the angular velocity of link 4.
- **(a)** Use the relative velocity method.
- **(b)** Use the instant center method.
- **(c)** Scale the figure to obtain numerical values of link lengths and directions, and use complex-number arithmetic.

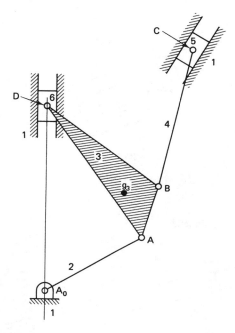

Figure P3.36

3.29. Given a slider velocity of $V_A = 20$ in/sec to the left, find V_B, V_C, and ω_5 for the mechanism in Fig. P3.37.
- **(a)** Use the relative velocity method.
- **(b)** Use the instant center method.
- **(c)** Scale the figure to obtain numerical values of link lengths and directions, and use complex-number arithmetic.

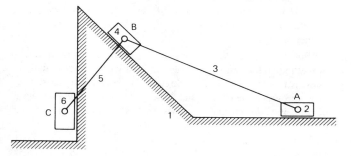

Figure P3.37

3.30. The inverted slider crank mechanism of Fig. P3.38 has a given angular velocity of $\omega_2 =$ 10 rad/sec ccw. Find ω_4.
   (a) Use the relative velocity method.
   (b) Use the instant center method.
   (c) Scale the figure to obtain numerical values of link lengths and directions, and use complex-number arithmetic.

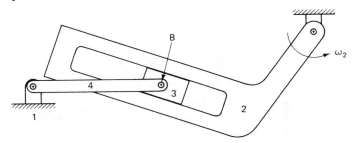

Figure P3.38

3.31. If the angular velocity of the crank of Fig. P3.39 is $\omega_2 = 100$ rpm cw, find ω_4.
   (a) Use the relative velocity method.
   (b) Use the instant center method.
   (c) Use complex-number arithmetic.

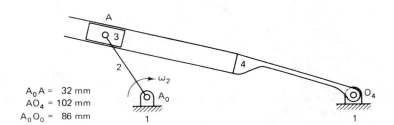

$$A_0 A = 32 \text{ mm}$$
$$AO_4 = 102 \text{ mm}$$
$$A_0 O_0 = 86 \text{ mm}$$

Figure P3.39

3.32. Determine the relative slider velocity $\mathbf{V}_{(C5)(C6)}$ and ω_6 of the Stephenson III six-bar linkage shown in Fig. P3.40. The linkage dimensions and inputs are $A_0 A = 2''$, $AB = 1''$, $B_0 B = 2.8''$, $A_0 B_0 = 1.11''$, $AC = 3.5''$, $B_0 C_0 = 6.9''$; $A_0 A$ is at $-30°$; $\omega_2 = 2$ rad/sec cw.
   (a) Use the relative velocity method.
   (b) Use the complex-number method.

3.33. In the eight-link mechanism of Fig. P3.41, find $\mathbf{V}_C$, $\mathbf{V}_E$, ω_5, and ω_8 given $\omega_2 = 100$ rpm cw.
   (a) Use the relative velocity method. [*Hint:* In some cases the relative velocity equations do not contain enough knowns to solve the problem directly. However, one can start at two ends of the mechanism (arbitrarily choosing one "test" velocity) and work toward the middle. If the middle velocities do not match, scale the "test" velocity up or down so that they do match.]
   (b) Use the instant center method.
   (c) Scale the figure to obtain numerical values of link lengths and directions, and use complex-number arithmetic.

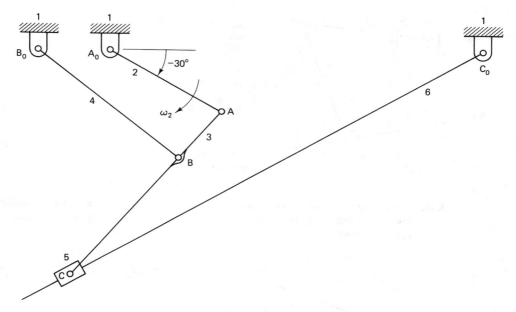

Figure P3.40

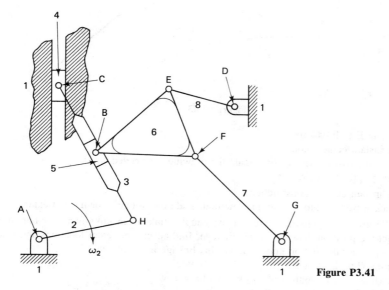

Figure P3.41

3.34. If the angular velocity of link 2 of Fig. P3.42 is 1 rad/sec cw:
   (a) Find ω_6 and V_B by the relative velocity method.
   (b) Determine the location of all the instant centers.
   (c) Find ω_6 and V_B by the instant center method and compare your results with part (a).
   (d) Use complex-number arithmetic.

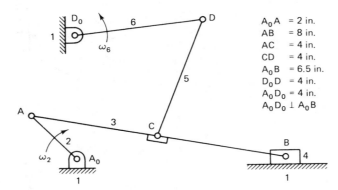

A_0A = 2 in.
AB = 8 in.
AC = 4 in.
CD = 4 in.
A_0B = 6.5 in.
D_0D = 4 in.
A_0D_0 = 4 in.
$A_0D_0 \perp A_0B$

Figure P3.42

3.35. Two known velocities for link n are shown in Fig. P3.43. Find the center of rotation of the link. If you had set up this problem, could you have arbitrarily drawn two vectors $\mathbf{V}_A$ and $\mathbf{V}_B$? Why or why not?

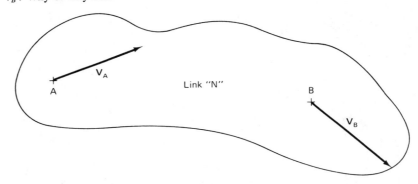

Figure P3.43

3.36. Given $\mathbf{V}_X$ in Fig. P3.44 find $\mathbf{V}_B$.
   (a) Use instant centers construction.
   (b) Use complex-number algebra (scale the drawing for needed data).

3.37. Given $\mathbf{V}_X$ in Fig. P3.45 find $\mathbf{V}_B$.
   (a) Use instant centers construction.
   (b) Check it by the complex-number method (scale the drawing for needed data).

3.38. Figure P3.46 is the velocity image of the coupler link of an unknown four-bar linkage. Construct the position diagram of this linkage, find all instant center locations and find ω_2, ω_3 and ω_4. The information known about this linkage is as follows.
   Input crank (link 2) A_0A length is 2.5".
   Coupler (link 3) AB length is 3", ω_3 known to be ccw.
   Output link (link 4) BB_0 length is 5".
   Ground link (link 1) A_0B_0 unknown length but smaller than 5".

3.39. Given $\omega_2 = 1$ rad/sec as shown in Fig. P3.47, find $\mathbf{V}_B$ in each case (use instant centers).
   (a) Assume rolling contact between link 1 and the carrier roller of point B.
   (b) Check your results by the complex-number method. (Scale the drawing for needed data.)

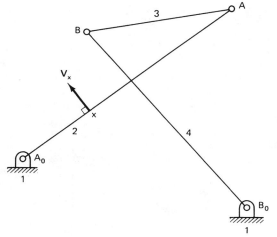

Figure P3.44

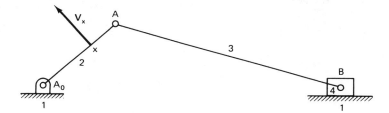

Figure P3.45

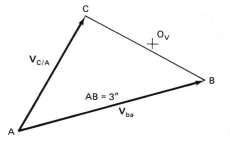

Figure P3.46

3.40. (a) Find the 10 instant centers of the five-link mechanism in Fig. P3.48.
  (b) For the position shown, represent the instantaneous angular velocity relationships between links 2 and 4 by a pair of gears pivoted at points O_2 and O_4.
  (c) For the position shown, represent the instantaneous angular velocity relationships between links 4 and 5 by a pair of gears pivoted at points O_4 and O_5.

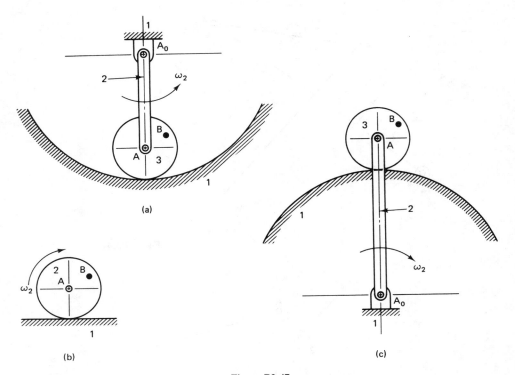

Figure P3.47

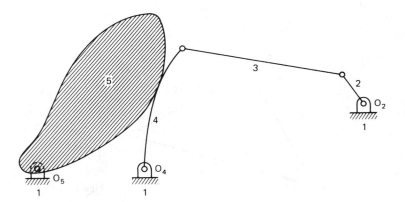

Figure P3.48

3.41. The fixed link (F), the input link (1), and the output link (N) of a mechanical system are shown in Fig. P3.49 with the revolute joints between 1 and F and N and F indicated.
   (a) Pick an acceptable instant center for 1 and N.
   (b) Write an equation relating the angular velocity ratio ω_N/ω_1 in terms of vectors connecting the instant centers.

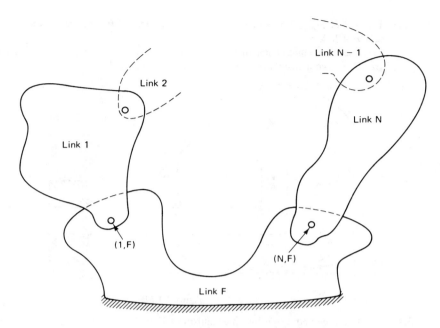

Figure P3.49

3.42. Find ω_5/ω_2 for the mechanism in Fig. P3.50.
   (a) Use the instant center method.
   (b) Check your results by the complex-number method.*

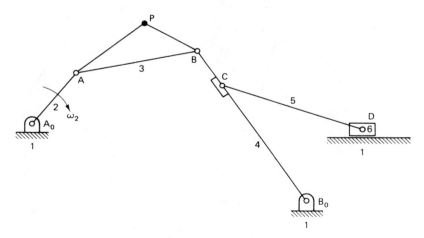

Figure P3.50

* Scale the figure for needed data.

3.43. If link 3 of Fig. P3.51 slides to the left at the rate of 1 m/sec, determine the velocity of the slider of link 5.

(a) Use the instant center method. Assume rolling contact between links 1 and 2.*

(b) Check your results by the complex-number method.[†]

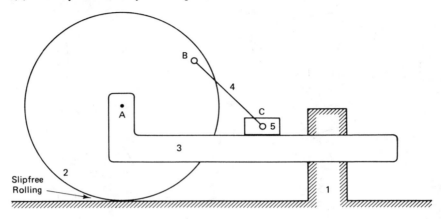

Figure P3.51

3.44. (a) In the geared seven-link mechanism of Fig. P3.52, find all the instant centers.[†]

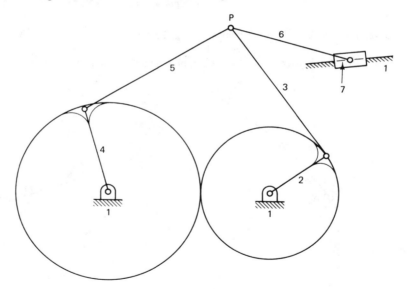

Figure P3.52

* As will be shown later, equivalence of higher and lower pair mechanisms prevails for degrees of freedom, displacements and velocities, but not for accelerations or higher-order motion derivatives.

[†] Scale the figure for needed data.

(b) What is the angular velocity ratio ω_6/ω_2?

(c) Check part (b) by the relative velocity method.

(d) What is the relative angular velocity of link 5 with respect to link 3?

(e) Check parts (b) and (d) by the complex-number method.

3.45. In the quick-return mechanism of Fig. P3.53

   (a) Determine the ratio V_C/ω_{in} by instant centers.*

   (b) Find F_{out}/F_{in}.

   (c) Check parts (a) and (b) by the complex-number method.

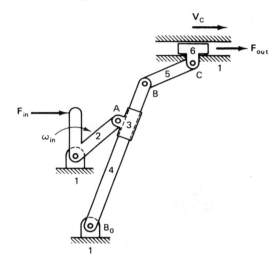

Figure P3.53

3.46. **(a)** Find the location of all the instant centers of the six-link mechanism in Fig. P3.54.*

   (b) Determine the angular velocity ratio ω_4/ω_2.

   (c) Determine the angular velocity ratio ω_6/ω_2.

   (d) Check the location of I_{13} by complex numbers.

   (e) Check the results of parts (b) and (c) by complex numbers.

3.47. **(a)** Determine the location of all the instant centers of the six-link mechanism in Fig. P3.55.

   (b) What is the angular velocity ratio ω_4/ω_2?

   (c) What is the angular velocity ratio ω_5/ω_2?

   (d) Find the velocity of point C given $\omega_2 = 1$ rad/sec cw by two methods: (1) by using instant center 2,5 and (2) instant center 2,6.

   (e) Check the results of parts (b), (c) and (d) by the complex-number method.*

3.48. Determine the torque ratio T_6/T_2 of the mechanism in Fig. P3.56.

   (a) Use the instant center method.

   (b) Check it by the complex-number method.

3.49. Find the ratio V_D/ω_2 for the mechanism in Fig. P3.57.

   (a) Use the instant center method.

   (b) Check it by the complex-number method.*

---

 * Scale the figure for needed data.

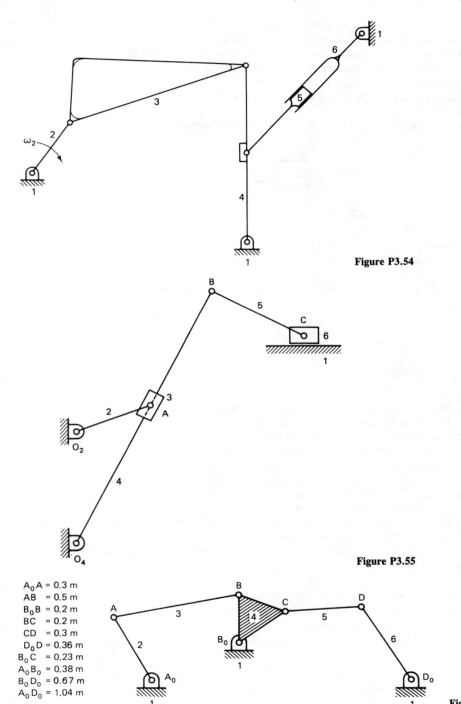

Figure P3.54

Figure P3.55

$A_0 A = 0.3$ m
$AB = 0.5$ m
$B_0 B = 0.2$ m
$BC = 0.2$ m
$CD = 0.3$ m
$D_0 D = 0.36$ m
$B_0 C = 0.23$ m
$A_0 B_0 = 0.38$ m
$B_0 D_0 = 0.67$ m
$A_0 D_0 = 1.04$ m

Figure P3.56

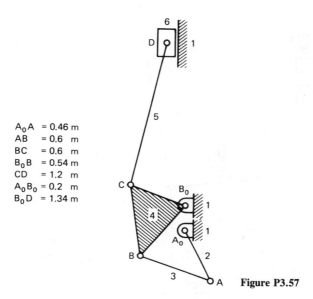

A_0A = 0.46 m
AB = 0.6 m
BC = 0.6 m
B_0B = 0.54 m
CD = 1.2 m
A_0B_0 = 0.2 m
B_0D = 1.34 m

Figure P3.57

3.50. What force ($\mathbf{F}_{in}$) is required at the piston in the mechanism of Fig. P3.58 to balance the weight $\mathbf{W}$ on link 5?
   (a) Use the instant center method.
   (b) Check your result by the complex-number method.*

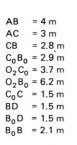

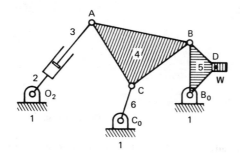

AB = 4 m
AC = 3 m
CB = 2.8 m
C_0B_0 = 2.9 m
O_2C_0 = 3.7 m
O_2B_0 = 6.2 m
C_0C = 1.5 m
BD = 1.5 m
B_0D = 1.5 m
B_0B = 2.1 m

Figure P3.58

3.51. (a) What is the relationship between the input force $\mathbf{F}_{in}$ and the resisting force $\mathbf{R}$ in Fig. P3.59? Use instant centers.
   (b) What is ω_5/ω_2?
   (c) Check your results in parts (a) and (b) by the complex-number method.

3.52. (a) What is the relationship between the input torque T_{in} and the resistance of the cutting tool $\mathbf{R}$ in the shaper mechanism of Fig. P3.60?
   (b) What is ω_2/V_6?
   (c) Check your results in parts (a) and (b) by the complex-number method.*

* See footnote on page 206.

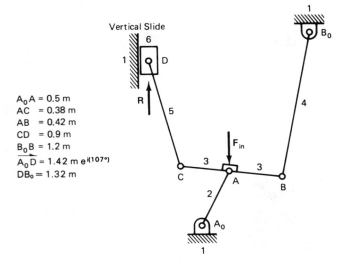

Vertical Slide

A_0A = 0.5 m
AC = 0.38 m
AB = 0.42 m
CD = 0.9 m
B_0B = 1.2 m
$\overrightarrow{A_0D}$ = 1.42 m $e^{i(107°)}$
DB_0 = 1.32 m

Figure P3.59

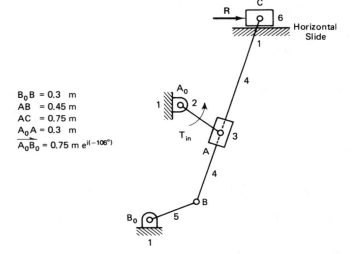

B_0B = 0.3 m
AB = 0.45 m
AC = 0.75 m
A_0A = 0.3 m
$\overrightarrow{A_0B_0}$ = 0.75 m $e^{i(-106°)}$

Horizontal
Slide

Figure P3.60

3.53. Determine the mechanical advantage (F_{out}/F_{in}) in the position shown of the inverted slider-crank in Fig. P3.61.

(a) Use instant centers.

(b) If friction exists in the slider generating a resisting force F_r, what is the mechanical advantage of the linkage?

(c) Check your results in parts (a) and (b) by the complex-number method.*

* Scale the figure for needed data.

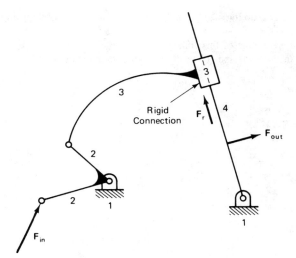

Figure P3.61

3.54. The linkage used in problem 3.9 is redrawn in Fig. P3.62. ω_2 is 2 rad/sec cw.
  (a) Find all instant center locations.
  (b) Determine the mechanical advantage.*

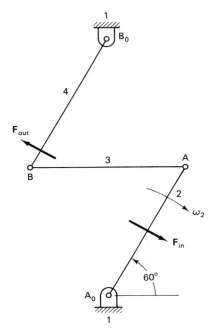

Figure P3.62

* Scale the drawing for the location of $\mathbf{F}_{in}$ & $\mathbf{F}_{out}$. Note that $\mathbf{F}_{out}$ is a resistance force.

3.55. **(a)** Find the angular velocity ratio (ω_4/ω_6) in the position shown of the mobile lifting mechanism in Fig. P3.63.

(b) Find the mechanical advantage of this linkage.

(c) Check your results in parts (a) and (b) by the complex-number method.*

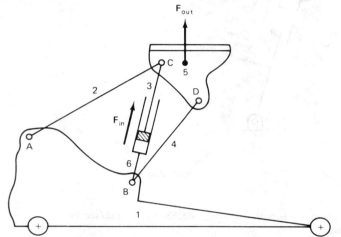

Figure P3.63

3.56. The input link of the Stephenson III six-bar shown in Fig. P3.64 has angular velocity $\omega_2 = 5$ rad/sec cw. Find the mechanical advantage if $\mathbf{F}_{in}$ acts on link 3 and $\mathbf{F}_{out}$ acts on link 6 as shown. Let A_0A be 3″ for scaling purposes; AB is at $+30°$.*

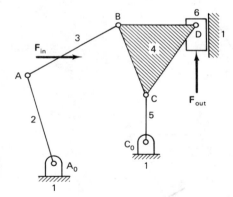

Figure P3.64

3.57. Determine the mechanical advantage of the dumping mechanism shown in Fig. P3.65. The hydraulic cylinder (2) provides the input force $(\mathbf{F}_{in})$. The payload is $(\mathbf{W})$.

(a) Use the method of your choice.

(b) Check your results with another method.*

* Scale the figure for needed data.

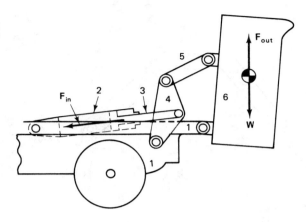

Figure P3.65

3.58. The velocity of slider 2 of the hydraulic actuator linkage is 10 in/sec to the right shown in Fig. P3.66. Link lengths are $AB_0 = 5''$, $BB_0 = 6''$.
(a) Find the angular velocity of link 4.
(b) Determine the mechanical advantage.*

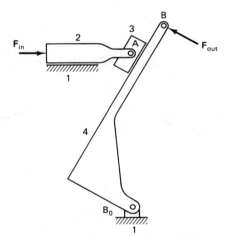

Figure P3.66

3.59. A cam surface is driving a four-bar linkage whose movement is resisted by a spring at point D (Fig. P3.67).
(a) What is the mechanical advantage of this mechanism?
(b) What is ω_5/ω_2?
(c) What is ω_5/ω_3?
(d) Check your results in parts (a) and (b) by the complex-number method.*

* Scale the figure for needed data.

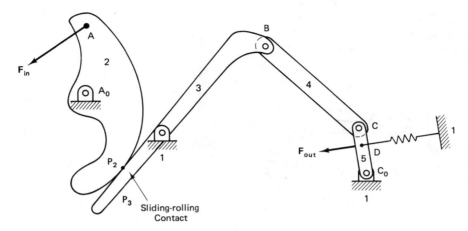

Figure P3.67

3.60. (a) Find the ratio of F_{out}/T_{in} for the riveting mechanisms in both positions shown in Fig. P3.68.

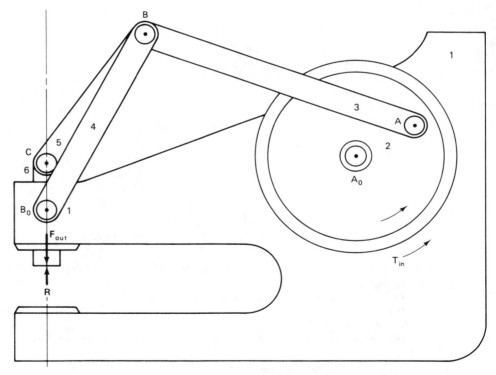

(a)

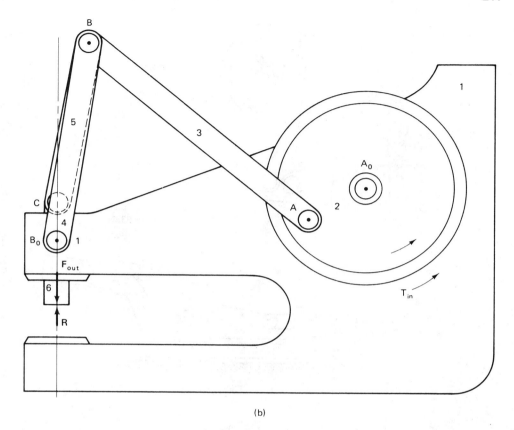

(b)

Figure P3.68

(b) Which position has the greater mechanical advantage?
(c) What is V_6/ω_2 in each case?
(d) Check your results in parts (a) and (b) by the complex-number method.*

3.61. In the rock crusher linkage of Fig. P3.69 determine the ratio of F_{out}/T_{in} by two methods:
    (a) Use instant centers (1,2), (1,6), and (2,6).
    (b) Use instant centers (1,2), (1,4), (2,4), (1,6), and (4,6).
    (c) Check your results in parts (a) and (b) by the complex-number method.*

3.62. **(a)** Find the mechanical advantage of the straight-line clamp of Fig. P3.70.
    (b) What is V_4/ω_2?
    (c) Check your results in parts (a) and (b) by the complex-number method.*

3.63. A hold-down clamp is shown in two positions in Fig. P3.71.
    (a) Find the mechanical advantage in both the solid and dashed positions.
    (b) Check your results by the complex-number method.*

* Scale the figure for needed data.

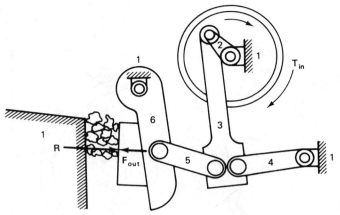

Figure P3.69

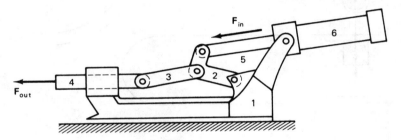

Figure P3.70

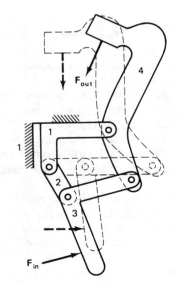

Figure P3.71

3.64. A four-bar linkage has been synthesized to remove the cap from a beverage bottle (see Fig. P3.72).

(a) What is the ratio of the torque applied to the bottle cap to the input force ($\mathbf{F}_{in}$) in the two positions shown?

(b) Determine ω_3/ω_2 in the positions shown.

(c) Check your results in parts (a) and (b) by the complex-number method.*

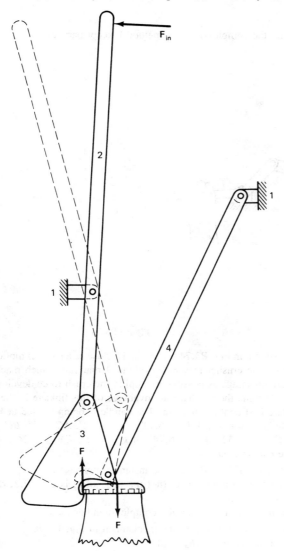

Figure P3.72

* Scale the figure for needed data.

3.65. After several unsuccessful attempts to open a can of liquid refreshment, a linkage was designed to assist in the task (Fig. P3.73). Determine the mechanical advantage of this "thirst-while" mechanism.

(a) Use instant centers (2,5), (1,2), and (1,5).

(b) Use instant centers (2,6), (1,2), and (1,6).

(c) Use instant centers (1,2), (2,4), (1,4), and (4,6).

(d) Use all of the above.

(e) Use a can opener.

(f) Check all the above by the complex-number method [except part (e)].*

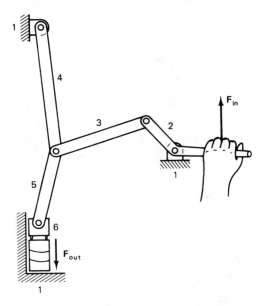

Figure P3.73

3.66. The Watt I six-bar linkage shown in Fig. P3.74 is designed for use as a parallel motion super-market based hand-operated can crusher to spur recycling of materials. Such machines require as large a mechanical advantage as possible to amplify the crush force developed from a limited human input. Determine the mechanical advantage of this linkage in the position shown when the crush plate first contacts the can. The handle is being rotated at 1 rad/sec cw. Positional information is $A_0A = 4''$, $A_0D = 6.13''$, $AB = 5.2''$, $AC = 7''$, $BC = 7.97''$, $BB_0 = 5''$, $A_0B_0 = 13''$, $CE = 17.52''$, $DE = 6.74''$, $A_0F_{in} = 12''$, $EF_{out} = 24''$; A_0A is at $+80°$, CE is at $-25°$, the handle is at $+40°$.

3.67. (a) What is the ratio of F_{out}/F_{in} of Fig. P3.75. Use instant centers.

(b) If the coefficient of friction in both sliders (not the fork joint) is $\mu = 0.2$, determine the new mechanical advantage.

(c) Check your results in parts (a) and (b) by the complex-number method.*

3.68. (a) Find the angular velocity ratio (ω_4/ω_6) of the mechanism in Fig. P3.76.

(b) Determine the mechanical advantage (F_{out}/F_{in}).

(c) Check your results in parts (a) and (b) by the complex-number method.*

* Scale the figure for needed data.

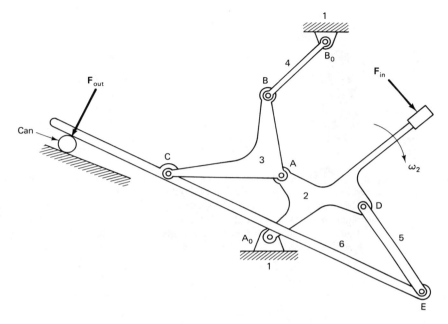

Figure P3.74

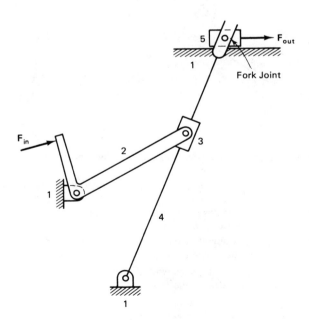

Figure P3.75

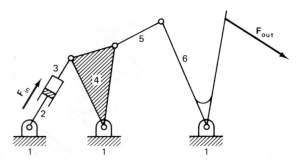

Figure P3.76

3.69. Links 1, 2, 3 and 4 have been designed for the Watt II six-bar shown in Fig. P3.77. It is desired to have link 6 perpendicular to the ground and F_{out} parallel and 20″ from the ground when the linkage is in the position shown. Link 5 is not yet designed and is shown in it's conceptual coupler position only.* If the overall mechanical advantage of the linkage is to be 3, how long should link 6 be? Known linkage dimensions are $A_0 B_0 = A_0 A = 36.334″$, $AB_0 = BB_0 = 20.881″$, $AB = B_0 C_0 = 12″$.

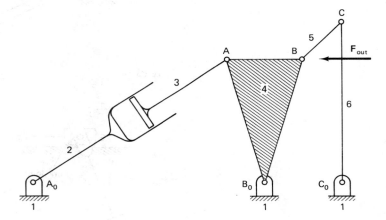

Figure P3.77

3.70. A typical automobile hood linkage is shown in Fig. P3.78. Notice the placement of the counterbalancing spring.
   (a) Determine (in terms of **W**) the force exerted by the spring (if other links are assumed to have negligible weight as compared with the hood) to balance **W**.
   (b) Does the effectiveness of the spring improve if its attachment point (P) is moved up vertically? Why or why not?
   (c) Does the placement of the spring make sense keeping in mind the entire range of motion of the mechanism? Why or why not?

3.71. A standard toggle press is shown in Fig. P3.79. Determine its mechanical advantage (force of the plunger divided by the force of the cylinder). Scale the drawing.

* It will connect link 4 in this current position to link 6 in its final position spanning points B and C.

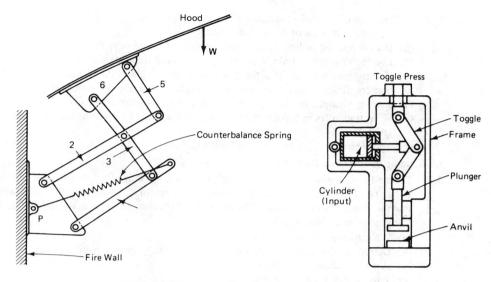

Figure P3.78   **Figure P3.79**

3.72. Polycentric motion is said to occur when the center of rotation of a moving body appears to move in plane space. This type of motion behavior is desirable for design situations in which it is desired to have a continuously changing mechanical advantage or in which it is desired to have self-locking behavior. For example [150], the polycentric hinge shown in Fig. P3.80 has a shifting instant center.

   (a) Regarding the door as the input, which inversion of the slider-crank is this linkage — similar to Fig. 3.8, 3.9, 3.10, or 3.11?

   (b) Plot the trajectory of the instant center of the door with respect to the frame as the door opens.

   (c) Describe in practical terms the advantage of these moving instant centers.

   (d) Spot-check your results for part (b) with complex numbers.*

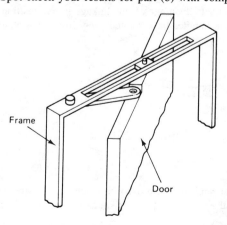

Figure P3.80 A polycentric door hinge.

* Scale the figure for needed data.

3.73. Referring to the casement window linkage of Fig. 1.4a,

    (a) Find the locations of the instant centers.

    (b) If T_2 is applied to link 2, how much torque (T_4) results at the window?

    (c) List the changes in linkage geometry that would improve part (b).

    (d) If a frictional force exists at the shoe (due to the weight of the window and a nonzero friction coefficient), find the transmission angle considering the sash (link 4) as the output.

    (e) List the changes in linkage geometry that would improve part (d).

    (f) What trade-offs in linkage performance will have to be made to improve both parts (b) and (c)?

    (g) Spot-check part (a) by complex numbers.*

3.74. Same questions as Prob. P3.73 but refer to Fig. 1.4b.

3.75. Figure P3.81 shows a Rongeur, which is used by orthopedic surgeons for cutting away bone. The leaf-type springs between the handles return the linkage to the open position so that the Rongeur can be operated by one hand. Note that the holes in the handles connecting to the jaws must be elongated.

    (a) What type of linkage is this?

    (b) Determine the mechanical advantage of this linkage in the position shown as well as the closed position (disregard the spring).*

    (c) Why is this device designed this way?

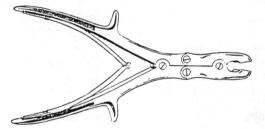

Figure P3.81

3.76. A pair of squeeze clamps are shown in Fig. P3.82.

    (a) What type of linkage is this?

    (b) Determine the mechanical advantage of this linkage in the position shown as well as in a position such that the handles have rotated away from each other 10° each.*

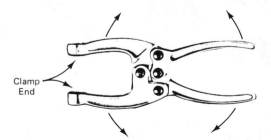

Clamp
End

Figure P3.82

3.77. Figure P3.83 shows a mechanical clamp used on machine tools for work-holding fixtures. Determine the mechanical advantage of this linkage. Why is it designed this way?*

* Scale the figure for needed data.

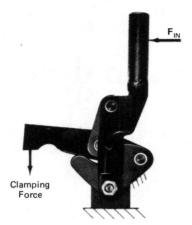

Clamping
Force

Figure P3.83

EXERCISES

3.1. In Fig. 3.43, let the following quantities be given: $R_{Ax} = -14.3$ mm, $R_{Ay} = 21.0$ mm, $R_{BAx} = 14.3$ mm, $R_{BAy} = 23.6$ mm, and $\omega_2 = 1.4$ rad/sec. Using the polar complex number form of Eq. (3.33), calculate $\mathbf{V}_A$, $\mathbf{V}_{BA}$, and $\mathbf{V}_B$.

3.2. Establish the values of the link vectors of the six-link mechanism of Fig. 3.49 by scaling the drawing and solve Exer. 3.5 by complex-number analysis.

3.3. Using the link vectors determined in Exer. 3.2, separate Eq. (3.49) into real and imaginary parts and solve for λ_2 and λ_4. Then use $\mathbf{I}_{1,3} = \lambda_2 \mathbf{Z}_2$ to find the instant center 1,3. Check your result graphically.

3.4. Prove Eq. (3.51b) by regarding $\mathbf{F}$ and $\mathbf{V}$ as space vectors confined to the xy plane of xyz space and forming their scalar [dot ($\cdot$)] product.

3.5. Regard link 3 as the output of the slider-crank mechanism of Fig. 3.77 and derive the expression for the mechanical advantage, (F_{out}/F_{in}), using the instant center method. Does your result verify Eq. (3.62)? It should!

3.6. Continuing as demonstrated in the text in connection with Fig. 3.78, draw and analyze free-body diagrams for links 3 and 4 of Fig. 3.77, leading up to Eq. (3.62), thus verifying both that equation and the graphical (geometric) construction demonstrated in connection with Fig. 3.77 for finding the M.A.

chapter four
Acceleration Analysis

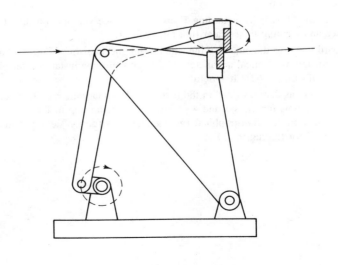

4.1 INTRODUCTION

In Chap. 3 several methods were described for velocity analysis of mechanisms. In this chapter acceleration analysis is discussed. Since the ultimate objective is inertia-force analysis of mechanisms and machines (Chap. 5), all acceleration components should be expressed in one and the same coordinate system: the inertial frame of reference of the fixed link of the mechanism.

It is extremely important to make sure that enough is known about the path of a point of interest in this coordinate system. If such a path is not obvious, acceleration must be determined in several point-to-point steps as introduced in Chap. 3 (see Table 3.1). In each of these steps the "relative" path must be known. We will find that in a typical linkage example, many more terms appear in acceleration equations than in the velocity equations of Chap. 3. Again, a careful study of the difference of motion between points of the same link and relative motion between different links will help avoid difficulty in solving acceleration problems.

Table 3.1 showed the four possible cases that are applicable when examining the motion of various points in a mechanism. The 2×2 matrix in this table represents combinations of the same or different points on the same or different links.

Figure 4.1 (which is a repeat of Fig. 3.36) will help review the four cases of Table 3.1. The motion of any point (say R) with respect to itself as part of the same link (say link 3) is trivial (*case 1*). The movement of point S as part of link 3 with respect to R of the same link is *case 2* motion. Case 3 analysis (same point–different member) can be trivial if there happens to be a revolute joint between those two members at that point (e.g., point T as part of link 4 and link 5). If there is no revolute joint (point U as part of links 4 and 5), case 3 analysis can be done only if the relative path is known.* A case 4

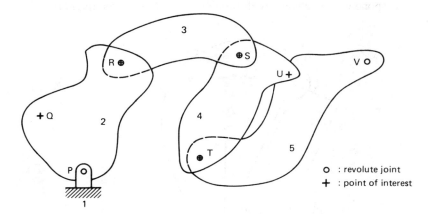

Figure 4.1 Acceleration difference occurs between points of the same link, say, point Q with respect to point R, both of link 2. Relative acceleration occurs between points on two different links, such as points R on link 2 and T on link 5.

* For acceleration analysis, the center of curvature of the relative path must be known.

analysis — different point–different member (say point T as part of link 5 and point R as part of 3) — can be solved directly only if the relative path is known, but usually this is not the case. A combined step-by-step analysis through a mechanism chain using case 2 and 3 analyses will usually solve a case 4 problem.

4.2 ACCELERATION DIFFERENCE*

As in Chap. 3, let us begin by describing the acceleration of a point on a body that is pinned to ground (Fig. 4.2). Point A on link 2 can be located with respect to the origin A_0 of the fixed complex plane in polar form by the vector $\mathbf{R}_A$, made up of angle θ_A measured positively counterclockwise from the fixed real x axis and the radius R_A:

$$\mathbf{R}_A = R_A e^{i\theta_A} \tag{4.1}$$

The velocity of point A was obtained in Chap. 3 by taking the derivative of the position vector with respect to time. If link 2 is a rigid member, only θ_A is changing with time, so the absolute linear velocity of point A on link 2 is (recall that $\omega_{21} \equiv d\theta_A/dt$)

$$\mathbf{V}_A = R_A \omega_{21} i e^{i\theta_A} = i\omega_{21}\mathbf{R}_A \tag{4.2}$$

The magnitude of the velocity vector is $R_A\omega_{21}$ and the direction is perpendicular to the vector $\mathbf{R}_A$ in the sense of the angular velocity ω_{21}. From now on, to simplify the notation, in most cases we write ω_2 to mean ω_{21} (i.e., angular velocity with respect to the fixed frame of reference). If there is another second index, say ω_{23}, it means the angular velocity of link 2 with respect to link 3.

The acceleration of point A may be found by taking the derivative of the velocity [Eq. (4.2)] with respect to time. Again, since all links are assumed to be rigid, the only components that change with time are θ_A and ω_2. If the *angular acceleration* is defined

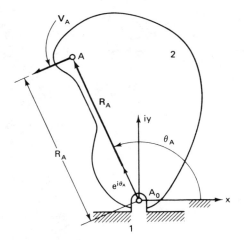

Figure 4.2   Velocity difference of A with respect to A_0.

* Case 2, Table 3.1.

as the rate of change of the angular velocity with respect to time ($\alpha_2 \equiv d\omega_2/dt$), *the absolute linear acceleration* of point A on link 2 with respect to ground is

$$\mathbf{A}_A = R_A \alpha_2 i e^{i\theta_A} - R_A \omega_2^2 e^{i\theta_A}$$

$$\mathbf{A}_A = R_A(-\omega_2^2 + \alpha_2 i)e^{i\theta_A} \qquad (4.3)$$

$$= (-\omega_2^2 + i\alpha_2)\mathbf{R}_A$$

Notice that in general there are two components of acceleration of a point on a rigid body rotating about a ground pivot (Fig. 4.3). One component has a magnitude of $R_A \alpha_2$ and a direction tangent to the path of A pointed in the sense of the angular acceleration. This component is called the *tangential acceleration*. Its presence is due solely to the rate of change of the angular velocity. The other component has a magnitude of $R_A \omega_2^2$ and, because of its $-$ sign, a direction opposite to that of the original position vector $\mathbf{R}_A$. This component, which always points toward the center of rotation because R_A and ω_2^2 are always positive, is called the *normal* or *centripetal acceleration*. This component is present due to the changing direction of the velocity vector (because point A is moving along a circular arc). An equivalent form for the magnitude of the normal acceleration is V_A^2/R_A. A special case occurs when point A moves in a straight line; then the radius of curvature of the path of A is infinite ($R_A = \infty$) and the normal acceleration is equal to zero.

Let us determine the total absolute acceleration of point B on the same link (see Fig. 4.4).

$$\mathbf{A}_B = R_B(-\omega_2^2 + i\alpha_2)e^{i\theta_B} = (-\omega_2^2 + i\alpha_2)\mathbf{R}_B \qquad (4.4)$$

What is the acceleration difference of B with respect to A ($\mathbf{A}_{BA}$)? Another way you might ask the same question: Suppose that I were sitting at point A (as if on a merry-go-

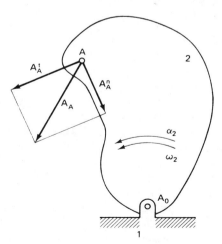

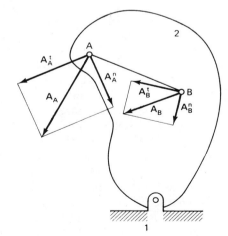

Figure 4.3 Acceleration difference of A with respect to A_0.

Figure 4.4 Accelerations of two different points on one rigid link.

round) in a swivel chair that always faces east, in the direction of the inertial x axis. If I were watching point B, what would be the acceleration of point B with respect to my fixedly oriented, orbiting frame of reference? This would be the acceleration difference of point B with respect to point A, or $\mathbf{A}_{BA}$. The total absolute acceleration of point B may be written in terms of the acceleration difference as

$$\mathbf{A}_B = \mathbf{A}_A + \mathbf{A}_{BA} \tag{4.5}$$

Each term of Eq. (4.5) has two possible components since link 2 is a rigid member, so that the full expression is

$$\mathbf{A}_B^n + \mathbf{A}_B^t = \mathbf{A}_A^n + \mathbf{A}_A^t + \mathbf{A}_{BA}^n + \mathbf{A}_{BA}^t$$

or   (4.6)

$$\mathbf{A}_{BA}^n + \mathbf{A}_{BA}^t = \mathbf{A}_B^n + \mathbf{A}_B^t - \mathbf{A}_A^n - \mathbf{A}_A^t$$

where the superscripts n and t refer to the normal and tangential components, respectively, described above. Initially, a graphical solution of the acceleration problem will be employed. To keep track of which components are known or unknown, an M will be written below a term if the magnitude is known and a D written if the direction is known. Thus from Fig. 4.4,

$$
\begin{array}{ccccc}
\mathbf{A}_{BA}^n & + \mathbf{A}_{BA}^t & = \mathbf{A}_B^n & + \mathbf{A}_B^t & - \mathbf{A}_A^n & - \mathbf{A}_A^t \\
\text{D}\swarrow & \text{D}\nwarrow & \text{D}\searrow & \text{D}\swarrow & & \\
\text{M} & \text{M} & \text{M} & \text{M} & &
\end{array}
\tag{4.7}
$$

A short line segment with an arrow pointing in the direction of the component is also helpful. Note that the respective signed magnitudes on the right side of Eq. (4.7) are

$$(R_B \omega_2^2), \qquad (R_B \alpha_2), \qquad (R_A \omega_2^2), \quad \text{and} \quad (R_A \alpha_2)$$

This complex equation [Eq. (4.7)] has four real unknowns: the direction and magnitude of $\mathbf{A}_{BA}^n$ and $\mathbf{A}_{BA}^t$. Unless two of these four unknowns can be found some other way, Eq. (4.7) cannot be solved since it represents only two scalar equations.

The directions of the normal and tangential components of the relative acceleration of B with respect to A are found by inspection, however. The normal component must be directed toward A (since the reference position is A). The tangential component is perpendicualr to the normal acceleration in the sense of the angular acceleration α_2.

Figure 4.5 shows the graphical solution to Eq. (4.7) yielding the normal and tangential acceleration difference components. A convenient center (O_A) of the acceleration diagram is located arbitrarily and all the components drawn to scale. Both $\mathbf{A}_A$ and $\mathbf{A}_B$ (combined normal and tangential components) are drawn from O_A since they are total absolute accelerations. Equation (4.7) tells us that the difference between the total acceleration of B and the total acceleration of A is the unknown $\mathbf{A}_{BA}$ (normal plus tangential). Intersections of the line of the direction of the normal acceleration with the line of the direction of the tangential acceleration yield the proper magnitudes of these components. Thus the graphical solution of Eq. (4.7) is complete.

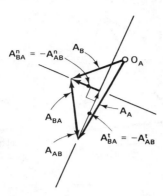

Figure 4.5 Geometric construction of the acceleration difference between points A and B of link 2 in Fig. 4.4.

Suppose that Eq. (4.7) had been written as

$$\mathbf{A}_{AB}^{n} + \mathbf{A}_{AB}^{t} = \mathbf{A}_{A}^{n} + \mathbf{A}_{A}^{t} - \mathbf{A}_{B}^{n} - \mathbf{A}_{B}^{t} \qquad (4.8)$$

Figure 4.5 demonstrates that $\mathbf{A}_{BA}^{t} = -\mathbf{A}_{AB}^{t}$ and $\mathbf{A}_{BA}^{n} = -\mathbf{A}_{AB}^{n}$ in this case 2 problem (see Table 3.1).

Example 4.1

Link J in Fig. 4.6 is in planar motion with respect to ground. Accelerometers located at points A and B measure total absolute linear accelerations as depicted by the vectors $\mathbf{A}_{A}$ and $\mathbf{A}_{B}$ in the Fig. 4.6.   (a) Find the angular acceleration of this link.   (b) Find the angular velocity of the link.

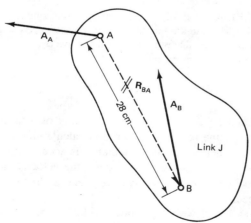

Figure 4.6 Accelerations of points A and B of link J in general planar motion.

Solution   Referring to Eq. (4.7), the components of the acceleration difference $\mathbf{A}_{BA}^{n}$ and $\mathbf{A}_{BA}^{t}$ may be obtained from the total absolute accelerations $\mathbf{A}_{A}$ and $\mathbf{A}_{B}$ as in Fig. 4.7. The resulting magnitudes of the tangential and normal components of the acceleration difference yield the answers to (a) and (b):

$$|\alpha_{J}| = \frac{A_{BA}^{t}}{R_{BA}} = \frac{459}{28} = 16.40 \text{ rad/sec}^2 \text{ ccw} \qquad (4.9)$$

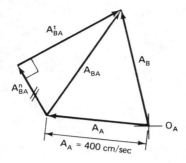

Figure 4.7 Geometric construction of the acceleration difference of point B with respect to point A of link J in Fig. 4.6.

$$|\omega_J| = \left|\sqrt{\frac{A_{BA}^n}{R_{BA}}}\right| = \left|\sqrt{\frac{234}{28}}\right| = 2.89 \text{ rad/sec} \qquad (4.10)$$

The sense of ω_j is not known from the information given. The sense of α_j is found by observing the directions of A_{BA}^t in Fig. 4.7. In digital computation, the algebraic sign of α_j (and thus its sense) is found by using the vector forms of A_{BA}^t and R_{BA} as follows:

$$A_{BA}^t = i\alpha_j R_{BA} \qquad \text{[from Eq. (4.3)]}$$

where R_{BA} is the vector locating point B with respect to point A (i.e., pointing from A to B). Therefore

$$\alpha_j = \frac{A_{BA}^t}{i R_{BA}} \qquad (4.10a)$$

Figures 4.6 and 4.7 show that A_{BA}^t and $i R_{BA}$ are collinear. Visualizing A_{BA}^t from Fig. 4.7 placed at point B on Fig. 4.6, one can see that this vector imposes a ccw angular acceleration to link J. Equation (4.10a) bears this out, yielding a positive α_j.

4.3 RELATIVE ACCELERATION*

In order to find linear or angular acceleration of an output link of a mechanism when input link acceleration is given, some relative acceleration calculations are usually performed. Suppose that we wish to determine the instantaneous acceleration of the slider of the slider-crank mechanism of Fig. 4.8 (see Sec. 3.5 for the velocity analysis of this mechanism) with a counterclockwise angular velocity (ω_2) and a clockwise angular acceleration of (α_2) of link 2 specified.

The slider is identified as link 4 while the known velocity and acceleration information is given for link 2. This is apparently a case 4 analysis. The path of point B on link 4 relative to point A on link 2 is not easily visualized without including link 3. To make the analysis simpler this example will be solved by superimposing several case 2 and case 3 solutions as follows:

---
* Case 3, Table 3.1.

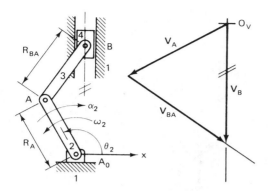

Figure 4.8 Geometric construction of slider velocity. Note that $\mathbf{V}_A$ is perpendicular to link 2 and $\mathbf{V}_{BA}$ is perpendicular to link 3.

Step 1.   Determine $\mathbf{A}_{(A2)}$ *(a case 2 analysis).* The acceleration of point A as part of 2 (signified as $\mathbf{A}_{(A2)}$) is

$$\mathbf{A}_{(A2)} = \mathbf{A}_{(A2)}^n + \mathbf{A}_{(A2)}^t = R_A(-\omega_2^2 + i\alpha_2)e^{i\theta_2} \tag{4.11}$$

Step 2.   Determine $\mathbf{A}_{(A3)}$. This is a case 3 analysis, but since point A is a revolute joint joining links 2 and 3, then

$$\mathbf{A}_{(A3)} = \mathbf{A}_{(A2)} \tag{4.12}$$

Step 3.   Determine $\mathbf{A}_{(B3)}$. Recognizing that point B can be thought of as a point on the slider (link 4) as well as a point of link 3, we will first find the total acceleration of B as part of 3 (case 2) by using the equation

$$\mathbf{A}_{(B3)} = \mathbf{A}_{(A3)} + \mathbf{A}_{BA}$$

or

$$
\begin{array}{cccccc}
\mathbf{A}_{(B3)}^n & + \mathbf{A}_{(B3)}^t & = \mathbf{A}_{(A3)}^n & + \mathbf{A}_{(A3)}^t & + \mathbf{A}_{BA}^n & + \mathbf{A}_{BA}^t \\
0 & D\updownarrow & D\searrow & D\nearrow & D\swarrow & D\nwarrow \\
0 & & M & M & M &
\end{array} \tag{4.13}
$$

The normal and tangential components of $\mathbf{A}_{(A3)}$ are known from steps 1 and 2. The direction of $\mathbf{A}_{BA}^n$ must be toward A while the magnitude is determined from the velocity analysis performed in Sec. 3.5: $A_{BA}^n = V_{BA}^2/R_{BA} = \omega_3^2 R_{BA}$. In vector form,

$$\mathbf{A}_{BA}^n = -\omega_3^2 \mathbf{R}_{BA} \tag{4.14}$$

The tangential acceleration $\mathbf{A}_{BA}^t$ has a known line of action represented by a double-headed arrow along a line perpendicular to the normal component.

On the left side of Eq. (4.13), the normal acceleration is zero since point B is constrained to move in a straight line with respect to the fixed inertial frame of reference. (Recall that the normal acceleration exists due to changing in direction of the linear velocity.) The absolute tangential acceleration of B (tangent instantaneously to the path of B relative to ground) is represented as a doubleheaded vertical arrow (see Fig. 4.9).

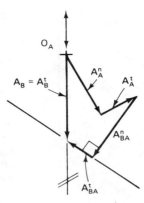

Figure 4.9 Geometric construction of slider acceleration in the mechanism of Fig. 4.8.

Thus Eq. (4.13) has two unknowns and can be solved graphically as illustrated in Fig. 4.9 for a typical ω_2 and α_2.

Step 4.   Determine $\mathbf{A}_{(B4)}$. This is a case 3 analysis but since point B is a revolute joint,

$$\mathbf{A}_{(B4)} = \mathbf{A}_{(B3)} \tag{4.15}$$

which completes our analysis.

Notice that the relative acceleration method follows the same technique as the relative velocity method (Sec. 3.5), but more terms are included. The slider-crank analysis above was presented as four separate steps in order to emphasize the different cases of motion of one point with respect to another represented in Table 3.1. In subsequent examples, these steps may not be specifically itemized, but one should include consideration of this step-by-step analysis concept when attacking any problem. This will help avoid errors.

Example 4.2

Determine the instantaneous angular acceleration of the coupler and output links of the four-bar linkage shown in Fig. 4.10, given $\omega_2 = 600$ rpm or 62.8 rad/sec cw and $\alpha_2 = 2000$ rad/sec² cw. Also find the linear acceleration of points B and C. Results of the velocity analysis, required in the forthcoming acceleration analysis, are shown in the velocity polygon of Fig. 4.10.

Solution   The total acceleration of point A as part of link 2 is

$$\mathbf{A}_A = \mathbf{A}_A^n + \mathbf{A}_A^t, \qquad \text{where } |\mathbf{A}_A^n| = (A_0A)\omega_2^2, \qquad |\mathbf{A}_A^t| = (A_0A)\alpha_2$$

and where $\mathbf{A}_A^n$ is directed from A to A_0 and $\mathbf{A}_A^t$ is perpendicular to $\mathbf{A}_A^n$ in the sense of α_2. Since the accelerations of A as part of links 2 and 3 are equal, the acceleration of B as part of 3 can be determined from

$$
\begin{array}{cccccc}
\mathbf{A}_B^n & + \mathbf{A}_B^t & = & \mathbf{A}_A^n & + & \mathbf{A}_A^t & + & \mathbf{A}_{BA}^n & + \mathbf{A}_{BA}^t \\
\text{D}\swarrow & \text{D}\searrow & & \text{D}\downarrow & & \text{D}\nearrow & & \text{D}\leftarrow & \text{D}\updownarrow \\
\text{M} & & & \text{M} & & \text{M} & & \text{M} & \\
[(B_0B)\omega_4^2] & & & [(A_0A)\omega_2^2] & & [(A_0A)\alpha_2)] & & [(BA)\omega_3^2)] &
\end{array} \tag{4.16}
$$

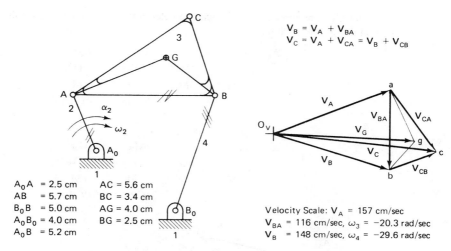

$$\mathbf{V}_B = \mathbf{V}_A + \mathbf{V}_{BA}$$
$$\mathbf{V}_C = \mathbf{V}_A + \mathbf{V}_{CA} = \mathbf{V}_B + \mathbf{V}_{CB}$$

A_0A	= 2.5 cm	AC	= 5.6 cm
AB	= 5.7 cm	BC	= 3.4 cm
B_0B	= 5.0 cm	AG	= 4.0 cm
A_0B_0	= 4.0 cm	BG	= 2.5 cm
A_0B	= 5.2 cm		

Velocity Scale: $\mathbf{V}_A$ = 157 cm/sec
$\mathbf{V}_{BA}$ = 116 cm/sec, ω_3 = −20.3 rad/sec
$\mathbf{V}_B$ = 148 cm/sec, ω_4 = −29.6 rad/sec

Figure 4.10 Geometric velocity analysis of four-bar mechanism.

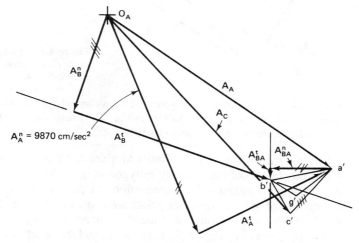

Figure 4.11 Geometric construction of accelerations in the four-bar mechanism of Fig. 4.10.

Figure 4.11 shows the acceleration polygon, which represents a graphical solution to Eq. (4.16). The intersection of the lines of $\mathbf{A}_B^t$ and $\mathbf{A}_{BA}^t$ marks the junction of these vectors and determines their magnitudes. Also,

$$\mathbf{A}_B = \mathbf{A}_B^n + \mathbf{A}_B^t = 9600 \text{ cm/sec}^2 \searrow$$

$$\alpha_3 = \frac{\mathbf{A}_{BA}^t}{i(\overrightarrow{AB})} = -\frac{500 \text{ cm/sec}^2}{5.7 \text{ cm}} = -87.7 \text{ rad/sec}^2 \text{ (cw)}$$

$$\alpha_4 = \frac{\mathbf{A}_B^t}{i(\overrightarrow{B_0B})} = -\frac{8500 \text{ cm/sec}^2}{5 \text{ cm}} = -1700 \text{ rad/sec}^2 \text{ (cw)}$$

The total absolute acceleration of C can be determined from either of the following acceleration difference equations:

$$\mathbf{A}_C = \mathbf{A}_A^n + \mathbf{A}_A^t + \mathbf{A}_{CA}^n + \mathbf{A}_{CA}^t$$

$$\begin{array}{cccc}
\text{D}\searrow & \text{D}\nearrow & \text{D}\swarrow & \text{D}\searrow \\
\text{M} & \text{M} & \text{M} & \text{M} \\
& & (\overrightarrow{CA})\omega_3^2 & (\overrightarrow{AC})i\alpha_3
\end{array}$$

(4.17)

or

$$\mathbf{A}_C = \mathbf{A}_B^n + \mathbf{A}_B^t + \mathbf{A}_{CB}^n + \mathbf{A}_{CB}^t$$

$$\begin{array}{cccc}
\text{D}\downarrow & \text{D}\searrow & \text{D}\rightarrow & \text{D}\nearrow \\
\text{M} & \text{M} & \text{M} & \text{M} \\
& & (\overrightarrow{CB})\omega_3^2 & (\overrightarrow{BC})i\alpha_3
\end{array}$$

(4.18)

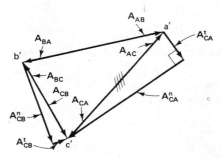

Figure 4.12 Acceleration image $\Delta a'b'c'$ of the coupler triangle ΔABC of the mechanism in Fig. 4.10.

Thus $\mathbf{A}_C = 11,100 \text{ cm/sec}^2 \searrow$. Figure 4.12 shows an enlarged drawing of the *acceleration image triangle* of the coupler ($\Delta a'b'c'$ in Fig. 4.11) (explained in the next paragraph) to help illustrate the solution of these equations.

Triangle $a'b'c'$ of Fig. 4.11 is similar to triangle ABC of the coupler link and also similar to triangle abc of the velocity polygon (Fig. 4.10). Therefore, a more rapid method of finding the total absolute acceleration of a point on a link when the total absolute accelerations of at least two other points are known is to utilize the *acceleration image triangle technique*. For example, if the absolute acceleration of the center of mass of the coupler (say point G of Fig. 4.10) is required, and the acceleration polygon has proceeded as far as solving Eq. (4.16), then image triangle $a'b'g'$ can be drawn similar to ABG. One must be careful to make sure that the triangles are not "flipped" on the wrong side of $a'b'$. This will be avoided if the velocity image and the acceleration image maintain the same directionally cyclic order as the coupler triangle. Here ABG, abg, and $a'b'g'$ are all directionally counterclockwise.

The similarity of the acceleration image to the original coupler can be proved by noting from Eq. (4.3) that

$$A_{GA} = (GA)|\sqrt{\omega_3^4 + \alpha_3^2}|$$

$$A_{GB} = (GB)|\sqrt{\omega_3^4 + \alpha_3^2}|$$

$$A_{AB} = (AB)|\sqrt{\omega_3^4 + \alpha_3^2}|$$

(4.19)

Thus the magnitudes of the acceleration differences of points on the coupler are proportional to the lengths between those points on the coupler link and therefore form a triangle similar to the coupler triangle.

Also notice that the velocity image triangle is turned 90° clockwise from the coupler triangle, and that the acceleration image has been turned through an angle of $(\pi - \beta)$ with respect to the coupler triangle, where

$$\beta = \arg(-\omega_3^2 + i\alpha_3) \tag{4.20}$$

See Prob. 4.1 at the end of the chapter for a proof of Eq. (4.20).

Analytical Expressions for Angular Accelerations in a Four-Bar Mechanism

In Sec. 3.10 the analytical expressions for the angular velocities of a four-bar mechanism (Fig. 3.82) were given [Eqs. (3.72) and (3.73)]. Recall that these were derived by taking derivatives of position vectors and doing some algebraic manipulation.

The expressions for the angular accelerations for the coupler link (α_3) and the output link (α_4) of this same four-bar mechanism (Fig. 3.82) are similarly derived (see Prob. 4.47):

$$\alpha_3 = \frac{-r_2\alpha_2 \sin(\theta_4 - \theta_2) + r_2\omega_2^2 \cos(\theta_4 - \theta_2) + r_3\omega_3^2 \cos(\theta_4 - \theta_3) - r_4\omega_4^2}{r_3 \sin(\theta_4 - \theta_3)}$$

$$\tag{4.21}$$

$$\alpha_4 = \frac{r_2\alpha_2 \sin(\theta_3 - \theta_2) - r_2\omega_2^2 \cos(\theta_3 - \theta_2) + r_4\omega_4^2 \cos(\theta_3 - \theta_4) - r_3\omega_3^2}{r_4 \sin(\theta_3 - \theta_4)}$$

$$\tag{4.22}$$

Note that these equations are in terms of α_2 and known angular velocity and position data. Equations (4.21) and (4.22) are easily programmed for automatic computation, yielding angular acceleration information of the entire four-bar mechanism for any number of positions θ_2 during the cycle of motion.

4.4 CORIOLIS ACCELERATION

Thus far acceleration analysis has been restricted to examples in which acceleration differences and relative accelerations have been between two points that have a fixed distance between them. In many mechanisms, however, lengths between points on different members do not remain constant (although the members themselves are rigid). Figure 4.13a shows the path of point P as it moves with respect to a reference coordinate system x,iy. In order to derive the acceleration of point P, this path* must be known (this is a case 4 analysis described in Table 3.1). If the path of P were traced by a coupler point of the four-bar mechanism shown in Fig. 4.13b, we are already prepared to

---

* Its slope, and radius of curvature at each point.

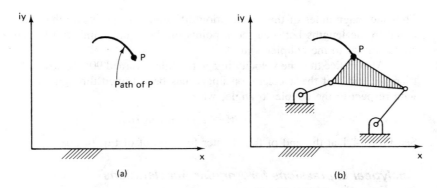

Figure 4.13 Point **P** moving with respect to fixed frame x,iy; (a) as a free point; (b) as a coupler point of a four-bar mechanism.

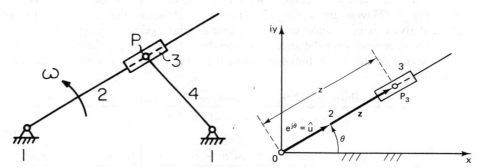

Figure 4.14 Inverted slider-crank mechanism.

Figure 4.15 Vector representation of the position of point P on the slider of Fig. 4.14.

find the acceleration of point P (as was done in Example 4.2) by case 2 and case 3 analyses. Let us now consider the case where a similar path might be generated by point P of the inverted slider-crank mechanism of Fig. 4.14 with link 2 as the input. Figure 4.15 shows this linkage without link 4. Let us determine the absolute acceleration of point P on the slider. The position P is defined by the vector **z**:

$$\mathbf{z} = ze^{i\theta} \tag{4.23}$$

The absolute velocity, acceleration, and further derivatives of the changing position vector of point P may be obtained by taking successive time derivatives of Eq. (4.23). In this example, both the angle θ and the length z change as slider 3 moves along link 2. If the following terms are defined:

$$v = \frac{dz}{dt}, \qquad \omega = \frac{d\theta}{dt}$$

$$a = \frac{dv}{dt}, \qquad \alpha = \frac{d\omega}{dt}$$

$$\dot{a} = \frac{da}{dt}, \qquad \dot{\alpha} = \frac{d\alpha}{dt}$$

then the terms of Fig. 4.16 are derived [144].

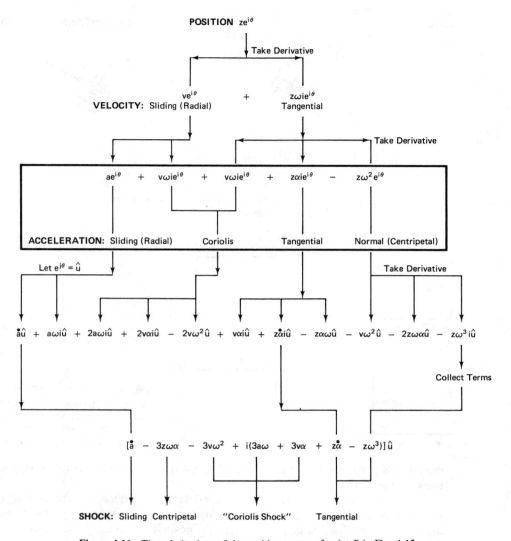

Figure 4.16 Time derivatives of the position vector of point P in Fig. 4.15.

There are two components of velocity (Fig. 4.17a), the *tangential velocity* with magnitude $z\omega$ and direction $\pm ie^{i\theta}$, the $\pm$ depending on the sense of ω (recall that ccw $= +$); the *sliding* or *radial velocity* has magnitude of v and a direction along link 2 in the sense of v: positive outward (z increasing) and negative inward (z decreasing). The derivatives of the velocity terms yield several acceleration terms (see Figs. 4.16 and 4.17b). Two familiar terms are the *normal (or centripetal) acceleration* (magnitude $z\omega^2$ and direction $-e^{i\theta}$) and the *tangential acceleration* (magnitude $z\alpha$ and the direction perpendicular to $e^{i\theta}$ in the sense of α). The *sliding* or *radial acceleration* is one new term that has magnitude a and is directed along the link, outward if v is increasing (positive acceleration), and vice versa. The other two components are equal and are combined to form the *Coriolis acceleration:* $2v\omega ie^{i\theta}$. The magnitude of this component is $2v\omega$ and its

233

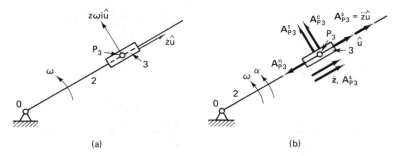

Figure 4.17 Vector representation of (a) the velocity and (b) the acceleration components of point P of Figs. 4.14 and 4.15.

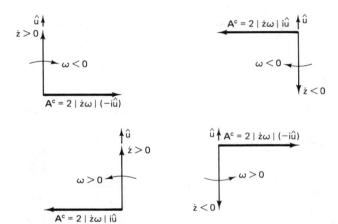

Figure 4.18 Determining the direction and sense of the Coriolis acceleration. In all cases, $\mathbf{A}^c = 2v\omega i_u$, where v is positive if it points in the outward direction from the center of rotation and ω is positive ccw.

direction is perpendicular to $e^{i\theta}$, directed one way or the other depending on the directions of v and ω.

Figure 4.18 can be used as a guide for determining which way the Coriolis component is directed. For example, if the velocity vector is directed outward (with respect to the center of rotation), which means that it is positive, and the angular velocity counterclockwise (which is also positive), the Coriolis component has the unit vector $i\hat{u}$, where $\hat{u} \equiv e^{i\theta}$, and therefore points to the left in Fig. 4.18. Another popular method is to rotate the velocity v 90° in the direction of ω; then the resulting direction is that of the Coriolis acceleration.

Coriolis acceleration* is usually believed to be conceptually difficult but a "rule of thumb" will help: *There is a Coriolis component of acceleration when the vector that locates the point is both rotating and changing length with respect to the fixed inertial frame of reference.* This is true of the example of Figs. 4.14 and 4.15, since the vector $\mathbf{z}$ is changing length as well as rotating as the mechanism moves.

---

*Coriolis acceleration affects our daily lives through its influence on both weather and ocean currents. For example, as air attempts to move from a high-pressure area to a "low," it is deflected by Coriolis acceleration so as to move in a circular path, counterclockwise in the northern hemisphere and clockwise in the southern hemisphere.

The Coriolis rule of thumb may be validated by inspection of Fig. 4.16. Both $v = dz/dt$ and $\omega = d\theta/dt$ must exist in order for the Coriolis component to appear in the acceleration expression. If $dz/dt = 0$, only the rigid-link tangential and normal acceleration components survive the two derivatives. In the case $d\theta/dt = 0$, only the sliding acceleration is present.

Figure 4.16 also includes a "shock" derivative development. The authors suggest that the terms

$$[(-3v\omega^2) + i(3a\omega + 3v\alpha)]\hat{u}$$

form a group that can be referred to as "Coriolis shock" or "Coriolis jerk," since they depend on simultaneous nonzero derivatives of both the angular and radial motions, just like the Coriolis acceleration. Chapter 6 discusses the importance of minimizing the derivatives of acceleration for cam applications.

An alternative proof for the existence of Coriolis acceleration is as follows. The slider of Fig. 4.15 is shown in two positions in Fig. 4.19. The movement of point P is separated into several submovements. Point P as part of the slide bar (link 2) moves from P_2 to P_2'. The accelerations associated with this movement are the rigid link normal and tangential accelerations. Then point P as part of the slider (link 3) moves from P_3 to A (which, for a small angle $d\theta$, is equal in distance to the movement from P_2' to P_3').

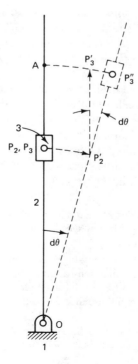

Figure 4.19 An aid to visualization of the origin of the Coriolis acceleration: arc $P_3'P_3''$ exists only if there is both sliding along the bar and rotation with the bar.

The sliding acceleration is associated with this movement. With only these three acceleration components, we are left short of our goal, namely P_3''.

$$\text{arc } P_3' P_3'' = \text{arc } AP_3'' - \text{arc } AP_3' = \text{arc } AP_3'' - \text{arc } P_2 P_2'$$

$$= (OA)\, d\theta - (OP_2)\, d\theta \qquad\qquad (4.24)$$

$$= (P_2' P_3')\, d\theta$$

But we know that $P_2' P_3' \simeq P_2 A = v\, dt$ and $d\theta = \omega_2\, dt$, so that

$$\text{arc } P_3' P_3'' = v\omega_2 (dt)^2 \qquad\qquad (4.25)$$

The right side of Eq. (4.25) has the units of acceleration times $(dt)^2$. In general, an infinitesimal displacement of a point moving along a path,

$$ds = \frac{1}{2} \ddot{s} (dt)^2 \qquad\qquad (4.26)$$

where $\ddot{s}$ is the acceleration along the path. Thus

$$\text{arc } P_3' P_3'' = \frac{1}{2} \ddot{s} (dt)^2 \qquad\qquad (4.27)$$

Combining Eqs. (4.25) and (4.27), the missing acceleration has the scalar value

$$\ddot{s} = 2v\omega_2 = \text{Coriolis contribution} \qquad\qquad (4.28)$$

whose direction must be determined on the basis of Fig. 4.18.

Note that, in the Coriolis term, *the velocity difference factor is a velocity along the original position vector of point P (representing a change in length of that vector), and the angular velocity is that of the position vector with respect to the fixed inertial frame of reference.*

Mechanism designs have been known to fail due to the mistaken omission of the Coriolis component in an acceleration analysis leading to incorrect inertia forces. The following examples contain analyses in which the Coriolis component is present. It is noteworthy that the complex-number treatment provides the magnitude and direction of the Coriolis term "automatically," without resorting to a rule of thumb for determining its direction and whether or not it is present. The graphical approach requires determination of direction as illustrated in Figure 4.16.

Example 4.3

Let us determine the angular acceleration of link 4 of the inverted slider crank mechanism of Fig. 4.20a, given the peripheral speed of the crank pin, $V_{P2} = 40$ cm/sec (constant angular velocity), as shown.

Solution   Three methods* for solving this example will be described; the first two are graphical (but in different directions around the loop), and the third is analytical. Both Table 3.1 and Fig. 4.16 will be useful as guides in setting up the correct equations. Figure 4.21 shows how we might visualize the relative and difference motion for this example.

---

* The student may want to concentrate on only one method in this and other examples of Coriolis acceleration to avoid possible undue confusion in the early stages of the learning process.

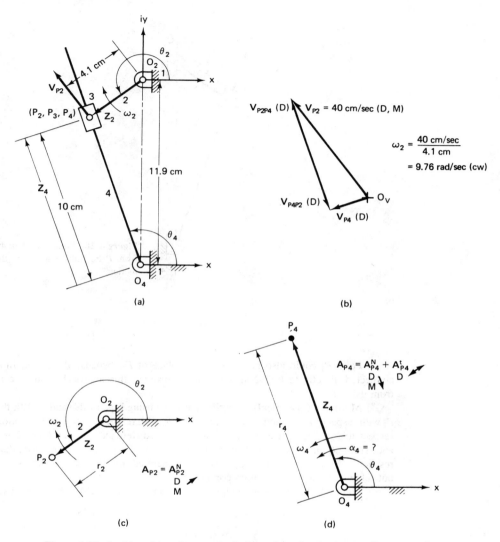

Figure 4.20 In (b) and in subsequent velocity and acceleration vector diagrams and equations, it is useful to enter (D, M) for a vector known both in magnitude and direction, and (D) where only the direction is known prior to the construction of the diagram.

Consider point P, the center of the crank pin, in the position shown in Fig. 4.20. At that instant, three points coincide at P: P_2 on link 2, P_3 on link 3, and P_4 on link 4 (Fig. 4.21). As the mechanism moves, P_2 and P_3 remain coincident, but P_4 moves apart from these, with relative velocity $\mathbf{V}_{P4P2}$ and relative acceleration $\mathbf{A}_{P4P2}$. The velocity and acceleration equations can be written in two ways:

$$\mathbf{V}_{P2} = \mathbf{V}_{P4} + \mathbf{V}_{P2P4} \tag{4.29}$$

$$\mathbf{A}_{P2} = \mathbf{A}_{P4} + \mathbf{A}_{P2P4} \tag{4.30}$$

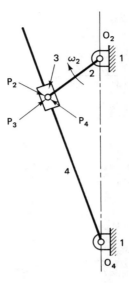

Figure 4.21 Point P on link 4 is separated from P on links 2 and 3 in the course of motion.

or

$$V_{P4} = V_{P2} + V_{P4P2} \qquad (4.31)$$

$$A_{P4} = A_{P2} + A_{P4P2} \qquad (4.32)$$

Point P_3 could also be used here in place of P_2, because the slider, link 3, is pin-connected directly to link 2 at this point. However, this approach would be no different from using P_2.

Method 1 (graphical):   Notice that when using Eqs. (4.29) and (4.30), the motion of P_2 with respect to O_2 is equal to the motion of P_4 with respect to O_4 plus the motion of P_2 with respect to P_4. The first and second motions are difference motions (case 2 in Table 3.1) and, since O_2 and O_4 are fixed, they are absolute motions. The third motion, that of P_2 with respect to P_4, is relative (case 3). The path generated in difference motion is a circular arc, but relative motion can be more complex.

Figure 4.20b shows the graphical solutions of Eq. (4.29), where we observe that

$$V_{P4} = Z_4\omega_4 i e^{i\theta_4} = (12 \text{ cm/sec}) i e^{i\theta_4}$$

$$V_{P4P2} = (38 \text{ cm/sec}) (-e^{i\theta_4})$$

and

$$\omega_4 = \frac{V_{P4}}{i Z_4} = \frac{12 \text{ cm/sec}}{10 \text{ cm}} = 1.2 \text{ rad/sec ccw}$$

Solution of Eq. (4.30) will take more thought. Expanding the total acceleration terms in Eq. (4.30) *may yield up to four components* (derived in Fig. 4.16): the *normal, tangential, sliding,* and the *Coriolis* accelerations for each term. Since Eq. (4.30) is made up of terms that represent only case 1, 2, and 3 motions, none of the terms will contain all four acceleration components. (The complex-number analysis that yielded the equations in Fig. 4.16 involved case 4 motion.)

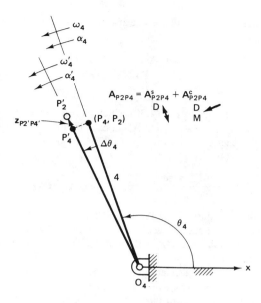

Figure 4.22 Visualization of how the Coriolis acceleration originates in the motion of P as a point of link 2 with respect to P as a point of link 4 of Fig. 4.21.

The motion of point P as part of 2 is constrained to a fixed radius at constant peripheral speed, so that only the normal term is present in $\mathbf{A}_{P2}$ (Fig. 4.20c). As for $\mathbf{A}_{P4}$, since $(O_4 P_4)$ is constant, the sliding and Coriolis terms are zero (Fig. 4.20d). In the tangential term, the direction is known, but the magnitude is not, because α_4 is as yet unknown. The normal term is known.

Solving for $\mathbf{A}_{P2P4}$, however, is more complicated. In Fig. 4.22, P_4 and P_2 coincide initially. In order to visualize the motion of P_2 with respect to P_4, we apply a small "virtual" displacement represented by the rotation $\Delta\theta_4$, after which P_4' and P_2' are separated by $z_{P_2'P_4'}$. Focusing on the $\theta_4 + \Delta\theta_4$ position, we have

$$z_{P_2'P_4'} = r_{P_2'P_4'}e^{i(\theta_4+\Delta\theta_4)}$$

$$\dot{z}_{P_2'P_4'} = \dot{r}_{P_2'P_4'}e^{i(\theta_4+\Delta\theta_4)} + r_{P_2'P_4'}\omega_4'ie^{i(\theta_4+\Delta\theta_4)}$$

$$\ddot{z}_{P_2'P_4'} = \ddot{r}_{P_2'P_4'}e^{i(\theta_4+\Delta\theta_4)} + 2\dot{r}_{P_2'P_4'}\omega_4'ie^{i(\theta_4+\Delta\theta_4)} + r_{P_2'P_4'}\alpha_4'ie^{i(\theta_4+\Delta\theta_4)} - r_{P_2'P_4'}\omega_4'^2e^{i(\theta_4+\Delta\theta_4)}$$

Now, to refer these results back to position P_4, P_2, we let the virtual displacement $\Delta\theta_4 \to 0$. Then

$$\ddot{z}_{P2P4} = \ddot{r}_{P2P4}e^{i\theta_4} + 2\dot{r}_{P2P4}\omega_4ie^{i\theta_4} + \underset{0}{r_{P2P4}\alpha_4ie^{i\theta_4}} - \underset{0}{r_{P2P4}\omega_4^2e^{i\theta_4}}$$

or

$$\mathbf{A}_{P2P4} = \underset{\substack{D\searrow\\M}}{\mathbf{A}^s_{P2P4}} + \underset{\substack{D\swarrow\\0}}{\mathbf{A}^c_{P2P4}} + \underset{0}{\mathbf{A}^t_{P2P4}} + \underset{0}{\mathbf{A}^n_{P2P4}} \tag{4.33}$$

In the limit $\Delta\theta_4 \to 0$ the last two terms also go to zero, because $r_{P2P4} \to 0$ (i.e., there is zero distance between P_2 and P_4 after the virtual displacement is set to zero). Equation (4.30) can now be expanded by inserting the results of our analysis thus far (entering on the right side only the nonzero components).

$$\mathbf{A}_{P2}^{n} \quad + \quad \mathbf{A}_{P2}^{t} \quad = \quad \mathbf{A}_{P4}^{n} \quad + \quad \mathbf{A}_{P4}^{t} \quad + \quad \mathbf{A}_{P2P4}^{s} \quad + \quad \mathbf{A}_{P2P4}^{c}$$

$$D \nearrow \qquad 0 \qquad D \searrow \qquad D \nearrow \qquad D \updownarrow \qquad D \swarrow \qquad \text{(4.33a)}$$

$$M \qquad 0* \qquad M \qquad\qquad\qquad M$$

Equation (4.33a) is the expanded form of Eq. (4.30) which has only two unknowns, the *directed magnitudes* of $\mathbf{A}_{P4}^{t}$ and $\mathbf{A}_{P2P4}^{s}$ (i.e., A_{P4}^{t} and A_{P2P4}^{s}, each is either positive or negative).

Figure 4.23 shows the vector solution to Eq. (4.33a). From this figure the angular acceleration of link 4 is $\mathbf{A}_{P4}^{t} = (461 \text{ cm/sec}^2)(-ie^{i\theta_4}) = i\alpha_4 \mathbf{Z}_4$.

$$\alpha_4 = \frac{\mathbf{A}_{P4}^{t}}{i\mathbf{Z}_4} = \frac{-461 \text{ cm/sec}^2}{10 \text{ cm}} = -46.1 \text{ rad/sec}^2 \text{ (cw)} \qquad \text{(4.33b)}$$

For the analytical version of this solution, see Prob. 4.44(a).

Method 2 (graphical): The angular acceleration α_4 may also be derived by using Eqs. (4.31) and (4.32). The velocity analysis will yield the same results as above except that $\mathbf{V}_{P4P2} = -\mathbf{V}_{P2P4}$ (see Fig. 4.20b).

The components of the relative acceleration expression (4.32) should be expanded using the same thought process as described in the first method. The absolute acceleration components $\mathbf{A}_{P2}$ and $\mathbf{A}_{P4}$ will be identical to those of Eq. (4.33a).

To find the relative acceleration $\mathbf{A}_{P4P2}$, it is only necessary to reverse the sense of the relative position vector in Fig. 4.22, thus obtaining $\mathbf{z}_{P'4P'2}$. Then, following through the derivation similar to that for $\mathbf{A}_{P2P4}$, we obtain

$$\ddot{\mathbf{z}}_{P4P2} = \ddot{r}_{P4P2}e^{i(\theta_4+\pi)} + 2\dot{r}_{P4P2}\omega_4 ie^{i(\theta_4+\pi)} + r_{P4P2}\alpha_4 ie^{i(\theta_4+\pi)} - r_{P4P2}\omega_4^2 e^{i(\theta_4+\pi)}$$

$$0 \qquad\qquad 0$$

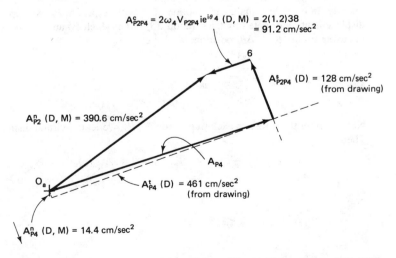

$$A_{P2P4}^{c} = 2\omega_4 V_{P2P4}ie^{i\theta_4} \text{ (D, M)} = 2(1.2)38$$
$$= 91.2 \text{ cm/sec}^2$$

6

A_{P2P4}^{s} (D) = 128 cm/sec^2
(from drawing)

A_{P2}^{n} (D, M) = 390.6 cm/sec^2

A_{P4}

A_{P4}^{t} (D) = 461 cm/sec^2
(from drawing)

O_a

A_{P4}^{n} (D, M) = 14.4 cm/sec^2

Figure 4.23 Vector diagram of the solution of Eq. (4.33a) (see Figs. 4.20–4.22).

* Because $\alpha_2 = 0$.

or

$$\mathbf{A}_{P4P2} = \mathbf{A}^s_{P4P2} \; + \; \mathbf{A}^c_{P4P2} \; + \; \mathbf{A}^t_{P4P2} \; + \; \mathbf{A}^n_{P4P2}$$

$$\qquad\qquad \mathrm{D}\searrow \qquad \mathrm{D}\nearrow \qquad 0 \qquad\quad 0$$

$$\qquad\qquad \mathrm{M} \qquad\quad 0 \qquad\quad 0$$

It is to be noted that,* as in Fig. 4.22 for $\mathbf{A}_{P2P4}$, we used the original configuration of the mechanism for finding the relative acceleration of point P_4 with respect to P_2. Using an inversion of the mechanism (as is done in some texts) would be incorrect, because accelerations are to be used for finding dynamic (inertial) forces in a mechanism, and must therefore be expressed in the fixed coordinate system, embedded in the fixed inertial frame of reference, which in this case is link 1. This means, for example, that it would be wrong to fix link 2 and express $\mathbf{A}_{P4P2}$ in a coordinate system embedded in link 2. The reason for this is that, although the magnitudes of the relative displacement vector $\mathbf{Z}_{P4P2}$ are invariant under inversion, the rotations of this vector are not. Since acceleration terms (other than the sliding acceleration) depend on the angular velocity and angular acceleration of the rotation of this vector, the resulting values would not be applicable in the original configuration of the mechanism, in which the fixed link is link 1.

Getting back to our method 2, using the results obtained above for $\mathbf{A}_{P4P2}$, we can expand Eq. (4.32) as follows:

$$\mathbf{A}^n_{P4} \; + \; \mathbf{A}^t_{P4} \; = \; \mathbf{A}^n_{P2} \; + \; \mathbf{A}^t_{P2} \; + \; \mathbf{A}^n_{P4P2} \; + \; \mathbf{A}^t_{P4P2} \; + \; \mathbf{A}^c_{P4P2} \; + \; \mathbf{A}^s_{P4P2}$$

$$\quad \mathrm{D}\searrow \quad\;\; \mathrm{D}\swarrow \quad\;\; \mathrm{D}\nearrow \qquad 0 \qquad\quad 0 \qquad\quad 0 \qquad\;\; \mathrm{D}\nearrow \quad\;\; \mathrm{D}\searrow \qquad (4.34)$$

$$\quad \mathrm{M} \qquad\qquad\quad\; \mathrm{M} \qquad\; 0^\dagger \qquad\quad 0 \qquad\quad 0 \qquad\;\; \mathrm{M}$$

For the analytical version of method 2, see Prob. 4-44(b).

Proof of Equivalence of Methods 1 and 2

A comparison of Eq. (4.33a) and (4.34) is in order here to prove equivalence of these two approaches. Since the normal and tangential components of the $\mathbf{A}_{P2}$ and $\mathbf{A}_{P4}$ terms appear in both equations, they can be eliminated by adding Eqs. (4.33a) and (4.34). Omitting zero terms, this yields

$$\mathbf{A}^s_{P2P4} \; + \; \mathbf{A}^c_{P2P4} \; + \; \mathbf{A}^s_{P4P2} \; + \; \mathbf{A}^c_{P4P2} = 0 \tag{4.35}$$

or

$$\ddot{r}_{P2P4}e^{i\theta_4} + 2\dot{r}_{P2P4}\omega_4 i e^{i\theta_4} + \ddot{r}_{P4P2}e^{i(\theta_4+\pi)} + 2\dot{r}_{P4P2}\omega_4 i e^{i(\theta_4+\pi)} = 0 \tag{4.36}$$

The fact that Eq. (4.36) is true can be verified as follows: $\dot{r}_{P2P4}$ is the relative speed of P_2 with respect to P_4. Since P_2 is separating from P_4, the distance is increasing, and therefore $\dot{r}_{P2P4}$ is positive. By the same token, $\dot{r}_{P4P2}$ is also positive. It can be shown similarly that $\ddot{r}_{P2P4}$ and $\ddot{r}_{P4P2}$ are either both positive or both negative. Furthermore, $\dot{r}_{P2P4}$ and $\dot{r}_{P4P2}$ are equal in magnitude, which is also true of $\ddot{r}_{P2P4}$ and $\ddot{r}_{P4P2}$. Thus

$$\dot{r}_{P2P4} = \dot{r}_{P4P2} \quad \text{and} \quad \ddot{r}_{P2P4} = \ddot{r}_{P4P2}$$

However, $e^{i(\theta_4+\pi)} = -e^{i\theta_4}$.

---

* The student may omit this paragraph at first reading.

$\dagger$ Because $\alpha_2 = 0$.

This completes our proof that Eq. (4.36) is true, and therefore methods 1 and 2 are equivalent. Q.E.D.

The example above provides additional insight into the thought process that one must follow in order to attack a linkage acceleration problem correctly. The steps may be summarized as follows:

1. Initially, one must keep Table 3.1 in mind when writing the velocity and acceleration equations so that *only case 2 and case 3 analyses need be performed.* The equations should be written so that the relative accelerations *take advantage of the physical constraints of the mechanism* — avoid the situation where the path of one point with respect to the other point's reference frame is not known by inspection.

2. Second, with each total acceleration term one should check to see which component (the normal, tangential, sliding, and Coriolis acceleration) should be included in the analysis. In general, the existence of these components follows the following rules: The normal acceleration vanishes when the radius of curvature of the path is infinite; the tangential acceleration is zero when the angular acceleration is zero; and the sliding acceleration component depends on straight-line motion, while the Coriolis component depends on both a lengthening of the radius vector (describing the path with respect to the fixed, inertial coordinate system) and a rotation of that vector also with respect to the fixed, inertial coordinate system. When analyzing lower-pair connections, one at a time, at least two of the four acceleration components are zero.

Method 3 (analytical): It should also be pointed out for the sake of computer-oriented engineers, that the problem of Example 4.3 can be solved by way of vector loop equations: the equation of closure and its first and second derivatives. This method is more readily programmed for digital computation and it is more accurate than the foregoing semigraphical methods, especially in the acceleration terms (although the visual feedback of the graphical methods for full initial understanding of these types of complex problems is very helpful in the learning process).

The procedure goes as follows.

Step 1.   Set up the vector model of the mechanism (Fig. 4.24).

Step 2.   Write the equation of closure

$$\mathbf{z}_4 = \mathbf{z}_1 + \mathbf{z}_2, \tag{4.37}$$

or in exponential form

$$z_4 e^{i\theta_4} = i z_1 + z_2 e^{i\theta_2} \tag{4.38}$$

Step 3.   Calculate θ_2 and θ_4 by the cosine rule. The result is

$$\theta_2 = 216.78°$$

$$\theta_4 = 109.17°$$

Step 4.   Take the time derivative of Eq. (4.38):

$$\dot{z}_4 e^{i\theta_4} + z_4 i \omega_4 e^{i\theta_4} = z_2 i \omega_2 e^{i\theta_2} \tag{4.39}$$

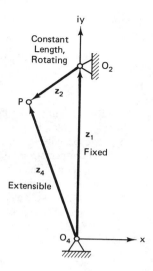

Figure 4.24 Vector model of the mechanism of Fig. 4.20a.

You will recognize the first term as the sliding velocity of the slider, the second term as its tangential velocity, and the third term as the tangential velocity of the crank pin.

Step 5.   Separate the real and imaginary parts of Eq. (4.39):

$$\mathscr{R}[\text{Eq. (4.39)}]: \dot{z}_4 \cos \theta_4 - z_4 \omega_4 \sin \theta_4 = -z_2 \omega_2 \sin \theta_2$$

$$\mathscr{I}[\text{Eq. (4.39)}]: \dot{z}_4 i \sin \theta_4 + z_4 i \omega_4 \cos \theta_4 = i z_2 \omega_2 \cos \theta_2$$

Step 6.   Cancel i in the imaginary part of Eq. (4.39). Now we have a system of two equations, linear and nonhomogeneous in the two unknown reals, $\dot{z}_4$ and ω_4. Solve the system simultaneously for $\dot{z}_4$ and ω_4, say, by way of determinants. The results are:

$$\dot{z}_4 = 38.13 \text{ cm/sec},$$

$$\omega_4 = 1.21 \text{ rad/sec, positive, therefore } (ccw)$$

These agree with the results of *Method 1*.

Step 7.   Next, simply take the time derivative of Eq. (4.39).

$$\ddot{z}_4 e^{i\theta_4} + \dot{z}_4 i \omega_4 e^{i\theta_4} + \dot{z}_4 i \omega_4 e^{i\theta_4} + z_4 i \alpha_4 e^{i\theta_4} - z_4 \omega_4^2 e^{i\theta_4} = -z_2 \omega_2^2 e^{i\theta_2} \qquad (4.40)$$

You will recognize the first term as the sliding acceleration of the slider (link 3) with respect to the slide bar (link 4), the second and third terms together as the Coriolis acceleration of the slider with respect to the slide bar, the fourth term as the tangential acceleration of the slide bar, the fifth term as the centrifugal acceleration of the slide bar, and the sixth term, which is on the right side of the equation, as the centrifugal acceleration of the crank pin of link 2.

Step 8.   Separate the real and imaginary parts of Eq. (4.40).

$$\mathscr{R}[\text{Eq. (4.40)}]: \ddot{z}_4 \cos \theta_4 - 2\dot{z}_4 \omega_4 \sin \theta_4 - z_4 \alpha_4 \sin \theta_4 - z_4 \omega_4^2 \cos \theta_4 = -z_2 \omega_2^2 \cos \theta_2$$

$$\frac{1}{i} \mathscr{I}[\text{Eq. (4.40)}]: \ddot{z}_4 \sin \theta_4 + 2\dot{z}_4 \omega_4 \cos \theta_4 + z_4 \alpha_4 \cos \theta_4 - z_4 \omega_4^2 \sin \theta_4 = -z_2 \omega_2^2 \sin \theta_2$$

This is a system of two equations, linear and nonhomogeneous in the two unknown reals, $\ddot{z}_4$ and α_4. Solving by determinants yields

$$\ddot{z}_4 = 132.66 \text{ cm/sec}^2, \quad \text{and}$$

$$\alpha_4 = -46.42 \text{ rad/sec}^2 \text{ (cw)}.$$

Thus the problem is solved.

Note that the value of α_4 differs slightly from the result of method 1. This is due to the greater accuracy of the computer-oriented loop-equation procedure, as compared to the previous semigraphical approaches, especially in the acceleration terms. The reader is encouraged to follow along this procedure performing the arithmetic, and convince himself/herself of the accuracy of the loop-equation method, which avoids the pitfalls of picking the wrong angular or linear velocity for the calculation of the Coriolis term.

It is interesting to note that Fig. 4.20a could also represent the geneva-wheel* problem if ω_4 were a given constant, and the angular acceleration of link 2 were sought. In that case, the forms of Eqs. (4.37), (4.38) and (4.39) would remain unchanged, but the unknowns in Eq. (4.39) would be $\dot{z}_4$ and ω_2. However, Eq. (4.40) would become:

$$\ddot{z}_4 e^{i\theta_4} + 2\dot{z}_4 i\omega_4 e^{i\theta_4} - z_4\omega_4^2 e^{i\theta_4} = z_2 i\alpha_2 e^{i\theta_2} - z_2\omega_2^2 e^{i\theta_2}, \tag{4.41}$$

containing the two different unknown reals, $\ddot{z}_4$ and α_2.

Example 4.4

The mechanism shown in Fig. 4.25 has been employed as a marine steering linkage called a Rapson's slide. O_2B represents the tiller and AC is the actuating rod. If the absolute velocity of the actuating rod is 30 cm/sec (constant) to the left, find the angular acceleration of the tiller in the position shown.

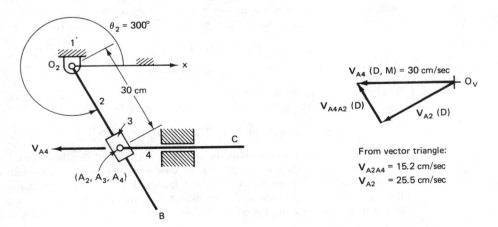

Figure 4.25 Graphical velocity analysis of marine steering linkage called Rapson's slide.

---

* See Fig. P4.40 representing a Geneva indexing mechanism.

Graphical Solution*   Again, the relative velocity and acceleration equation will be written relating A_4 and A_2 since the paths of these points are known. The velocity equation will be written so that the relative velocity of A_4 is with respect to A_2 (which is a point on the tiller):

$$\mathbf{V}_{A4} = \mathbf{V}_{A2} + \mathbf{V}_{A4A2} \tag{4.42}$$

Equation (4.42) is solved graphically in Fig. 4.25. The total acceleration equation is

$$\mathbf{A}_{A4} = \mathbf{A}_{A2} + \mathbf{A}_{A4A2} \tag{4.43}$$

Expanding each total acceleration to include appropriate components yields (see Figs. 4.26 and 4.27)

$$
\underbrace{\cancel{\mathbf{A}_{A4}^{n}} + \cancel{\mathbf{A}_{A4}^{t}}}_{(a)} = \underbrace{\mathbf{A}_{A2}^{n} + \mathbf{A}_{A2}^{t}}_{(b)} + \underbrace{\cancel{\mathbf{A}_{A4A2}^{n}} + \mathbf{A}_{A4A2}^{s} + \mathbf{A}_{A4A2}^{c}}_{(c)} \tag{4.44}
$$

$$
\begin{array}{ccccc}
0 & 0 & D\nwarrow & D\nearrow & 0 & D\searrow & D\nearrow \\
R=\infty & \alpha_4=0 & M & & r_{A4A2}=0 & & M \\
& & \dfrac{V_{A2}^2}{O_2A} & & & & 2V_{A4A2}\omega_2
\end{array}
$$

(a) Also, $\mathbf{A}_{A4} = 0$, because $\mathbf{V}_{A4} = $ constant both in magnitude and direction.
(b) Referring to Fig. 4.25, we have

$$\ddot{\mathbf{z}}_2 = \underset{0}{\ddot{r}_2 e^{i\theta_2}} + \underset{0}{2\dot{r}_2\omega_2 i e^{i\theta_2}} + r_2\alpha_2 i e^{i\theta_2} - r_2\omega_2^2 e^{i\theta_2} \tag{4.44a}$$

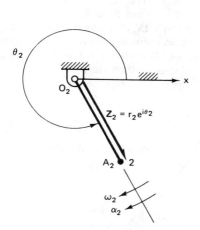

Figure 4.26 Vector notation associated with point A on link 2 of the mechanism in Fig. 4.24.

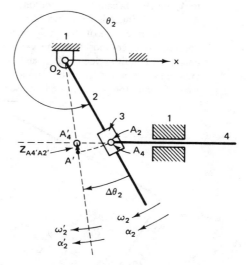

Figure 4.27 Relative motion of A_4 with respect to A_2 in marine steering linkage.

* See Example 4.4a for the analytical solution.

where $\omega_2 = \mathbf{V}_{A2}/i\mathbf{Z}_2$.

$$\mathbf{A}_{A2} = \mathbf{A}_{A2}^{s} + \mathbf{A}_{A2}^{c} + \mathbf{A}_{A2}^{t} \quad + \mathbf{A}_{A2}^{n}$$
$$\begin{array}{cccc} 0 & 0 & D\nearrow & D\nwarrow \\ 0 & 0 & M & \end{array} \qquad (4.44\text{b})$$

$$\omega_2 = -\frac{25.5}{30} = -0.85 \text{ rad/sec}, \qquad \omega_2^2 = 0.72 \text{ rad/sec}^2$$

$$\mathbf{A}_{A2}^{n} = -(30)(0.72)e^{i\theta_2} = (-21.68 \text{ cm/sec}^2)e^{i\theta_2}\nwarrow$$

(c) Referring to Fig. 4.26, we have

$$\ddot{\mathbf{Z}}_{A'4A'2} = \ddot{r}_{A'4A'2}(-e^{i(\theta_2+\Delta\theta_2)}) + 2\dot{r}_{A'4A'2}\omega_2'(-ie^{i(\theta_2+\Delta\theta_2)})$$
$$- r_{A'4A'2}\alpha_2'ie^{i(\theta_2+\Delta\theta_2)} + r_{A'4A'2}\omega_2'^2 e^{i(\theta_2+\Delta\theta_2)} \qquad (4.44\text{c})$$

Letting $\Delta\theta_2 \to 0$, then $r_{A'4A'2} \to r_{A4A2} = 0$, we get

$$\ddot{z}_{A4A2} = -\ddot{r}_{A4A2}e^{i\theta_2} - 2\dot{r}_{A4A2}\omega_2 ie^{i\theta_2} \qquad (4.44\text{d})$$

$$\mathbf{A}_{A4A2} = \mathbf{A}_{A4A2}^{s} + \mathbf{A}_{A4A2}^{c}$$
$$\begin{array}{cc} D\searrow & D\nearrow \\ & M \end{array} \qquad (4.44\text{e})$$

$$\mathbf{A}_{A4A2}^{c} = -2(+15.2)(-0.85)ie^{i\theta_2}$$
$$= (25.84 \text{ cm/sec}^2)ie^{i\theta_2}$$

There are several zero terms in the acceleration expression of Eq. (4.44), most notably: The normal accelerations of point A as a point of link 4, A_4, and as a point of link 2, A_2, relative to A_4 are zero because of the reasons given under Eq. (4.44); the tangential acceleration of A as part of link 4 is zero since the actuating rod has constant velocity.

The direction of the Coriolis component is shown under Eq. (4.44). The Coriolis component is composed of the velocity difference ($\mathbf{V}_{A4A2}$), which expresses the rate of change in length of vector $\mathbf{Z}_{A4A2}$ and ω_2, which identifies the rate of change in angular orientation of vector $\mathbf{Z}_{A4A2}$. The solution to Eq. (4.44) is shown in Fig. 4.28. Since $\mathbf{A}_{A2}^{t} = r_2\alpha_2 ie^{i\theta_2}$, therefore $\alpha_2 = (25.84 \text{ cm/sec}^2)/-30 \text{ cm} = -0.86 \text{ rad/sec}^2$ (cw).

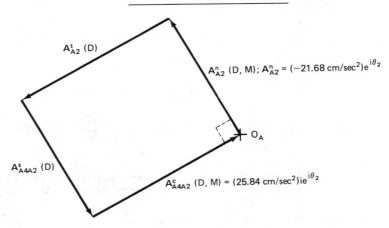

A_{A2}^{t} (D)

A_{A2}^{n} (D, M); $A_{A2}^{n} = (-21.68 \text{ cm/sec}^2)e^{i\theta_2}$

O_A

A_{A4A2}^{s} (D)

A_{A4A2}^{c} (D, M) = $(25.84 \text{ cm/sec}^2)ie^{i\theta_2}$

Figure 4.28   Graphical solution of vector Eq. (4.44).

Some of the foregoing essential considerations in Exs. 4.3 and 4.4 may seem difficult to visualize to some readers. However, they can be followed by careful analysis of the point-to-point position vectors and their rates of change in the course of motion of the mechanism. This is illustrated in Figs. 4.20c, 4.20d, 4.22, 4.26, and 4.27, together with their accompanying complex-number equations.

Also, while the sense of acceleration components is difficult to visualize by inspection, the complex-number approach makes it easy and foolproof. For example, rewriting Eq. (4.44) in complex vector form clearly displays the directions of all terms:

$$\mathbf{A}_4 = 0 = \mathbf{A}_{A2}^n \quad + \quad \mathbf{A}_{A2}^t \quad + \quad \mathbf{A}_{A4A2}^s \quad + \quad \mathbf{A}_{A4A2}^c$$

$$\mathbf{A}_4 = 0 = r_2\omega_2^2(-e^{i\theta_2}) + r_2\alpha_2(ie^{i\theta_2}) + \ddot{r}_{A4A2}(e^{i\theta_2}) + 2\dot{r}_{A4A2}\omega_2(-ie^{i\theta_2})$$

$$(4.45)$$

Since r_2 is an absolute value, $r_2\omega_2^2 > 0$, and $(-e^{i\theta_2})$ points from A_2 to O_2. Refer to Figs. 4.26 and 4.28 [the latter a graphical solution of Eqs. (4.44) and (4.45)].

Sense depends on sign of α_2, which is unknown; direction is perpendicular to link 2 (the slide). From Fig. 4.28: sense is $(-ie^{i\theta_2})$; therefore, $\alpha_2 < 0$ (i.e., cw).

Sense depends on sign of $\ddot{r}_{A4A2}$ which is unknown; direction is along link 2. From Fig. 4.28: sense is $e^{i\theta_2}$.

Sense is $ie^{i\theta_2}$ because $\omega_2 < 0$ (i.e., cw) and $\dot{r}_{A4A2} > 0$.

For full analytical verification of this example, see Example 4.4a below.

If we reverse the arrows on the point-to-point position vectors in Fig. 4.28 and analyze the components of $\mathbf{A}_{A2A4}$ (see Prob. 4.2), we find that

$$\mathbf{A}_{A2A4} = -\mathbf{A}_{A4A2} \qquad (4.46)$$

which shows that this method is equivalent to that of Eq. (4.43).

Example 4.4a: Analytical Solution of Example 4.4

Step 1. To draw the vector model of the mechanism of Fig. 4.25, establish the frame vector connecting O_2 with the fixed slide, say, $\mathbf{z}_1 = r_1 e^{i\theta_1}$ where θ_1 is, say, $-45°$ (see Fig. 4.29). Therefore

Step 2. The loop equation of closure is

$$\mathbf{z}_2 = \mathbf{z}_1 + \mathbf{z}_4$$

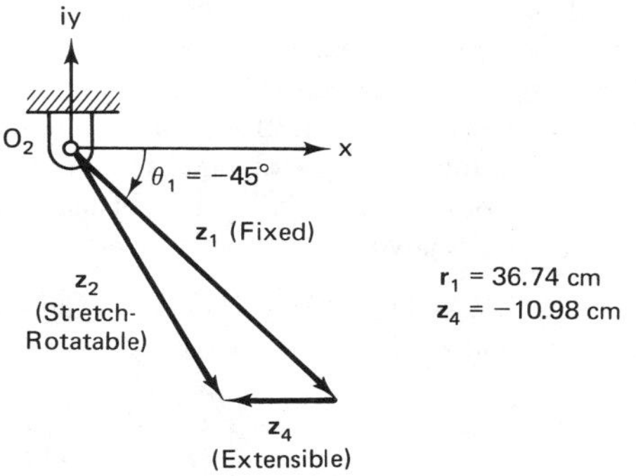

iy

O_2

$\theta_1 = -45°$

x

$\mathbf{z}_1$ (Fixed)

$\mathbf{z}_2$
(Stretch-Rotatable)

$\mathbf{z}_4$
(Extensible)

$r_1 = 36.74$ cm
$z_4 = -10.98$ cm

Figure 4.29 Vector model of the mechanism of Figure 4.25.

or in exponential form

$$\dot{r}_2 e^{i\theta_2} = r_1 e^{i\theta_1} + r_4 e^{i\theta_4} \tag{4.47}$$

Step 3.   The time derivative of Eq. (4.47) is

$$\dot{r}_2 e^{i\theta_2} + r_2 i \omega_2 e^{i\theta_2} = 0 + \dot{r}_4 e^{i\pi} + 0 = -\dot{r}_4 \tag{4.47a}$$

Since $\dot{r}_4$ is given as the constant 30 cm/sec leftward, Eq. (4.47a) can be solved for the two unknown reals, $\dot{r}_2$ and ω_2.

Step 4.   To this end, we separate the real and imaginary parts.

$$\mathcal{R}[\text{Eq. (4.47a)}: \dot{r}_2 \cos \theta_2 - r_2 \omega_2 \sin \theta_2 = -\dot{r}_4$$

$$\frac{1}{i} \mathcal{I}[\text{Eq. (4.47a)}: \dot{r}_2 \sin \theta_2 + r_2 \omega_2 \cos \theta_2 = 0$$

Substituting known values, we see that this is a system of two equations, linear and nonhomogeneous in $\dot{r}_2$ and ω_2. Solving simultaneously, say, by determinants, we get

$$\dot{r}_2 = -15 \text{ cm/sec}, \quad \text{and} \quad \omega_2 = -0.866 \text{ rad/sec}.$$

Step 5.   Now take the time derivative of Eq. (4.47a).

$$\ddot{r}_2 e^{i\theta_2} + \dot{r}_2 i \omega_2 e^{i\theta_2} + \dot{r}_2 i \omega_2 e^{i\theta_2} + r_2 i \alpha_2 e^{i\theta_2} - r_2 \omega_2^2 e^{i\theta_2} = 0 \tag{4.47b}$$

There are now two unknown reals $\ddot{r}_2$ and α_2.

Step 6.   Separating real and imaginary parts.

$$\mathcal{R}[\text{Eq. (4.47b)}: \ddot{r}_2 \cos \theta_2 - 2\dot{r}_2 \omega_2 \sin \theta_2 - r_2 \alpha_2 \sin \theta_2 - r_2 \omega_2^2 \cos \theta_2 = 0$$

$$\frac{1}{i} \mathcal{I}[\text{Eq. (4.47b)}: \ddot{r}_2 \sin \theta_2 + 2\dot{r}_2 \omega_2 \cos \theta_2 + r_2 \alpha_2 \cos \theta_2 - r_2 \omega_2^2 \sin \theta_2 = 0$$

This is a system of two equations, linear and nonhomogeneous in the above two unknown reals. Solving simultaneously, we get the results

$$\ddot{r}_2 = -33.75 \text{ cm/sec}^2 \quad \text{and}$$

$$\alpha_2 = -0.866 \text{ rad/sec}^2.$$

Example 4.5

Figure 4.30 shows a Whitworth quick-return mechanism which is used in shapers. It is required that the tool holder (link 6) maintain a nearly constant velocity during the cutting stroke but accelerate on the return stroke in order to reduce idle time.

Variations of link geometry will of course affect the timing and performance of this mechanism. For the geometry shown in Fig. 4.30 and the given velocity polygon (with $V_{A2} = 6$ m/sec, $\alpha_2 = 0$ as inputs), find the acceleration of the tool holder. The variation in the magnitude of A_C is a measure of compliance with the constant-velocity requirement of this mechanism. Both graphical and analytical solutions follow.

Graphical Solution   There are two kinematic loops in this six-link mechanism which will require two sets of velocity and acceleration equations.

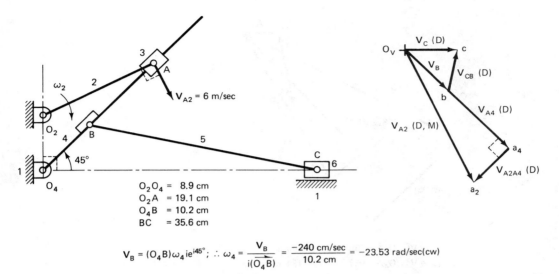

$$V_B = (O_4B)\omega_4\,ie^{i45°}; \quad \therefore \omega_4 = \frac{V_B}{i(\overrightarrow{O_4B})} = \frac{-240 \text{ cm/sec}}{10.2 \text{ cm}} = -23.53 \text{ rad/sec(cw)}$$

Figure 4.30 Six-link quick-return mechanism and its velocity vector polygon.

From the given velocity polygon of Fig. 4.30, the following velocity magnitudes are observed:

$$V_B = 2.40 \text{ m/sec}, \qquad V_{A2A4} = 2.0 \text{ m/sec}$$

$$V_C = 2.0 \text{ m/sec}, \qquad V_{CB} = 1.7 \text{ m/sec}$$

The total acceleration equation for the first loop of links 1, 2, 3, and 4 is

$$\mathbf{A}_{A2} = \mathbf{A}_{A4} + \mathbf{A}_{A2A4} \tag{4.48}$$

The sense of $\mathbf{A}_{A2A4}^C$ is determined by complex numbers as follows (see Fig. 4.31):

$$\ddot{\mathbf{Z}}_{A2A4} = \ddot{r}_{A2A4}e^{i(5\pi/4)} + 2\dot{r}_{A2A4}^{(+)}\overset{(-)}{\omega}_4 ie^{i(5\pi/4)}$$

or

$$\mathbf{A}_{A2A4} = \mathbf{A}_{A2A4}^s + \mathbf{A}_{A2A4}^c$$
$$\qquad\qquad\quad D\nearrow \qquad D\nwarrow$$
$$\qquad\qquad\qquad\qquad\ M$$

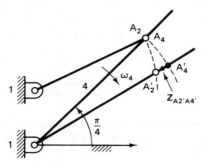

Figure 4.31 Visualization of relative motion of A_2 with respect to A_4.

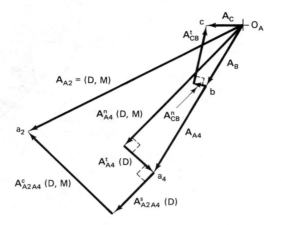

$$A_{A2} = V_{A2}^2/O_2A = 36/0.191 = 188 \text{ m/sec}^2$$
$$A_{A4}^n = V_{A4}^2/O_4A = (5.7)^2/0.243 = 134 \text{ m/sec}^2$$
$$A_{A2A4}^c = 2V_{A2A4}\omega_4 = 2(2.0)(23.53) = 94.12 \text{ m/sec}^2$$

Figure 4.32 Acceleration polygon of the mechanism in Fig. 4.29.

With this, expanding Eq. (4.48) yields

$$\mathbf{A}_{A2}^n \quad + \quad \mathbf{A}_{A2}^t \quad = \mathbf{A}_{A4}^n \quad + \mathbf{A}_{A4}^t \quad + \mathbf{A}_{A2A4}^s \quad + \mathbf{A}_{A2A4}^c$$

$$\text{D}\swarrow \qquad 0 \qquad \text{D}\swarrow \qquad \text{D}\searrow \qquad \text{D}\leftrightarrow \qquad \text{D}\nwarrow$$

$$\text{M} \qquad \alpha_2 = 0 \qquad \text{M} \qquad\qquad\qquad \text{M} \tag{4.49}$$

$$2V_{A2A4}\omega_4 ie^{i225°} = 2(2.0)(-23.53)ie^{i225°}$$

With $A_{A2}^n = V_{A2}^2/O_2A_2 = 188 \text{ m/sec}^2$, $A_{A4}^n = V_{A4}^2/O_4A_4 = 134 \text{ m/sec}^2$, and $A_{A2A4}^c = 94.12 \text{ m/sec}^2$, the acceleration polygons of Fig. 4.32 show that

$$A_B = 56.8 \text{ m/sec}^2$$

The acceleration equation for the second loop is

$$\mathbf{A}_C = \mathbf{A}_B + \mathbf{A}_{CB} \tag{4.50}$$

or

$$\mathbf{A}_C^s \quad = \mathbf{A}_B \quad + \mathbf{A}_{CB}^n + \mathbf{A}_{CB}^t \tag{4.51}$$

$$\text{D}\leftrightarrow \qquad \text{D}\swarrow \quad \text{D}\searrow \qquad \text{D}\nearrow$$

$$\text{M} \qquad \text{M}$$

where $A_{CB}^n = V_{CB}^2/CB = (1.7)^2/0.356 = 8.1 \text{ m/sec}^2$. With this, Fig. 4.32 shows that

$$\mathbf{A}_C = 28.5 \text{ m/sec}^2 \leftarrow$$

Analytical Solution

We start by drawing the vector model of the mechanism of Fig. 4.30, as shown on Fig. 4.33.

Step 1. Since $\mathbf{z}_2$ is the input crank, we begin by solving loop 1. The equation of loop 1 closure is

$$r_2 e^{i\theta_2} = r_4 e^{i\theta_4} - r_1 e^{i\theta_1} \tag{4.52}$$

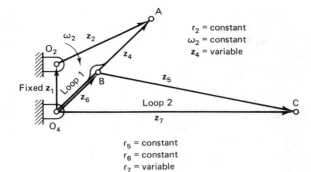

r_2 = constant
ω_2 = constant
z_4 = variable

r_5 = constant
r_6 = constant
r_7 = variable

Figure 4.33 Vector diagram of the mechanism of Figure 4.30.

Separating real and imaginary parts we obtain

$$\mathscr{R}: r_2 \cos \theta_2 = r_4 \cos \theta_4 - r_1 \cos \theta_1$$

$$(1/i)\mathscr{I}: r_2 \sin \theta_2 = r_4 \sin \theta_4 - r_1 \sin \theta_1$$

Squaring both equations we get

$$r_2^2 \cos^2\theta_2 = r_4^2 \cos^2\theta_4 - 2r_4r_1 \cos \theta_4 \cos \theta_1 + r_1^2 \cos^2\theta_1$$

$$r_2^2 \sin^2\theta_2 = r_4^2 \sin^2\theta_4 - 2r_4r_1 \sin \theta_4 \sin \theta_1 + r_1^2 \sin^2\theta_1$$

Adding the two squared equations eliminates θ_2.

$$r_2^2 = r_4^2 - 2r_4r_1(\cos \theta_4 \cos \theta_1 + \sin \theta_4 \sin \theta_1) + r_1^2$$

Rearranging, we obtain a quadratic in r_4, which yields two roots: $r_4 = 24.327$ and -11.740 cm. The negative root applies when link 4 has rotated 180°; therefore, we use the positive root. Substituting this value back into the real part of Eq. (4.52) and solving for $\cos \theta_2$, we get $\cos \theta_2 = 0.90061$, and $\theta_2 = 25.762°$.

Step 2.   Now we take the time derivative of Eq. (4.52)

$$r_2i\omega_2 e^{i\theta_2} = \dot{r}_4 e^{i\theta_4} + r_4 i\omega_4 e^{i\theta_4} \tag{4.52a}$$

There are two unknown reals in this equation: $\dot{r}_4$ and ω_4. The real and imaginary parts of Equation (4.52a) constitute a system which is linear and nonhomogeneous in the two unknown reals, $\dot{r}_4$ and ω_4. Solving simultaneously, we get

$$\dot{r}_4 = -197.69 \text{ cm/sec} \quad \text{and} \quad \omega_4 = -23.287 \text{ rad/sec}.$$

Step 3.   Next, we take the second time derivative of the loop 1 closure equation.

$$e^{i\theta_4}\ddot{r}_4 + ir_4 e^{i\theta_4}\alpha_4 = -r_2\omega_2^2 e^{i\theta_2} - 2\dot{r}_4 i\omega_4 e^{i\theta_4} + r_4\omega_4^2 e^{i\theta_4} \tag{4.52b}$$

This equation contains two new unknown reals: $\ddot{r}_4$, and α_4. Solving the real and imaginary parts simultaneously yields:

$$\ddot{r}_4 = 2,995.48 \text{ cm/sec}^2 \quad \text{and} \quad \alpha_4 = -123.20 \text{ rad/sec}^2.$$

Step 4.   Turning now to loop 2, the equation of closure is

$$r_7 e^{i\theta_7} - r_6 e^{i\theta_6} = r_5 e^{i\theta_5} \tag{4.52c}$$

This equation is linear in the real unknown r_7, and transcendental in the real unknown θ_5. To solve, we separate the real and imaginary parts of the equation, and eliminate θ_5 by squaring and adding, as was demonstrated in Step No. 1 above. The resulting quadratic in r_7 has the positive root $r_7 = 41.51$ cm. This applies to the mechanism position shown in the figures. Substituting this value back, say, in the imaginary part of Eq. (4.52c), we get $\sin \theta_5 = -0.202598$ and $\theta_5 = -11.69°$.

  Step 5. The first time derivative of Eq. (4.52c) is

$$\dot{r}_7 e^{i\theta_7} - r_5 i\omega_5 e^{i\theta_5} = r_6 i\omega_6 e^{i\theta_6} \tag{4.52d}$$

Separating the real and imaginary parts of this equation, the resulting system is linear and nonhomogeneous in the two unknown reals, $\dot{r}_7$ and ω_5. Solving simultaneously, we get $\dot{r}_7 = 202.71$ cm/sec and $\omega_5 = 4.818$ rad/sec.

  Step 6. Finally, we take the second time derivative of Eq. (4.52c).

$$\ddot{r}_7 e^{i\theta_7} - r_5 i\alpha_5 e^{i\theta_5} + r_5 \omega_5^2 e^{i\theta_5} = r_6 i\alpha_6 e^{i\theta_6} - r_6 \omega_6^2 e^{i\theta_6} \tag{4.52e}$$

Separating real and imaginary parts yields a system linear and nonhomogeneous in the two unknown reals, $\ddot{r}_7$ and α_5. Solving simultaneously, we get: $\alpha_5 = 132.88$ rad/sec^2 and $\ddot{r}_7 = -2{,}873$ cm/sec^2, or 28.73 m/sec^2 leftward, which agrees closely with the result obtained in the graphical solution.

4.5 MECHANISMS WITH CURVED SLOTS AND HIGHER-PAIR CONNECTIONS*

The examples above illustrate in part the complex number approach to solving acceleration problems with pin and straight slide connections. One case that has not been covered by example is a curved-slot case (see, e.g., Figs. P4.32, P4.33, and P4.37). This seemingly difficult problem is made simple by replacing the curved member (with known constant curvature) with an equivalent pin-jointed link, which results in a non-Coriolis problem. The validity of acceleration (as well as velocity) equivalence of such linkage can be shown by way of path curvature theory (Chap. 4 of Vol. 2) (see Example 4.6, Probs. 4.39 and 4.41 and Ref. [144]).

Example 4.6

  In the mechanism of Fig. 4.34, let $O_A O_B = 6.4$ cm, $O_B B = 5.3$ cm, $O_A B = 2.3$ cm, and $\rho = 4.8$ cm. Find α_4 and the linear acceleration of the slider along the curved path on link

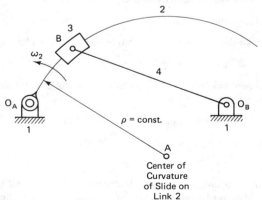

Center of
Curvature
of Slide on
Link 2

Figure 4.34 Four-link mechanism based on link with curved slide.

---

* This section has been expanded at the suggestion of Dr. Harold Johnson of Georgia Tech.

2, as well as the total acceleration of point B with respect to the ground, if $\omega_2 = 1$ rad/sec, constant.

Solution

Step 1. Since ρ is a constant, slider 3 can be replaced by a bar of length ρ, pivoted at A on link 2. Thus the problem is reduced to that of a four-bar linkage (Fig. 4.35). The vector model of the mechanism is shown in Fig. 4.36.

The loop-closure equation is $\sum_{j=1}^{4} \mathbf{z}_j = 0$; here $\mathbf{z}_1$ is fixed, and $\mathbf{z}_2$ is the input. In exponential form:

$$r_1 e^{i\theta_1} + r_2 e^{i\theta_2} + r_3 e^{i\theta_3} + r_4 e^{i\theta_4} = 0 \tag{4.53}$$

Here θ_2 is the input. Since the lengths of all links are given, we proceed to find θ_3 and θ_4 as follows.

Step 2. Let $-r_1 \cos \theta_1 + r_2 \cos \theta_2 = D_x$, and $r_1 \sin \theta_1 + r_2 \sin \theta_2 = D_y$. Then $O_B A = D = |\sqrt{D_x^2 + D_y^2}|$, and $\theta_D = \tan^{-1}(D_y/D_x)$. The quadrant of θ_D can be determined from the vector diagram of Fig. 4.36 or by using the Cartesian-to-polar conversion feature available in most hand calculators and computer languages. Thus, we have $\mathbf{D} = D e^{i\theta_D}$. With this, the equation of closure reduces to

$$D e^{i\theta_d} + r_3 e^{i\theta_3} = -r_4 e^{i\theta_4} \tag{4.53a}$$

We separate the real and imaginary parts

$$\mathcal{R}: D \cos \theta_d + r_3 \cos \theta_3 = -r_4 \cos \theta_4$$

$$(1/i)\mathcal{I}: D \sin \theta_d + r_3 \sin \theta_3 = -r_4 \sin \theta_4$$

and square both equations

$$D^2 \cos^2\theta_d + 2Dr_3 \cos \theta_d \cos \theta_3 + r_3^2 \cos^2 \theta_3 = r_4^2 \cos^2\theta_4$$

$$D^2 \sin^2\theta_d + 2Dr_3 \sin \theta_d \sin \theta_3 + r_3^2 \sin^2 \theta_3 = r_4^2 \sin^2\theta_4$$

Then we add the two equations.

$$D^2 + 2Dr_3(\cos \theta_d \cos \theta_3 + \sin \theta_d \sin \theta_3) + r_3^2 = r_4^2$$

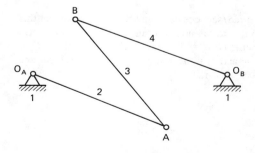

Figure 4.35 Four-bar linkage, equivalent to the curved-slide mechanism for its motion and all motion derivatives provided the radius of curvature is constant. For the treatment of variable-curvature curved-slot mechanism see Example 4.7.

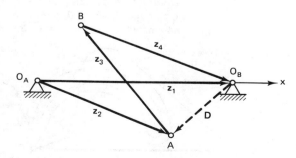

Figure 4.36 Vector model of the mechanism of Figure 4.35.

Using a well-known trigonometric identity and rearranging, we obtain

$$\cos(\theta_d - \theta_3) = (r_4^2 - r_3^2 - D^2)/2Dr_3;$$

and

$$\theta_3 = \theta_d - \cos^{-1}[(r_4^2 - r_3^2 - D^2)/2Dr_3].$$

This yields two values for θ_3. Decide by inspection of Fig. 4.36 which to use. Substitute this value back into the real part of Eq. (4.53a). This will yield two values for θ_4. Again decide by inspection which to use.

   Step 3.   The first time derivative of Eq. (4.53) is

$$r_2 i\omega_2 e^{i\theta_2} + r_3 i\omega_3 e^{i\theta_3} + r_4 i\omega_4 e^{i\theta_4} = 0 \qquad (4.53b)$$

Cancel i, separate the real and imaginary parts, and solve the resulting system of two linear nonhomogeneous equations simultaneously for the two unknown reals, ω_3 and ω_4.

   Step 4.   Take the second derivative of Eq. (4.53),

$$-r_2 \omega_2^2 e^{i\theta_2} + r_3 i\alpha_3 e^{i\theta_3} - r_3 \omega_3^2 e^{i\theta_3} + r_4 i\alpha_4 e^{i\theta_4} - r_4 \omega_4^2 e^{i\theta_4} = 0$$

and rearrange it to

$$(r_3 ie^{i\theta_3})\alpha_3 + (r_4 ie^{i\theta_4})\alpha_4 = r_2 \omega_2^2 e^{i\theta_2} + r_3 \omega_3^2 e^{i\theta_3} + r_4 \omega_4^2 e^{i\theta_4}$$

Separate the real and imaginary parts and solve the resulting system of two equations, linear and nonhomogeneous in the two unknown reals, for the two angular accelerations, α_3 and α_4.

   Step 5.   The linear acceleration of the slider along the curved path of link 2, is

$$\mathbf{A}_{BA}^t = \rho\alpha_3 ie^{i\theta_3} \text{ cm/sec}^2 \text{ (note: } \rho = r_3\text{)}.$$

The total acceleration of point B on slider 3 with respect to ground is

$$\mathbf{A}_B = \mathbf{A}_A + \mathbf{A}_{BA} = \mathbf{A}_A^t + \mathbf{A}_A^N + \mathbf{A}_{BA}^t + \mathbf{A}_{BA}^n$$

$$= 0 - r_2 \omega_2^2 e^{i\theta_2} + \rho\alpha_3 ie^{i\theta_3} - \rho(\omega_3 - \omega_2)^2 e^{i\theta_3}$$

The actual arithmetic work is left as an exercise.

Example 4.7

   For the mechanism of Figure 4.37, let $\omega_2 = 1$ rad/sec, constant, and find α_3. Notice that this example is more difficult than Example 4.6 due to the non-constant curvature of the slot!

   Data: $\mathbf{z}_1 = 5.2$ cm, $O_A A = 4.5$ cm, $O_B B = 1.1$ cm, $O_B C = 0.9$ cm, $BC = 1.1$ cm, $\mathbf{z}_4 = r_4 e^{i\theta_4}$, where $\theta_4 = \theta_3 - \delta + \psi$ and $r_4 = a_0 + a_1\psi$, where $a_0 = 1.1$ cm and $a_1 = 0.4$ cm/rad. Note that $\mathbf{z}_3 = \overline{O_B B}$. $\delta = -\frac{3}{4}\pi$ rad.

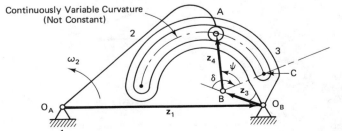

Figure 4.37 Three-link higher-pair mechanism with curved slot having continuously varying curvature.

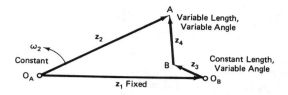

Figure 4.38 Vector model of the mechanism of Fig. 4.37. z_1 is fixed, z_2 is given, θ_2 is the input rotation, z_3 is given, θ_3 is variable, z_4 and θ_4 are variable.

Solution We start with the vector model of the mechanism, shown in Fig. 4.38. The loop equation of closure is

$$z_2 = z_1 + z_3 + z_4, \quad \text{or}$$

$$r_2 e^{i\theta_2} = r_1 e^{iO} + r_3 e^{i\theta_3} + r_4 e^{i\theta_4}$$

where r_1, r_2 and r_3 are given constants, r_4 is unknown and is a function of ψ (see data above), θ_2 is input, θ_3 is unknown as is also θ_4, but $\theta_4 = \theta_3 - \delta + \psi$, where ψ is also unknown. Thus the loop equation of closure contains too many unknowns. We proceed as follows.

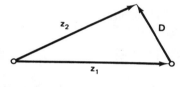

(a) Loop 1   (b) Loop 2   **Figure 4.39**

Step 1.   We separate the vector diagram into two loops, as shown in Fig. 4.39. The equation of closure for loop 1 is $z_2 = z_1 + D$, where D is unknown. In exponential form this is

$$r_2 e^{i\theta_2} = r_1 e^{iO} + D e^{i\theta_d} \tag{4.54}$$

Solve this equation for D and θ_d (see step 4 in the analytical solution of Example 4.5).
Step 2.   The loop 2 closure equation is $z_3 + z_4 = D$, or in exponential form is

$$r_3 e^{i\theta_3} + r_4 e^{i(\theta_3 - \delta + \psi)} - D e^{i\theta_d} = 0 \tag{4.54a}$$

Note that $\theta_4 = \theta_3 - \delta + \psi$, and $r_4 = a_0 + a_1\psi$. Thus, the unknowns are θ_3 and ψ, both in transcendental form. Solve for these two reals numerically, say, by way of a two-dimensional optimization algorithm. (See, for example, T. Shoup and F. Mistree, *Optimization Methods with Applications for Personal Computers*, Englewood Cliffs, Prentice-Hall, 1987; pp 110–118).
Step 3.   The first time derivative of the Loop 1 equation of closure is

$$\dot{z}_2 = \dot{z}_1 + \dot{D}$$

or

$$r_2 i\omega_2 e^{i\theta_2} = Di\omega_d e^{i\theta_d} + \dot{D} e^{i\theta_d} \tag{4.54b}$$

Here we have two unknown reals, ω_d and $\dot{D}$, for which the system of real and imaginary parts can be solved linearly simultaneously.

Step 4. The first time derivative of the Loop 2 Equation of Closure is

$$r_3 i \omega_3 e^{i\theta_3} + \dot{r}_4 e^{i\theta_4} + r_4 i \omega_4 e^{i\theta_4} = Di\omega_d e^{i\theta_d} + \dot{D} e^{i\theta_d} \tag{4.54c}$$

Here,

$$\dot{\theta}_4 = \omega_4 = \frac{d}{dt}(\theta_3 - \delta + \psi) = \omega_3 + \dot{\psi}$$

and

$$\dot{r}_4 = \frac{d}{dt}(a_0 + a_1\psi) = a_1\dot{\psi}$$

Using these, Eq. (4.54c) becomes

$$r_3 i \omega_3 e^{i\theta_3} + a_1 \dot{\psi} e^{i\theta_4} + r_4 i(\omega_3 + \dot{\psi}) e^{i\theta_4} - \dot{D} e^{i\theta_d} + Di\omega_d e^{i\theta_d} = 0$$

Separate real and imaginary parts, collect ω_3 and $\dot{\psi}$ terms, and solve the system linearly, simultaneously for ω_3 and $\dot{\psi}$.

Step 5. The second time derivative of the Loop 1 closure equation is (note that α_2 is zero):

$$-r_2\omega_2^2 e^{i\theta_2} = \dot{D}i\omega_d e^{i\theta_d} + Di\alpha_d e^{i\theta_d} - D\omega_d^2 e^{i\theta_d} + \ddot{D} e^{i\theta_d} + \dot{D}i\omega_d e^{i\theta_d} \tag{4.54d}$$

Separate real and imaginary parts and solve for $\ddot{D}$ and α_d.

Step 6. The second time derivative of the Loop 2 closure equation is

$$r_3 i \alpha_3 e^{i\theta_3} - r_3 \omega_3^2 e^{i\theta_3} + \dot{r}_4 i \omega_4 e^{i\theta_4} + \ddot{r}_4 e^{i\theta_4} + \dot{r}_4 i \omega_4 e^{i\theta_4} + r_4 i \alpha_4 e^{i\theta_4} - r_4 \omega_4^2 e^{i\theta_4} = \ddot{D} e^{i\theta_d}$$
$$+ \dot{D}i\omega_d e^{i\theta_d} + \dot{D}i\omega_d e^{i\theta_d} + Di\alpha_d e^{i\theta_d} - D\omega_d^2 e^{i\theta_d} \tag{4.54e}$$

Here,

$$\alpha_4 = \dot{\omega}_4 = \frac{d}{dt}(\dot{\theta}_3 + \dot{\psi}) = \alpha_3 + \ddot{\psi}$$

and

$$\ddot{r}_4 = \frac{d^2}{dt^2}(a_0 + a_1\psi) = a_1\ddot{\psi}$$

Substitute these in Eq. (4.54e), separate real and imaginary parts, and solve linearly simultaneously for α_3 and $\ddot{\psi}$. This completes the solution. The arithmetic work is left as an exercise.

PROBLEMS

4.1. Prove Eq. (4.20). [*Hint:* Let link AB be represented by the vector $\mathbf{R}_{BA} = (AB)e^{i\theta}$. Then derive the components of $\mathbf{A}_{BA}$ in the $e^{i\theta}$ and the $ie^{i\theta}$ directions and find the angle of rotation from $\mathbf{R}_{BA}$ to $\mathbf{A}_{BA}$.]

4.2. Starting with Eq. (4.43) and using the results of the velocity triangle in Fig. 4.25, perform all steps of the derivations and find the solutions to Example 4.4. [*Hint:* Your results should be the same as those based on Eq. (4.44).]

4.3. For the four-bar mechanism shown in Fig. P4.1, knowing that $O_2A = 10$ in., $AB = 12$ in., $O_4B = 12$ in., $O_2O_4 = 22.5$ in., $\omega_2 = -20$ rad/sec, and $\alpha_2 = -160$ rad/sec², and also knowing the velocity polygon, draw the acceleration polygon, and find the acceleration of point B.

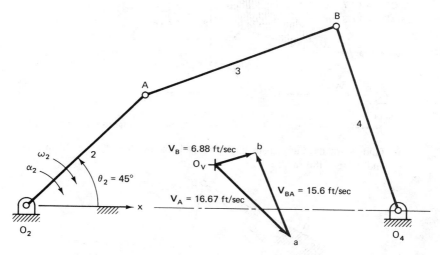

Figure P4.1

4.4. Draw the velocity and acceleration polygons for the four-bar mechanism shown in Fig. P4.2. Given that length of $O_2A = 2$ in., $O_2O_4 = 4.5$ in., $O_4B = 2.5$ in., $AB = 3.6$ in., $\omega_2 = -60$ rad/sec, and $\alpha_2 = -900$ rad/sec², find the velocity and acceleration of point B.

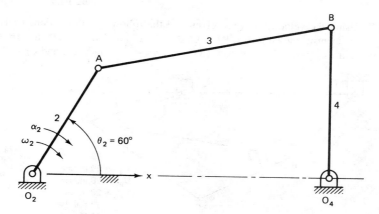

Figure P4.2

4.5. Knowing that the angular velocity of link 2 is a constant in the mechanism of Fig. P4.3, determine A_B, A_D, and α_4.

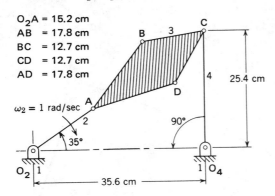

O₂A = 15.2 cm
AB = 17.8 cm
BC = 12.7 cm
CD = 12.7 cm
AD = 17.8 cm

ω_2 = 1 rad/sec

35°

90°

25.4 cm

35.6 cm

Figure P4.3

4.6. Find A_C of the mechanism in Fig. P4.4 knowing that $O_2O_4 = 15$ in., $\omega_2 = -25$ rad/sec, and $\alpha_2 = -180$ rad/sec^2.

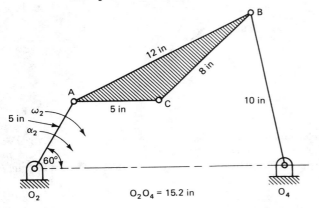

12 in

8 in

A

5 in

C

10 in

5 in
ω_2
α_2

60°

O_2 O_2O_4 = 15.2 in O_4 **Figure P4.4**

4.7. Using the velocity polygon given and the following information, find A_C and α_4 for the mechanism shown in Fig. P4.5. $O_4B = 1.125$ in., $O_2A = 1.25$ in., $AB = 2.5$ in., $O_2O_4 = 2$ in., $BC = 1.25$ in., $CA = 1.667$ in., $\omega_2 = -40$ rad/sec, and $\alpha_2 = 20$ rad/sec^2.

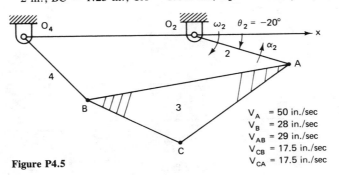
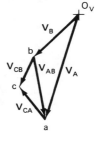

O_4 O_2 ω_2 $\theta_2 = -20°$ x

2   α_2 A

4

B   3

C

V_A = 50 in./sec
V_B = 28 in./sec
V_{AB} = 29 in./sec
V_{CB} = 17.5 in./sec
V_{CA} = 17.5 in./sec

O_V

V_B

b

V_{CB} V_{AB} V_A

c

V_{CA}

a

Figure P4.5

4.8. Referring to Fig. P4.6,

    (a) Determine the velocity of point P by the method of instant centers.

    (b) Determine the angular velocity of the connecting rod (CP).

    (c) Find the acceleration of point P with respect to the ground when OC is rotating at a constant speed of 120 rpm ccw.

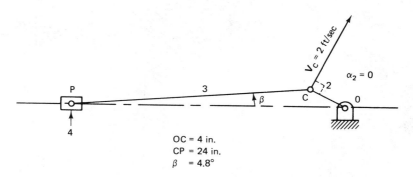

OC = 4 in.
CP = 24 in.
β = 4.8°

Figure P4.6

4.9. Given the velocity polygon for the mechanism in Fig. P4.7 and knowing that $V_A = 14$ m/sec, find $\mathbf{A}_B$ ($\alpha_2 = 0$).

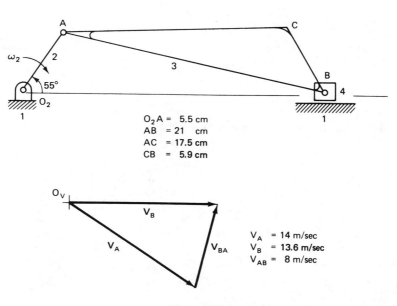

O_2A = 5.5 cm
AB = 21 cm
AC = 17.5 cm
CB = 5.9 cm

V_A = 14 m/sec
V_B = 13.6 m/sec
V_{AB} = 8 m/sec

Figure P4.7

4.10. Knowing that $\alpha_2 = 100$ rad/sec² and $\omega_2 = 15$ rad/sec, find $\mathbf{A}_C$ in the mechanism of Fig. P4.8.

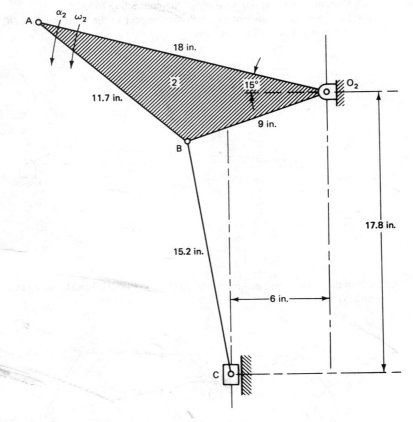

18 in.

11.7 in.

9 in.

17.8 in.

15.2 in.

6 in.

Figure P4.8

4.11. Find $\mathbf{A}_B$ for the mechanism shown in Fig. P4.9 ($\alpha_2 = 0$).

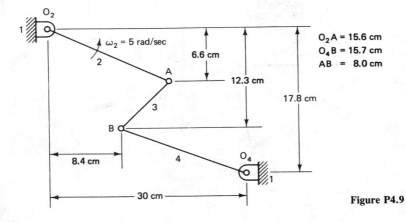

$\omega_2 = 5$ rad/sec

6.6 cm

12.3 cm

17.8 cm

8.4 cm

30 cm

$O_2A = 15.6$ cm
$O_4B = 15.7$ cm
$AB = 8.0$ cm

Figure P4.9

4.12. Link 2 rotates at a constant angular velocity in Fig. P4.10. Determine the velocity and acceleration of point D.

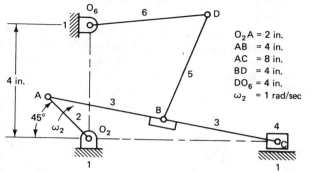

O_2A = 2 in.
AB = 4 in.
AC = 8 in.
BD = 4 in.
DO_6 = 4 in.
ω_2 = 1 rad/sec

Figure P4.10

4.13. In Fig. P4.11, slide 2 moves to the left at a constant velocity of 1 cm/sec. Find α_3 and α_6 in the position shown.

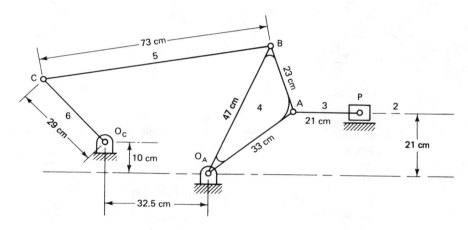

Figure P4.11

4.14. For the mechanism shown in Fig. P4.12, find the acceleration of point D with respect to point B.

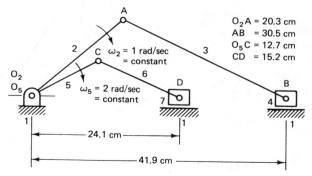

O_2A = 20.3 cm
AB = 30.5 cm
O_5C = 12.7 cm
CD = 15.2 cm

Figure P4.12

4.15. Find the location of a point C on link 3 of the linkage shown in Fig. P4.13 which at this instant has zero relative acceleration with respect to ground. The link lengths are $A_0A = 5$ in at $135°$, $AB = 6$ in at $0°$, $B_0B = 3.064$ in at $80°$, $A_0B_0 = 2$ in at $15°$. The angular velocity of link 2 is constant at $\omega_2 = 2$ rad/sec ccw.

4.16. The partial acceleration polygon of an unknown 4-bar linkage is shown in Fig. P4.14. The linkage has constant-length links. Complete the acceleration polygon, determine α_3 and α_4, and sketch the linkage clearly showing the directions of all angular velocities and accelerations. The only other information known of this linkage is $A_0B_0 = 2.534$ in at $-75°$ and is the ground link, $\omega_2 = 5$ rad/sec ccw, $\omega_4 = 2.887$ rad/sec ccw.

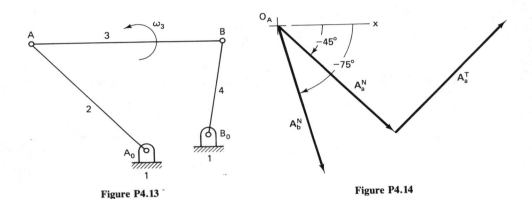

Figure P4.13   Figure P4.14

4.17. Referring to Fig. P4.15, given $\alpha_2 = 1$ rad/sec^2, determine A_{D6}, α_4, and α_5 given the velocity polygon.

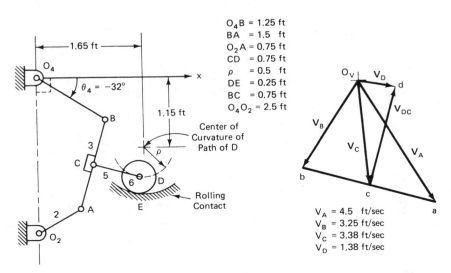

$O_4B = 1.25$ ft
$BA = 1.5$ ft
$O_2A = 0.75$ ft
$CD = 0.75$ ft
$\rho = 0.5$ ft
$DE = 0.25$ ft
$BC = 0.75$ ft
$O_4O_2 = 2.5$ ft

Center of
Curvature of
Path of D

Rolling
Contact

$V_A = 4.5$ ft/sec
$V_B = 3.25$ ft/sec
$V_C = 3.38$ ft/sec
$V_D = 1.38$ ft/sec

Figure P4.15

4.18. Find the absolute acceleration of point A on the slider ($\mathbf{A}_{A3}$) of the 3 link system of Fig. P4.16. Known information is: $A_0A = 2''$, $\omega_2 = 3$ rad/sec ccw, $\alpha_2 = 3$ rad/sec^2 cw, $\mathbf{V}_{A3/A2} = 1$ in/sec, $\mathbf{A}_{A3/A2}^S = 18$ in/sec^2.

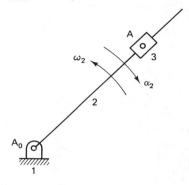

Figure P4.16

4.19. For the mechanism shown in Fig. P4.17 determine:
  (a) The angular velocity ω_3.
  (b) The magnitude of the sliding velocity of the pin in the slot.
  (c) The angular acceleration α_3.

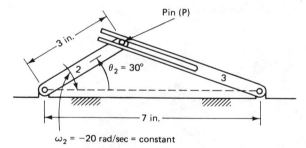

$\omega_2 = -20$ rad/sec = constant   **Figure P4.17**

4.20. Do an acceleration analysis of the mechanism shown in Fig. P4.18 given the velocity polygon.

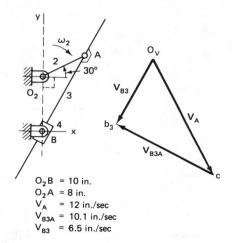

$O_2B = 10$ in.
$O_2A = 8$ in.
$V_A = 12$ in./sec
$V_{B3A} = 10.1$ in./sec
$V_{B3} = 6.5$ in./sec

Figure P4.18

4.21. Refer to Prob. 3.17 and Fig. P3.25 for the velocity analysis of the match lighting mechanism redrawn as Fig. P4.19. Determine α_3 and α_4 and whether link 3 is angularly and linearly speeding up or slowing down. $\omega_2 = 4$ rad/sec ccw and constant. The link lengths are
$A_0A = 2.5$ in at $-30°$, $A_0B = 4.33$ in, $AB = 2.5$ in.

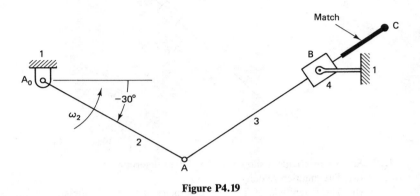

Figure P4.19

4.22. Complete the acceleration polygon for the mechanism of Fig. P4.20.

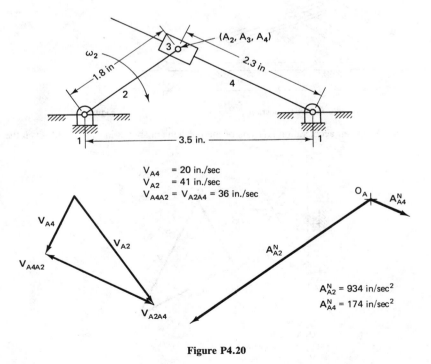

$V_{A4} = 20$ in./sec
$V_{A2} = 41$ in./sec
$V_{A4A2} = V_{A2A4} = 36$ in./sec

$A_{A2}^N = 934$ in/sec^2
$A_{A4}^N = 174$ in/sec^2

Figure P4.20

4.23. Determine the angular acceleration of link 2 in Fig. P4.21. Also find $\mathbf{A}_B$.

Lengths

O_2A = 10.2 cm
O_4A = 15.2 cm
AB   = 7.62 cm
O_2B = 10.2 cm

ω_4 = −1 rad/sec
     = constant

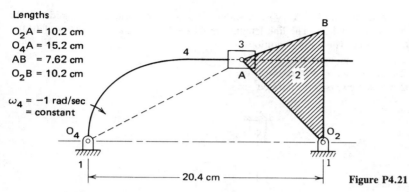

Figure P4.21

4.24. Link 2 in Fig. P4.22 rotates at a constant $\omega_2 = 1$ rad/sec ($\theta_2 = 60°$). Find V_{A4} and A_{A4} if $O_2A = 2.4$ cm.

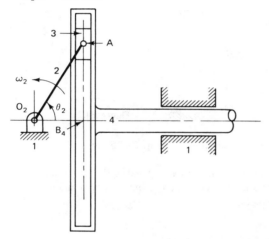

Figure P4.22

4.25. Determine the linear acceleration of A_{P4} in Fig. P4.23 if $\alpha_2 = 0$.

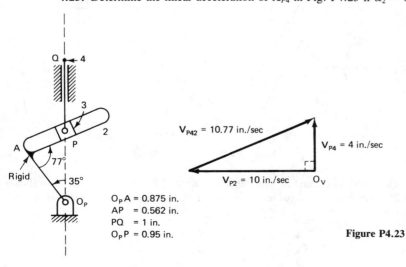

V_{P42} = 10.77 in./sec

V_{P4} = 4 in./sec

V_{P2} = 10 in./sec

O_PA = 0.875 in.
AP = 0.562 in.
PQ = 1 in.
O_PP = 0.95 in.

Figure P4.23

4.26. Determine α_4 and the coriolis ($A_{B4/B2}^C$) and sliding ($A_{B4/B2}^S$) accelerations of the slider in the
linkage shown in Fig. P4.24. The link lengths are $A_0A = 1.061$ in at $135°$, $AB = 1.061$ in
at $45°$, $BB_0 = 1.5$ in. The input conditions are
(a) $\omega_2 = 2$ rad/sec cw, constant
(b) $\omega_2 = 2$ rad/sec cw, $\alpha_2 = 4$ rad/sec^2 cw
(c) $\omega_2 = 2$ rad/sec cw, $\alpha_2 = 8$ rad/sec^2 cw

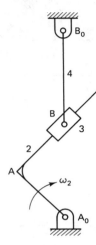

Figure P4.24

4.27. The dimensions of the mechanism drawn in Fig. P4.25 are $A_0A = 2$ in at $0°$, $B_0A = 3$ in
at $195°$.
(a) Find the sliding acceleration of link 3 ($A_{A4/A2}^S$) and α_4 if $\omega_2 = 3$ rad/sec ccw and constant.
(b) Find α_2 and α_4 if the sliding acceleration of link 3 is zero and $\omega_2 = 3$ rad/sec ccw but
not necessarily constant.

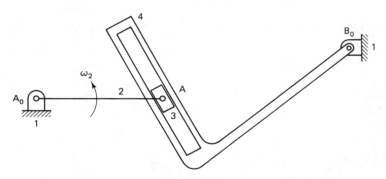

Figure P4.25

4.28. Find the sliding acceleration and the angular acceleration of link 4 in the mechanism shown
in Fig. P4.26. The linkage dimensions are $A_0A = 1$ ft, $B_0B = 2$ ft, $AB = 2$ ft. ω_2 is constant at 15 rad/sec.

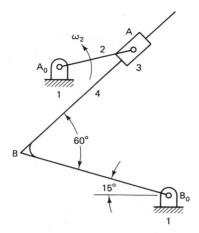

Figure P4.26

4.29. Find the acceleration of link 4 (A_{P4}) of the Scotch yoke in Fig. P4.27. Write the acceleration equation, indicating known magnitudes and directions.

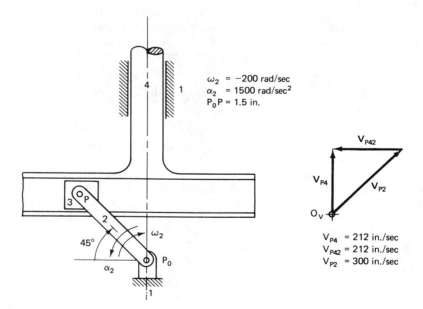

Figure P4.27

4.30. In the device of Fig. P4.28, the actuator (link 3) moves to the right with a constant acceleration ($A_{B3} = 500$ m/sec^2). Given the velocity polygon, solve for A_{B2}, A_C, α_2, and α_{23}. Write the vector equations that must be used to solve for these values.

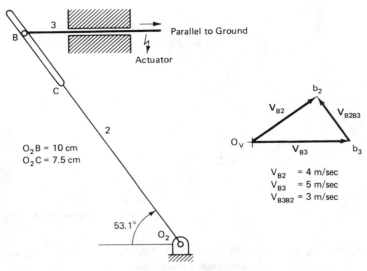

Figure P4.28

4.31. The linkage shown in Fig. P4.29, drawn when link 2 is in a general position $\theta°$, is a Rapson slide rudder mechanism used to steer large ships. Links 3 and 4 are sliders. Link 4 slides horizontally on ground link 1 and link 3 slides along link 2. There is a turning pair joining links 3 and 4. The general acceleration equation describing this mechanism is

$$\mathbf{A}_4^N + \mathbf{A}_4^T + \mathbf{A}_4^C + \mathbf{A}_4^S = \mathbf{A}_2^N + \mathbf{A}_2^T + \mathbf{A}_2^C + \mathbf{A}_2^S + \mathbf{A}_{42}^N + \mathbf{A}_{42}^T + \mathbf{A}_{42}^C + \mathbf{A}_{42}^S$$

(a) Show that the relative acceleration equation describing this mechanism can at most consist of the following terms.

$$\mathbf{A}_4^S = \mathbf{A}_2^N + \mathbf{A}_2^T + \mathbf{A}_{42}^C + \mathbf{A}_{42}^S$$

(b) At what value of $\theta°$ does $\mathbf{A}_{42}^C = 0$ and what are the relationships between the remaining terms?

(c) Show that if $\theta \neq 0°$ and $\mathbf{A}_4^S = 0$ then both $\mathbf{A}_{42}^S$ and $\mathbf{A}_2^T$ must exist.

(d) Suppose $\theta = 30°$ and $\mathbf{A}_{42}^S = 0$. If $\mathbf{V}_4^S = V$ directed to the left, show that $A_4^S = \sqrt{3}V\omega_2$ in/sec^2 and $\alpha_2 = (5V\omega_2)/(2r_2)$ rad/sec^2 clockwise.

(e) Suppose $\theta = 30°$ and $\alpha_2 = 0$. If $\mathbf{V}_4^S = V$ directed to the right, show that $A_4^S = (2\sqrt{3}V\omega_2)/3$ and $A_{42}^S = (5\sqrt{3}\,V\omega_2)/6$.

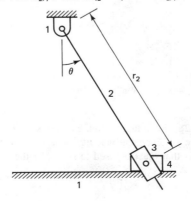

Figure P4.29

4.32. In the double-slider mechanism of Fig. P4.30, find the acceleration of point P on 4 ($\mathbf{A}_{P4}$), when $\alpha_2 = 0$.

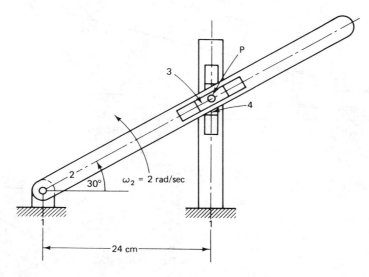

Figure P4.30

4.33. Fig. P4.31 redepicts the slider linkage of Prob. 3.58 and Fig. P3.66.* The velocity of the slider link 2 is 10 in/sec to the right. Find the sliding acceleration of link 3 and the angular acceleration of link 4 if the magnitude of the link 2 sliding acceleration equals the magnitude of the tangential acceleration of point A on link 4, $|\mathbf{A}_{A2}^S| = |\mathbf{A}_{A4}^T|$.

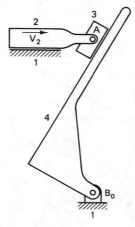

Figure P4.31

4.34. If the slider (2) of Fig. P4.32 accelerates at 1 in/sec^2 in the direction shown, find α_4. (*Hint:* With a known center of curvature, an equivalent pin-jointed dyad can replace the slider 3 and curved slider 4.)

* Scale the figure for needed data.

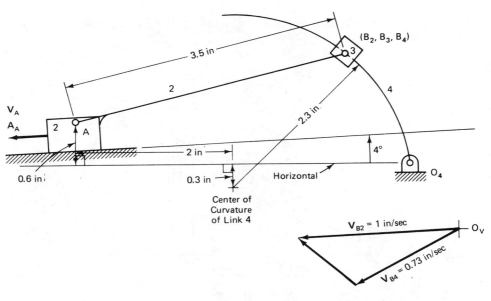

Figure P4.32

4.35. Find the acceleration of point B on the inverted slider-crank bucket mechanism drawn in Fig. P4.33. The input angular velocity is constant at 2 rad/sec cw. The linkage dimensions are $A_0A = (4\sqrt{3})/3$ ft (or 2.31 ft) at $-150°$, $AB = .75$ ft, and curved link 4 has a 2 ft radius.

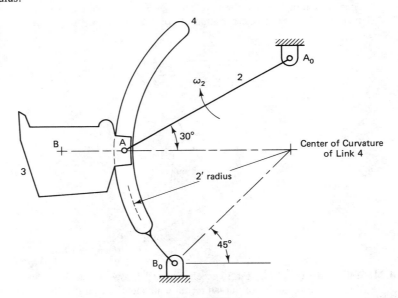

Figure P4.33

4.36. The Stephenson III linkage of Prob. 3.32 and Fig. P3.40 is shown in Fig. P4.34. Determine the relative sliding acceleration of slider link 5 and α_6 if ω_2 is constant at 2 rad/sec cw.

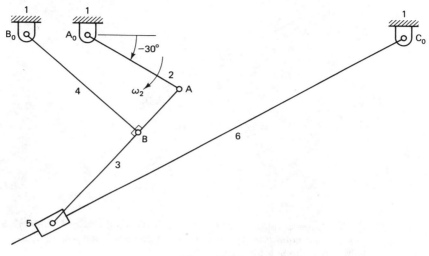

Figure P4.34

4.37. Determine the acceleration of the link-5 slider block in the metal stamping mechanism shown in Fig. P4.35. At this instant, ω_2 is constant at 2 rad/sec cw. The link lengths are $A_0A = 5$ in at 150°, $A_0B_0 = 6.83$ in, $B_0A = 3.53$ in at 45°, and $B_0B = 7.07$ in.

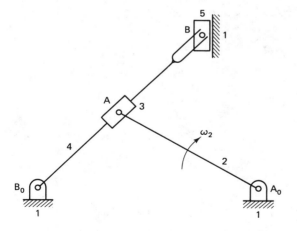

Figure P4.35

4.38. Find α_3, α_4, and α_5 and linear accelerations $\mathbf{A}_{B4}$ and $\mathbf{A}_D$ for the Stephenson III turning block mechanism sketched in Fig. P4.36. The input angular velocity is constant at $\omega_2 = -5$ rad/sec cw. The link lengths are $A_0A = 2$ in at 135°, $AB = 4.83$ in at $-30°$, $AC = 7.07$ in, $B_0B = 4$ in at 60°, $CD = 3.796$ in at 90°, and slider 6 operates against a plane inclined at 60°.

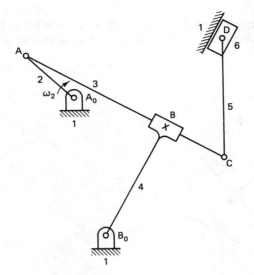

Figure P4.36

4.39. For the flat-face follower and for the curved-face follower of Fig. P4.37 determine V_Q, A_Q, ω_2, and α_2 if point P is the point of contact on body 3 and point Q is the point of contact on body 2 (see the hint for Prob. 4.34).

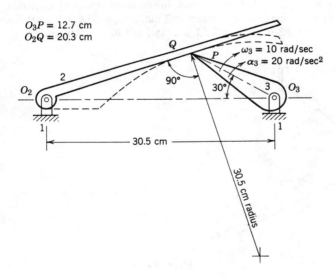

$O_3P = 12.7$ cm
$O_2Q = 20.3$ cm

$\omega_3 = 10$ rad/sec
$\alpha_3 = 20$ rad/sec^2

90°

30°

30.5 cm

30.5 cm radius

Figure P4.37

4.40. Determine α_4 of the 2-degree-of-freedom 5-bar linkage shown in Fig. P4.38 if $\omega_2 = -5$ rad/sec cw, $\alpha_2 = -25$ rad/sec^2 cw, and the sliding velocity of link 5 is constant at $V_{B5}^S = 2.5$ in/sec upwards. The link lengths are $A_0A = 2$ in at 135° and $BA = 3.38$ in at 105°.

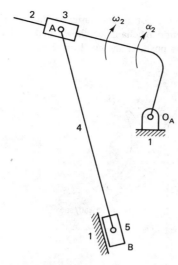

Figure P4.38

4.41. Write the equation of the acceleration of P on 4, indicating all the known quantities (magnitudes and directions), and find α_4 for the mechanism of Fig. P4.39. Note that the relative center of curvature of link 3 (slider) with respect to link 4 is at C and the curved link 4 pivots at O_4 (see the hint for Prob. 4.34).

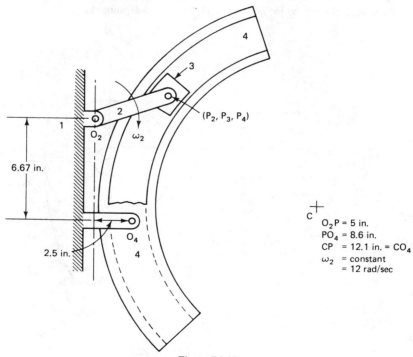

$O_2P = 5$ in.
$PO_4 = 8.6$ in.
$CP = 12.1$ in. $= CO_4$
$\omega_2 =$ constant
$\quad = 12$ rad/sec

Figure P4.39

4.42. Determine and plot ω_3 and α_3 for a one-quarter revolution of link 2 for the Geneva mechanism (an intermittent motion mechanism) shown in Fig. P4.40. Let θ_2 be taken in increments of 5° starting in the position shown. Link 2 rotates at a constant angular velocity in the direction shown.

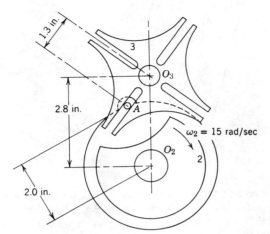

Figure P4.40

4.43. Referring to the mechanism of Fig. P4.41, determine $\mathbf{A}_P$.

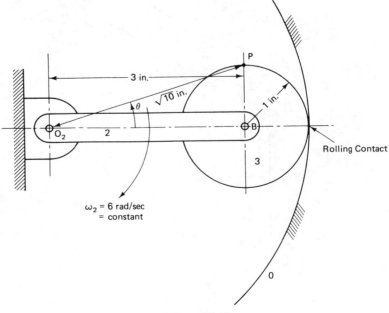

Figure P4.41

4.44. **(a)** Assume a convenient coordinate system for the mechanism shown in Fig. 4.20a, say, in which $\overrightarrow{O_4 O_2} = 0 + i11.9$ cm. Let $\overrightarrow{O_2 P_2} = \mathbf{z}_{P2}$, $\overrightarrow{O_4 P_4} = \mathbf{z}_{P4}$ and solve the triangle $\Delta O_4 O_2 (P_2 \equiv P_4)$ to obtain the arguments $\theta_2 = \arg \mathbf{z}_{P2}$ and $\theta_4 = \arg \mathbf{z}_{P4}$. Write Eq. (4.29) in exponential form, convert to Cartesian form, separate real and imaginary parts, and solve the two resulting real equations simultaneously for the directed (signed) magnitudes $V_{P2P4} = \dot{z}_{P2P4}$ and $V_{P4} = \dot{z}_{P4}$. Similarly, write Eq. (4.33) in exponential form, convert to Cartesian form, separate real and imaginary parts, and solve the resulting two real equations simultaneously for the directed (signed) magnitudes A_{P4}^t and A_{P2P4}^s. Then calulate α_4 as in Eq. (4.33b).

(b) Similar to part (a), except use Eqs. (4.31) and (4.34).

4.45. Following a complex-number analytical method outlined in Prob. 4.44, solve Eq. (4.45) analytically for $(r_2 \alpha_2)$ and $\ddot{r}_{A4A2}$ and thus verify the statements in the table below Eq. (4.45).

4.46. Write the complex-number expression for the linear acceleration of point P of the four-bar mechanism of Fig. 3.82 using Eqs. (4.21) and (4.22).

4.47. Derive Eqs. (4.21) and (4.22) from Fig. 3.82.

4.48. Find $\mathbf{A}_C$ of Fig. P4.4 using the results of Prob. 4.46. Compare this answer to the graphical solution of Prob. 4.6.

4.49. Use the results of Prob. 4.46 to solve Prob. 4.5.

4.50. Use the results of Prob. 4.46 to solve Prob. 4.7.

4.51. Write the complex-number expression for the linear acceleration of point P of the linkage in Fig. 3.82 if point B is on a slider moving horizontally, in other words, if link 4 is vertical of infinite length.

4.52. For the mechanism and the partially completed velocity and acceleration polygons shown in Fig. P4.42:

(a) Outline the solution procedure to find the acceleration of point P on link 6. Write the vector equations necessary, indicating known magnitudes and directions; indicate known directions by appropriate arrows.

(b) Find $\mathbf{A}_{P3}$, $\mathbf{A}_{P6}$, α_6, and α_{54}.

4.53. In Fig. P4.43, a block (link 5) slides outward on a branch of link 3 at a uniform relative velocity of 10 cm/sec, while link 2 of the four-bar mechanism $A_0 ABB_0$ rotates counter-clockwise at the constant angular velocity of $\omega_2 = 10$ rad/sec. The linkage dimensions are $A_0 A = 1.5$ cm, $AB = 4$ cm, $AC = 1$ cm, $CP = 1$ cm, $A_0 B_0 = 2$ cm, $\theta_2 = 120°$, $\theta_3 = 9°$, and $\measuredangle BCP = 51°$. Determine

(a) The absolute velocity of point P on link 3 (point P_3).

(b) The absolute velocity of point P on link 5 (point P_5).

(c) The absolute acceleration of point P on link 3.

(d) The absolute acceleration of point P on link 5.

4.54. An additional link 6 of length 2 cm, oriented at 30° from the positive x axis ccw, measured at P, is connected to the slider (link 5) of Fig. P4.43 at P and is pivoted to ground at P_0 as shown in Fig. P4.44. The rest of the mechanism is the same as that of Fig. P4.43. Determine

(a) The relative velocity of point P_5, with respect to point P_3, $\mathbf{V}_{P5P3}$.

(b) The absolute velocity of point P on link 5, $\mathbf{V}_{P5}$.

(c) The angular velocity of link 6, ω_6.

(d) The relative acceleration of point P_5 of link 5, with respect to point P_3, $\mathbf{A}_{P5P3}$.

(e) The absolute acceleration of point P on link 5, $\mathbf{A}_{P5}$.

(f) The angular acceleration of link 6, α_6.

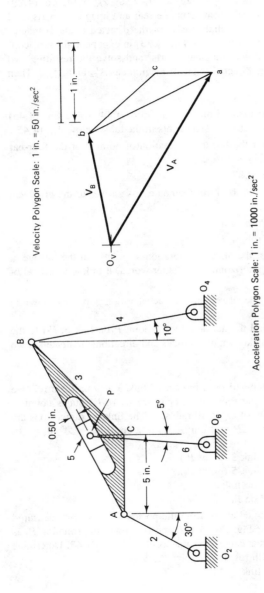

Velocity Polygon Scale: 1 in. = 50 in./sec²

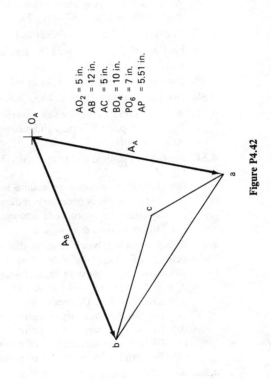

Acceleration Polygon Scale: 1 in. = 1000 in./sec²

AO_2 = 5 in.
AB = 12 in.
AC = 5 in.
BO_4 = 10 in.
PO_6 = 7 in.
AP = 5.51 in.

Figure P4.42

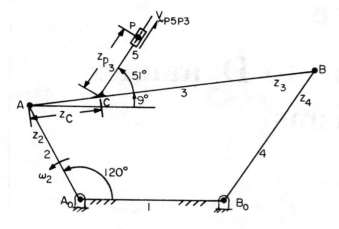

Figure P4.43

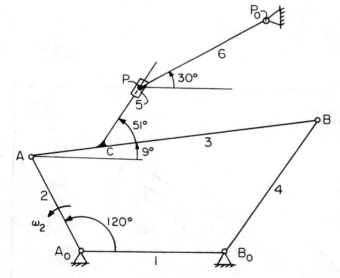

Figure P4.44

4.55. Derive Eqs. (4.21) and (4.22). Start by taking the time derivative of Eq. (3.68), use Euler's equation, separate real and imaginary parts to obtain two real equations, and then eliminate first α_3 to solve for α_5 and vice versa. Finally, use trigonometric identities to obtain the sines and cosines of $(\theta_3 - \theta_1)$, and so on.

chapter five

Introduction to Dynamics of Mechanisms

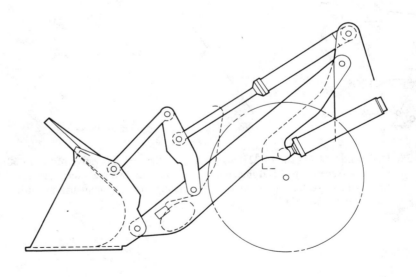

5.1 INTRODUCTION

The phrases "dynamics," "dynamics of machinery," and "mechanism dynamics" are over-used and appear to have different meanings in different contexts. We make an effort in the following sections to be specific in describing the various subcategories of mechanism dynamics (see Table 5.1).

The major ingredients of what is called dynamics of mechanical systems are *force* and *motion*. The degree of difficulty of a dynamic analysis rests on what is known and unknown in a problem and assumptions that can be made about the unknowns. Several levels of dynamics of mechanisms are described in this chapter. The advanced topics appear in Chap. 5 of Vol. 2.

Newtonian Mechanics

Let us begin with Newton's laws. We will not proceed, however, as one might in a particle physics or elementary mechanics problem, because examples in mechanism dynam-

TABLE 5.1 METHODS OF ANALYSIS OF FORCES AND TORQUES IN MECHANISMS*

	Method		
	Statics (mechanical advantage)	Kinetostatics	Dynamics (time response)
Input information, assumptions (given)			
Masses	Zero†	Specified	Specified
Loading	Specified or parameterized, as in input/output ratio	Specified at each position	Specified in terms of position, velocity, and/or time
Motion	Positions specified	Position, velocity, and acceleration specified	Unknown
Output information (sought)	Required input force to balance the load Mechanical advantage at each position Pin reactions	Required input force to sustain assumed motion Pin reactions	Position, velocity, and acceleration of each member as a function of time: that is, the *actual* motion
Required analytical tools	Statics, linear algebra	D'Alembert's principle, statics, linear algebra	Writing differential equations of motion, solution by computer

* Arranged in order of increasing complexity from left to right.
† The weight of the links may play a part in the analysis but not the inertia.
Source: Ref. [18].

ics are posed inversely to the typical physics problem. In elementary mechanics, the force (or torque) is a known quantity (such as gravity causing a block to slide down an inclined plane) and it is required to determine the resulting motion (displacement, velocity, and/or acceleration). In mechanism analysis, however, the motion is usually known (or assumed known) either by experimentation or analytical predictions based on kinematic analysis. Physical constraints at the joints of the mechanism help predict motion while the forces (and torques) that cause these motions are to be determined. Newton's laws certainly still apply but they are used in a different manner. Newton's laws of motion may be expressed as follows:

1. A particle will remain at rest or will continue in straight-line motion at a constant speed unless acted on by a force.
2. The time rate of change of momentum of a particle is equal to the magnitude of the applied force and acts in the direction of the force.

Static forces.   The laws of statics are also fundamental to mechanism analysis:

1. A rigid body acted on by two forces is in static equilibrium only if the two forces are collinear and equal in magnitude but opposite in sense (Fig. 5.1).
2. A rigid body acted upon by three forces can be in static equilibrium only if the lines of action of the three forces are concurrent at some point and if the force vectors form a closed triangle (Fig. 5.2).
3. A rigid body acted on by a couple is in static equilibrium only if it is also acted on by another couple equal in magnitude and opposite in sense (Fig. 5.3).

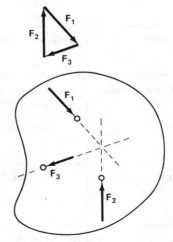

Figure 5.2   Rigid body in static equilibrium under the action of three forces.

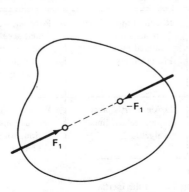

Figure 5.1   Rigid body in static equilibrium.

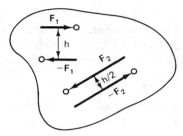

Figure 5.3 Rigid body in static equilibrium under the action of two couples of equal magnitude but opposite sense ($F_2 = 2F_1$).

If more than three forces are acting on a body in static equilibrium or if combinations of forces and couples are acting, the *principle of superposition* may be used in conjunction with the three laws of statics; that is, the effect of each force or moment may be analyzed independently, and the effect of all the forces and moments is the vector summation of the results of all individual analyses. (The principle of superposition is valid for any linear system — where effects of the components of the system are independent and thus do not influence one another.) Static analysis of mechanisms composed of rigid links involve free-body diagrams and application of the laws of statics.

Most engineers feel comfortable with statics. In fact in Sec. 3.9 we performed static force analyses of several linkages using the instant center approach for mechanical advantage determination. That technique is a shortcut method which yields the same results as applying the laws of statics to each linkage member. It would be nice to be able to solve a "dynamic" problem with the same degree of simplicity and comfort. Newton's laws will help to further explore the relationship between statics and dynamics.

Dynamic forces.   Newton's laws describe the relationship between the motion of and forces acting on a particle. Mechanisms are made up of rigid links which are composed of millions of particles forming the solid body. In order to build on Newton's laws, it becomes desirable to further develop them to the point where analogous statements can also be made about the motion of solid links.

Since the great majority of this text is concerned with planar motion, let us look at the rigid-body planar motion of link k (Fig. 5.4) and express vector quantities in com-

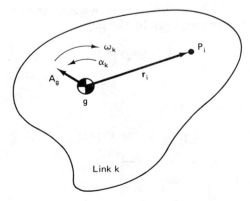

Figure 5.4 Link **k** in general planar motion.

plex number form. At a particular instant in time link k has a known angular velocity ω_k and angular acceleration α_k. The center of mass is determined to be located at g and is accelerating at $\mathbf{A}_g$. Any particle (say P_i) within this link must obey Newton's laws. The acceleration of P_i may be found by the acceleration-difference technique (Sec. 4.2).

$$\mathbf{A}_{P_i} = \mathbf{A}_g + \mathbf{A}_{(P_i)g} \tag{5.1}$$

Or we can express $\mathbf{A}_{P_i}$ and $\mathbf{A}_{(P_i)g}$ in their two acceleration components (there is no Coriolis acceleration because the vector $\mathbf{r}_i$ does not change length).

$$\mathbf{A}^n_{P_i} + \mathbf{A}^t_{P_i} = \mathbf{A}_g + \mathbf{A}^n_{(P_i)g} + \mathbf{A}^t_{(P_i)g} \tag{5.2}$$

where $\mathbf{A}^n_{(P_i)g} = -\mathbf{r}_i\omega_k^2$, and is directed from P_i toward g; and $\mathbf{A}^t_{(P_i)g} = \mathbf{r}_i\alpha_k e^{i\pi/2}$ and is perpendicular to the normal component in the sense of α_k. Newton's second law may be applied to particle P_i to determine the force applied to particle P_i in a planar link:

$$\frac{d\mathbf{M}_{P_i}}{dt} = m_i\frac{d\mathbf{V}_{P_i}}{dt} = m_i\mathbf{A}_{P_i} = m_i\mathbf{A}_g - m_i\mathbf{r}_i\omega_k^2 + m_i\mathbf{r}_i\alpha_k e^{i\pi/2} = \mathbf{F}_i \tag{5.3}$$

where m_i is particle mass, and $\mathbf{M}$ is momentum expressed in complex vector form, as are also $\mathbf{V}$, $\mathbf{A}$, and $\mathbf{r}$.

The resultant force applied to the entire link may be found by summing the contributions of all particles P_i

$$\mathbf{F} = \sum_i \mathbf{F}_i = \sum_i m_i\mathbf{A}_g - \sum_i m_i\mathbf{r}_i\omega_k^2 + \sum_i m_i\mathbf{r}_i\alpha_k e^{i\pi/2} \tag{5.4}$$

or

$$\mathbf{F} = \sum_i \mathbf{F}_i = \mathbf{A}_g\sum_i m_i - \omega_k^2\sum_i m_i\mathbf{r}_i + \alpha_k e^{i\pi/2}\sum_i m_i\mathbf{r}_i \tag{5.5}$$

Equation (5.5) was derived from Eq. (5.4) based on the following observations: (1) in the first term, the acceleration of the center of gravity remains constant over the summation; (2) in the second and third terms, ω_k^2 and α_k may be brought out of the summation since they are constant for the particular position in time in which the forces on link k are being analyzed. (3) The minus sign on the ω_k^2 terms is present since the force is directed from each particle toward g, the center of mass, while $\mathbf{r}_i$ points from g to P_i.

Equation (5.5) may be further simplified by noticing that

$$\sum_i m_i = M \qquad \text{the total mass of the link } k \tag{5.6}$$

and

$$\sum_i m_i\mathbf{r}_i = 0 \tag{5.7}$$

since g is the center of gravity. Thus Eq. (5.5) simplifies to

$$\boxed{\mathbf{F} = M\mathbf{A}_g} \tag{5.8}$$

If a summation of moments about the center of gravity of all points P_i is made, then the normal terms disappear (because they have no moments about the CG) and the resulting torque becomes

$$T = \sum_i m_i r_i \alpha_k r_i = \alpha_k \sum_i m_i r_i^2 \qquad (5.9)$$

By definition the summation on the right side of the equation is the *mass moment of inertia about the center of gravity* (I_g); so Eq. (5.9) may be rewritten as

$$\boxed{T = I_g \alpha_k} \qquad (5.10)$$

Thus treating the rigid link k of Fig. 5.4 as a conglomeration of particles will lead to Eq. (5.5), which can then be simplified to a force $\mathbf{F} = M\mathbf{A}_g$ through the center of gravity in the direction of the acceleration and a torque $T = I_g \alpha_k$ in the sense of the angular acceleration. So if we observe that the link is moving with $\mathbf{A}_g$, ω_k, and α_k, then the force $\mathbf{F}$ and the torque T must be acting on that link. Since Eq. (5.8) has two components in the planar case, there will result three independent equations of *dynamic equilibrium* which may be written for any link k

$$\sum F_x = MA_{gx} \qquad (5.11)$$

$$\sum F_y = MA_{gy} \qquad (5.12)$$

$$\sum T = I_g \alpha_k \qquad (5.13)$$

where the summation of forces in the x direction in Eq. (5.11) and the y direction in Eq. (5.12) are parallel to the axes of any convenient fixedly oriented x,iy system.

Equations (5.11) to (5.13) represent the laws of dynamics of rigid bodies moving in planar motion. They may be arranged into convenient extensions of the laws of statics. A preferred arrangement results by applying D'Alembert's principle, which can be stated as follows: *the sum of the inertial or body forces and torques and the external forces and torques together produce equilibrium, or*

$$\sum F_x + (-MA_{gx}) = 0 \qquad (5.14)$$

$$\sum F_y + (-MA_{gy}) = 0 \qquad (5.15)$$

$$\sum T + (-I_g \alpha_k) = 0 \qquad (5.16)$$

Here Eqs. (5.11) to (5.13) have been rewritten to express dynamic equilibrium, in which $(-MA_{gx})$ and $(-MA_{gy})$ are known as *inertia force coordinates* in the x and y directions, or as their real and imaginary parts, and $(-I_g \alpha_k)$ is known as the *inertia torque, inertia couple,* or *inertia moment.* The inertia force and inertia torque concepts have been devised to enable handling a dynamic problem similarly to a static problem: by way of equations of equilibrium.

5.2 INERTIA FORCES IN LINKAGES

Suppose that a mechanism has constrained motion in a horizontal plane, so that gravity has no effect. Furthermore, all joints are free from friction, and there are no forces or moments transmitted through link k. Assume further that the accelerations are known and the pin forces are to be determined.

Referring to link k of the mechanism (Fig. 5.5a) in light of the foregoing discussion, the linear acceleration A_g and the angular acceleration α_k are known from a kinematic analysis. The pin joints at A and B are points where unknown forces F_A and F_B act on link k. These two forces add vectorially to yield a resultant force R. If, for convenience, the y axis is set in the direction of the acceleration A_g, Eqs. (5.14) and (5.15) become

$$R_y - mA_{gy} = 0 \tag{5.17}$$

and

$$R_x = 0 \tag{5.18}$$

where m is the mass of the link. Hence the resultant R must be parallel to and point in the same direction as A_g. The inertia force F_0 is a measure of the inertia of the link (which represents a resistance to linear translational acceleration) and is defined as

$$F_0 = -mA_g \tag{5.19}$$

Therefore,

$$R + F_0 = 0 \tag{5.20}$$

and the forces on link k are in dynamic equilibrium.

Equation (5.16) must also be satisfied. There are two approaches in general use which assure dynamic equilibrium in the presence of rotary inertia. The *first* is to replace the effects of A_g and α_k by F_0 [defined by Eq. (5.19)], but which is offset from the CG by ε_k (see Fig. 5.5a). Offset ε_k is found by summing moments of the applied forces about the center of gravity.

Here we digress to define the moment of a force in complex notation. In three-dimensional coordinates, $M_R = r \times F$, where F is the force and r is a vector from the fulcrum to any point on the line of action of F. In determinant form this is

$$M_R = \begin{vmatrix} \mathbf{i} & \mathbf{j} & \mathbf{k} \\ r_x & r_y & r_z \\ F_x & F_y & F_z \end{vmatrix}$$

Specializing this to our case yields

$$\mathbf{M}_R = \begin{vmatrix} \mathbf{i} & \mathbf{j} & \mathbf{k} \\ \varepsilon_{kx} & \varepsilon_{ky} & 0 \\ 0 & R_y & 0 \end{vmatrix} = (\varepsilon_{kx} R_y)\mathbf{k}$$

In the coordinate system of Fig. 5.5a, $\varepsilon_{ky} = 0$, and hence, ε_k is real, which means that it is perpendicular to R, turned 90° cw from the direction of R.

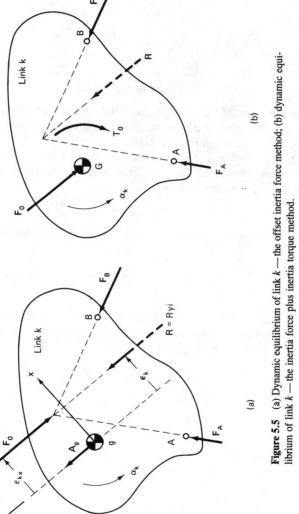

Figure 5.5 (a) Dynamic equilibrium of link k — the offset inertia force method; (b) dynamic equilibrium of link k — the inertia force plus inertia torque method.

Applying this to express the dynamic equilibrium of link k of Fig. 5.5a, we get

$$R_y \varepsilon_{kx} - I_g \alpha_k = 0 \tag{5.21}$$

where R_y and ε_{kx} are signed reals, giving the correct sign of the moment of **R** about g. Since I_g, α_k, and $\mathbf{R} = R_y i$ are known, the eccentricity vector of the inertia force is

$$\varepsilon_k = \varepsilon_{kx} = \frac{I_g \alpha_k}{R_y} = \frac{I_g \alpha_k}{m A_{gy}} \tag{5.22}$$

(y axis in the $\mathbf{A}_g$ direction)

which gives the value of the offset ε_{kx} with the correct sign along the x axis. Notice from Eq. (5.22) that the *offset of the inertia force is purely a function of the linear and angular accelerations and the mass and inertia.*

It is to be noted that it is not necessary to take the y axis in the direction of $\mathbf{A}_g$. Regardless of the choice of the coordinate system, the following vector relationships prevail, expressed in vector notation,

$$\boldsymbol{\varepsilon}_k \times \mathbf{R} - I_g \boldsymbol{\alpha}_k = 0$$

or in three-dimensional notation applied to our two-dimensional case:

$$\begin{vmatrix} \mathbf{i} & \mathbf{j} & \mathbf{k} \\ \varepsilon_{kx} & \varepsilon_{ky} & 0 \\ R_x & R_y & 0 \end{vmatrix} = (\varepsilon_{kx} R_y - \varepsilon_{ky} R_x)\mathbf{k} = I_g \alpha_k \mathbf{k} \tag{5.23}$$

Canceling **k** and recalling that the direction of $\boldsymbol{\varepsilon}_k$ (which is a vector from g to any point on the line of action of **R**) is arbitrary, we let $\boldsymbol{\varepsilon}_k$ be perpendicular to **R**; then it can be shown (see Prob. 5.1b and Fig. 5.6).

$$\frac{\varepsilon_{kx}}{\varepsilon_{ky}} = -\frac{R_y}{R_x} \tag{5.24}$$

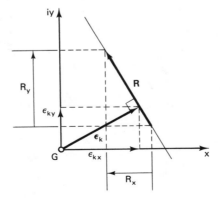

Figure 5.6 Graphical proof of Eq. (5.24) by way of similar triangles turned 90° to one another.

Solving Eqs. (5.23) and (5.24) simultaneously for ε_{kx} and ε_{ky}, we obtain

$$\varepsilon_{kx} R_y - \varepsilon_{ky} R_x = I_g \alpha_k$$
$$\varepsilon_{kx} R_x + \varepsilon_{ky} R_y = 0 \tag{5.25}$$

$$\varepsilon_{kx} = \frac{\begin{vmatrix} I_g \alpha_k & -R_x \\ 0 & R_y \end{vmatrix}}{\begin{vmatrix} R_y & -R_x \\ R_x & R_y \end{vmatrix}} = \frac{R_y I_g \alpha_k}{R^2} \tag{5.26}$$

and

$$\varepsilon_{ky} = \frac{\begin{vmatrix} R_y & I_g \alpha_k \\ R_x & 0 \end{vmatrix}}{R^2} = \frac{-R_x I_g \alpha_k}{R^2} \tag{5.27}$$

Equations (5.26) and (5.27) give the real and imaginary parts of ε_k with the correct algebraic sign, thus eliminating the need for visual inspection of the geometric configuration of the link and forces acting on it, and are therefore suitable for automatic digital computation.

The *second* method for assuring dynamic equilibrium (see Fig. 5.5b) uses the inertia force $\mathbf{F}_0$ acting through the center of gravity and an inertia torque $\mathbf{T}_0$ acting in the opposite sense as the angular acceleration where

$$T_0 = -I_g \alpha_k \tag{5.28}$$

These representations of the inertia force and inertia torque will be used shortly for the purpose of mechanism analysis.

Example 5.1

Figure 5.7a shows a coupler link 3 of a four-bar linkage with $m_3 = 18$ kg and $I_{g3} = 4.5$ kg·m². The forces of the input and output links acting on the coupler ($\mathbf{F}_{23}$ and $\mathbf{F}_{43}$) balance the inertia force $\mathbf{F}_{O3}$ since no other forces are acting on the coupler. If $\mathbf{A}_{g3} = 10e^{i60°}$ m/sec² and $\alpha_3 = 4$ rad/sec² ccw, determine ε_3, $\mathbf{F}_{23}$, and $\mathbf{F}_{43}$ by both methods described above. Notice that the line of action of the force of link 4 on 3 ($\mathbf{F}_{43}$) is known.

From Eq. (5.22), changing the y axis to coincide with $\mathbf{A}_{g3}$,

$$\varepsilon_{3x} = \frac{(4.5)(4)}{(18)(10)} = 0.1 \text{ m}$$

From Eqs. (5.19) and (5.28),

$$F_{O3y} = -180 \text{ N} \qquad (\text{N} \equiv \text{newton})$$

and

$$T_{O3} = -18 \text{ N·m cw}$$

Method 1.   Figure 5.7b and c show a graphical solution to the three-force equilibrium ($\mathbf{F}_{23}$, $\mathbf{F}_{43}$, and $\mathbf{F}_{O3}$). Notice that the moment of $\mathbf{F}_{O3}$ is opposite the angular acceleration and that the line of action of $\mathbf{F}_{O3}$ is offset from g_3 by ε_3. (Fig. 5.7a and b). From

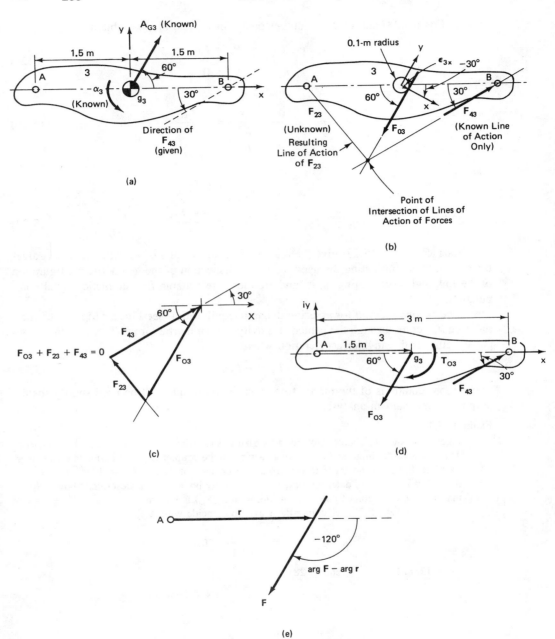

Figure 5.7 Dynamic equilibrium of coupler link 3 of four-bar mechanism.

Fig. 5.7b, the line of action of $\mathbf{F}_{23}$ is determined. Figure 5.7c shows the graphical solution for the magnitudes and sense of $\mathbf{F}_{23}$ and $\mathbf{F}_{43}$, from which:

$$\mathbf{F}_{23} = 93e^{i126°} \text{ N}$$

$$\mathbf{F}_{43} = 167.8e^{i30°} \text{ N}$$

Method 2.   Figure 5.7d shows the inertia force $\mathbf{F}_{O3}$ and the inertia torque $\mathbf{T}_{O3}$. Here a note about the moment of a force $\mathbf{F}$ about a point A, expressed in terms of complex numbers, is in order. In Fig. 5.7e, $\mathbf{r}$ is a vector from A to any point on the line of action of $\mathbf{F}$. Then the moment of $\mathbf{F}$ about A is

$$M_A = rF \, \sin(\arg \mathbf{F} - \arg \mathbf{r}) \tag{5.29}$$

Notice that Eq. (5.29) gives the moment as a signed scalar: positive ccw and negative cw. Referring now to Fig. 5.7d and using Eq. (5.29), summation of moments about A yields

$$T_{O3} - 1.5F_{O3} \sin 60° + 3F_{43} \sin 30° = -18 - (1.5)(180)(0.87) + 3F_{43}(0.5) = 0$$

from which

$$F_{43} = 167.8 \text{ N}$$

From the summation of forces in the x and y directions, now taking the x axis along the line AB, we obtain

$$\mathbf{F}_{23} = 93e^{i126°} \text{ N}$$

Center of Percussion*

When a link is rotating about a fixed axis, an interesting property of the offset $\boldsymbol{\varepsilon}_k$ can be found by analysis of Fig. 5.8a. The acceleration components $\mathbf{A}_g^t$ and $\mathbf{A}_g^n$ form a triangle with $\mathbf{A}_g$ that is similar to triangle gpE, where g is the center of gravity, p is the intersection of the inertia force with the centerline of link B_0B, and E is the intersection of the line of action of the inertia force $\mathbf{F}_0$ and the perpendicular to the line of $\mathbf{F}_0$ through g. Now

$$\boldsymbol{\varepsilon}_k \times \mathbf{F}_0 = -I_g \boldsymbol{\alpha}_k$$

or

$$\boldsymbol{\varepsilon}_k \times (-m\mathbf{A}_g) = -I_g \boldsymbol{\alpha}_k$$

Expressing the cross product trigonometrically,

$$-m\varepsilon_k A_g \, \sin(\arg \mathbf{A}_g - \arg \boldsymbol{\varepsilon}_k)\mathbf{k} = -I_g \alpha_k \mathbf{k}$$

or canceling $-\mathbf{k}$ and solving for $\boldsymbol{\varepsilon}_k$, we obtain

$$\varepsilon_k = \frac{I_g \alpha_k}{mA_g \, \sin(\arg \mathbf{A}_g - \arg \boldsymbol{\varepsilon}_k)}$$

---

* This and the next subsection may be skipped without loss of continuity.

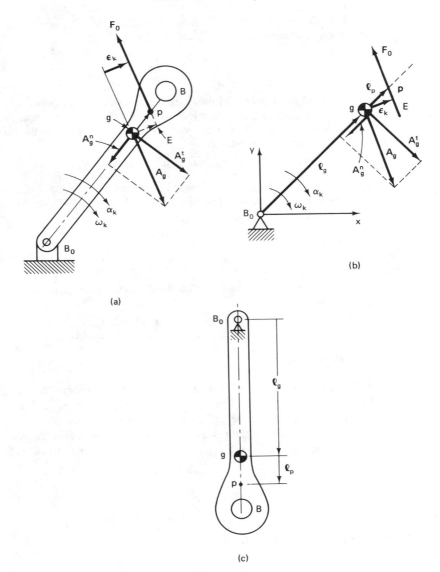

Figure 5.8 Center of percussion p of a link pivoted at B_0.

Note that $\sin(\arg \mathbf{A}_g - \arg \boldsymbol{\varepsilon}_k) = \operatorname{signum}(\alpha_k) = \alpha_k/|\alpha_k|$, since $(\arg \mathbf{A}_g - \arg \boldsymbol{\varepsilon}_k) = \pm 90°$ according as α_k is positive or negative, and $\operatorname{signum}(\alpha_k)$ is $+1$ for ccw α_k and -1 for cw α_k. Therefore,

$$\varepsilon_k = \frac{I_g \alpha_k}{mA_g(\alpha_k/|\alpha_k|)} = \frac{I_g}{mA_g}|\alpha_k| \tag{5.30}$$

The direction of ε_k is $(\arg \mathbf{A}_g) \pm \pi/2$ according as α_k is cw or ccw. Therefore,

$$\boldsymbol{\varepsilon}_k = \varepsilon_k e^{i(\arg \mathbf{A}_g - (\alpha_k/|\alpha_k|)(\pi/2))} \tag{5.31}$$

From Fig. 5.8b,

$$\frac{\ell_p}{\varepsilon_k} = \frac{l_p}{\varepsilon_k} e^{i(\arg \ell_p - \arg \varepsilon_k)} = \frac{A_g}{A_g^t} e^{i(\arg \mathbf{A}_g^t - \arg \mathbf{A}_g)}$$

$$\ell_p = \varepsilon_k \frac{A_g}{A_g^t} e^{i(\arg \mathbf{A}_g^t - \arg \mathbf{A}_g)}$$

$$\ell_p = \left[\frac{I_g}{mA_g}|\alpha_k|e^{i[\arg \mathbf{A}_g - (\alpha_k/|\alpha_k|)(\pi/2)]}\right]\frac{A_g}{A_g^t} e^{i(\arg \mathbf{A}_g^t - \arg \mathbf{A}_g)}$$

$$\ell_p = \frac{I_g}{ml_g|\alpha_k|}|\alpha_k|e^{i[\arg \mathbf{A}_g^t - (\alpha_k/|\alpha_k|)(\pi/2)]}$$

but

$$e^{i[\arg \mathbf{A}_g^t - (\alpha_k/|\alpha_k|)(\pi/2)]} = e^{i(\arg \ell_g)}$$

Therefore,

$$\ell_p = \frac{I_g}{ml_g} e^{i(\arg \ell_g)} \tag{5.32}$$

It can be observed from Eq. (5.32) that the vector ℓ_p is not a function of the acceleration components and in fact is a *constant*. Thus the inertia force $\mathbf{F}_0$ always passes through point p, which is called the *center of percussion* of link k with respect to its fixed axis of rotation B_0. Another property of the center of percussion (from which its name is derived) is that when a pendulum is struck by a force in the plane of oscillation and perpendicular to the B_0g axis such that the line of action of the force goes through the center of percussion, there will be no reaction force at the point of suspension. Sensible evidence of this is the "sting" effect produced on the hands when batting a ball if the ball strikes the bat at other than the center of percussion. Note that the center of percussion p is always farther from the pivot B_0 than the center of gravity g is from B_0.

Finding Mass-Moment of Inertia by Experiment

Links of mechanisms are often designed with varied geometries and thicknesses due to interference, stress, or other considerations. Calculation of the moment of inertia about the center of gravity can be tedious in these cases. A simple experimental procedure can be used instead.

Figure 5.8c shows the rigid link of Fig. 5.8a, which is set in oscillation on a knife-edge at B_0 so that it is permitted to swing freely as a pendulum. If k is the principle *radius of gyration** (radius of gyration about the center of gravity), then from Eq. (5.32),

---

* The radius of gyration is that radius at which the entire distributed mass of the body could be concentrated in a thin ring around the (now massless) center of gravity and yield the same mass moment of inertia as the actual distributed mass.

$$l_p = \frac{mk^2}{ml_g} = \frac{k^2}{l_g}$$

Thus

$$k^2 = l_p l_g \tag{5.33}$$

The magnitude of ℓ_p and thus the location of the center of percussion can be derived more simply from the following consideration:

Where should the line of action of a force **F**, perpendicular to $(B_0 g)$, intersect the axis $B_0 g$ in order to impart angular acceleration α_k to link k without causing a reaction force at B_0?

To this end, the force must accelerate the center of mass g by $\alpha_k l_g$ and, in addition, impart angular acceleration α_k about g by exerting a moment about g, thus leaving B_0 momentarily stationary:

$$F = \alpha_k l_g m \tag{5.34}$$

$$F l_p = \alpha_k I_g = \alpha_k m k^2 \tag{5.35}$$

From Eq. (5.34),

$$\frac{F}{m\alpha_k} = l_g \tag{5.36}$$

and from Eq. (5.35),

$$\frac{F}{m\alpha_k} = \frac{k^2}{l_p} \tag{5.37}$$

Equating the right sides of Eqs. (5.36) and (5.37) and solving for l_p yields

$$l_p = \frac{k^2}{l_g} \tag{5.33a}$$

which agrees with Eq. (5.33).

As known from elementary physics, a property of the center of percussion is that if a rigid body swings freely as a physical pendulum, its period of oscillation is the same as that of a simple pendulum having a length equal to the distance between the center of percussion and the axis of oscillation (distance $l_g + l_p$ in Fig. 5.8c). This suggests an experimental technique for determining the moment of inertia of a link of a mechanism.

The period of oscillation T of the rigid link in Fig. 5.8c can be determined from elementary physics using

$$T = 2\pi \sqrt{\frac{l_g + l_p}{g}} \tag{5.38}$$

where g is the acceleration of gravity. Therefore,

$$l_p = \frac{T^2 g}{4\pi^2} - l_g \tag{5.39}$$

Also, from Eq. (5.33), we have

$$I_g = mk^2 = ml_g l_p \tag{5.40}$$

Thus, from (5.39) and (5.40),

$$I_g = ml_g \left(\frac{T^2 g}{4\pi^2} - l_g \right) \tag{5.41}$$

Once the period T is measured, the moment of inertia may be obtained from Eq. (5.41).

A property of the mass moment of inertia for a link rotating about a fixed center is important in linkage force analysis. If in Fig. 5.7 the summation of torques is made about a fixed center, and the moment of inertia about the center of gravity is used, terms in the moment equation will include $I_g \alpha_k$ and $\overrightarrow{Ag_k} \times \mathbf{F}_0$. If, however, the moment of inertia about the axis of rotation is used, the inertia force $\mathbf{F}_0$ drops out and only the $I_0 \alpha_k$ term appears, where $I_0 = I_g + m(l_g)^2$.

Example 5.2

Where would you recommend a horizontal blow be applied to the free-hanging bar (Fig. 5.9) to minimize the reactions at the bearing at B_0 if the angular acceleration is to be -20 rad/sec^2 cw?

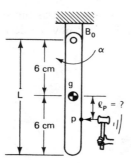

Figure 5.9   When a rod is struck at the center of percussion, the point of suspension B_0 is free from lateral reaction: the batter's hand feels no "sting" if the center of percussion of the bat strikes the ball.

Solution   The moment of inertia of the (prismatic) bar about its center of gravity is

$$I_g = \tfrac{1}{12} mL^2 = mk_g^2$$

Thus

$$k_g^2 = \tfrac{1}{12} L^2 = 12 \text{ cm}^2$$

From Eq. (5.33), $l_p = 2$ cm or the blow should be 8 cm from B_0.

5.3 KINETOSTATIC ANALYSIS OF MECHANISMS

Table 5.1 shows that inertia forces can be accounted for in mechanisms in a kinetostatic analysis yielding algebraic equations. Kinetostatic methods are discussed in Secs. 5.3 to 5.7. In Fig. 5.10, a typical four-bar mechanism is shown at a particular instant of the motion cycle. For the following analysis, linkages are assumed to be comprised of rigid links connected by frictionless joints in continuous contact (no clearance in joints). Let us assume that the linear accelerations of the mass centers and the angular accelerations of the moving links have been determined by the methods of Chap. 4. The three inertia forces $\mathbf{F}_{O2}$, $\mathbf{F}_{O3}$, and $\mathbf{F}_{O4}$ represent kinetostatic loading (see Table 5.1) of the mechanism. Assume that their magnitudes and locations have been determined by Eqs. (5.19) and (5.22). The objective of the kinetostatic analysis is the determination of the bearing forces and the required input shaft torque at the particular instant in the position shown in Fig. 5.10 by using the concept of dynamic equilibrium applied to the inertia forces and moments. Once analytical expressions are found for a single position, magnitudes of bearing forces and input torque for additional positions are easily determined. There are three techniques in general use for accomplishing these force analysis objectives: the graphical or analytical superposition method and the matrix method. We will describe all these methods. Since the methods presented may be difficult initially for the reader, a number of problems and case studies will be worked out in great detail within the chapter.

5.4 THE SUPERPOSITION METHOD [110] (GRAPHICAL AND ANALYTICAL)

For a mechanism in a certain position with assumed (or known) velocity and acceleration conditions, the equations that will now be derived are *algebraic* and *linear* in the

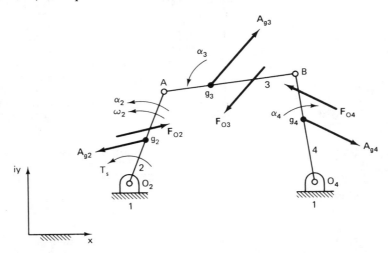

Figure 5.10 Inertia forces F_{Ok} ($k = 2, 3, 4$) arising in a four-bar mechanism due to link instantaneous accelerations.

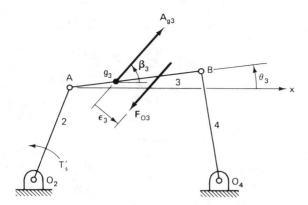

Figure 5.11 In linear superposition, we take the effect of one inertia force at a time.

inertia forces $\mathbf{F}_{O2}$, $\mathbf{F}_{O3}$, and $\mathbf{F}_{O4}$. Therefore, the principle of superposition applies: The effects of the individual inertia forces can be treated separately and then superposed to determine their combined effect. The equations of motion are based on individual free-body diagrams, one for each moving link. The resulting solutions are simple and easily solved graphically, or analytically, say, on a hand calculator, or a personal computer.

Here is how we get started. Figure 5.11 shows the four-bar linkage with inertia force $\mathbf{F}_{O3}$ as the only load. Thus the bearing forces and shaft torque to be determined are those related to $\mathbf{F}_{O3}$ alone. Similar independent force analyses will follow with $\mathbf{F}_{O2}$ and $\mathbf{F}_{O4}$ each acting alone, and the resultant bearing forces and shaft torque will be obtained by superposition of all three sets of results.

Step 1: The Effect of F_{O3}

With $\mathbf{F}_{O3}$ acting alone, the free body to be considered first is that of link 3 (Figure 5.12b). Since the acceleration $\mathbf{A}_{g3} = (A_{g3}e^{i\beta_3})$ and the angular acceleration α_3 are known, the inertia force vector $\mathbf{F}_{O3}$ may be determined by

$$\mathbf{F}_{O3} = (m_3 A_{g3})e^{i(\beta_3 + \pi)} \tag{5.42}$$

where $(\beta_3 + \pi)$ indicates that the sense of $\mathbf{F}_{O3}$ is opposite to that of $\mathbf{A}_{g3}$, whose sense is given by β_3. Because of the presence of angular acceleration α_3, the line of action of $\mathbf{F}_{O3}$ is offset by ($\varepsilon_3 = |I_3\alpha_3/F_{O3}|$) (Eq. (5.22)) from the line of action of $\mathbf{A}_{g3}$. In complex vector form,

$$\varepsilon_3 = -\frac{\alpha_3}{|\alpha_3|}\varepsilon_3 i e^{i\beta_3}$$

where $ie^{i\beta_3}$ gives the direction of ε_3 perpendicular to $\mathbf{A}_{g3}$ and $-\alpha_3/|\alpha_3|$ yields the algebraic sign which determines the sense of ε_3 along the $ie^{i\beta_3}$ direction. Substituting for ε_3, we have

$$\boldsymbol{\varepsilon}_3 = -\frac{\alpha_3}{|\alpha_3|}\left|\frac{I_3\alpha_3}{F_{O3}}\right|ie^{i\beta_3} = \frac{-I_3\alpha_3}{F_{O3}}ie^{i\beta_3}$$

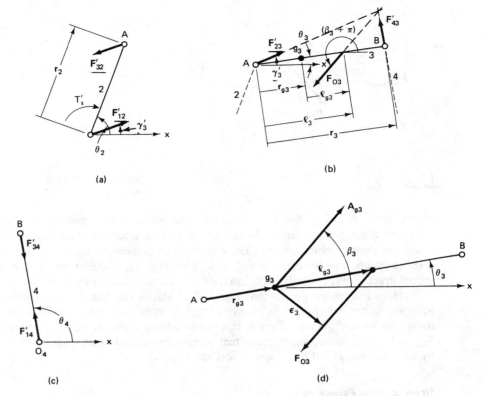

Figure 5.12 The effect of inertia force F_{03} on pin forces. Note that γ_3' in (b) is not to scale: it is unknown, to be found by graphical or analytical means.

For convenience in making calculations, the location of the line of action of $\mathbf{F}_{03}$ may be given by the vector ℓ_3 from the moving pivot A, as shown in Fig. 5.12b, where

$$\ell_3 = \mathbf{r}_{g3} + \ell_{g3}$$

From Fig. 5.12d,

$$\varepsilon_3 = \ell_{g3} \cos\left(\theta_3 + \frac{\pi}{2} - \beta_3\right)e^{-i[\theta_3+(\pi/2)-\beta_3]}$$

$$\varepsilon_3 = -\ell_{g3} \sin(\theta_3 - \beta_3)e^{-i[\theta_3+(\pi/2)-\beta_3]}$$

$$\ell_{g3} = \frac{-\varepsilon_3}{\sin(\theta_3 - \beta_3)}e^{i[\theta_3+(\pi/2)-\beta_3]} = +\frac{I_3\alpha_3 i e^{i\beta_3}e^{i[\theta_3+(\pi/2)-\beta_3]}}{F_{03}\sin(\theta_3 - \beta_3)}$$

$$\ell_{g3} = \frac{-I_3\alpha_3 e^{i\theta_3}}{F_{03}\sin(\theta_3 - \beta_3)}$$

or

$$\ell_3 = \left[r_{g3} - \frac{I_3\alpha_3/F_{O3}}{\sin(\theta_3 - \beta_3)} \right] e^{i\theta_3} \tag{5.43}$$

where I_3 and F_{O3} are absolute values and the sign of α_3 takes care of the sense of the vector ℓ_{g3}.

Of the three forces acting on link 3, $\mathbf{F}_{O3}$ is the known dynamic load and $\mathbf{F}'_{23}$ and $\mathbf{F}'_{43}$ are the unknown joint forces. Link 4, which for the moment is considered massless, is acted on by only two forces as shown in Fig. 5.12c; therefore, link 4 can be considered as a two-force member and the direction of the line of action of $\mathbf{F}'_{43}$ is θ_4. Thus the forces acting on the coupler are identical to those in the situation proposed in Ex. 5.1. Summing the forces on link 3 expresses their kinetostatic equilibrium

$$\mathbf{F}'_{23} + \mathbf{F}'_{43} + \mathbf{F}_{O3} = 0 \tag{5.44}$$

which can be solved graphically as shown in Figs. 5.13 and 5.14.

To solve Eq. (5.44) analytically, we rewrite it in exponential form:

$$F'_{23}(e^{i\gamma'_3}) + F'_{43}(e^{i\theta_4}) + F_{O3}[e^{i(\beta_3+\pi)}] = 0 \tag{5.45}$$

Separating real and imaginary parts of Eq. (5.45), we have

$$F'_{23} \cos\gamma'_3 + F'_{43} \cos\theta_4 + F_{O3} \cos(\beta_3 + \pi) = 0 \tag{5.46}$$

$$F'_{23} \sin\gamma'_3 + F'_{43} \sin\theta_4 + F_{O3} \sin(\beta_3 + \pi) = 0 \tag{5.47}$$

To determine all three unknowns, however, another equation is required in addition to Eqs. (5.46) and (5.47). The additional equation is one of equilibrium of moments about some point on the coupler. Choosing point A, we get

$$F'_{43}r_3 \sin(\theta_4 - \theta_3) + F_{O3}l_3 \sin(\beta_3 + \pi - \theta_3) = 0 \tag{5.48}$$

or

$$F'_{43} = -F_{O3} \frac{l_3 \sin(\beta_3 + \pi - \theta_3)}{r_3 \sin(\theta_4 - \theta_3)}$$

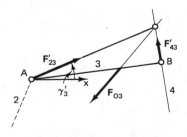

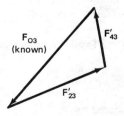

Figure 5.13 Graphical determination of the line of action of $\mathbf{F}'_{23}$.

Figure 5.14 Graphical determination of the magnitude and sense of forces $\mathbf{F}'_{23}$ and $\mathbf{F}'_{43}$.

The real and imaginary components of $\mathbf{F}'_{23}$ (F'_{23x} and F'_{23y}) may now be determined from Eqs. (5.46) and (5.47).

$$F'_{23x} = F'_{23} \cos \gamma'_3 = -F'_{43} \cos \theta_4 - F_{O3} \cos(\beta_3 + \pi) \qquad (5.49)$$

$$F'_{23y} = F'_{23} \sin \gamma'_3 = -F'_{43} \sin \theta_4 - F_{O3} \sin(\beta_3 + \pi) \qquad (5.49a)$$

Note that F'_{23x} and F'_{23y} each may turn out to be positive or negative. The magnitude and direction of $\mathbf{F}'_{23}$ are computed as follows.

$$F'_{23} = \sqrt{(F'_{23x})^2 + (F'_{23y})^2} \qquad (5.50)$$

$$\gamma'_3 = \arg(F'_{23x} + iF'_{23y}) \qquad (5.51)$$

The free-body diagram of link 4 (Fig. 5.12c), shows that the bearing force $\mathbf{F}'_{14}$ at O_4 is identical to the force $\mathbf{F}'_{43}$, because only two forces act on link 4. Similarly, there are only two forces acting on link 2, as shown in Fig. 5.12a, because it is considered momentarily massless and otherwise unloaded. Thus the bearing force $\mathbf{F}'_{12}$ at O_2 is identical to $\mathbf{F}'_{23}$. The final step of determining the shaft torque required to maintain the assumed state of motion of link 3, T'_s, may be derived from the equilibrium of moments acting on link 2.

To find T'_s graphically, redraw Fig. 5.12a to scale and measure the perpendicular distance between the parallel forces, say "d." Then $|T'_s| = |F'_{32}|d$, and the sense must counteract the effects of F'_{12} and F'_{32} and, therefore, is clockwise.

To find T'_s analytically, using Eq. (5.29), we obtain

$$\begin{aligned} T'_s &= -F'_{12}r_2 \sin(\gamma'_3 - \theta_2 + \pi) \\ &= -F'_{23}r_2 \sin(\theta_2 - \gamma'_3) \end{aligned} \qquad (5.52)$$

Thus we have determined the bearing forces and shaft torque due to the load $\mathbf{F}_{O3}$. A similar analysis with only $\mathbf{F}_{O4}$ acting and a third analysis yielding the influence of F_{O2} will now be outlined.

Step 2: The Effect of F_{O4}

Following the same procedure as step 1, the inertia force of link 4 is described in complex-number notation (refer to Fig. 5.15).

$$\mathbf{F}_{O4} = (m_4 A_{g4})e^{i(\beta_4 + \pi)} \qquad (5.53)$$

$\mathbf{F}_{O4}$ is offset from the center of mass by

$$\varepsilon_4 = -\frac{I_4 \alpha_4}{F_{O4}} i e^{i\beta_4} \qquad (5.54)$$

and is located by

$$\ell_4 = \mathbf{r}_{g4} + \frac{\varepsilon_4(\alpha_4/|\alpha_4|)e^{i\theta_4}}{\sin(\beta_4 - \theta_4)} = \mathbf{r}_{g4} + \frac{I_4\alpha_4/F_{O4}}{\sin(\beta_4 - \theta_4)}e^{i\theta_4} \qquad (5.55)$$

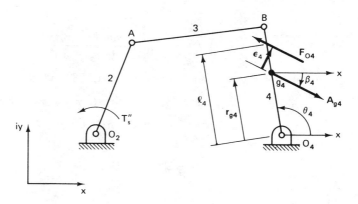

Figure 5.15 To evaluate the effect of the inertia of link 4 upon input torque requirement, consider all other links massless and unloaded.

Summation of the forces on link 4 yields

$$\mathbf{F}''_{34} + \mathbf{F}''_{14} + \mathbf{F}_{O4} = 0 \tag{5.56}$$

To solve Eq. (5.56) graphically we proceed as shown in Fig. 5.16.

The analytical solution of Eq. (5.56) begins by writing it in exponential form:

$$F''_{34}e^{i\theta_3} + F''_{14}e^{i\gamma''_4} + F_{O4}e^{i(\beta_4+\pi)} = 0 \tag{5.57}$$

and then separating real and imaginary parts.

$$F''_{34}\cos\theta_3 + F''_{14}\cos\gamma''_4 + F_{O4}\cos(\beta_4 + \pi) = 0 \tag{5.58}$$

$$F''_{34}\sin\theta_3 + F''_{14}\sin\gamma''_4 + F_{O4}\sin(\beta_4 + \pi) = 0 \tag{5.59}$$

where the unknowns are F''_{34}, F''_{14}, and γ''_4. Summation of moments about O_4 gives us the additional required equation

$$F''_{34}r_4\sin(\theta_3 - \theta_4) + F_{O4}l_4\sin(\theta_4 - \beta_4) = 0 \tag{5.60}$$

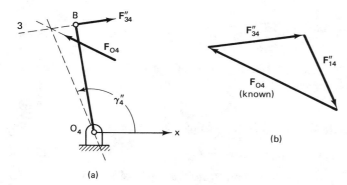

Figure 5.16 (a) Finding the line of action of $\mathbf{F}''_{14}$; (b) finding the magnitudes and sense of $\mathbf{F}''_{34}$ and $\mathbf{F}''_{14}$ (graphical method).

so that

$$F''_{34} = -F_{O4} \frac{l_4 \sin(\theta_4 - \beta_4)}{r_4 \sin(\theta_3 - \theta_4)} \tag{5.61}$$

Using this value of F''_{34}, we can now solve Eqs. (5.58) and (5.59) for F''_{14}:

$$F''_{14x} = F''_{14} \cos \gamma''_4 = -F_{O4} \cos(\beta_4 + \pi) - F''_{34} \cos \theta_3$$
$$F''_{14y} = F''_{14} \sin \gamma''_4 = -F_{O4} \sin(\beta_4 + \pi) - F''_{34} \sin \theta_3 \tag{5.62}$$

so that

$$F''_{14} = \sqrt{(F''_{14x})^2 + (F''_{14y})^2} \tag{5.63}$$

and

$$\gamma''_4 = \arg(F''_{14x} + iF''_{14y}) \tag{5.64}$$

From Fig. 5.17b we see that

$$\mathbf{F}''_{43} = -\mathbf{F}''_{34}$$
$$\mathbf{F}''_{32} = -\mathbf{F}''_{23} = -\mathbf{F}''_{34} \tag{5.65}$$

Summing moments about O_2 on link 2 yields

$$T''_s = -F''_{32}r_2 \sin(\theta_3 - \theta_2) \tag{5.66}$$

The graphical method for finding T''_s is based on a scale drawing of Fig. 5.17a, and is similar to that outlined for T'_s above. Thus the bearing forces and the shaft torque are determined due to the inertia of link 4.

Step 3: The Effect of F_{O2}

The inertia force of link 2 is (see Figs. 5.18 and 5.19)

$$\mathbf{F}_{O2} = (m_2 A_{g2})e^{i(\beta_2 + \pi)} \tag{5.67}$$

and

$$\ell_2 = \mathbf{r}_{g2} + \frac{\varepsilon_2(\alpha_2/|\alpha_2|)e^{i\theta_2}}{\sin(\beta_2 - \theta_2)} = \mathbf{r}_{g2} + \frac{I_2\alpha_2/F_{O2}}{\sin(\beta_2 - \theta_2)}e^{i\theta_2} \tag{5.68}$$

This analysis is much simpler since all bearing forces except F'''_{12} are zero, since both link 3 and link 4 are now considered massless and unloaded.

$$\mathbf{F}'''_{12} = -\mathbf{F}_{O2} \tag{5.69}$$

The required shaft torque to overcome the inertia of link 2 is

$$T'''_s = F_{O2}l_2 \sin(\beta_2 - \theta_2) \tag{5.70}$$

Graphical determination of T'''_s can be based on a scale drawing like Fig. 5.19.

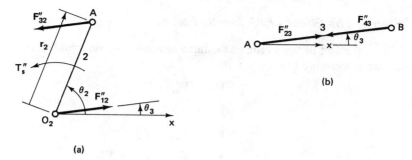

(a)

(b)

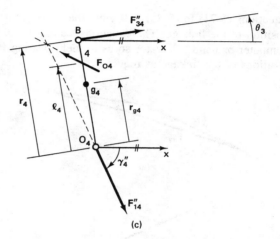

(c)

Figure 5.17   Effect of the inertia of link 4 on pin forces and input torque.

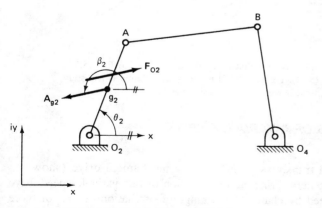

Figure 5.18   Inertia of the input link 2 considered alone.

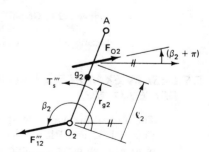

Figure 5.19   Balance of forces and torque on link 2 from the inertia of link 2 only.

Step 4: The Effect of All Inertia Forces

The total effect of the inertia of the three moving links on joint forces and required input torque is therefore (see Fig. 5.20)

$$\mathbf{F}_{14} = \mathbf{F}'_{14} + \mathbf{F}''_{14} \tag{5.71}$$

$$\mathbf{F}_{34} = -\mathbf{F}'_{43} + \mathbf{F}''_{34} \tag{5.72}$$

$$\mathbf{F}_{23} = \mathbf{F}'_{23} + \mathbf{F}''_{23} \tag{5.73}$$

$$\mathbf{F}_{12} = \mathbf{F}'_{12} + \mathbf{F}''_{12} + \mathbf{F}'''_{12} \tag{5.74}$$

$$T_s = T'_s + T''_s + T'''_s \tag{5.75}$$

Equations (5.71) to (5.75) will yield the bearing forces and required torque for a single position of the linkage. The equations leading up to step 4 are easily programmed on a computer or hand calculator so as to yield bearing forces and input torque for as many positions of the linkage as required.

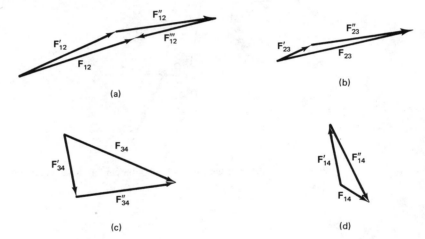

Figure 5.20 Graphical solutions of Equations: (a) Eq. (5.74); (b) Eq. (5.73); (c) Eq. (5.72); and (d) Eq. (5.71). For Eq. (5.75) no graphical solution is necessary, because in this planar work, torques are real signed scalar numbers.

5.5 DESIGN EXAMPLE: ANALYSIS OF A VARIABLE-SPEED DRIVE [41, 103]

The Zero-Max variable-speed transmission is a purely mechanical drive (shown in Figs. P1.32 and P1.33). The general principle of operation of this variable-speed drive is to give stepless variable speed by changing the range of oscillation of four or more one-way clutches which successively drive the output shaft (see Figs. 5.21 and 5.22). The drive has four sets of out-of-phase linkages (with input cranks $\mathbf{Z}_2$ 90° apart) which

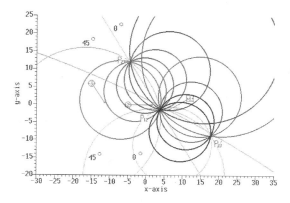

Figure 1: Ground pivot (green) and moving pivot (blue) circles for three prescribed motion positions of Example 8.9 (generated on an IBM AT by Elizabeth Logan, University of Minnesota).

Figure 3: A dump truck with a dumpster loading four bar is shown in partial animation in Macintosh LINCAGES. The cubic curves on the screen represent possible location of ground and moving pivots for four prescribed positions. (See Vol. 2, courtesy of Ken Chin-Purcell, University of Minnesota.)

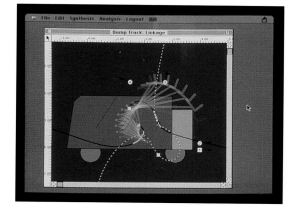

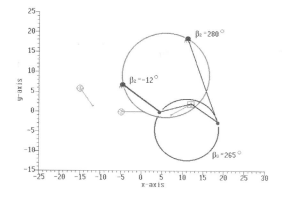

Figure 2: A single pair of ground and moving pivot circles for $\beta_3 = 265°$ for Example 8.9. Two dyads for $\beta_2 = -12°$ and $\beta_2 = 280°$ are shown.

Figure 4: A front end loader designed in the E&S version of LINCAGES. A four bar controls the near-straight line path of the bucket (generated by Dave Rosen, University of Minnesota).

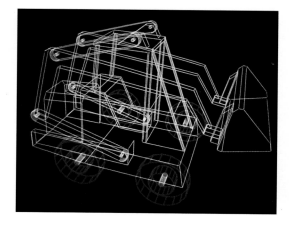

Figure 5: A credit card embossing mechanism similar to the mechanism in Problem 1.15 (courtesy of Data Card Corporation).

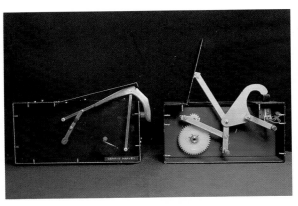

Figure 6: Two different types of mechanisms that turn themselves off. (See problem 1.16, courtesy of Dennis Harvey, MTS Corporation, and Tom Bjorklund, 3M Corporation.)

Figure 7: An agitator mechanism for a washing machine. (See Problem 1.18.)

Figure 8: A casement window and operating mechanism. (See Figures 1.4, 1.5, and the Appendix of Chapter 8.)

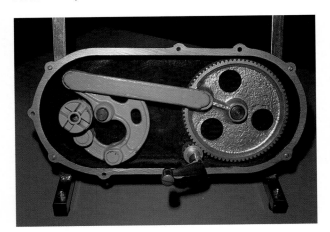

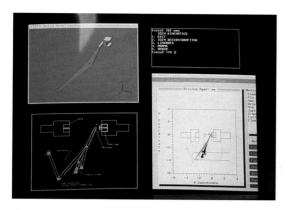

Figure 9: Design of an extractor mechanism to remove a casting from a die using CDC's ICEM KINEMATICS. The four windows from the upper right clockwise are: the main menu, the four bar being designed in LINCAGES-4, the completed six bar, and the coupler link solid model ready for stress analysis.

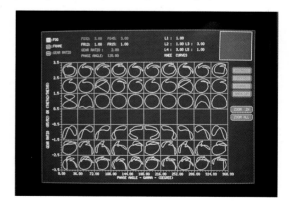

Figure 10: A solid model of the synthesized six bar of Figure 9 (both generated on an IRIS workstation and courtesy of CDC, MDI, and the University of Minnesota).

Figure 11: Sample geared five-bar coupler curve (with input (2) = output (3) = ground (1) = 1 in; the other links (3, 4) = 3 in; the gear ratio between links 2 and 5 = 2) (courtesy of S. Kota, University of Michigan).

Figure 12: One geared five-bar coupler curve design chart for geometry of Figure 11 but varying gear ratio and phase angle (angle between pivot centers and path tracer point) (courtesy of S. Kota, University of Michigan).

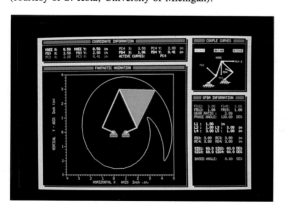

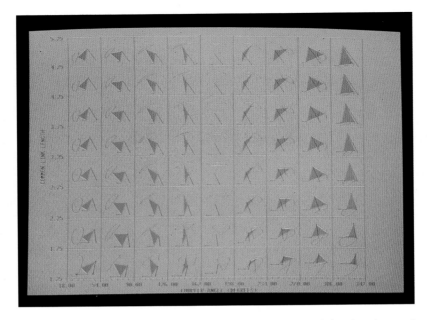

Figure 13: Four-bar symmetric coupler curve design chart with crank-length = 1, ground length = 2, and all the other lengths equal. Therefore, the common length and the coupler angle are the only design parameters. Here, the two parameters form the design chart.

Figure 14: Same as Figure 13, except the vertical axis represents the ground length and a fixed common length = 2.5. (Both of these figures are courtesy of S. Kota, University of Michigan.) [99].

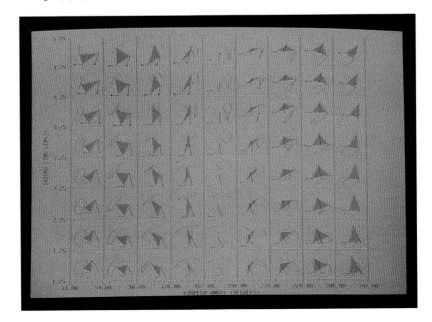

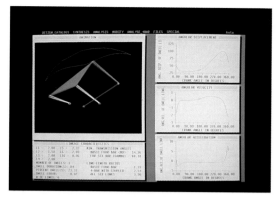

Figure 15: Apollo screen copy from MINN-DWELL [6, 97, 98]. The four bar (white and yellow links) generates a symmetric coupler curve with a circular arc portion. The attached dyad (red link) dwells since its moving pivot is essentially motionless at the center of the circular arc; see position, velocity, and acceleration plots.

Figure 17: The task is to design a mechanism to move a monitor from under a desk to a view position without interfering with the blue region [62].

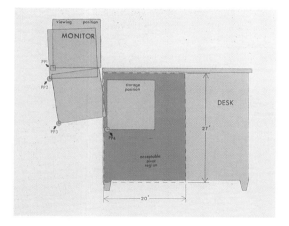

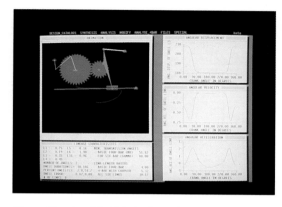

Figure 16: An eight link dwell mechanism where the four bar (white, yellow, and small gear radius) traces a "J"-shaped curve. The slider-bar (red) dwells when the slider (blue) slides and returns along the near straight line. The input dyad (blue) drives the gear pair to improve transmission characteristics. (Both figures are courtesy of S. Kota, University of Michigan.)

Figure 18: Six-bar linkage solution to task of Figure 17 found in the LINCAGES-6 software package [23, 24] (courtesy of Tom Chase, University of Minnesota).

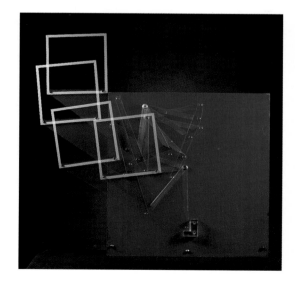

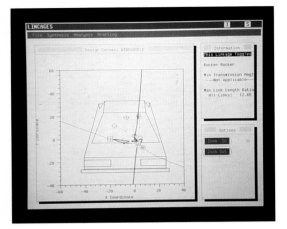

Figure 19: The design task is to design a window wiper mechanism for the rear hatchback of an automobile. The Apollo version of LINCAGES-4 shows the cubic design curves for the desired four wiper positions shown [124].

Figure 20: A solution for the task of Figure 19. A four-bar motion generation from Figure 19 is driven with a dyad function generator.

Figure 21: A model of an elastic four-bar mechanism with light reflectors along the length of the moving links. A high speed camera monitors the location of these reflectors [55, 106, 107].

Figure 22: Graphics display of digitized results from Figure 21 (courtesy of F. W. Liou, University of Missouri, Rolla).

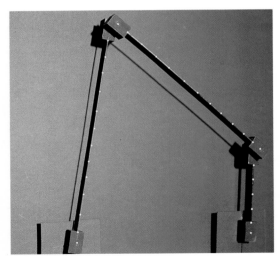

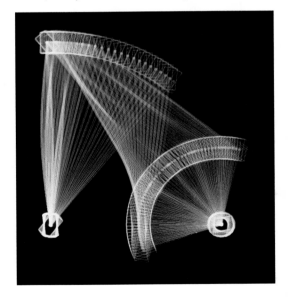

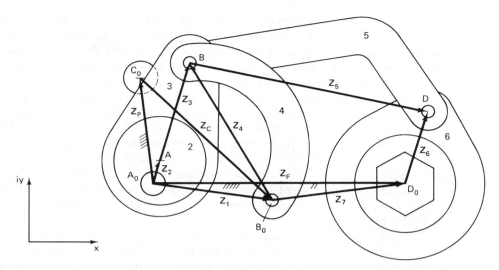

Figure 5.21 Schematic drawing of one lamination of the variable-speed drive, showing the vectors that represent each of the links.

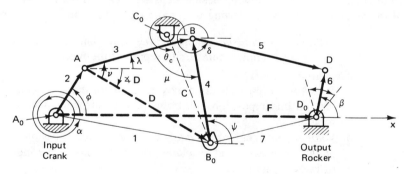

Figure 5.22 Vector representation of one lamination of the variable-speed drive (not to scale).

use three common shafts, A_0, B_0, and D_0. Figure 5.16 shows only one of these linkages, which is referred to as a single lamination.

Referring to Fig. 5.22, which is a not-to-scale schematic of the linkage in Fig. 5.21, the rotation of the input $A_0 A$ causes the output link $D_0 D$ to oscillate, thus rotating the output shaft D_0 in one direction (due to the one-way clutch mounted in the hub of link 6). The position of pivot B_0 is adjusted by rotating the speed control arm C about C_0 to change the output speed of the drive. As B_0 approaches the moving pivot D, the output speed decreases since the center of curvature of the trajectory of B will approach D reducing the displacement of the latter.

The objective of this case study is to report on the analysis involved in designing a larger integral-horsepower drive adapting the same linkage type designed for a fractional-horsepower unit. Preliminary prototypes in which smaller units were "scaled up" to meet higher power ratings resulted in high and varying torque required to maintain or to change speed (the torque on the speed control arm $C_0 B_0$ was excessive). An optimal linkage configuration was sought which would minimize the required control torque while retaining an acceptable performance level of the drive.

Notice that the dynamics problem was *perceived here* as undesirable torques on a linkage member. The reader is referred back to Chap. 2 (Table 2.3) for an outline of symptoms that may be perceived as a problem with a machine or mechanism. Translating those symptoms into a "prescription" for alleviating the problem is sometimes a difficult task. In this case there was a problem in the transmissibility area (although contrary to the figure in Table 2.3, where the lack of transmissibility is the issue, here there is too much torque transmitted to the speed control link). Notice that under "possible causes" in Table 2.3 "excessive inertia forces" are included. Each of the four possible causes listed there were considered in the analysis.

The Seven Stages of Engineering Design (Fig. 2.1) are also recalled to the reader since this case study required consideration of the stages of the design process as well as the assumptions made in each design category. Since the major objective was the redesign of a mechanism in which the type was fixed, stages 5 and 6, modeling and analysis, were emphasized. The following design steps were carried out to accomplish a redesign of the variable-speed drive (starting with more elementary analyses — see Table 2.2).

1. *Modeling of mechanism.* Determine the equivalent linkage and then model the linkage using complex numbers to represent the links.
2. *Kinematic analysis*
   a. *Position analysis.* Determine expressions for the dependent variables (angular and linear positions) in terms of the independent variable, input crank rotation, ϕ.
   b. *Velocity and acceleration analysis.* Determine expressions for the linear and angular velocities and accelerations of all links of the mechanism in terms of known parameters.
3. *Inertia-force analysis.* Determine expressions for and the effects of the inertia of the various members of the mechanism on the speed control arm.
4. *Dynamic synthesis.* Determine methods of synthesizing new mechanisms (of the same type) to yield the desired dynamic performance. These methods are based on experimental and intuitive considerations and rely on the computer for rapid solutions to assess the effect of intuitive changes.
5. *Experimental verification.* Test by building a prototype of a mechanism derived from the process outlined above. The validity of all assumptions and intuitive choices made above would be determined in this step.

Several assumptions (Table 2.2) were made in order to simplify the analysis, making it adaptable to modest time-sharing computation. The links in the drive are assumed to be rigid throughout their motion and the input angular velocity regarded as constant over a cycle (at a particular speed setting and output load). This important timesaving assumption seemed valid since several units were tested and no variation of input angular velocity could be detected. Joints are assumed to be frictionless and the viscous damping of the lubricating oil is assumed to be negligible. Also, the engaging and disengaging of the overrunning clutches is assumed to be smooth, and therefore the dynamics of these clutches are considered to have little effect on the performance of the drive.

Step 1: Modeling of the Mechanism

For any one speed setting, with B_0 stationary, each lamination of the drive (Figs. 5.21 and 5.22) is equivalent to an adjustable Watt six-link mechanism where the primary four-bar linkage (A_0, A, B, B_0) is of the crank-rocker type and the secondary four-bar (B_0, B, D, D_0) is a double-rocker. The speed control arm C, the seventh link, is pivoted at C_0. Figure 5.21 shows a vector diagram of the adjustable six-link mechanism where the links, including the seventh, the adjusting link, are represented by complex numbers $(\mathbf{Z}_1 \rightarrow \mathbf{Z}_7)$.

Step 2: Kinematic Analysis

Position analysis.   Referring to Fig. 5.22 (which is an unscaled schematic kinematic diagram of the drive), with ϕ the input, the dependent position variables in the primary four-bar linkage are λ and ψ. Expressions for λ and ψ in terms of Z_1, Z_2, Z_3, Z_4, and α were developed in Sec. 3.3. They are redeveloped here.

$$\mathbf{D} = Z_1 e^{i\alpha} - Z_2 e^{i\phi}$$

$$\cos \nu = \frac{D^2 + Z_3^2 - Z_4^2}{2DZ_3}, \qquad \sin \nu = |\sqrt{1 - \cos^2 \nu}|$$

$$\nu = \arg(\cos \nu + i \sin \nu) \tag{5.76}$$

$$\lambda = \nu + \sphericalangle \mathbf{D}$$

$$Z_4 e^{i\psi} = Z_2 e^{i\phi} + Z_3 e^{i\lambda} - Z_1 e^{i\alpha}$$

Therefore,

$$\psi = \arg(Z_2 e^{i\phi} + Z_3 e^{i\lambda} - Z_1 e^{i\alpha}) \tag{5.77}$$

$$\mu = \psi - \lambda \tag{5.78}$$

Since this lamination of the variable-speed drive is made up of two four-bar mechanisms in series, expressions for δ and β in terms of ψ are obtained by proper substitution of the variables of the secondary four-bar linkage into Eqs. (5.76), (5.77), and (5.78).

Velocity and acceleration analysis.   Looking again at the primary four-bar linkage, expressions are desired for the angular velocities and accelerations of $\mathbf{Z}_3$ and $\mathbf{Z}_4$ in terms of the input angular velocity and angular acceleration and the position of the linkage. From Eqs. (3.72), (3.73), (4.21), and (4.22),

$$\dot{\lambda} = -\frac{Z_2}{Z_3} \dot{\phi} \frac{\sin(\psi - \phi)}{\sin(\psi - \lambda)} \tag{5.79}$$

$$\dot{\psi} = \frac{Z_2}{Z_4}\dot{\phi}\frac{\sin(\lambda - \phi)}{\sin(\lambda - \psi)} \tag{5.80}$$

$$\ddot{\lambda} = \frac{-Z_2\ddot{\phi}\sin(\psi - \phi) + Z_2\dot{\phi}^2\cos(\psi - \phi) + Z_3\dot{\lambda}^2\cos(\psi - \lambda) - Z_4\dot{\psi}^2}{Z_3\sin(\psi - \lambda)}$$

$$\tag{5.81}$$

$$\ddot{\psi} = \frac{Z_2\ddot{\phi}\sin(\lambda - \phi) + Z_2\dot{\phi}^2\cos(\lambda - \phi) + Z_4\dot{\psi}^2\cos(\lambda - \psi) - Z_3\dot{\lambda}^2}{Z_4\sin(\lambda - \psi)}$$

$$\tag{5.82}$$

Expressions for $\dot{\delta}$, $\dot{\beta}$, $\ddot{\delta}$, and $\ddot{\beta}$ in terms of $\dot{\psi}$ and $\ddot{\psi}$ are obtained by proper substitutions of the variables of the secondary four-bar linkage into Eqs. (5.79) to (5.82).

Step 3: Inertia-Force Analysis

The objective of the inertia-force analysis is to determine the effect of inertia forces $\mathbf{F}_{O2}$, $\mathbf{F}_{O3}$, $\mathbf{F}_{O4}$, $\mathbf{F}_{O5}$, and $\mathbf{F}_{O6}$ (of links 2, 3, 4, 5, and 6 respectively) on the speed control arm C. The method of determining the magnitude and direction of the inertia force will be demonstrated by looking first at the secondary four-bar linkage.

Figure 5.23 shows a schematic drawing of the secondary four-bar linkage with mass centers g_4, g_5, and g_6. The linear accelerations $\mathbf{A}_{g4}$, $\mathbf{A}_{g5}$, and $\mathbf{A}_{g6}$ can be computed for a particular configuration of the mechanism knowing the angular velocities and accelerations from above. For example,

$$\mathbf{A}_{g5} = Z_4(i\ddot{\psi} - \dot{\psi}^2)e^{i\psi} + r_{g5}(i\ddot{\delta} - \dot{\delta}^2)e^{i\delta} \tag{5.83}$$

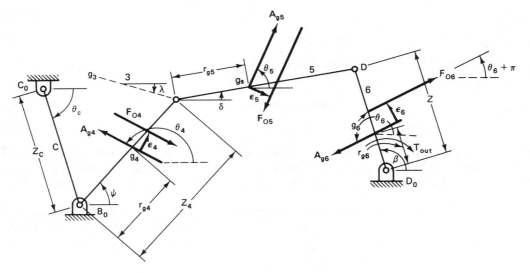

Figure 5.23 Unscaled kinematic diagram of a portion of one lamination of the variable-speed drive, the secondary four-bar linkage of Fig. 5.22, showing location of mass centers, vectors representing acceleration of mass centers, inertia-force vectors and their offsets from the mass centers.

The inertia forces $\mathbf{F}_{O4}$, $\mathbf{F}_{O5}$, and $\mathbf{F}_{O6}$ represent the "dynamic loading" of this four-bar. Using Eqs. (5.19) and (5.22), the offsets of these forces are determined. For example

$$\varepsilon_5 = -\frac{I_5\ddot{\delta}}{F_{O5}} ie^{i(\arg A_{g5})} \tag{5.84}$$

where I_5 is the mass moment of inertia of link 5 about its center of gravity.

As described previously, the principle of superposition allows us to compute the effect of each inertia force independently and sum their effects to obtain the total torque required for each position. The laws of statics were used to obtain the following expression.

Effect of $\mathbf{F}_{O6}$.   The torque T_{c6} on arm C due to $\mathbf{F}_{O6}$ is

$$T_{c6} = -\left\{\frac{F_{O6}\sin(\beta - \theta_6)[r_{g6} + \varepsilon_6/\sin(\theta_6 - \beta)]}{Z_6\sin(\beta - \delta)}\right\}\frac{\sin(\lambda - \delta)}{\sin(\lambda - \psi)}[Z_c\sin(\psi - \theta_c)] \tag{5.85}$$

The first expression on the right transfers $\mathbf{F}_{O6}$ to a force $\mathbf{F}'_{O6}$ along link 5, the second expression transfers $\mathbf{F}'_{O6}$ to a force $\mathbf{F}''_{O6}$ along link 4, and the final expression multiplied by the foregoing expressions gives the torque of $\mathbf{F}''_{O6}$ about C_0. A positive T_{c6} means that the control arm C, if released, will tend toward high speed.

Effect of the output load (T_{out}) on the drive.   The torque T_{cL} on arm C due to the load torque T_{out} is similar to T_{c6}:

$$T_{cL} = \left[\frac{T_{out}}{Z_6\sin(\beta - \delta)}\frac{\sin(\lambda - \delta)}{\sin(\lambda - \psi)}\right][Z_c\sin(\psi - \theta_c)] \tag{5.86}$$

Notice that Eq. (5.86) is valid only when the clutch of a particular lamination is engaged.

Effect of $\mathbf{F}_{O5}$.   The torque T_{c5} on arm C due to $\mathbf{F}_{O5}$ is

$$T_{c5} = -F_{O5}\left[\frac{\left(r_{g5} + \dfrac{\varepsilon_5}{\cos(\theta_5 - \delta)}\right)\sin(\theta_5 - \delta)}{Z_5\sin(\beta - \delta)}\right.$$
$$\left. \times \frac{\sin\beta - \cos\beta\tan\lambda - \sin\theta_5 + \cos\theta_5\tan\lambda}{\sin(\lambda - \psi)}\right][Z_c\sin(\psi - \theta_c)] \tag{5.87}$$

Effect of $\mathbf{F}_{O4}$.   The torque T_{c4} on arm C due to $\mathbf{F}_{O4}$ is

$$T_{c4} = -F_{c4}\sin(\Sigma - \theta_c)Z_c \tag{5.88}$$

where

$$\Sigma = \arg(B + iA) \quad \text{and} \quad F_{c4} = \sqrt{A^2 + B^2}$$

where

$$A = F_{O4} \sin \theta_4 - F_{34} \sin \lambda$$

$$B = F_{O4} \cos \theta_4 - F_{34} \cos \lambda$$

$$F_{34} = F_{O4} \frac{r_{g4} \sin(\theta_4 - \psi) + \varepsilon_4}{r_4 \sin(\psi - \lambda)}$$

Effect of F_{O3}.   The torque T_{c3} on arm C due to F_{O3} is

$$T_{c3} = -F_{O3} \left[\frac{r_{g3} \sin(\theta_3 - \lambda) + \varepsilon_3}{r_3 \sin(\psi - \lambda)} \right] [Z_c \sin(\psi - \theta_c)] \qquad (5.89)$$

Effect of F_{O2}.   Since the input angular velocity is assumed constant, there will only be a centripetal acceleration on link 2, and therefore no forces will be transmitted to the control arm.

Step 4: Dynamic Synthesis

The objective of the dynamic synthesis is to develop a means or a strategy for arriving at one or more mechanisms to satisfy best certain dynamic criteria. In the redesign of this drive, the major concern is minimizing the torque requirements on the control arm to make the drive more easily controllable. Minimizing the angular accelerations of all the moving links in the mechanism, however, turned out to be the major design strategy since angular accelerations are easily monitored. The magnitudes of angular accelerations directly affect the level of inertia force and torques in the mechanism. Also, minimization of angular accelerations will tend to reduce bearing forces, and should result in a cooler running unit as well as decreasing wear and increasing unit life.

Steps 2 and 3 of the design procedure were programmed on a time-sharing computer to yield positions, velocities, and accelerations of links of the mechanism. For example, Fig. 5.24 shows a graph of the angular accelerations of each link of unit A, which is a scaled-up version of a fractional horsepower unit (shown in Fig. 5.25) plotted for one complete revolution of the input link.

The function of the primary four-bar linkage of the drive is to act as an adjustable input to the secondary four-bar linkage (see Fig. 5.22). This input is in the form of an oscillation of point B on link 4 about center B_0.

Experimenters report that link 3 of unit A "thrashes around" substantially. Figure 5.24 bears out this fact—notice the variation in magnitude of the angular acceleration $\ddot{\lambda}$ of link 3. Since link 3 need not have high angular accelerations, the strategy of "uncramping" the primary four-bar, that is, moving A_0 away from B_0 (see Fig. 5.21), for the purpose of reducing the angular acceleration of link 3 (as well as the rest of the links) was pursued. This was chosen as the major means for optimization.

Starting with unit A shown in Fig. 5.25, the primary four-bar linkage is slowly "uncramped." An interactive computer terminal was utilized to optimize the variable-speed-drive dimensions with a step-by-step procedure. For each possible linkage configuration a kinematic and an inertia force analysis was performed. The printout time was

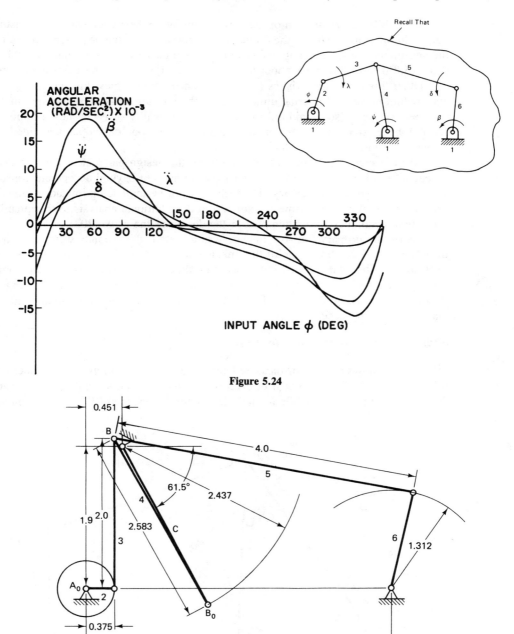

Figure 5.24

Figure 5.25   Scaled kinematic diagram of unit A. For simplicity, the center of gravity of each link is assumed to be at each geometric center. The weight of each link is as follows: 3:2.1 oz; 4:2.1 oz; 5:1.25 oz; and 6:3.0 oz. Dimensions are in inches.

long enough for an evaluation of the new design by the best engineer, the "human computer," in order to decide on the direction for proceeding in the next step (choice of new linkage dimensions). The angular velocities, angular accelerations, and control torque [Eqs. (5.85) to (5.89)] throughout the cycles of motion of the mechanism were printed out. Other linkage characteristics (e.g., two transmission angles, linkage size and clearance within prescribed internal dimensions of the housing, maximum and minimum output speeds) were also monitored to check the practicality of each "new prototype." A "feel" for the effect of changes in each parameter was soon developed and aided the designer in arriving at an acceptable solution.

As is usually the case, optimizing one of the design parameters tends to decrease acceptability of one or more of the others or drive them beyond acceptable constraints, so that compromise is necessary. The techniques of systematic optimization would have been very useful here, but time did not permit the additional analysis. Subsequently, optimization techniques were attempted in the design of the drive [103].

After the primary four-bar linkage was adjusted, the secondary linkage was optimized. Figure 5.26 shows one of the linkages which showed considerable improvement over unit A. This unit (unit B) is the result of about 150 interactions with the computer. For example, one of the important sets of parameters monitored is displayed in Fig. 5.27. A comparison of Fig. 5.27 with Fig. 5.24 shows improvement in the magnitude of the maximum angular accelerations.

Step 5: Experimental Verification

Figure 5.28 shows a graph of measured control torques which represent the inertial contribution [Eqs. (5.85) to (5.89)] of each link of the lamination while that lamination is driving the output load. While a single lamination is engaged (90° of input crank rota-

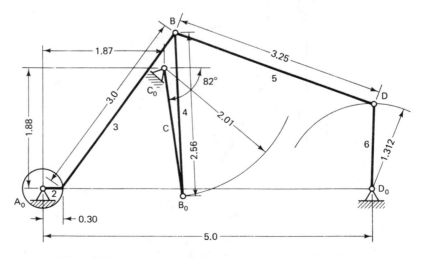

Figure 5.26   Scaled kinematic diagram of an optimized unit (unit B).

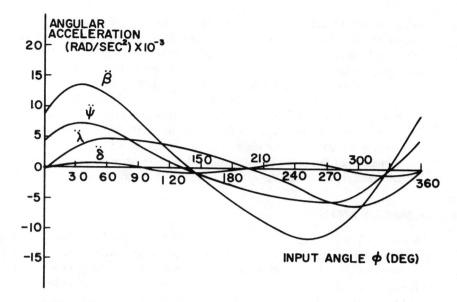

Figure 5.27   Angular accelerations of the links of unit B (see, for comparison, Fig. 5.24).

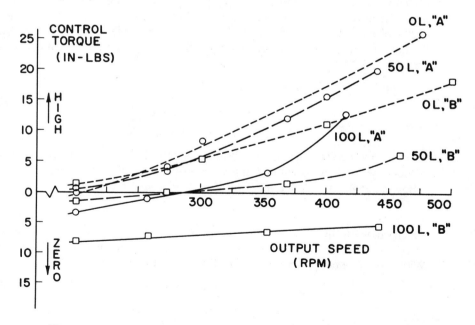

Figure 5.28   Comparison of control torque requirements between unit A (circles) and unit B (squares) versus output speed. Three different output load settings are compared: 100 in·lb (solid lines), 50 in·lb (long dashes), and 0 in·lb (short dashes).

tion), the inertia forces change in magnitude and direction so that the experimentally measured control torque is the mechanically damped average of the contributions (over 90° of input rotation) of each of the links in each lamination. One can see that the control torques have been reduced in unit B as well as shifted toward zero (if one lets go of the control arm, the unit would tend to go to zero speed by itself) side of the graph. For the higher output speed range, the control torque is considerably reduced. Unit B also exhibits a nearly linear control torque versus output speed relationship for a 100 in·lb load. The experimental results were within 10% of the analytical model. If more accurcy were required, a model including elastic effects (see Sec. 5.17 of Vol. 2) clearances and/or the dynamics of the one-way clutch would have to be used, and the link models would have to take into account the offset of the center of gravity of some of the links.

5.6 THE MATRIX METHOD

Another approach to the force analysis of a linkage (kinetostatic analysis in this section) is the matrix method. This technique will be demonstrated by referring to the four-bar linkage of Fig. 5.10. Rather than using the principle of superposition, this method considers all the inertia forces together. The advantage of this method is that the equations of motion are quickly derived. The disadvantage is the need for matrix manipulation in order to solve the equations. Another change from the previous method is the way that the inertia torques are expressed (although this change is not required to use the matrix method). In each free-body diagram of a link the inertia forces are assumed to act at the center of mass and an inertia torque is added (as suggested in Fig. 5.5b). Referring to the three free-body diagrams of Fig. 5.29, three static equilibrium equations are written for each link ($\Sigma F_x = 0$, $\Sigma F_y = 0$, and $\Sigma T_g = 0$).

Link 2 (Fig. 5.29a)

$$F_{12x} + F_{32x} + F_{O2x} = 0 \qquad (5.90)$$

$$F_{12y} + F_{32y} + F_{O2y} = 0 \qquad (5.91)$$

In order to write the moment-equilibrium equations in a uniform, programmable manner, without the need for visual determination of the sense of each moment, we will use the following expression of the moment of a force around a fulcrum

$$M_F = r_x F_y - r_y F_x \qquad (5.92)$$

which comes from the determinant form of the vector cross product $\mathbf{M}_F = \mathbf{r} \times \mathbf{F}$ and where $\mathbf{r}$ is a vector from the fulcrum to the point of application of the force (Fig. 5.29d). With this, for example in Fig. 5.29a, the moment of $\mathbf{F}_{12}$ about g_2 is

$$M_{F_{12/g}} = (-\mathbf{r}_{g2}) \times \mathbf{F}_{12} = (-r_{g2x})F_{12y} - (-r_{g2y})F_{12x}$$

$$= F_{12x}r_{g2} \sin \theta_2 - F_{12y}r_{g2} \cos \theta_2$$

Accordingly, the moment equation for link 2 becomes

$$T_{O2} + T_s + F_{12x}r_{g2} \sin \theta_2 - F_{12y}r_{g2} \cos \theta_2 - F_{32x}(r_2 - r_{g2})\sin \theta_2$$
$$+ F_{32y}(r_2 - r_{g2})\cos \theta_2 = 0 \qquad (5.93)$$

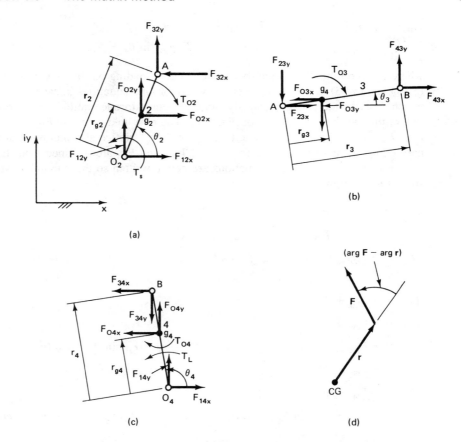

Figure 5.29 (a), (b), and (c) show free-body diagrams of individual links used in deriving the matrix equation of kinetostatic equilibrium; (d) illustrates the forming of the cross product $\mathbf{r} \times \mathbf{F}$ in complex numbers.

where T_s is the driving torque.

Link 3 (Fig. 5.29b)

$$F_{23x} + F_{43x} + F_{O3x} = 0 \tag{5.94}$$

$$F_{23y} + F_{43y} + F_{O3y} = 0 \tag{5.95}$$

$$T_{O3} + F_{23x}r_{g3}\sin\theta_3 - F_{23y}r_{g3}\cos\theta_3 - F_{43x}(r_3 - r_{g3})\sin\theta_3$$
$$+ F_{43y}(r_3 - r_{g3})\cos\theta_3 = 0 \tag{5.96}$$

where we used joint forces F_{23} and F_{43} acting on link 3.

Link 4 (Fig. 5.29c)

$$F_{34x} + F_{14x} + F_{O4x} = 0 \tag{5.97}$$

$$F_{34y} + F_{14y} + F_{O4y} = 0 \tag{5.98}$$

$$T_{O4} - F_{34x}(r_4 - r_{g4})\sin\theta_4 + F_{34y}(r_4 - r_{g4})\cos\theta_4 + T_L$$
$$+ F_{14x}r_{g4}\sin\theta_4 - F_{14y}r_{g4}\cos\theta_4 = 0 \qquad (5.99)$$

where T_L represents the torque due to the external loading on the unit. Equations (5.90), (5.91) and (5.93) to (5.99) represent a system of nine equations linear in nine unknowns which describe the instantaneous dynamic force and torque equilibrium on each moving link of the four-bar mechanism of Fig. 5.10. The nine unknowns are F_{12x}, F_{12y}, F_{23x}, F_{23y}, F_{34x}, F_{34y}, F_{14x}, F_{14y}, and T_s. Note that $F_{jkx} = -F_{kjx}$. Notice also that the real and imaginary components of the inertia forces $\mathbf{F}_{O2}$, $\mathbf{F}_{O3}$, and $\mathbf{F}_{O4}$ are obtained from Eqs. (5.42), (5.53), and (5.67). The nine equations are now rewritten so that the unknown terms are shifted to the right side:

$$F_{O2x} = -F_{12x} + F_{23x}$$
$$F_{O2y} = -F_{12y} + F_{23y}$$
$$T_{O2} = -F_{12x}r_{g2}\sin\theta_2 + F_{12y}(r_{g2}\cos\theta_2) - T_s - F_{23x}(r_2 - r_{g2})\sin\theta_2$$
$$+ F_{23y}(r_2 - r_{g2})\cos\theta_2$$

$$F_{O3x} = -F_{23x} + F_{34x}$$
$$F_{O3y} = -F_{23y} + F_{34y} \qquad (5.100)$$
$$T_{O3} = -F_{23x}r_{g3}\sin\theta_3 + F_{23y}r_{g3}\cos\theta_3 - F_{34x}(r_3 - r_{g3})\sin\theta_3 + F_{34y}(r_3 - r_{g3})\cos\theta_3$$
$$F_{O4x} = -F_{34x} - F_{14x}$$
$$F_{O4y} = -F_{34y} - F_{14y}$$
$$T_{O4} = F_{34x}(r_4 - r_{g4})\sin\theta_4 - F_{34y}(r_4 - r_{g4})\cos\theta_4 - F_{14x}r_{g4}\sin\theta_4 + F_{14y}r_{g4}\cos\theta_4 - T_L$$

This system of equations is linear in the unknowns (the pin forces plus T_s) since the sines and cosines are known for each position of the linkage. Let us define

$$R_2 \equiv r_2 - r_{g2}$$
$$R_3 \equiv r_3 - r_{g3}$$
$$R_4 \equiv r_4 - r_{g4}$$

The system of Eqs. (5.100) is expressed in matrix form [Eq. (5.101)], or in symbolic form,

$$[F_I] = [L][F_B] \qquad (5.102)$$

where

$[F_I]$ = column matrix of known external load plus inertia forces and torques
$[L]$ = square matrix of known linkage parameters and position angles
$[F_B]$ = column matrix of unknown bearing forces and input torque

The system may be solved for the unknowns by matrix computation techniques. The right and left sides of Eq. (5.102) are premultiplied by $[L]^{-1}$ (the inverse of $[L]$):

$$[L]^{-1}[F_I] = [L]^{-1}[L][F_B] \qquad (5.103)$$

Since $[L]^{-1}[L] = [I]$, the identity matrix,

$$[F_B] = [L]^{-1}[F_I] \qquad (5.104)$$

The solution of Eq. (5.104) requires calculation of the inverse of the linkage parameter matrix and the multiplication of the two matrices on the right-hand side of Eq. (5.104). In most instances, these matrix operations are available as standard functions on computer systems. Therefore, one may adopt a computer program (see the flowchart of Fig. 5.47) to perform a kinetostatic analysis of any four-bar linkage. The card-feeder mechanism example, presented later, uses the matrix technique.

It should be noted that taking the inverse in Eq. (5.104) can be a time-consuming step on the computer. If minimizing computer time is important, a large reduction in the number of calculations can be accomplished by using the Gauss–Jordan reduction to put the array of coefficients of Eq. (5.101) into upper-right-triangle form and then solving by direct back substitution [82]. This generally requires $[(3m)^3/3 + (3m)^2/2]$ multiplications (where $m = 3$ is the number of moving links), a reduction of two-thirds from the matrix inversion method. Partial pivoting of the array is usually performed to avoid dividing by small coefficients and to retain accuracy [36].

Example 5.3*

The crank of the four-bar linkage shown in Fig. 5.30 is in the $\theta_2 = 150°$ position and the motion is given by $\omega_2 = 5$ rad/sec ccw and $\alpha_2 = 5$ rad/sec^2 cw. The lengths, weights and mass moments of inertia of the links are given in the chart below.

Link	Length (in)	Weight (lbf)	I_g (lbf in sec^2)
1	12	—	—
2	4	.8	.012
3	12	2.4	.119
4	7	1.4	.038

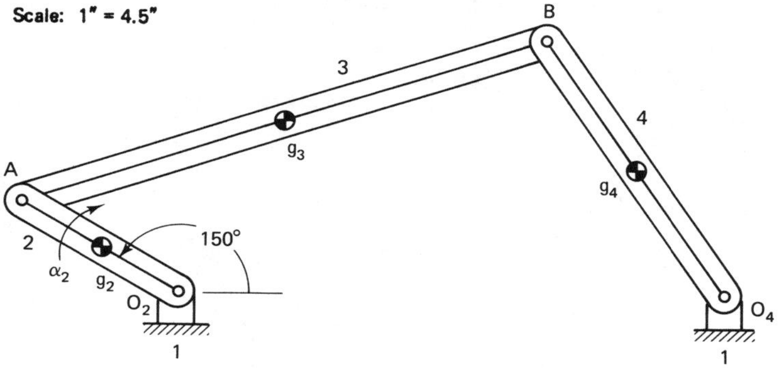

Figure 5.30 Diagram of four-bar linkage in desired position.

* This example has been contributed by John Titus, University of Minnesota.

$$
\begin{bmatrix}
F_{O2x} \\
F_{O2y} \\
T_{O2} \\
F_{O3x} \\
F_{O3y} \\
T_{O3} \\
F_{O4x} \\
F_{O4y} \\
T_{O4} + T_L
\end{bmatrix}
=
\begin{bmatrix}
-1 & 0 & 0 & 1 & 0 & 0 & 0 & 0 & 0 \\
0 & -1 & 0 & 0 & 1 & 0 & 0 & 0 & 0 \\
-r_{g2}\sin\theta_2 & r_{g2}\cos\theta_2 & -1 & -R_2\sin\theta_2 & R_2\cos\theta_2 & 0 & 0 & 0 & 0 \\
0 & 0 & 0 & -1 & 0 & 1 & 0 & 0 & 0 \\
0 & 0 & 0 & 0 & -1 & 0 & 1 & 0 & 0 \\
0 & 0 & 0 & -r_{g3}\sin\theta_3 & r_{g3}\cos\theta_3 & -R_3\sin\theta_3 & R_3\cos\theta_3 & 0 & 0 \\
0 & 0 & 0 & 0 & 0 & -1 & 0 & -1 & 0 \\
0 & 0 & 0 & 0 & 0 & 0 & -1 & 0 & -1 \\
0 & 0 & 0 & 0 & 0 & R_4\sin\theta_4 & -R_4\cos\theta_4 & -r_{g4}\sin\theta_4 & r_{g4}\cos\theta_4
\end{bmatrix}
\begin{bmatrix}
F_{12x} \\
F_{12y} \\
T_s \\
F_{23x} \\
F_{23y} \\
F_{34x} \\
F_{34y} \\
F_{14x} \\
F_{14y}
\end{bmatrix}
\tag{5.101}
$$

The mass centers are located at the middle of each link. Find the pin reaction forces and the input shaft torque required to maintain this motion.

Solution   We will demonstrate the matrix and superposition solution methods for this example. Both solutions will require the inertia forces and torques for each link; so the first step is to construct the acceleration polygon following the techniques of Chap. 4. The mass center linear accelerations are then found using the acceleration image. The results are listed below and graphically displayed in Fig. 5.31.

$$\alpha_2 = 5 \text{ rad/sec}^2 \text{ cw} \qquad \mathbf{A}_{g2} = 51 \text{ in/sec}^2 \text{ at } 341°$$

$$\alpha_3 = 4.01 \text{ rad/sec}^2 \text{ ccw} \qquad \mathbf{A}_{g3} = 86.98 \text{ in/sec}^2 \text{ at } 353°$$

$$\alpha_4 = 9.76 \text{ rad/sec}^2 \text{ cw} \qquad \mathbf{A}_{g4} = 38.34 \text{ in/sec}^2 \text{ at } 8°$$

The inertia forces and torques are calulated next.

$$\mathbf{F}_{O2} = .106 \text{ lbf at } 161° \qquad T_{O2} = .06 \text{ in·lbf ccw}$$

$$\mathbf{F}_{O3} = .54 \text{ lbf at } 173° \qquad T_{O3} = .477 \text{ in·lbf cw}$$

$$\mathbf{F}_{O4} = .14 \text{ lbf at } 188° \qquad T_{O4} = .371 \text{ in·lbf ccw}$$

At this point the two solution methods diverge. We will follow the matrix method solution presented in Sec. 5.6 first and follow up with the superposition technique discussed in Sec. 5.4.

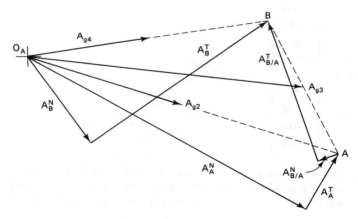

Figure 5.31   Acceleration polygon of four-bar linkage showing mass center accelerations.

Matrix Method Solution   The equations corresponding to equation set (5.100) can be immediately written as

$$-.1002 = -F_{12x} + F_{23x}$$

$$.0345 = -F_{12y} + F_{23y}$$

$$.06 = -F_{12x} - 1.732F_{12y} - T_s - F_{23x} - 1.732F_{23y}$$

$$-.536 = -F_{23x} + F_{34x}$$

$$.066 = -F_{23y} + F_{34y}$$

$$-.477 = -1.854F_{23x} + 5.706F_{23y} - 1.854F_{34x} + 5.706F_{34y}$$

$$-.139 = -F_{34x} - F_{14x}$$

$$-.019 = -F_{34y} - F_{14y}$$

$$.371 = 2.856F_{34x} + 2.027F_{34y} - 2.856F_{14x} - 2.027F_{14y}$$

The matrices of Eq. (5.102) can now be constructed.

$$[L] = \begin{bmatrix} -1.0 & 0 & 0 & 1.0 & 0 & 0 & 0 & 0 & 0 \\ 0 & -1.0 & 0 & 0 & 1.0 & 0 & 0 & 0 & 0 \\ -1.0 & -1.732 & -1.0 & -1.0 & -1.732 & 0 & 0 & 0 & 0 \\ 0 & 0 & 0 & -1.0 & 0 & 1.0 & 0 & 0 & 0 \\ 0 & 0 & 0 & 0 & -1.0 & 0 & 1.0 & 0 & 0 \\ 0 & 0 & 0 & -1.854 & 5.706 & -1.854 & 5.706 & 0 & 0 \\ 0 & 0 & 0 & 0 & 0 & -1.0 & 0 & -1.0 & 0 \\ 0 & 0 & 0 & 0 & 0 & 0 & -1.0 & 0 & -1.0 \\ 0 & 0 & 0 & 0 & 0 & 2.856 & 2.027 & -2.856 & -2.027 \end{bmatrix}$$

$$[F_l] = \begin{bmatrix} -.1002 \\ .0345 \\ .06 \\ -.536 \\ .066 \\ -.477 \\ -.139 \\ -.019 \\ .371 \end{bmatrix}$$

The linkage parameter matrix must be inverted and then must premultiply the known loading column matrix to form the bearing force and input torque solution column matrix; see Eq. (5.104). At this point a computer program was used to find the final solutions to be

$$F_{14x} = .069 \text{ lbf}$$

$$F_{14y} = -.081 \text{ lbf}$$

$$F_{34x} = .070 \text{ lbf}$$

$$F_{34y} = .101 \text{ lbf}$$

$$F_{32x} = -.605 \text{ lbf}$$

$$F_{32y} = -.035 \text{ lbf}$$

$$F_{12x} = .706 \text{ lbf}$$

$$F_{12y} = .001 \text{ lbf}$$

$$T_s = -1.432 \text{ in·lbf}$$

Superposition Method   We will demonstrate the graphical technique of this method. For this reason, the inertia effect on each link must be resolved into a force-couple system. This results in a single offset inertia force. This offset distance was introduced as $\boldsymbol{\varepsilon}$, the eccentricity vector, in Sec. 5.2. The eccentricity vectors for this example are

$$\boldsymbol{\varepsilon}_2 = .566 \text{ inches at } 71°$$

$$\boldsymbol{\varepsilon}_3 = .883 \text{ inches at } 263°$$

$$\boldsymbol{\varepsilon}_4 = 2.65 \text{ inches at } 98°$$

A diagram of the offset inertia forces due to the motion of this linkage is shown in Fig. 5.32. This is the total loading on the linkage and it can not be solved graphically.

The effects on the input of the three inertial loadings shown in Fig. 5.33 will be superposed to obtain the effect of the total loading shown in Fig. 5.32. There are now three loading subproblems, however they are relatively easy to solve.

Solution to subproblem 1. On link 4, three forces are acting. Therefore the force system has a point of concurrency (intersection of the lines of action of the three forces). Two such lines of action are required to fix the location of this point. The line of action of $\mathbf{F}_{O4}$ is known from the acceleration analysis. Note that link 3 is a two force member. This supplies the line of action of $\mathbf{F}'_{34}$. This fixes the location of the point of concurrency and the line of action of $\mathbf{F}'_{14}$. The free-body diagram of this link is shown in Fig. 5.34. The kinetostatic force equation for this link is

$$\mathbf{F}'_{34} + \mathbf{F}_{O4} + \mathbf{F}'_{14} = 0 \tag{5.105}$$

This equation is solved graphically by constructing a force polygon in the same manner as was done for the velocity and acceleration polygons. The polygon for this case is shown in Fig. 5.35. Measurement of the vectors yields

$$\mathbf{F}'_{34} = .13 \text{ lbf at } 18°$$

$$\mathbf{F}'_{14} = .03 \text{ lbf at } 313°$$

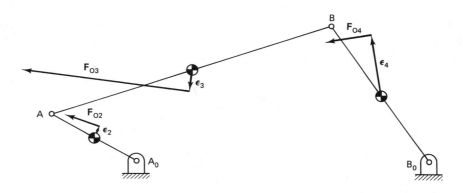

Figure 5.32   Inertial loading diagram of the linkage.

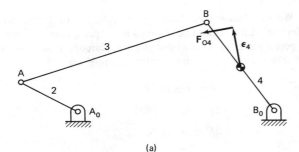

(a)

Figure 5.33a   Subproblem 1: Link 4 is considered massive, links 2 and 3 are assumed massless.

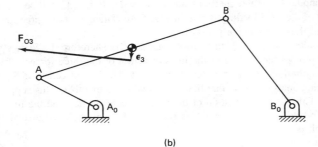

(b)

Figure 5.33b   Subproblem 2: Link 3 is considered massive, links 2 and 4 are assumed massless.

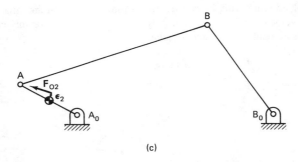

(c)

Figure 5.33c   Subproblem 3: Link 2 is considered massive, links 3 and 4 are assumed massless.

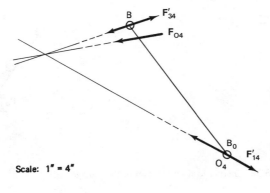

Scale: 1″ = 4″

Figure 5.34   Free-body diagram of link 4 of Fig. 5.33a showing point of concurrency.

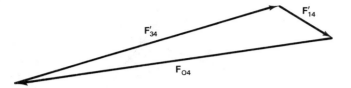

Figure 5.35 Force polygon of link 4 of Fig. 5.33a.

Newton's third law dictates that $\mathbf{F}'_{34} = -\mathbf{F}'_{32}$. We can skip the free-body diagram of link 3 and move directly to the free-body diagram of link 2 shown in Fig. 5.36. The kinetostatic equations are

$$\mathbf{F}'_{32} + \mathbf{F}'_{12} = 0$$

$$\mathbf{T}'_s + \mathbf{d}' \times \mathbf{F}'_{32} = 0 \qquad (5.106)$$

Here $\mathbf{d}'$ is the moment arm vector and is 3 inches long. The results for the first subproblem are

$$\mathbf{F}'_{14} = .03 \text{ lbf at } 313°$$

$$\mathbf{F}'_{34} = .13 \text{ lbf at } 18°$$

$$\mathbf{F}'_{32} = .13 \text{ lbf at } 198°$$

$$\mathbf{F}'_{12} = .13 \text{ lbf at } 18°$$

$$T'_s = .39 \text{ in·lbf cw}$$

Solution to subproblem 2. In this case link 3 is the only link with mass and the free-body diagram of this link should be considered first. Note that link 4 is now a two force member which determines the line of action of $\mathbf{F}''_{43}$. This now known line of action of $\mathbf{F}''_{43}$ along with $\mathbf{F}_{O3}$ fixes the location of the point of concurrency. This allows the line of action of $\mathbf{F}''_{23}$ to be determined as shown in Fig. 5.37. The dynamical equilibrium equation for this free-body diagram is:

$$\mathbf{F}''_{43} + \mathbf{F}''_{23} + \mathbf{F}_{O3} = 0 \qquad (5.107)$$

The force polygon is shown in Fig. 5.38. The results of this analysis are:

$$\mathbf{F}''_{43} = .08 \text{ lbf at } 305°$$

$$\mathbf{F}''_{23} = .49 \text{ lbf at } 0°$$

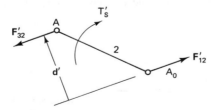

Figure 5.36 Free-body diagram of link 2 of Fig. 5.33a.

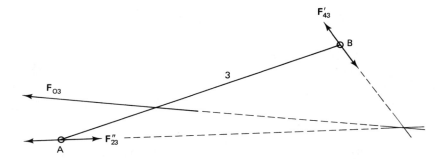

Figure 5.37 Free-body diagram of link 3 of Fig. 5.33b showing point of concurrency.

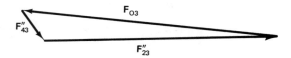

Scale: 1″ = 0.2 lb

Figure 5.38 Force polygon for link 3 of Fig. 5.33b.

A free-body diagram of link 4 would show that $\mathbf{F}''_{41} = -\mathbf{F}''_{43}$ because it is a two force member. The free-body diagram of link 2 is shown in Fig. 5.39. Again, $\mathbf{F}''_{32}$ is known from equation (5.107) and an application of Newton's third law. The kinetostatic equations for $\mathbf{F}''_{12}$ and $\mathbf{T}''_s$ are similar in form to those listed in equation (5.106). The final results for this subproblem are given in the same form as those for subproblem 2.

$$\mathbf{F}''_{14} = .08 \text{ lbf at } 305°$$

$$\mathbf{F}''_{34} = .08 \text{ lbf at } 125°$$

$$\mathbf{F}''_{32} = .49 \text{ lbf at } 180°$$

$$\mathbf{F}''_{12} = .49 \text{ lbf at } 0°$$

$$T''_s = .97 \text{ in·lbf cw}$$

Solution to subproblem 3. The final subproblem is actually the easiest. A set of free-body diagrams for links 3 and 4 are shown in Fig. 5.40. For the purposes of this subproblem, both of these links are two force members, having mechanical force application points at joints A, B and B_0. Therefore, the line of action of forces $\mathbf{F}'''_{23}$ and $\mathbf{F}'''_{43}$ must be along link 3. Also, the line of action of forces $\mathbf{F}'''_{34}$ and $\mathbf{F}'''_{14}$ must be along link 4. Now consider joint B.

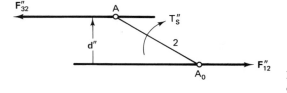

Figure 5.39 Free-body diagram of link 2 of Fig. 5.33b.

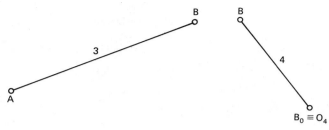

Figure 5.40 Proposed free-body diagrams for links 3 and 4 of Fig. 5.33c.

Scale: 1″ = 6″

Newton's third law maintains that $\mathbf{F}'''_{43} = -\mathbf{F}'''_{34}$. This is impossible, the directions of the forces do not align except at the change point position (if it is possible for this linkage to reach a change point configuration). The force system must be satisfied for all positions of the linkage. The only possible solution to this dilemma is that links 3 and 4 are zero force members; certainly a special case of a two force member system. This makes working with the free-body diagram of link 2 extremely easy. The diagram is shown in Fig. 5.41. The kinetostatic equations for this system are

$$\mathbf{F}_{O2} + \mathbf{F}'''_{12} = 0$$

$$\mathbf{T}'''_s + \mathbf{d}''' \times \mathbf{F}_{O2} = 0 \qquad\qquad (5.108)$$

The results of this subproblem are

$$\mathbf{F}'''_{14} = 0 \text{ lb}$$

$$\mathbf{F}'''_{34} = 0 \text{ lb}$$

$$\mathbf{F}'''_{32} = 0 \text{ lb}$$

$$\mathbf{F}'''_{12} = .106 \text{ lb at } 341°$$

$$T'''_s = .102 \text{ in·lb cw}$$

A note concerning the results of each subproblem is in order. The superposition technique clearly separates out the effect of the mass of each link upon the total required input torque. The input shaft torque calculated for each subproblem is the torque required to move the link possessing mass at the specified motion. It can be considered the cost incurred by the mass of each link. Changing the mass of a link will change the input torque required to move that link. The same realization is true for the pin reaction forces. More massive links give rise to higher pin forces as well as requiring greater input torques to sustain the motion. This concept is not as dramatically demonstrated by the matrix method which only yields the total values of the forces and torques required for the given motion.

Also note that the values of the required forces and torques were highly dependent upon the linkage position. Each position will give a different value of the input torque re-

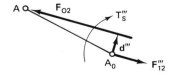

Figure 5.41 Free-body diagram of link 2 of Fig. 5.33c.

quired to sustain that motion. Kinetostatic analysis supplies the designer with the time varying torque requirements of the linkage.

The final answers to this problem are found by vectorally adding the answers from the subproblems as in Eqs. (5.71) to (5.75). The results are

$$\mathbf{F}_{14} = .11 \text{ lbf at } 307°$$

$$\mathbf{F}_{34} = .13 \text{ lbf at } 54°$$

$$\mathbf{F}_{32} = .61 \text{ lbf at } 184°$$

$$\mathbf{F}_{12} = .71 \text{ lbf at } .5°$$

$$T_s = 1.462 \text{ in·lbf cw}$$

Design Example: Kinetostatic Analysis of a Card Feeder [15]

Tabulating card punches and card readers are common types of peripheral equipment used in digital computer systems. The card feeder mechanisms, which feed cards from the input hopper into the machine, have to be capable to feed cards at a high rate. Typical rates are 300 to 1000 cards per minute. Techniques for card feeding, which are simpler and cheaper but reliable, are desirable.

Figure 5.42 shows one proposed design concept for a card feeder involving a four-bar linkage. A crank is driven by a suitable driving mechanism to produce oscillating

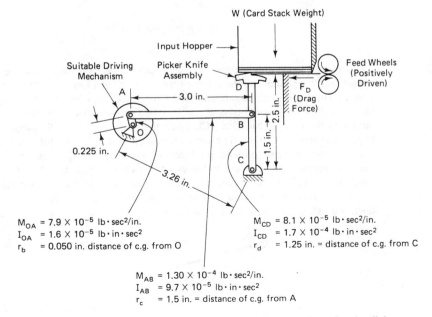

$M_{OA} = 7.9 \times 10^{-5} \text{ lb·sec}^2/\text{in.}$
$I_{OA} = 1.6 \times 10^{-5} \text{ lb·in·sec}^2$
$r_b = 0.050 \text{ in. distance of c.g. from O}$

$M_{CD} = 8.1 \times 10^{-5} \text{ lb·sec}^2/\text{in.}$
$I_{CD} = 1.7 \times 10^{-4} \text{ lb·in·sec}^2$
$r_d = 1.25 \text{ in.} = \text{distance of c.g. from C}$

$M_{AB} = 1.30 \times 10^{-4} \text{ lb·sec}^2/\text{in.}$
$I_{AB} = 9.7 \times 10^{-5} \text{ lb·in·sec}^2$
$r_c = 1.5 \text{ in.} = \text{distance of c.g. from A}$

Figure 5.42 Schematic diagram of card feeder mechanism using a four-bar linkage.

motion of a picker knife assembly. When the mechanism starts from rest, the picker knives engage a 0.007- to 0.009-inch-thick card. When the picker knives have reached maximum velocity, the card is caught by the feeder wheels, which are driven by another source. The card is then fed along its path through the machine. Finally, the mechanism is brought to rest back at its original position ready for the next card.

While the picker knives engage the card during the initial part of the cycle, the linkage has to overcome the card to card and the card-to-steel drag forces while supporting part of the weight of the card stack. After the card being fed is engaged by the feed wheels, only the card-to-steel drag force is acting. This drag force is small compared to the card-to-card drag force.

Figures 5.43 to 5.45 show the kinematic characteristics and the loading assumed for the mechanism. From Fig. 5.43 the angular position, velocity, and acceleration with respect to time are derived from the drive motor characteristics.

When $t \leq 0.01234$ sec, the average acceleration is

$$(\ddot{\theta}_2)_{\text{avg}} = \frac{-266.67 \text{ rad/sec}}{0.01234 \text{ sec}} = -21{,}610.2 \text{ rad/sec}^2;$$

$$\theta_2 = -\frac{21{,}610.2}{2} t^2 + 3.6183 \text{ rad}$$

$$\dot{\theta}_2 = -21{,}610.2t \text{ rad/sec}$$

$$\ddot{\theta}_2 = -21{,}610.2 \text{ rad/sec}^2$$

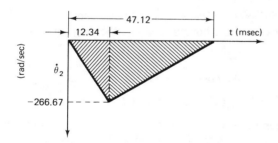

Figure 5.43 Angular velocity input to driving crank of sample four-bar linkage.

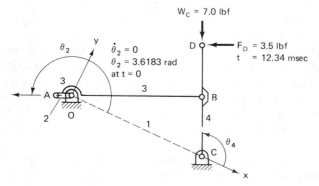

Figure 5.44 Definition of loads and angular coordinates at time equal to zero.

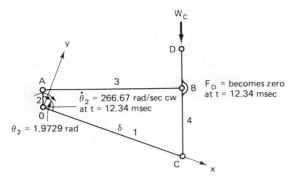

Figure 5.45 Definition of loads and angular coordinates at $T = 12.34$ msec.

When $t \geq 0.01234$ sec, the average acceleration is

$$\frac{266.67 \text{ rad/sec}}{(0.04712 - 0.01234)\text{sec}} = 7667.2 \frac{\text{rad}}{\text{sec}^2}$$

$$\theta_2 = \left[-266.67(t - 0.01234) + \frac{7667.2}{2}(t - 0.01234)^2 + 1.9729 \right] \text{rad}$$

$$\dot{\theta}_2 = [-266.67 + 7667.2(t - 0.01234)] \text{ rad/sec}$$

$$\ddot{\theta}_2 = 7667.2 \text{ rad/sec}^2$$

The area under the velocity curve in Fig. 5.43 represents one revolution or 2π radians of input crank rotation.

Figures 5.44 and 5.45 depict the loads imposed in this application, which are W_C, the weight of the stack of cards in the hopper, and F_D, the drag force on the card.

$$F_{L_x} = +W_C \sin \delta - F_D \cos \delta$$

$$F_{L_y} = -W_C \cos \delta - F_D \sin \delta$$

$$T_L = -F_{L_x}\overline{CD} \sin \theta_4 + F_{L_y}\overline{CD} \cos \theta_4$$

where δ is the angle between the y axis and link DC, $\mathbf{F}_L$ is the load force applied to the linkage at point D, and T_L is the load torque applied by $\mathbf{F}_L$ to link 4.

The solid line in Fig. 5.46 shows the input torque required for one complete revolution of the crank (link 2) based on Eqs. (5.100) to (5.104). Since F_{L_x}, F_{L_y}, and T_L are not functions of $\dot{\theta}_2$, $\ddot{\theta}_2$, or t, we can see the effects of inertia on input torque by setting $\dot{\theta}_2$ and $\ddot{\theta}_2$ equal to zero. The resulting torque should be the static torque required to maintain equilibrium. The dashed line in Fig. 5.46 shows the static torque required. The difference between the solid and dashed line may be thought of as being proportional to the equivalent inertia of the four-bar linkage referred to link 2. If a variable inertia load equal to the equivalent inertia were attached to the input link, the same variable input torque would be required. It should be noted that the equivalent inertia is not constant during the cycle.

Although the kinematic characteristics, as shown in Fig. 5.43, may be desirable, in practice they would be *difficult to achieve*. One would have to carefully measure the

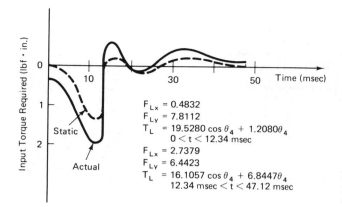

$F_{Lx} = 0.4832$
$F_{Ly} = 7.8112$
$T_L = 19.5280 \cos \theta_4 + 1.2080 \theta_4$
  $0 < t < 12.34$ msec
$F_{Lx} = 2.7379$
$F_{Ly} = 6.4423$
$T_L = 16.1057 \cos \theta_4 + 6.8447 \theta_4$
  12.34 msec $< t < 47.12$ msec

Figure 5.46   Given load and kinematic input, torque required to maintain dynamic equilibrium.

torque-speed characteristics of a motor while loading it with the presumed varying inertia of the linkage in order to obtain the true Fig. 5.43 for the motor. Therefore, the problem remains whether we can achieve the strict kinematic requirements of feeding cards without knowing the exact input torque characteristics of the driving mechanism. If more precision is required, a kinetostatic analysis is not appropriate and the time-response method of Sec. 5.3 of Vol. 2 must be considered.

5.7 DISCUSSION OF THE SUPERPOSITION AND MATRIX APPROACH TO KINETOSTATICS

The three kinetostatic techniques* described previously will of course yield the same results. They both provide numerical answers of the nine unknowns — eight pin reaction components and the required input torque. In both cases the technique must be repeated as many times as required to suitably describe the motion of the linkage (i.e., if the analysis is required for every 2° of input rotation for a crank-rocker, the analysis must be repeated 180 times). The flow diagram of Fig. 5.47 outlines computer programs which will perform a kinetostatic analysis of a four-bar mechanism by both methods.

All three of these kinetostatic techniques can be applied to any planar linkage. The same general procedure that is outlined in Fig. 5.47 would be used in applying these techniques to more complex linkages. Note that all of these analyses have been limited to constraining the center of masses along the link centerlines and no additional external loads. Generalizing the equations to include variable link geometry as well as many external loads is straightforward (see Sec. 5.10 of Vol. 2, on balancing).

Chapter 5 of Vol. 2 continues the development of mechanism dynamics. Time response, vibrations, balancing, and high-speed elastic mechanisms are subjects covered in that chapter.

Several software packages are available for dynamic analysis of planar mechanisms. These programs can perform static, kinetostatic, and in most cases, time re-

* Analytical and geometric superposition and the matrix method.

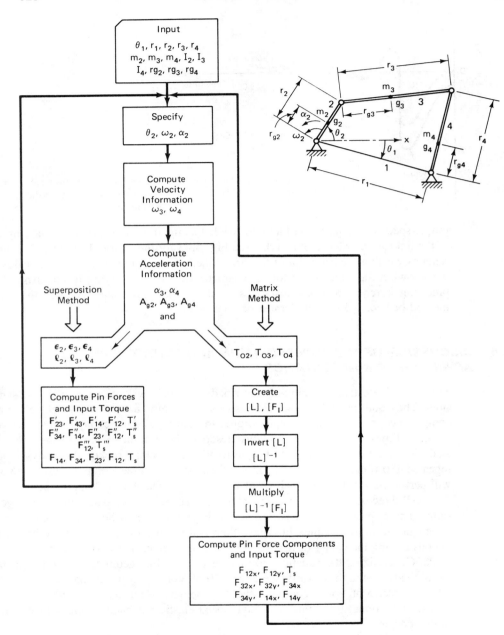

Figure 5.47 Computer program flowchart for kinetostatic analysis of a four-bar mechanism. Drawing on top right gives input notation. The superposition method (left branch of flowchart) can also be carried out graphically.

sponse analysis of multiloop planar mechanisms with lower pair and higher pair joints
[20-22, 85, 93, 121, 147, 159, 169-171].

PROBLEMS

5.1. Figure P5.1 shows link 2 rotating about A_0 at $\omega_2 = 12$ rad/sec and $\alpha_2 = 60$ rad/sec^2. If
$I_2 = 0.04$ slug·ft^2, $W_2 = 6$ lbf, $A_0A = 12$ in., and $A_0cg_2 = 6.6$ in.:
  (a) Find the magnitude and direction of the inertia force at the center of gravity.
  (b) Find the value of the offset vector ε_2 of the inertia force that would yield the correct
  inertia torque.
  (c) Draw a picture of the link with the inertia force in the correct location. (*Hint:* consider ε_2).

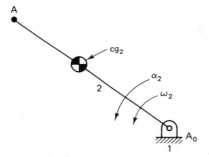

Figure P5.1

5.2. Figure P5.2 shows a four-bar linkage with inertia forces F_{O2}, F_{O3}, and F_{O4} already computed
and located on the figure. (The lengths of the vectors represent the correct magnitudes of
these forces. Scale the drawing for lengths and directions.)
  (a) Find the resultant of these inertia forces as seen from the ground.
  (b) Determine the magnitude and direction of pin forces at A and B.
  (c) Do (a) and (b) both analytically and graphically.

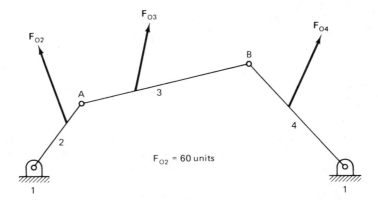

Figure P5.2

5.3. In Fig. P5.3 known external force **S** and inertia force **P** are applied to links 3 and 4, respectively. Links 2 and 3 are assumed to have very small inertia as compared to forces **P** and **S**. What torque must be applied to link 2 for equilibrium? Use free-body diagrams and solve graphically.

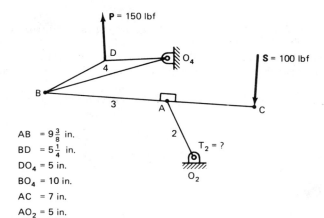

AB $= 9\frac{3}{8}$ in.

BD $= 5\frac{1}{4}$ in.

DO_4 = 5 in.

BO_4 = 10 in.

AC   = 7 in.

AO_2 = 5 in.

Figure P5.3

5.4. The four-bar linkage in Fig. P5.4 has two forces acting on it, an inertia force **P** = 100 lbf acting at point B and an external torque $T_4 = -600$ in·lbf acting on link 4. Find the required input torque T_2 and the pin forces at O_2, A, C, and O_4.
(a) Use the graphical method based on free-body diagrams.
(b) Use the complex-number method.

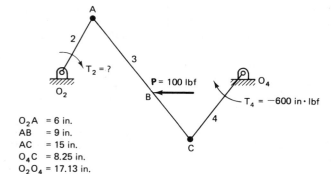

O_2A = 6 in.

AB   = 9 in.

AC   = 15 in.

O_4C = 8.25 in.

O_2O_4 = 17.13 in.

Figure P5.4

5.5. Figure P5.5 shows a double-piston mechanism which is assumed to be in equilibrium under the action of external forces **P** and **S** and an input torque T_2. Find the input torque T_2 and the pin forces at A, C, and O_2 as well as the contact force between the pistons and their guides.
(a) Use the graphical method based on free-body diagrams.
(b) Use the complex-number method.

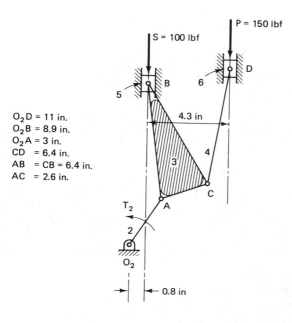

O_2D = 11 in.
O_2B = 8.9 in.
O_2A = 3 in.
CD = 6.4 in.
AB = CB = 6.4 in.
AC = 2.6 in.

Figure P5.5

5.6. Find the input torque T_2 applied to link 2 to maintain equilibrium of the linkage shown in Fig. P5.6 with external force **P** acting on it.
   (a) Use the graphical method based on free-body diagrams.
   (b) Use the complex-number method.

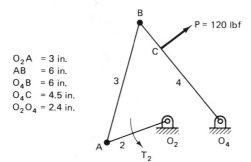

O_2A = 3 in.
AB   = 6 in.
O_4B = 6 in.
O_4C = 4.5 in.
O_2O_4 = 2.4 in.

Figure P5.6

5.7. The four-bar mechanism of Fig. P5.7 has one external force **P** = 200 lbf and one inertia force **S** = 150 lbf acting on it. The system is in dynamic equilibrium as a result of torque T_2 applied to link 2. Find T_2 and the pin forces.
   (a) Use the graphical method based on free-body diagrams.
   (b) Use the complex-number method.

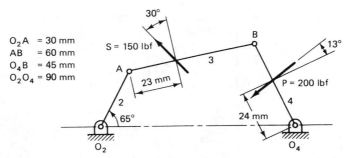

$$O_2A = 30 \text{ mm}$$
$$AB = 60 \text{ mm}$$
$$O_4B = 45 \text{ mm}$$
$$O_2O_4 = 90 \text{ mm}$$

Figure P5.7

5.8. The input crank of the four-bar linkage shown in Fig. P5.8 rotates at a constant speed $\omega_2 = 300$ rpm ccw. Only the coupler link is considered to have significant inertia. The velocity and acceleration diagrams are given in the figure. Calculate the values of velocities and accelerations shown in these diagrams. Then

   (a) Determine the linear acceleration of the center of gravity A_{g3} and the angular acceleration α_3 of link 3.

   (b) Find the inertia force F_{O3} of the coupler link.

   (c) Find the offset ε_3 of the inertia force F_{O3}.

   (d) Sketch the inertia force in its correct position on the linkage.

   (e) Find the directions and magnitudes of the pin forces at A and B.

   (f) Determine the required input torque to drive this mechanism in this position under the conditions described above.

   (g) Solve (e) and (f) graphically and analytically.

$$O_2A = 5 \text{ in.}$$
$$AB = 15 \text{ in.}$$
$$AG_3 = 7.5 \text{ in.}$$
$$O_4B = 10 \text{ in.}$$
$$O_2O_4 = 14 \text{ in.}$$
$$W_3 = 11.45 \text{ lbf}$$
$$I_3 = 0.451 \text{ lbf} \cdot \text{in} \cdot \text{sec}^2$$

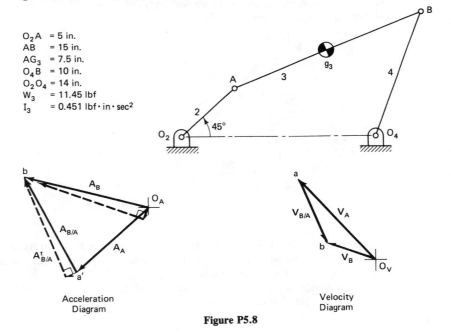

Acceleration
Diagram

Velocity
Diagram

Figure P5.8

5.9. The input crank of the four-bar linkage of Fig. P5.9 rotates at a constant speed of $\omega_2 = 500$ rad/sec cw. Each link has significant inertia. The velocity and acceleration diagrams are provided in the figure. Calculate the values of all velocities and accelerations in these diagrams. Then

(a) Determine the linear accelerations of each center of gravity and angular accelerations α_2, α_3, and α_4.

(b) Find the inertia forces $\mathbf{F}_{O2}$, $\mathbf{F}_{O3}$, and $\mathbf{F}_{O4}$.

(c) Find the offsets ε_2, ε_3, and ε_4 of the inertia forces.

(d) Sketch the inertia forces in their correct positions on the linkage.

(e) Find the directions and magnitudes of the pin forces at A and B.

(f) Determine the required input torque to drive this mechanism in this position under the conditions described above. Solve (e) and (f) by the matrix method.

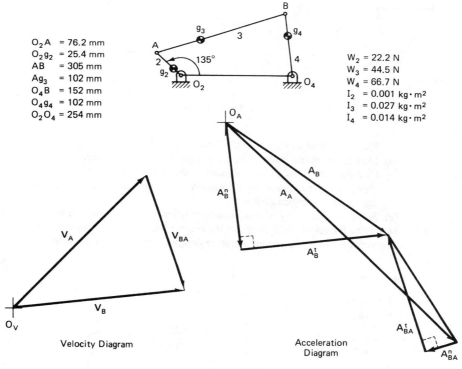

O_2A = 76.2 mm
O_2g_2 = 25.4 mm
AB = 305 mm
Ag_3 = 102 mm
O_4B = 152 mm
O_4g_4 = 102 mm
O_2O_4 = 254 mm

W_2 = 22.2 N
W_3 = 44.5 N
W_4 = 66.7 N
I_2 = 0.001 kg·m^2
I_3 = 0.027 kg·m^2
I_4 = 0.014 kg·m^2

Velocity Diagram

Acceleration Diagram

Figure P5.9

5.10. The slider-crank mechanism of Fig. P5.10 is to be analyzed to determine the effect of the inertia of the connecting rod (link 3). The velocity diagram is shown in the figure and the magnitude of $\mathbf{V}_A$ is given. Calculate the crank vector $\overrightarrow{O_2A}$ and the input angular velocity ω_2, and proceed to calculate the values of all vectors in the diagram. Then

(a) Determine the linear acceleration of the center of gravity of link 3 and the angular acceleration α_3.

(b) Find the inertia force $\mathbf{F}_{O3}$ of the coupler link.

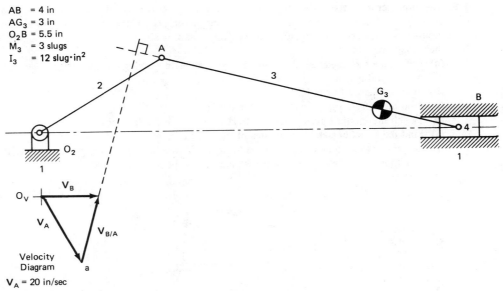

AB = 4 in
AG_3 = 3 in
O_2B = 5.5 in
M_3 = 3 slugs
I_3 = 12 slug·in²

V_A = 20 in/sec

Figure P5.10

(c) Find the offset ε_3 of the inertia force $\mathbf{F}_{O3}$.
(d) Sketch the inertia force in its correct position on the linkage.
(e) Find the directions and magnitudes of the pin forces at A and B.
(f) Determine the required input torque to drive this mechanism in this position under the conditions described above.
(g) Solve (e) and (f) by graphical and analytical superposition and by the matrix method.

5.11. Determine the effect of the inertia of the coupler link of the four-bar mechanism in Fig. P5.11. The pertinent data and the acceleration diagram are shown in the figure.

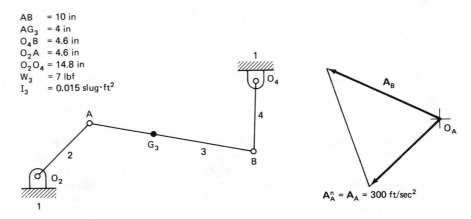

AB = 10 in
AG_3 = 4 in
O_4B = 4.6 in
O_2A = 4.6 in
O_2O_4 = 14.8 in
W_3 = 7 lbf
I_3 = 0.015 slug·ft²

$A_A^n = A_A$ = 300 ft/sec²

Figure P5.11

(a) Determine the linear acceleration of the center of gravity of link 3 and the angular acceleration α_3.

(b) Find the inertia force $\mathbf{F}_{O3}$ of the coupler link.

(c) Find the offset ε_3 of the inertia force $\mathbf{F}_{O3}$.

(d) Sketch the inertia force in its correct position on the linkage.

(e) Find the directions and magnitudes of the pin forces at A and B.

(f) Determine the required input torque to drive this mechanism in this position under the conditions described above.

(g) Solve (e) and (f) by graphical and analytical superposition and by the matrix method.

5.12. Let us determine the effect of considering the inertia of links 2 and 4 in Prob. 5.11. Refer again to Fig. P5.11, and add the following data:

$$O_2 G_2 = O_4 G_4 = 2.5 \text{ in}$$

$$W_2 = W_4 = 5 \text{ lbf}$$

$$I_2 = I_4 = 0.01 \text{ slug·ft}^2$$

(a) Determine the linear acceleration of G_2 and G_4 and the angular accelerations α_2 and α_4.

(b) Find the inertia forces $\mathbf{F}_{O2}$ and $\mathbf{F}_{O4}$.

(c) Find the offsets ε_2 and ε_4.

(d) Sketch the inertia forces in the correct positions on the linkage.

(e) With all inertias considered, find the required input torque.

(f) Solve (d) and (e) by graphical and analytical superposition and by the matrix method.

5.13. Figure P5.12 shows a four-bar mechanism with only the mass and inertia of link 3 considered pertinent. For the given data, determine θ_3 and θ_4 and then:

(a) Find the inertia force $\mathbf{F}_{O3}$ of the coupler link.

(b) Find the offset ε_3 of the inertia force $\mathbf{F}_{O3}$.

(c) Sketch the inertia force in its correct position on the linkage.

(d) Find the directions and magnitudes of the pin forces at A and B.

(e) Determine the required input torque to drive this mechanism in this position under the conditions described above.

(f) Solve (d) and (e) graphically and analytically.

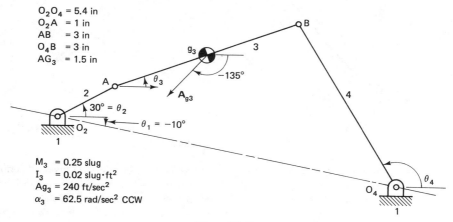

$O_2O_4 = 5.4$ in
O_2A = 1 in
AB = 3 in
O_4B = 3 in
AG_3 = 1.5 in

M_3 = 0.25 slug
I_3 = 0.02 slug·ft^2
Ag_3 = 240 ft/sec^2
α_3 = 62.5 rad/sec^2 CCW

Figure P5.12

5.14. The four-bar mechanism in Fig. P5.13 is to be analyzed instantaneously to determine the effect of the inertia of three moving links. If the input angular velocity and acceleration are $\omega_2 = 2\pi$ rad/sec ccw and $\alpha_2 = 2\pi$ rad/sec^2 ccw and the velocity and acceleration diagrams are given, then

(a) Determine the linear accelerations of each center of gravity and angular accelerations α_3 and α_4.

(b) Find the inertia forces $\mathbf{F}_{O2}$, $\mathbf{F}_{O3}$, and $\mathbf{F}_{O4}$.

(c) Find the offsets ε_2, ε_3, and ε_4 of the inertia forces.

(d) Sketch the inertia forces in their correct positions on the linkage.

(e) Find the directions and magnitudes of the pin forces at A and B.

(f) Determine the required input torque to drive this mechanism in this position under the conditions described above.

(g) Solve (e) and (f) by graphical and analytical superposition and by the matrix method.

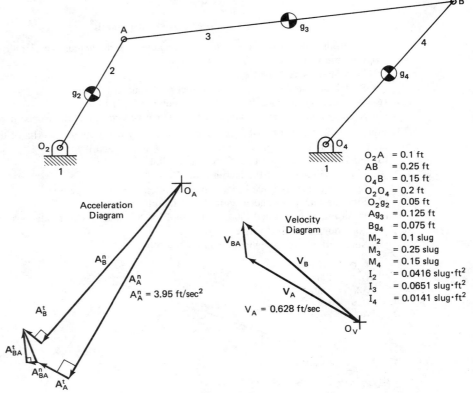

$$O_2A = 0.1 \text{ ft}$$
$$AB = 0.25 \text{ ft}$$
$$O_4B = 0.15 \text{ ft}$$
$$O_2O_4 = 0.2 \text{ ft}$$
$$O_2g_2 = 0.05 \text{ ft}$$
$$Ag_3 = 0.125 \text{ ft}$$
$$Bg_4 = 0.075 \text{ ft}$$
$$M_2 = 0.1 \text{ slug}$$
$$M_3 = 0.25 \text{ slug}$$
$$M_4 = 0.15 \text{ slug}$$
$$I_2 = 0.0416 \text{ slug} \cdot \text{ft}^2$$
$$I_3 = 0.0651 \text{ slug} \cdot \text{ft}^2$$
$$I_4 = 0.0141 \text{ slug} \cdot \text{ft}^2$$

Figure P5.13

5.15. In Fig. P5.14, link 2 rotates about fixed axis O_2. For the data given, determine the inertia force vector $\mathbf{F}_{O2}$ and draw it in its proper position on a scale drawing of link 2. Determine the external force vector $\mathbf{F}_A$ to produce the given angular motion. Data: $\omega_2 = -5$ rad/sec cw, $\alpha_2 = 200$ rad/sec^2 ccw, $W_2 = 44.5$ N, and $I_2 = 0.0212$ kg·m^2. Solve graphically and analytically.

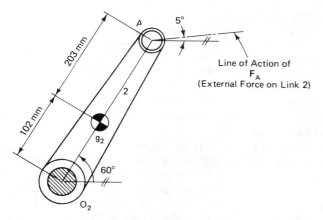

Figure P5.14

5.16. Same as Prob. 5.15, but with different data: $\omega_2 = -10$ rad/sec cw, $\alpha_2 = 2000$ rad/sec^2 ccw, $W_2 = 8$ lbf, $I_2 = 1/64$ slug·ft^2, $O_2 g_2 = 3$ in., and $g_2 A = 5$ in.

5.17. Figure P5.15 shows a four-bar mechanism that has a counterweight on link 2 so that the center of mass of link 2 is at O_2. The radius of gyration of link 3 about g_3 is 4.5 in., while the radius of gyration of link 4 about O_4 is 3.9 in. If the input angular velocity is constant at $\omega_2 = 40$ rad/sec ccw, use superposition to:

   (a) Determine the linear accelerations of each center of gravity and angular accelerations α_2, α_3, and α_4.

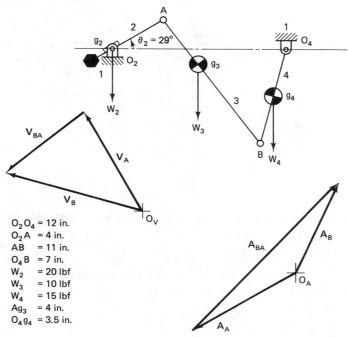

Figure P5.15

(b) Find the inertia forces F_{O2}, F_{O3}, and F_{O4}.

(c) Find the offsets ε_2, ε_3, and ε_4 of the inertia forces.

(d) Sketch the inertia forces in their correct positions on the linkage.

(e) Find the directions and magnitudes of the pin forces at A and B.

(f) Determine the required input torque to drive this mechanism in this position under the conditions described above.

(g) Use both graphical and analytical superposition for (e) and (f).

5.18. Same as Prob. 5.17 but use the matrix approach where the entire mechanism is treated at the same time [disregard parts (c) and (d)].

5.19. A slider-crank mechanism shown in Fig. P5.16 is the main drive of a compressor. The input crank is partially counterbalanced with a weight so that g_2 is $\frac{1}{2}$ in. from O_2 toward the balancing weight. The radius of gyration of link 3 about g_3 is 4.1 in. The only external force is the given gas force F_4 acting on the face of the piston. All four inertia forces are also given. (The mechanism is moving in the horizontal plane, so the link weights have no effect.) Check the given values of the inertia forces and inertia torque in view of the given input motion. Are they right? If not, correct them. Then:

(a) Find the directions and magnitudes of the pin forces at A and B.

(b) Determine the required input torque to drive this mechanism in this position under the conditions described above.

(c) Use both analytical and graphical superposition and also the matrix method for (a) and (b).

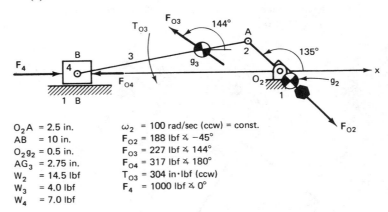

O_2A = 2.5 in.	ω_2 = 100 rad/sec (ccw) = const.
AB = 10 in.	F_{O2} = 188 lbf $\angle$ $-45°$
O_2g_2 = 0.5 in.	F_{O3} = 227 lbf $\angle$ 144°
AG_3 = 2.75 in.	F_{O4} = 317 lbf $\angle$ 180°
W_2 = 14.5 lbf	T_{O3} = 304 in·lbf (ccw)
W_3 = 4.0 lbf	F_4 = 1000 lbf $\angle$ 0°
W_4 = 7.0 lbf	

Figure P5.16

5.20. Same question as Prob. 5.19 but the position is top dead center (far left position).

5.21. A scoop mechanism is shown in Fig. P5.17. We assume that the inertia of link 3 has little effect in the force analysis. Link 2 has mass of 5 kg and mass moment of inertia about g_2 of 24 kg·cm². Both the direction of the input force F_{in} and its point of application are known. Verify the values given in the velocity and acceleration diagrams in view of the given input motion. Correct them if necessary. Then:

(a) Determine A_{g2} and α_2.

(b) Find the inertia force F_{O2} of the output link.

(c) Find the offset ε_2 of F_{O2}.

(d) Sketch the inertia force in its correct position on the linkage.

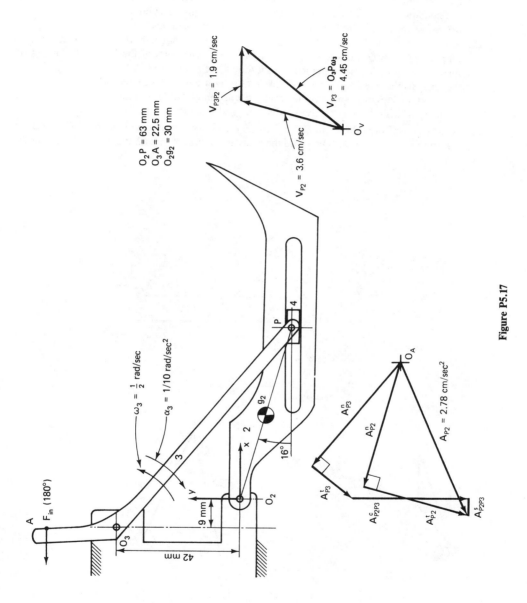

$O_2P = 63$ mm
$O_3A = 22.5$ mm
$O_2g_2 = 30$ mm

$V_{P_3P_2} = 1.9$ cm/sec

$V_{P_3} = O_3P\omega_3 = 4.45$ cm/sec

$V_{P_2} = 3.6$ cm/sec

$\omega_3 = \dfrac{1}{2}$ rad/sec

$\alpha_3 = 1/10$ rad/sec^2

F_{in} (180°)

16°

9 mm

42 mm

$A_{P_3}^n$

$A_{P_2}^n$

$A_{P_2} = 2.78$ cm/sec^2

$A_{P_3}^t$

$A_{P_2P_3}^c$

$A_{P_2}^t$

$A_{P_2P_3}^s$

Figure P5.17

(e) Find the directions and magnitudes of the pin forces at O_3 and P and the normal force between slider 4 and the slot in link 2.

(f) Determine the required input torque to drive this mechanism in this position under the conditions described above and find F_{in}. *Hint:* Ignore friction.

(g) Solve (e) and (f) by both analytical and graphical methods.

5.22. Same questions as Prob. 5.21, but in this case a load of dirt is in the scoop with center of mass 6 cm from O_2 on a line $0°$ to the right of O_2. The mass of the load is 10 kg and the mass moment of inertia of the load about the CG of the load is 20 kg·cm². Also include in the analysis the mass and inertia of link 3: $O_3g_3 = 4$ cm on the centerline O_3P, $m_3 = 2$ kg, and $I_3 = 15$ kg·cm². How do the resulting pin forces, slider-guide force and input force compare with Prob. 5.21? Solve analytically and graphically.

5.23. A lift mechanism is shown in Fig. P5.18. For this problem assume that only link 3 has significant mass and inertia. (Disregard the mass and inertia of links 2 and 4 and of the load W.)

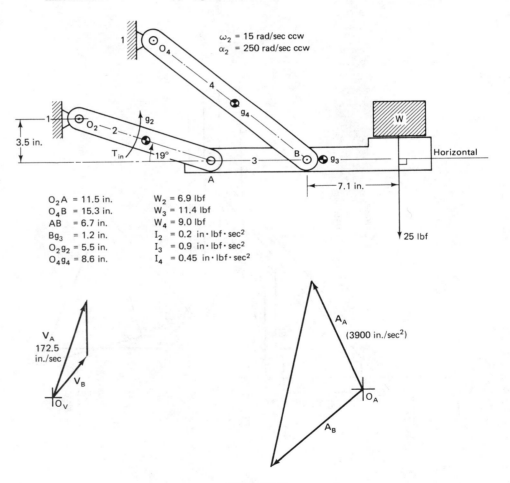

O_2A = 11.5 in.	W_2 = 6.9 lbf
O_4B = 15.3 in.	W_3 = 11.4 lbf
AB = 6.7 in.	W_4 = 9.0 lbf
Bg_3 = 1.2 in.	I_2 = 0.2 in·lbf·sec²
O_2g_2 = 5.5 in.	I_3 = 0.9 in·lbf·sec²
O_4g_4 = 8.6 in.	I_4 = 0.45 in·lbf·sec²

Figure P5.18

(a) Determine $\mathbf{A}_{g3}$ and α_3.
(b) Find the inertia force $\mathbf{F}_{O3}$.
(c) Locate the offset $\boldsymbol{\varepsilon}_3$ of $\mathbf{F}_{O3}$.
(d) Sketch the inertia force in its correct position on the linkage.
(e) Find the directions and magnitudes of the pin forces at A and B.
(f) Determine the required input torque to drive this mechanism in this position under the conditions described above. Solve by analytical and graphical superposition and by the matrix method.

5.24. Referring to Prob. 5.23, include the effects of the mass and inertia of links 2 and 4 as well as of the load W.
(a) Determine the linear accelerations of g_2 and g_4 and angular acceleration α_4.
(b) Find the inertia forces $\mathbf{F}_{O2}$ and $\mathbf{F}_{O4}$.
(c) Locate the offsets $\boldsymbol{\varepsilon}_2$ and $\boldsymbol{\varepsilon}_4$.
(d) Sketch the inertia forces in their correct positions on the linkage.
(e) Find the directions and magnitudes of the pin forces at A and B.
(f) Determine the required input torque to drive this mechanism in this position under the conditions described above. Solve by analytical and graphical superposition and by the matrix method.

chapter six
Cam Design

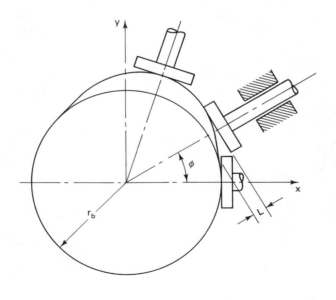

6.1 INTRODUCTION

A *cam* is a convenient device for transforming one motion into another. This machine element has a curved or grooved surface which mates with a *follower* and imparts motion to it. The motion of the cam (usually rotation) is transformed into follower oscillation, translation, or both. Because of the various cam geometries, and the large number of cam and follower combinations, the cam is an extremely versatile mechanical element. Although a cam and follower may be designed for motion, path, or function generation, the majority of applications utilize the cam and follower for function generation. Figure 6.1 illustrates all three types; a and b are self-explanatory. Figure 6.1c illustrates a car lift whereby similar cam mechanisms arranged in tandem impart circular translation to platform *p* and the cams are designed such that the velocity ratio between the speed of lead screw *b* and platform *p* will be constant. An electric motor turns lead screw *b* which moves crossheads *c* and *d* guided by rollers. Arms *e* and *f* pivoted to the crossheads carry rollers *g* and *h*, driving them between stationary cams *i*, *k*, and cam levers *m*, *n* which are pivoted to the frame. Pins *E* and *F* of levers *m* and *n* carry platform *p*. The linear velocity relationship is achieved by coordinating the design of the cam members on *m* with *i* and on *n* with *h*; the shape of one can be assumed and the other designed accordingly.

Figure 6.1d shows an aircraft drift meter in which the sight wire *f* is aligned to follow an object on the earth which passes through its center 0, thereby defining the direction of the aircraft relative to ground. While the instrument must rotate about 0, a physical pivot at this point would hinder vision; therefore, rotation is defined by two fixed guiding pins which guide a circular arc shaped cam slot with its center at 0.

The computer disc that accompanies this book has a cam design module CAM-SYN that will allow the reader to design cam profiles using the equations developed in this chapter. See section 6.8 for more details.

6.2 CAM AND FOLLOWER TYPES

The most common cam types according to cam shapes are *plate* or *disk, translating* (two-dimensional or planar), and *cylindrical* (three-dimensional or spatial) cams. Figure 6.2 shows examples of each of these cam types as well as some other three-dimensional cams: the *conical* and the *globoidal* cams.

Followers can be classified in several ways: according to follower motion, such as translation or oscillation; according to whether the translational (straight-line) follower motion may be radial or offset from the center of the cam shaft; and according to the shape of the follower contact surface (e.g., flat-face, roller, point [knife-edge], spherical, planar-curved, or spatial-curved surface). Figure 6.3 illustrates many of these follower classifications.

Disk Cam with Radial Roller Follower

Figure 6.4 shows a disk cam with a radial (in-line) roller follower and with the standard cam nomenclature: The smallest circle that can be drawn tangent to the cam surface con-

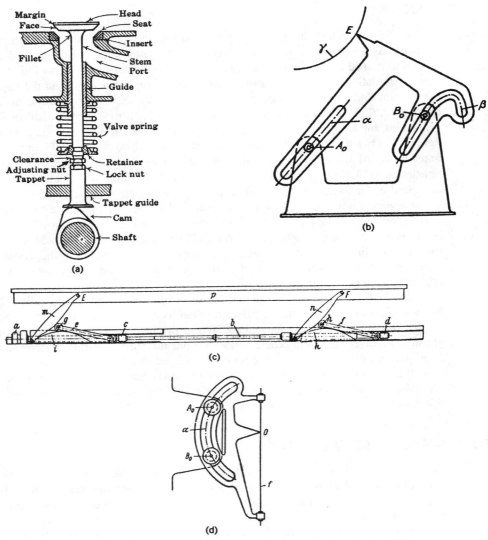

Figure 6.1 (a) Function-generator cam: the lift of the valve is a function of the cam-shaft position; (b) path-generator cams: guiding point E along path γ by means of moving cam slots α and β; (c) motion-generator cams: a stationary car lift with stationary and moving cams (Schroeder, Munich); (d) motion-generator cam: drift meter: example of a coupler cam-guided plane.

centric with the camshaft is the *base circle*.* The *tracer point* is a point at the center of the follower that generates the *pitch curve*. The *pressure angle* is similar to the deviation angle in linkage analysis (see Sec. 3.1) and is the complement of the transmission angle.

* Engineers in practice will more naturally think of a cam cut out of a blank (with a maximum *blank diameter*) rather than building up from the base circle.

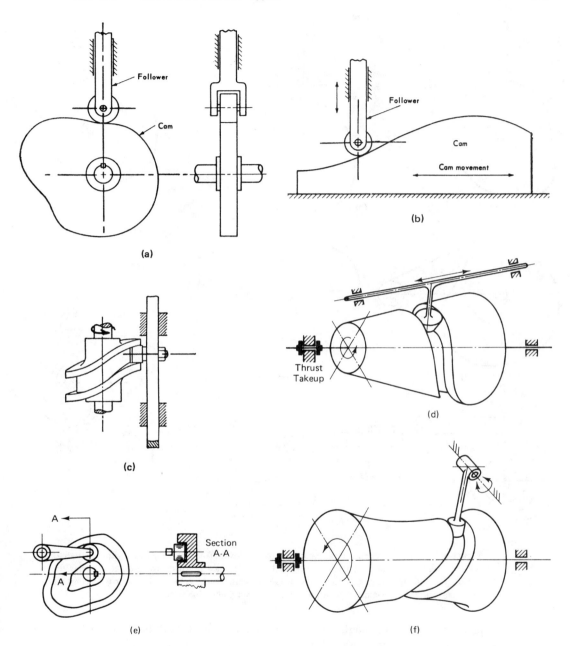

Figure 6.2 Cam types: (a) plate or disk cam with translating roller follower; (b) translating or wedge cam with translating roller follower; (c) cylindrical cam with translating roller follower; (d) conical cam with translating follower; (e) face cam with oscillating follower; (f) globoidal cam with oscillating follower [37].

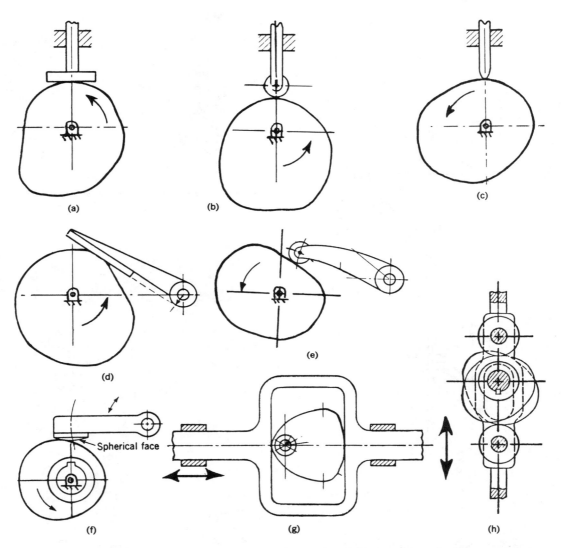

Figure 6.3 Follower types: (a) translating flat-face follower; (b) translating roller follower (radial); (c) translating point follower (radial); (d) oscillating flat-face follower; (e) oscillating roller follower; (f) oscillating spherical-face follower; (g) translating positive-return follower with constant-diameter cam; (h) translating double-roller follower and double-lobed cam.

The pressure angle is the angle between the direction of the path of the follower and the normal to the pitch curve through the center of the cam follower. Neglecting friction, this normal is collinear with the contact force between the cam and follower. As in a linkage, the pressure angle varies during the cycle and is a measure of the ability of the cam to transmit motion to the follower. In the case of a disk cam with a translating roller follower, a large pressure angle will produce an appreciable lateral force exerted on the

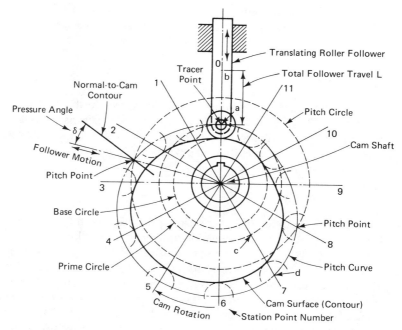

Figure 6.4 Disk cam and radial roller follower with appropriate nomenclature. Distance c-d is the rise of the follower in position 7.

stem of the follower, which, in the presence of friction, would tend to bind the follower in the guide. The *pitch point* is usually the location of the maximum pressure angle along the pitch curve. The *pitch circle* has a radius from the camshaft center to the pitch point, while the *prime circle* is the smallest circle from the camshaft center tangent to the *pitch curve,* the path traced by the tracer point relative to the cam.

A typical cam application would require a follower displacement such as the one shown in Fig. 6.5. In this example one full revolution of the cam (with the developed length of the circumference of the prime circle) is represented on the abscissa while the displacement of the follower is shown on the ordinate. The tracer point of the follower is

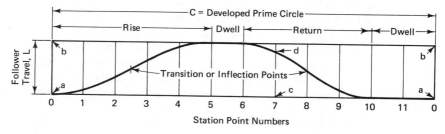

Figure 6.5 Follower displacement profile corresponding to Fig. 6.4. Distance c-d is the rise of the follower in position 7. Maximum follower travel L represents movement from point a on the prime circle to point b at stations 5 and 6.

required to rise off the prime circle by the rise L, to remain for a while (or "dwell") at height L, return to the prime circle, and remain at rest in a second dwell before repeating the cycle.

6.3 CAM SYNTHESIS

Numerous applications in automatic machinery require the kind of intermittent motion suggested in Fig. 6.5. A typical example will call for a rise–dwell–return and perhaps another dwell period of a specified number of degrees each, together with a required follower displacement measured in centimeters (inches) or degrees. The designer's job is to lay out the cam accordingly.

Type Synthesis

The first decision to be made is to choose the cam and follower types. The specific application may dictate the combination of the cam and follower. Some factors that should enter into the decision are *geometric* considerations — type of output (rotary or translational), distance between cam shaft and the center of the required oscillatory output, space allowed for cam and follower; *dynamic* considerations — rotational speed of cam, loading on cam and follower, and masses to be moved; *environmental* considerations — projected environmental conditions in which the cam will be required to operate, as well as environmental requirements of the cam system (e.g., noise, cleanliness); and *economic* matters — first and maintenance cost, number of duplicate systems, and so on.

Figures 6.2 and 6.3 show some of the options available for the type synthesis step. A certain degree of experience will help pick an appropriate cam and follower. Simplicity is always a governing factor in the choice of cam and follower. For this reason, most of this chapter is devoted to the simplest type, the plate cam, whose geometry and method of synthesis are the same as those of the face cam (Fig. 6.2). Section 6.10 does extend the discussion of type synthesis to cam-linkage mechanisms, referred to as cam-modulated linkages.

Follower Motion

Once a cam and follower pair has been chosen, the follower motion must be synthesized. In most cases a cam-follower is required to be displaced through a specified rise or fall (return). The shape of the displacement curve (such as the one shown in Fig. 6.5) may not initially seem important, but the cam and follower are just a segment of a dynamic mechanical system — one whose performance may well depend on the inertial (and impact) properties of the cam and follower. Therefore, the velocity, acceleration, and in some cases further derivatives of the displacement of the follower are of great importance. Sections 6.4 and 6.5 will be devoted to some standard displacement diagrams.

Dimensional Synthesis of the Cam Profile

The cam profile cannot be generated by just wrapping the proposed displacement diagram around the base circle of the cam — except in the case of a radially translating point follower. Interference of portions of a roller or flat-face follower adjacent to the contact point or line with adjacent portions of the cam surface requires a more detailed cam profile synthesis procedure. Sections 6.7 and 6.8 provide graphical and analytical procedures, respectively, for several different follower types.

It should be noted that in many cases, the key section of a machine in which inertia forces must be minimized may be many links removed from the cam and follower (see, e.g., Fig. 8.8). It is at this remote location from the follower that the derivatives of the displacement profile must be minimized, not at the follower. In these design situations the kinematics at the key portion of the machine must be traced back to the follower to create a compensated follower displacement profile before the cam profile may be synthesized.

Cam Analysis

Once the cam profile is synthesized, the design may not be acceptable for many different reasons. For example, (1) the pressure angles may be unacceptable; (2) the follower may not be able to follow the cam surface due to local curvature conditions; (3) too large a return spring may be required to keep the follower in contact with the cam surface during the cycle; or (4) excessive dimensions may be required (e.g., length of a flat-face follower or size of a spherical-face follower).

Usually, several parameters can be varied to help alleviate such problems. Some of these parameters may be: the cam base circle diameter, the offset of the follower travel if a straight-line translating follower is used or the relative locations of the fixed pivot of the follower arm to the center of the cam if an oscillating follower is employed (two independent parameters are needed here, such as two rectangular or two polar coordinates), the length of the follower arm and the radius of the follower if a roller or spherical follower is used. By modifying any of these arbitrary choices, a new cam profile will be synthesized without the need to pick another cam or follower type (although this too may become necessary).

6.4 DISPLACEMENT DIAGRAMS: GRAPHICAL DEVELOPMENT*

The task of a cam designer is to prescribe a follower displacement profile, the kind shown in Fig. 6.5. Let us concentrate on the rise portion of the follower motion and think through various possible displacement profiles. Each will have a total rise L in β degrees of cam rotation.

Constant Velocity

The simplest displacement profile one can think of is a straight line between zero follower displacement and the end of the rise (points a and b at stations 0 and 5 in Fig. 6.5). This *straight line* or *constant velocity* profile is shown in Fig. 6.6, where the position, velocity (slope of the displacement diagram), and acceleration diagrams ap-

---

* This section may be skipped for the reader only interested in the analytical approach.

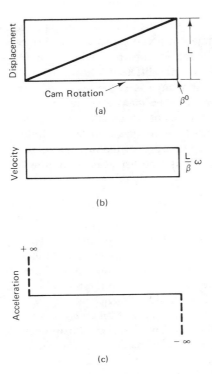

Figure 6.6 The constant velocity curve.

pear. The disadvantage of this simple profile is the infinite acceleration at the beginning and end of the rise. The consequent large inertia forces at these points in the cam's cycle will disqualify this profile for any application that requires moderate or high cam velocity. High inertia forces can also result in vibrations, noise, high stress levels, and wear.

To avoid the infinite accelerations at the beginning and the end of the rise, a *modified straight-line* curve is sometimes used (see Fig. 6.7a). The step changes in velocity are removed by smoothing the displacement by a convenient radius R. The shorter R, the nearer to the undesirable conditions of the constant velocity profile; the longer R, the more gradual the acceleration conditions at the ends, but the higher the velocity during the midsection of the rise. A radius equal to the follower displacement is chosen often in practice as shown in Fig. 6.7a.

Unfortunately, the modified straight-line profile does not exhibit very attractive characteristics either. In the constant-velocity case, the derivative of velocity was unacceptable. The derivative of acceleration called *jerk* or *pulse* will have infinite spikes in the modified straight-line case. This derivative is a measure of the time rate of change of inertia force and thus indicates impact levels. Impact in mechanical components contributes to noise and shortens life due to surface wear and fatigue of adjacent components.

The search for a better rise-and-fall profile naturally led cam designers to try some well-known functions such as *parabolic, simple harmonic,* and *cycloidal* profiles. Each of these will be described and their dynamic characteristics compared.

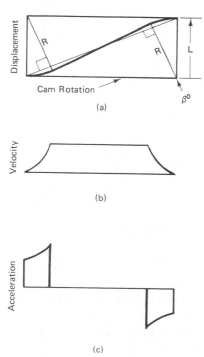

(a)

(b)

(c)

Figure 6.7 The modified constant velocity curve.

Constant Acceleration (Parabolic)

The next obvious choice of profiles to correct the disadvantages described above is constant acceleration, as shown in Fig. 6.8. Notice that the displacement and its three derivatives are shown together. The acceleration is a positive constant for half of the rise and a negative constant for the second half. Although there is no infinite value for follower acceleration, the jerk profile has three undesirable infinite spikes due to the step changes in the acceleration level. Thus, where vibrations, noise and/or wear are not tolerated, this design profile would not be a wise choice.

A graphical construction technique for parabolic motion is shown in Fig. 6.9a. A line at an arbitrary angle is constructed through the origin of the displacement diagram. The rise time (the horizontal axis) must be divided into an even number of equal divisions of six or larger. Six will be used here for illustrative purposes. (For an actual layout construction, many more divisions and a larger drawing scale would be required.) The arbitrary line should be divided into 1, 3, 5, 5, 3, 1 equal parts if six divisions are used on the abscissa, 1, 3, 5, 7, 7, 5, 3, 1 equal parts if eight divisions are used, etc. The end of the last division should be connected to the end of the ordinate representing the length of the rise. The remaining points should be connected to the ordinate by lines parallel to the first. These ordinate intersections should now be transferred by horizontal

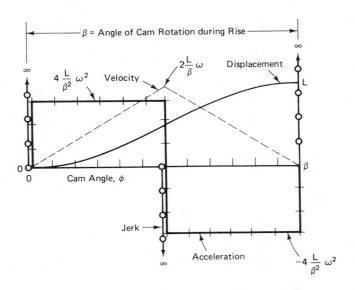

Figure 6.8 Displacement, velocity, acceleration, and jerk relations for parabolic motion.

lines to intersect the vertical lines through the corresponding division indicator on the abscissa. These intersections are points on the parabolic-rise curve.

Another simpler graphical method is shown in Fig. 6.9b, which is self-explanatory. A third method, based on the tangents of the displacement curve, is shown in Fig. 6.9c. Algebraic proofs for validity of the constructions in Fig. 6.9b and c are found in the next section.

In high-speed application of disk cams, it may be desirable to reduce the magnitude of the negative acceleration during the second part of the rise. The purpose of this is to reduce the tendency toward "follower jump," the separation of the follower from the cam surface followed by a "crashing" of the follower back to the surface. This phenomenon occurs when external contact-closing forces are insufficient to balance dynamic separation forces caused by follower-system inertia.

Assume that it has been determined that a 2:1 ratio is desirable between the magnitudes of follower acceleration and deceleration. In parabolic profiles, this is accomplished as depicted in Fig. 6.9d, which shows a nonsymmetric parabolic displacement curve. Construction by either of the foregoing geometric methods is a scaled-down (respectively, scaled-up) version of those shown in Fig. 6.9a, b, and c.

Simple Harmonic Motion

Another algebraic representation of an acceleration curve, that by its name might suggest continuous derivatives, is simple harmonic motion. For rise L in β rotation of the cam, the displacement, velocity, acceleration and jerk curves are shown in Fig. 6.10. Note that although the acceleration shape is harmonic in nature, at $\phi = 0$ and $\phi = \beta$ there are finite changes in acceleration that cause two theoretically infinite spikes in the jerk profile (recall that there are three such spikes in parabolic motion). Despite this defi-

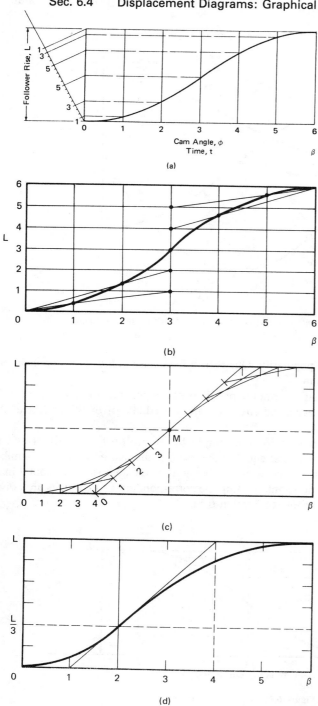

Figure 6.9 (a) Parabolic displacement diagram; (b) geometric construction of parabolic rise; (c) geometric construction of the parabolic displacement curve by way of its tangents; (d) nonsymmetric parabolic rise to reduce the magnitude of deceleration.

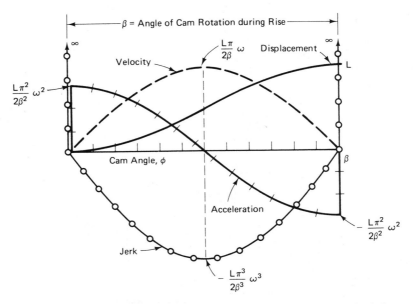

Figure 6.10 Displacement, velocity, acceleration, and jerk relations for simple harmonic motion.

ciency, this profile has some popularity. One reason for its use in low-speed applications is that it is easy to manufacture.

Worth noting here is that a harmonic profile is easily generated in the case of a radial flat face follower if an eccentric circular cam is used. Proof of this is left to the reader as an exercise.

The graphical construction for the simple harmonic profile is illustrated in Fig. 6.11. A semicircle of a diameter equal to the rise L is divided into the same number of equal angular increments as the equal linear divisions of the abscissa. Horizontal lines through the circumferential marks intersecting the corresponding vertical lines are points on the harmonic curve; (for proof see Section 6.5).

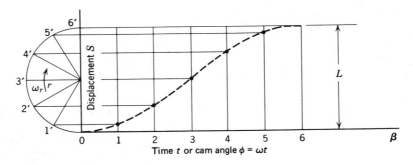

Figure 6.11 Simple harmonic motion.

Cycloidal Motion

Continuing our search for cam rise profiles with finite derivatives, we next look at the cycloidal displacement curve and its three derivatives (Fig. 6.12). Notice that, for the first time in this chapter, the jerk curve has finite magnitude throughout the cycle. Although the maximum acceleration looks higher than for the previous profiles, the finite jerk makes the cycloidal profile the best so far for high-speed applications. Graphically, the displacement for cycloidal motion may be generated by a point of a circle of radius $L/2\pi$, where L is the required rise, rolling on the zero ordinate (see the left side of Fig. 6.13a). To construct the displacement curve, divide the zero ordinate into the same number of equal parts as the abscissa. Let the generating point P of the rolling circle coincide with the origin O in the zero position of the circle. Then, when the generating circle rolls vertically up to tangency with the ordinate at, say, point number 2, project the generating point P horizontally to the correspondingly-numbered ordinate. This is a point on the cycloidal displacement curve; (for proof see Section 6.5).

For an alternative construction, draw a circle of the same diameter centered at B (as shown in the upper right corner of Fig. 6.13a). After dividing the circle into the same number of parts as the number of equal divisions on the abscissa, project the points from the periphery of the circle horizontally onto the vertical line through B. Then from these intersections, project each point parallel to the diagonal $O-B$ to intersect the correspondingly numbered ordinate, thus obtaining points on the cycloidal displacement curve; (mathematical proof is given in Section 6.5).

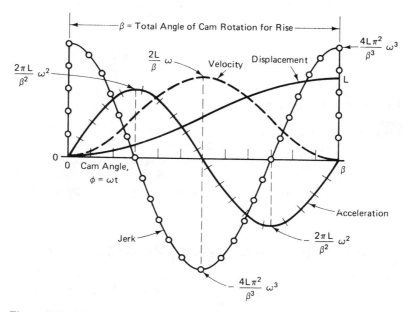

Figure 6.12 Displacement, velocity, acceleration, and jerk relations for cycloidal motion.

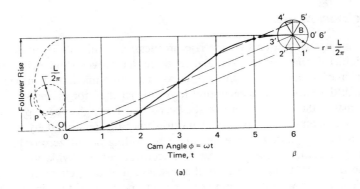

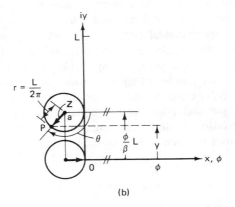

(b)

Figure 6.13   (a) A cycloidal motion; (b) generating the cycloidal rise by means of a rolling complex phasor.

Comparison of Basic Profiles

Figure 6.14* summarizes the characteristics of the constant velocity, constant acceleration, simple harmonic, and cycloidal motions. For the same input conditions (see next section for details), the maximum values for velocity, acceleration, and jerk are labeled. The cycloidal profile has the best overall characteristics for all three derivatives. Notice, however, that there is a dramatic increase in maximum magnitude of acceleration over the harmonic profile.

6.5 DISPLACEMENT DIAGRAMS: ANALYTICAL DEVELOPMENT

This section will parallel section 6.4 by analytically developing the basic cam profiles. The reader may wish to skip to Sec. 6.6 if the graphical presentation was sufficient.

* Source: Refs. [135, 148].

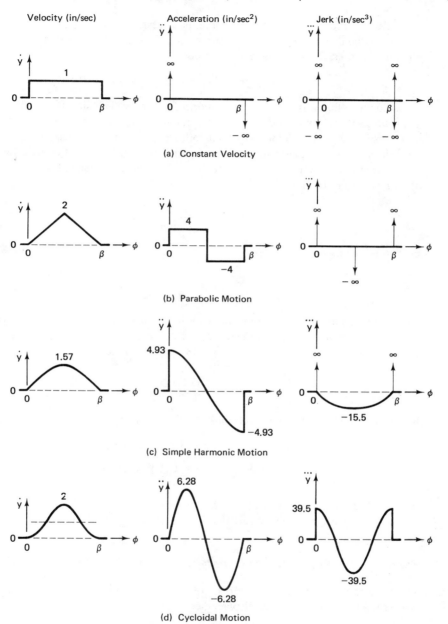

Figure 6.14 Comparison of the kinematic characteristics of four basic motions for angular velocity ω (deg/sec) $= \beta°/\text{sec}$ and rise $L = 1$ in. With these, dim $(\dot{y}) = $ in/sec, dim $(\ddot{y}) = $ in/sec^2, and dim $(\dddot{y}) = $ in/sec^3. (a) constant-velocity motion; (b) parabolic motion; (c) simple harmonic motion; (d) cycloidal motion.

During a dwell the follower remains at rest. Mathematically expressed: $y = k$, where y is the displacement and k is a constant. This may be regarded as a *zeroth-degree polynomial* in ϕ, the angle of cam rotation:

$$y = 0(\phi)^3 + 0(\phi)^2 + 0(\phi)^1 + k(\phi)^0 = P^0(\phi) \tag{6.1}$$

Suppose that a rise of L is required in β degrees. The simplest profile to utilize is a *constant velocity* or *straight-line* displacement, as shown in Fig. 6.6. If the cam rotates at a constant angular velocity, constant follower velocity means constant rate of change of position with respect to cam rotation. In other words, instead of expressing velocity as a time derivative, we replace it with the derivative of displacement with respect to angle of cam rotation. Thus

$$v = \frac{dy}{dt} = \frac{dy}{d\phi}\frac{d\phi}{dt} = \omega\frac{dy}{d\phi}$$

or if ϕ is expressed in degrees,

$$v = \frac{180}{\pi}\omega\frac{dy}{d\phi} \tag{6.2}$$

where ω is the angular velocity of the cam in radians per second. From the viewpoint of cam geometry, ω is irrelevant. Therefore, while dealing with the geometry of cam contour design, it is sufficient to use, say, $(180/\pi)\omega = 1$, and then

$$v = \frac{dy}{d\phi} = \frac{L}{\beta^\circ} \tag{6.3}$$

Expressed in the form of a polynomial, the displacement curve for constant velocity motion becomes

$$y = \frac{L}{\beta^\circ}(\phi^\circ)^1 + k(\phi^\circ)^0 = P^1(\phi^\circ) \tag{6.4}$$

or a first-degree polynomial in (ϕ°), whose first derivative with respect to ϕ°, $\dot{y} = (L/\beta^\circ)\omega(180/\pi)$, clearly satisfies Eq. (6.3).

Parabolic Displacement or Constant Acceleration

The next choice of profile is the parabolic (also called constant acceleration or gravity motion). The equation of the displacement s is

$$s = \tfrac{1}{2}At^2 \tag{6.5}$$

When A represents the acceleration of gravity, Eq. (6.5) represents the equation of motion of a freely falling body, starting from rest. Assuming constant angular velocity for the cam, the polynomial form of this profile is

$$\text{Displacement:} \quad y = k_2(\phi°)^2 + 0(\phi°)^1 + 0(\phi°)^0$$

$$\text{Velocity:} \quad \dot{y} = 2\omega k_2(\phi°) \tag{6.6}$$

$$\text{Acceleration:} \quad \ddot{y} = 2k_2\omega^2 \quad \text{(where } \omega \text{ is deg/sec)}$$

(Note that $\omega = $ const.; therefore $\dot{\omega} = 0$. Note also that superior dot(s) signify derivative(s) with respect to cam rotation $\phi°$.) The foregoing expressions are valid for the first half of the rise up to when $\phi = \frac{1}{2}\beta$, $y = \frac{1}{2}L$. Therefore, $L/2 = k_2(\beta^2/4)$, or $k_2 = 2L/\beta^2$, where L is in inches, β in degrees, and hence k_2 is in in./(deg)^2. For the second half of the rise, $y = L - k_2(\beta - \phi)^2$. At $\phi = \beta/2$, when $y = L/2$, $L/2 = L - (2L/\beta^2)(\beta/2)^2$, which is an identity and thus verifies continuity of the displacement curve. The follower velocity in the second half of the rise is

$$\dot{y} = \frac{d}{d\phi}[-k_2(\beta^2 - 2\beta\phi + \phi^2)]\omega = (2k_2\beta - 2k_2\phi)\omega = 2k_2(\beta - \phi)\omega \tag{6.7}$$

At $\phi = \beta/2$, $\dot{y} = (2k_2\beta/2)\omega = k_2\beta\omega$, which is the same as $\dot{y}$ at $\beta/2$ determined from the expression for the first half of the rise, and therefore shows that the velocity is also continuous. The acceleration in the second half of the rise is

$$\ddot{y} = -2k_2\omega^2 \tag{6.8}$$

a constant, equal and opposite to the acceleration in the first half of the rise. Figure 6.9 shows three methods of graphical construction of the constant acceleration displacement profile. We shall verify the second method.

From the construction, in the first half of the rise

$$y = \frac{\phi}{\beta/2}\left(\frac{\phi}{\beta/2}\right)\frac{L}{2} = \frac{L}{2}\frac{4\phi^2}{\beta^2} = \frac{2L}{\beta^2}\phi^2 = k_2\phi^2 \tag{6.9}$$

which is the desired parabolic function. Symmetry makes a separate proof of the second half of the construction unnecessary.

The third method of construction in Fig. 6.9 is based on the tangents of the displacement curve. At $\phi = \beta/2$,

$$\dot{y} = k_2\beta\omega = \frac{2L}{\beta^2}\beta\omega = \frac{2L}{\beta}\omega = \frac{L/2}{\beta/4}\omega \tag{6.10}$$

For geometric construction we can ignore ω by tacitly assuming that $\omega = 1$. Then it is easy to see that the tangent of the displacement curve at its halfway point M intersects the 0 abscissa at $\beta/4$, which leads to the construction of Fig. 6.9c. As shown there, the first $\beta/4$ on the horizontal axis and the tangent segment between $\beta/4$ and M are both divided into equal numbers of equal parts and numbered. Then corresponding division points are connected to obtain additional tangents to the displaced curve. Proof of this method is left to the reader as an exercise, as is also the determination of the point of tangency between each successive tangent and the parabolic curve. (*Hint:* Extend each tangent to the ordinate through M.)

In Fig. 6.9d, a 2:1 nonsymmetric parabolic displacement curve is suggested as a way to decrease the magnitude of the negative acceleration during the second part of the rise; this reduces the tendency of the cam follower to separate from the cam surface. Here in the first third of the rise: $0 \leq \phi \leq \beta/3$, $y = k_{2a}\phi^2$, $y_{\beta/3} = k_{2a}(\beta/3)^2 = L/3$, from which

$$k_{2a} = \frac{L}{3}\frac{9}{(\beta°)^2} = \frac{3L}{(\beta°)^2}\frac{\text{in.}}{(\text{deg})^2} \tag{6.11}$$

Velocity

$$\dot{y} = 2k_{2a}\phi\omega\left(\frac{\text{in.}}{(\text{deg})^2}\text{deg}\frac{\text{deg}}{\text{sec}^2} = \frac{\text{in.}}{\text{sec}^2}\right) \tag{6.12}$$

Acceleration

$$\ddot{y} = 2k_{2a}\omega^2\left(\frac{\text{in.}}{(\text{deg})^2}\frac{(\text{deg})^2}{\text{sec}^2} = \frac{\text{in.}}{\text{sec}^2}\right) = 2\frac{3L}{\beta^2}\omega^2$$

$$= \frac{6L}{\beta^2}\omega^2 \tag{6.13}$$

For the second part of the parabolic rise $\beta/3 \leq \phi \leq \beta$, $y = L - k_{2d}(\beta - \phi)^2$, $y_{\beta/3} = L/3 = L - k_{2d}(\frac{2}{3}\beta)^2$, from which

$$k_{2d} = \frac{2L}{3}\frac{9}{4}\frac{1}{\beta^2} = \frac{3L}{2\beta^2} = \tfrac{1}{2}k_{2a} \tag{6.14}$$

Velocity

$$\dot{y} = 2k_{2d}(\beta - \phi)\omega = 2\frac{3L}{2\beta^2}(\beta - \phi)\omega = \frac{3L}{\beta^2}(\beta - \phi)\omega \tag{6.15}$$

Acceleration

$$\ddot{y} = -2k_{2d}\omega^2 = \frac{-3L}{\beta^2}\omega^2 \tag{6.16}$$

whose magnitude is one-half that of the acceleration during the first part of the rise, and therefore follower-jump tendency has been reduced.

However, there is no such thing as a free lunch: we paid for this decrease in the magnitude of deceleration with higher acceleration in the first part of the rise, as compared to a symmetric parabolic rise of the same magnitude. Proof of this is left to the reader as an exercise.

Simple Harmonic Motion

The next profile considered is the simple harmonic motion depicted in Fig. 6.10. Let us derive the displacement, velocity, acceleration and jerk expressions, Eqs. (6.19) through

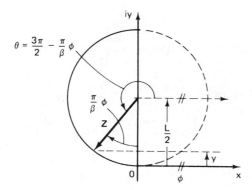

Figure 6.15

(6.22), by using a complex phasor pivoted at the center of a semicircle, as follows (See Fig. 6.15).

$$\mathbf{Z} = \frac{L}{2} e^{i\pi[(3/2)-(\phi/\beta)]} = \frac{L}{2} e^{i3\pi/2} e^{-i\pi\phi/\beta} \tag{6.17}$$

Let $\mathbf{Z}_0 = \dfrac{L}{2} e^{i3\pi/2} = -i\dfrac{L}{2}$

$$\text{Then } \mathbf{Z} = \mathbf{Z}_0 e^{-i\pi\phi/\beta} = \mathbf{Z}_0\left[\cos\left(-\frac{\pi\phi}{\beta}\right) + i\,\sin\left(-\frac{\pi\phi}{\beta}\right)\right] \tag{6.18}$$

The displacement

$$y = \frac{L}{2} + \mathscr{I}m(\mathbf{Z})$$

$$y = \frac{L}{2} + \mathscr{I}m(\mathbf{Z}_0 e^{-i\pi\phi/\beta}) = \frac{L}{2} + \mathscr{I}m\left\{-i\frac{L}{2}\left[\cos\left(-\frac{\pi\phi}{\beta}\right) + i\,\sin\left(-\frac{\pi\phi}{\beta}\right)\right]\right\}$$

$$y = \frac{L}{2}\left[1 - \cos\left(-\frac{\pi\phi}{\beta}\right)\right] = \frac{L}{2}\left(1 - \cos\frac{\pi\phi}{\beta}\right) \tag{6.19}$$

where $\mathscr{I}m(\ )$ means "the imaginary part of" ( ).

Checking this by substituting $\phi = 0$, $\frac{\beta}{2}$ and β:

$$y_0 = \frac{L}{2}(1 - \cos 0) = 0, \qquad y_{\beta/2} = \frac{L}{2}\left(1 - \cos\frac{\pi}{2}\right) = \frac{L}{2} \quad \text{and}$$

$$y_\beta = \frac{L}{2}(1 - \cos \pi) = L,$$

which verify the general expression for y_ϕ.

The velocity

$$\dot{y} = \frac{d}{dt}\left(\frac{L}{2} + \mathcal{I}m\mathbf{Z}\right) = \frac{d}{dt}\left(\mathcal{I}m(\mathbf{Z})\right) = \mathcal{I}m\left[\frac{d}{dt}\mathbf{Z}\right], \quad \text{or}$$

$$\dot{y} = \mathcal{I}m(\dot{\mathbf{Z}}) = \mathcal{I}m\left[\frac{d}{dt}\left(\mathbf{Z}_0 e^{-i\pi\phi/\beta}\right)\right] = \mathcal{I}m\left[\mathbf{Z}_o\left(-i\frac{\pi}{\beta}\omega\right)e^{-i\pi\phi/\beta}\right]$$

$$\dot{y} = \mathcal{I}m\left\{-i\frac{L}{2}\left(-i\frac{\pi}{\beta}\omega\right)\left[\cos\left(-\frac{\pi\phi}{\beta}\right) + i\,\sin\left(-\frac{\pi\phi}{\beta}\right)\right]\right\} \quad (6.20)$$

$$\dot{y} = -\frac{L\pi}{2\beta}\omega\left(-\sin\frac{\pi\phi}{\beta}\right) = \frac{L\pi}{2\beta}\omega\,\sin\frac{\pi\phi}{\beta}$$

Checking,

$$\dot{y}_0 = 0, \qquad \dot{y}_{\beta/2} = \frac{L\pi}{2\beta}\omega, \qquad \dot{y}_\beta = 0.$$

Similarly

$$\ddot{y} = \frac{L}{2}\frac{\pi^2}{\beta^2}\omega^2\cos\frac{\pi\phi}{\beta}; \qquad \ddot{y}_0 = \frac{L}{2}\frac{\pi^2}{(\beta^2)}\omega^2; \qquad \ddot{y}_{\beta/2} = 0; \qquad \ddot{y}_\beta = -\frac{L}{2}\frac{\pi^2}{\beta^2}\omega^2$$

$$(6.21)$$

$$\dddot{y} = -\frac{L}{2}\frac{\pi^3}{\beta^3}\omega^3\sin\frac{\pi\phi}{\beta}; \qquad \dddot{y}_{0+} = 0; \qquad \dddot{y}_{(\beta/2)} = -\frac{L}{2}\frac{\pi^3}{\beta^3}\omega^3; \qquad \dddot{y}_{\beta-} = 0$$

$$(6.22)$$

where the subscript 0+ indicates that $\ddot{y}$ is evaluated just after the start of the harmonic rise, and the subscript $\beta-$ shows that $\ddot{y}$ is evaluated just before the end of the harmonic rise. However, at the very instant of the start and finish of the rise, the acceleration takes a finite jump, therefore the jerk approaches infinity (see Fig. 6.10).

Cycloidal Motion

The next profile chosen to obtain continuous derivatives of displacement is the cycloidal rise.

   With the aid of the complex phasor, we can derive the equation of the cycloidal rise as follows (see Fig. 6.13b).

$$\mathbf{Z} = re^{-i\theta}$$

where

$$r = \frac{L}{2\pi} \quad \text{and} \quad \theta = \frac{\phi L}{\beta r} = \frac{\phi L}{\beta}\frac{2\pi}{L} = \frac{2\pi}{\beta}\phi$$

and

$$\mathbf{Z} = re^{-i(2\pi/\beta)\phi} \qquad (6.23)$$

From Fig. 6.13b

$$y = \frac{\phi}{\beta} L + \mathscr{I}m(\mathbf{Z})$$

$$y = \frac{\phi}{\beta} L + r\mathscr{I}m(e^{-i(2\pi/\beta)\phi})$$

$$y = \frac{\phi}{\beta} L + r\mathscr{I}m\left[\cos\left(-\frac{2\pi}{\beta}\phi\right) + i \sin\left(-\frac{2\pi}{\beta}\phi\right)\right]$$

$$y = \frac{\phi}{\beta} L - r \sin\left(\frac{2\pi}{\beta}\phi\right)$$

(6.24)

Checking this at $\phi = 0$, $\beta/2$, and β yields

$$y_0 = 0, \qquad y_{\beta/2} = \frac{L}{2} - r \sin \pi = \frac{L}{2} \qquad y_\beta = L - r \sin 2\pi = L$$

Also note that, for $0 \leq \phi \leq \beta/2$, the argument of the sine term is less than or equal to π. For example, at $\phi = \beta/3$, the argument is $(2\pi/\beta)(\beta/3) = \frac{2}{3}\pi < \pi$, and therefore y is below the diagonal (see Fig. 6.13a). Conversely, for $\beta/2 \leq \phi \leq \beta$, y is above the diagonal. The velocity of a cycloidal rise is

$$\dot{y} = \frac{L\omega}{\beta} - r\frac{2\pi}{\beta}\omega \cos\left(\frac{2\pi}{\beta}\phi\right) = \frac{L}{\beta}\omega\left[1 - \cos\left(\frac{2\pi}{\beta}\phi\right)\right]$$

(6.25)

Checking

$$\dot{y}_0 = 0, \qquad \dot{y}_{(\beta/4)} = \frac{L}{\beta}\omega\left(1 - \cos\frac{\pi}{2}\right) = \frac{L}{\beta}\omega,$$

$$\dot{y}_{(\beta/2)} = \frac{L}{\beta}\omega(1 - \cos \pi) = \frac{2L}{\beta}\omega$$

$$\dot{y}_{(3\beta/4)} = \frac{L}{\beta}\omega\left[1 - \cos\left(\frac{3}{2}\pi\right)\right] = \frac{L}{\beta}\omega \quad \text{and} \quad \dot{y}_\beta = \frac{L}{\beta}\omega[1 - \cos(2\pi)] = 0$$

Acceleration

$$\ddot{y} = \frac{2L\pi}{\beta^2}\omega^2 \sin\left(\frac{2\pi}{\beta}\phi\right)$$

(6.26)

Checking

$$\ddot{y}_0 = 0, \qquad \ddot{y}_{\beta/4} = \frac{2\pi L}{\beta^2}\omega^2, \qquad \ddot{y}_{\beta/2} = 0, \qquad \ddot{y}_{(3\beta/4)} = -\frac{2\pi L}{\beta^2}\omega^2, \qquad \ddot{y}_\beta = 0$$

Jerk

$$\dddot{y} = \frac{4L\pi^2}{\beta^3}\omega^3 \cos\left(\frac{2\pi}{\beta}\phi\right)$$

(6.27)

Checking

$$\dddot{y}_0 = \frac{4L\pi^2}{\beta^3}\omega^3, \qquad \dddot{y}_{\beta/4} = 0, \qquad \dddot{y}_{\beta/2} = -\frac{4L\pi^2}{\beta^3}\omega^3,$$

$$\dddot{y}_{3\beta/4} = 0, \qquad \dddot{y}_\beta = \frac{4L\pi^2}{\beta^3}\omega^3$$

Comparison of Basic Profiles

Figure 6.14 shows a comparison between the constant-velocity, parabolic, simple harmonic, and cycloidal motions, each for the same rise, $L = 1$ in., and angular velocity $\omega(\deg/\sec) = \beta°/\sec$. The maximum values shown in Fig. 6.14 can be verified as follows.

In the constant-velocity motion, $\dot{y} = (L/\beta)\omega$. If we let $\dot{y} = 1$ in./sec, then $L = \dot{y}\beta/\omega = \beta/\omega$ in., and if ω (deg/sec) $= \beta°/\sec$, the $L = 1$ in. Using this value of L and ω in the other motions as well, we have:

Constant acceleration

$$\dot{y}_{\max} = 2\frac{L}{\beta}\omega = 2\frac{\text{in}}{\sec}$$

$$\ddot{y}_{\max} = 4\frac{L}{\beta^2}\omega^2 = 4\frac{\beta}{\omega}\frac{\omega^2}{\beta^2} = 4\frac{\omega}{\beta} = 4 \text{ in/sec}^2$$

Harmonic motion

$$\dot{y}_{\max} = \frac{L\pi}{2\beta}\omega = \frac{\beta\pi}{2\omega\beta}\omega = \frac{\pi}{2} = 1.57 \text{ in/sec}$$

$$\ddot{y}_{\max} = \frac{L\pi^2}{2\beta^2}\omega^2 = \frac{\beta}{\omega}\frac{\pi^2}{2\beta^2}\omega^2 = \frac{\pi^2}{2}\frac{\omega}{\beta} = 4.93 \text{ in/sec}^2$$

$$\dddot{y}_{\max\text{(extreme in middle of rise)}} = \frac{-L\pi^3}{2\beta^3}\omega^3 = \frac{-\beta}{\omega}\frac{\pi^3}{2\beta^3}\omega^3 = \frac{-\pi^3}{2}\frac{\omega^2}{\beta^2} = -15.5 \text{ in/sec}^3$$

Cycloidal motion

$$\dot{y}_{\max} = \frac{2L}{\beta}\omega = \frac{2\beta}{\omega}\frac{\omega}{\beta} = 2 \text{ in/sec}$$

$$\ddot{y}_{\max} = \frac{2\pi L}{\beta^2}\omega^2 = \frac{2\pi}{\beta^2}\frac{\beta}{\omega}\omega^2 = 2\pi\frac{\omega}{\beta} = 6.28 \text{ in/sec}^2$$

$$\dddot{y}_{\max} = \frac{4L\pi^2}{\beta^3}\omega^3 = \frac{4\beta\pi^2}{\omega\beta^3}\omega^3 = 4\pi^2\frac{\omega^2}{\beta^2} = 39.5 \text{ in/sec}^3$$

6.6 ADVANCED CAM PROFILE TECHNIQUES

The last two sections provided us with the usual cam profiles for a basic rise or fall requirement. Situations will arise, however, in which more advanced techniques are required. Such is the case, for example, when a high-speed precision cam is required for a rise–dwell–fall displacement sequence. Also, sometimes displacement requirements differ from the conventional rise-dwell-fall sequence. In these cases none of the foregoing profiles exhibit satisfactory dynamic characteristics.

High-Speed Applications

The cycloidal profile was determined to be the best of the displacement curves discussed so far. Its disadvantage was the higher level of extreme accelerations. The *modified trapezoidal* profile (see Fig. 6.16) is designed to minimize extreme accelerations. The trapezoidal acceleration profile (not shown) consists of two trapezoids, one above the abscissa for acceleration and one below for deceleration. Note that the "modified trap" acceleration profile is without discontinuities, which yields a finite jerk profile throughout the rise cycle. Furthermore, to avoid discontinuities (sudden changes) in the value of jerk, the corners of the trapezoidal acceleration graph are rounded off in the "modified" trapezoidal profile.

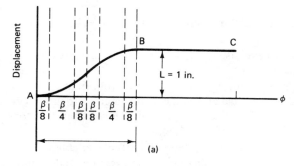

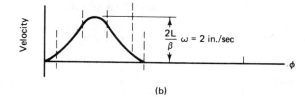

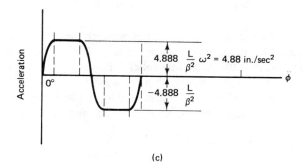

Figure 6.16   Modified-trapezoidal motion curve. ω (deg/sec) $= \beta°/$sec.

    A typical contour of the modified trapezoidal acceleration, like that shown in Fig. 6.16, is derived as a combination of cycloidal and parabolic (constant acceleration) displacement curves. Accordingly, the segments of the acceleration curve may be described as follows:

For $0 \leq \phi \leq \dfrac{\beta}{8}$ a sinusoidal quarter-wave rise from zero to $\ddot{y}_{max}$

For $\dfrac{\beta}{8} \leq \phi \leq \dfrac{3\beta}{8}$ constant at $\ddot{y}_{max}$

For $\dfrac{3\beta}{8} \leq \phi \leq \dfrac{5\beta}{8}$ a sinusoidal half-wave descent from $\ddot{y}_{max}$ to $\ddot{y}_{min} = -\ddot{y}_{max}$

For $\dfrac{5\beta}{8} \leq \phi \leq \dfrac{7\beta}{8}$ constant at $-\ddot{y}_{max}$

For $\dfrac{7\beta}{8} \leq \phi \leq \beta$ a sinusoidal quarter-wave rise from $-\ddot{y}_{max}$ to zero

    It is left to the reader as an exercise to derive the piecewise expressions for the acceleration, velocity, and displacement of this modified trapezoidal profile, thereby verifying the values given in Fig. 6.16 for L, $\dot{y}_{max}$, and $\ddot{y}_{max}$ (see Prob. 6.25).

    Note that for a rise of $L = 1$ in. and an angular cam velocity $\omega(\text{deg/sec}) = \beta°/\text{sec}$, the maximum acceleration of 4.888 in/sec in Fig. 6.16 is lower than that for both the cycloidal and simple harmonic profiles of Fig. 6.14.

    The *modified sine* is another popular acceleration profile available from vendors of "off the shelf" cams. This motion is a combination of the cycloidal and harmonic acceleration curves. This profile is known for its smooth transition from maximum acceleration to maximum deceleration. The segments may be described as follows:

For $0 \leq \phi \leq \dfrac{\beta}{8}$ a cycloidal quarter-wave rise from zero to $\ddot{y}_{max}$

For $\dfrac{\beta}{8} \leq \phi \leq \dfrac{7\beta}{8}$ a harmonic half-wave descent from $\ddot{y}_{max}$ to $\ddot{y}_{min} = -\ddot{y}_{max}$

For $\dfrac{7\beta}{8} \leq \phi \leq \beta$ a cycloidal quarter-wave rise from $-\ddot{y}_{max}$ to zero

    Figure 6.17* provides several interesting comparisons between the cycloidal, modified sine, and modified trapezoid for the same conditions: rise = 2 in, 60° of cam rotation, pitch circle of cam 2.375 in, radially translating roller follower radius of 0.625 in. Notice that the lowest maximum velocity occurs with the modified sine, the lowest maximum acceleration with the modified trapezoid curve, the lowest pressure angle (defined in Fig. 6.4 and accompanying text) with the modified sine, and the contact force

---

* Contributed by John Thoreson, 3M Co., St. Paul, MN.

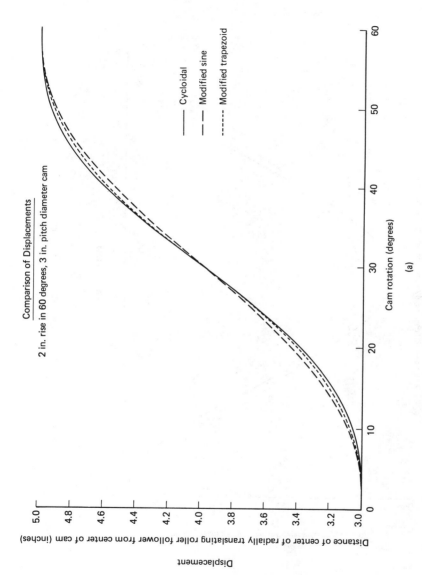

Figure 6.17(a) Courtesy of Jon Thoreson, 3M Company, St. Paul, MN.

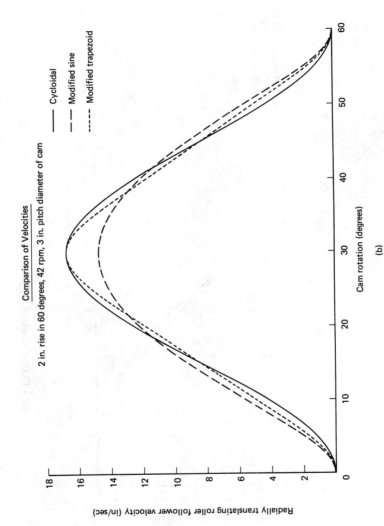

Figure 6.17(b) Courtesy of Jon Thoreson, 3M Company, St. Paul, MN.

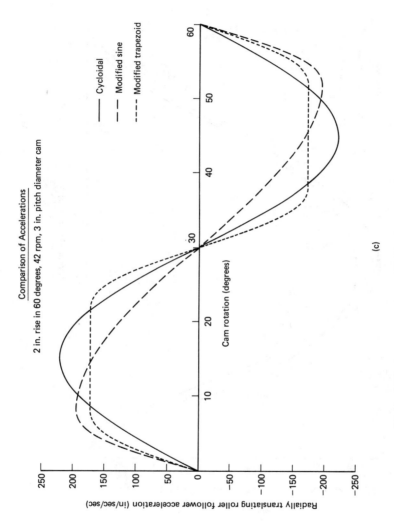

Figure 6.17(c) Courtesy of Jon Thoreson, 3M Company, St. Paul, MN.

369

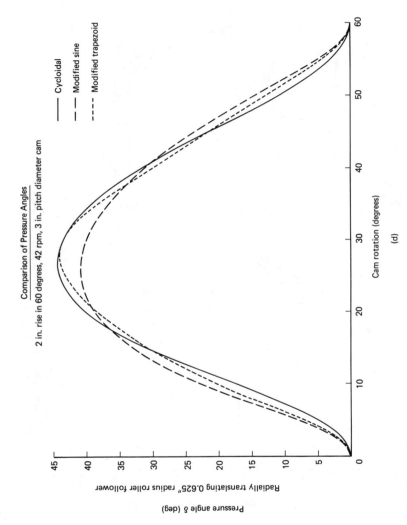

Figure 6.17(d) Courtesy of Jon Thoreson, 3M Company, St. Paul, MN.

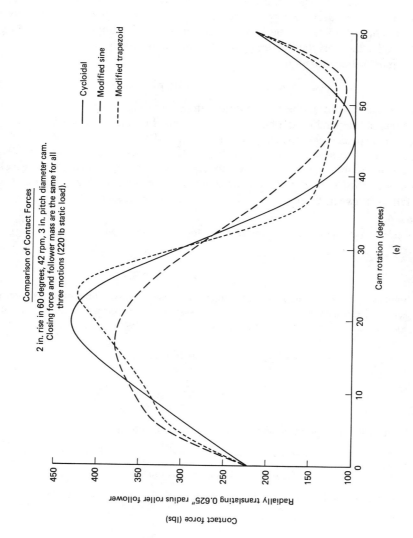

Figure 6.17(e) Courtesy of Jon Thoreson, 3M Company, St. Paul, MN.

371

between follower and cam surface is lowest with the modified sine. These types of comparisons help one choose the best profile for all or part of the entire cam based on a specific set of design constraints. Other profiles are sometimes used, such as a modified sine with a constant velocity portion in the middle of the rise.

Another profile that is increasingly being used is the *polynomial* curve. Since polynomials can be used to approximate arbitrary functions, it is not surprising to find that polynomial profiles can be "custom designed" to fit almost any required profile. Unfortunately, the analytical calculations become cumbersome in many instances. The standard polynomial equation is

$$s = a_n\phi^n + a_{n-1}\phi^{n-1} + \cdots + a_2\phi^2 + a_1\phi + a_0 = 0 \qquad (6.28)$$

where s is the rise, ϕ is the cam rotation angle, and a_0, a_1 and so on, are constants. These constants depend on the boundary conditions. For each boundary condition specified (either displacement, velocity, acceleration, and so on, at any point in the cycle), another term is required in Eq. (6.28). Figure 6.18 shows an eighth-degree polynomial displacement, with corresponding seventh-degree velocity and sixth-degree acceleration polynomials. This particular curve is designed to provide peak acceleration and pressure angle falling between those of the harmonic and cycloidal profiles (Fig. 6.14).

Example 6.1

Design a polynomial profile to satisfy the following conditions:

$$\text{At } \phi = 0: \qquad s = 0, \qquad \dot{s} = 0, \qquad \ddot{s} = 0$$

$$\text{At } \phi = \beta: \qquad s = L, \qquad \dot{s} = 0, \qquad \ddot{s} = 0$$

Solution   These are the same requirements as those satisfied by the cycloidal and modified trapezoidal profiles of Figs. 6.12 and 6.16. Based on six boundary conditions, a fifth-degree polynomial must be written with six constant coefficients:

$$s = \sum_n a_n\phi^n \qquad n = 0, 1, 2, \ldots, 5$$

or

$$s = a_0 + a_1\phi + a_2\phi^2 + a_3\phi^3 + a_4\phi^4 + a_5\phi^5 \qquad (6.29)$$

With $\omega = $ const., the first and second derivatives of Eq. (6.29) are

$$\dot{s} = \omega a_1 + 2\omega a_2\phi + 3\omega a_3\phi^2 + 4\omega a_4\phi^3 + 5\omega a_5\phi^4 \qquad (6.30)$$

$$\ddot{s} = 2\omega^2 a_2 + 6\omega^2 a_3\phi + 12\omega^2 a_4\phi^2 + 20\omega^2 a_5\phi^3 \qquad (6.31)$$

Substituting the boundary conditions into Eqs. (6.29), (6.30), and (6.31) yield

$$0 = a_0$$

$$L = a_0 + a_1\beta + a_2\beta^2 + a_3\beta^3 + a_4\beta^4 + a_5\beta^5$$

$$0 = \omega a_1$$

$$0 = \omega a_1 + 2\omega a_2\beta + 3\omega a_3\beta^2 + 4\omega a_4\beta^3 + 5\omega a_5\beta^4 \qquad (6.32)$$

$$0 = 2\omega^2 a_2$$

$$0 = 2\omega^2 a_2 + 6\omega^2 a_3\beta + 12\omega^2 a_4\beta^2 + 20\omega^2 a_5\beta^3$$

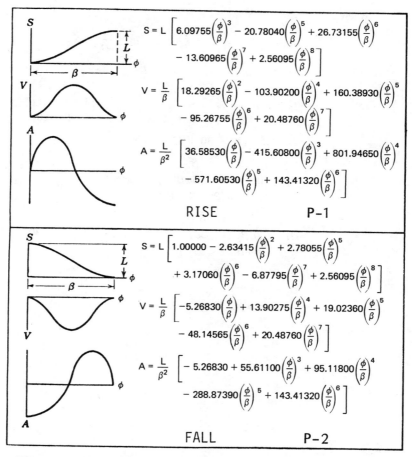

Figure 6.18 Eight-degree polynomial motion characteristics. S = displacement, inches; V = velocity, inches per degrees; A = acceleration, inches per degree squared; ϕ = angle of a cam rotation (in degrees) measured from start of the rise; β is the arc of total cam rotation for the rise as well as for the fall. P-1 (Polynomial-1) shows a rise, and P-2 (Polynomial-2) shows a descent or fall. (See note in caption of Fig. (6.20).)

Solving Eqs. (6.32) simultaneously results in

$$a_0 = 0$$

$$a_1 = 0$$

$$a_2 = 0$$

With these values, the remaining equations reduce to

$$
\begin{bmatrix}
\beta^3 & \beta^4 & \beta^5 \\
3\omega\beta^2 & 4\omega\beta^3 & 5\omega\beta^4 \\
6\omega^2\beta & 12\omega^2\beta^2 & 20\omega^2\beta^3
\end{bmatrix}
\begin{bmatrix}
a_3 \\
a_4 \\
a_5
\end{bmatrix}
=
\begin{bmatrix}
L \\
0 \\
0
\end{bmatrix}
\tag{6.33}
$$

where

$$a_3 = \frac{D_3}{D}, \qquad a_4 = \frac{D_4}{D}, \qquad a_5 = \frac{D_5}{D}$$

where

$$D = \beta^3\beta^2\beta\omega\omega^2 \begin{vmatrix} 1 & \beta & \beta^2 \\ 3 & 4\beta & 5\beta^2 \\ 6 & 12\beta & 20\beta^2 \end{vmatrix} = \beta^9\omega^3 \begin{vmatrix} 1 & 1 & 1 \\ 3 & 4 & 5 \\ 6 & 12 & 20 \end{vmatrix}$$

$$D = \beta^9\omega^3(20 - 30 + 12) = \omega^3\beta^9 2$$

$$D_3 = \begin{vmatrix} L & \beta^4 & \beta^5 \\ 0 & 4\omega\beta^3 & 5\omega\beta^4 \\ 0 & 12\omega^2\beta^2 & 20\omega^2\beta^3 \end{vmatrix} = L\omega^3\beta^2\beta^3 \begin{vmatrix} 4\beta & 5\beta \\ 12 & 20 \end{vmatrix} = \omega^3\beta^6(20)L$$

$$D_4 = \begin{vmatrix} \beta^3 & L & \beta^5 \\ 3\omega\beta^2 & 0 & 5\omega\beta^4 \\ 6\omega^2\beta & 0 & 20\omega^2\beta^3 \end{vmatrix} = -L\omega^3\beta^5 \begin{vmatrix} 3 & 5 \\ 6 & 20 \end{vmatrix} = -30\omega^3\beta^5 L$$

$$D_5 = \begin{vmatrix} \beta^3 & \beta^4 & L \\ 3\omega\beta^2 & 4\omega\beta^3 & 0 \\ 6\omega^2\beta & 12\omega^2\beta^2 & 0 \end{vmatrix} = L\omega^3\beta^4 \begin{vmatrix} 3 & 4 \\ 6 & 12 \end{vmatrix} = 12\omega^3\beta^4 L$$

from which

$$a_3 = \frac{10L}{\beta^3}$$

$$a_4 = \frac{-15L}{\beta^4}$$

$$a_5 = \frac{6L}{\beta^5}$$

Substituting these constants into Eq. (6.29) yields the displacement equation

$$s = \frac{10L}{\beta^3}\phi^3 - \frac{15L}{\beta^4}\phi^4 + \frac{6L}{\beta^5}\phi^5 \tag{6.34}$$

or letting $\phi/\beta = x$, then for $0 \le \phi \le \beta$, we will have $0 \le x \le 1$. Also, dividing through by L and letting $s/L = y$, so that for $0 \le s \le L$, we have $0 \le y \le 1$, our fully normalized equation will become

$$y = 10x^3 - 15x^4 + 6x^5 \tag{6.35}$$

Equation (6.35) is called a 3–4–5 polynomial based on the degrees of the remaining terms.

Let us verify our resulting equation against the prescribed boundary conditions and determine the extreme values of the motion parameters.

$$\dot{y} = (30x^2 - 60x^3 + 30x^4)\dot{x} \qquad \left(\dot{x} = \frac{\dot{\phi}}{\beta} = \frac{\omega}{\beta} = \text{const.}\right)$$

$$\ddot{y} = (60x - 180x^2 + 120x^3)\dot{x}^2$$

$$\dddot{y} = (60 - 360x + 360x^2)\dot{x}^3$$

At $\phi = 0$: $x = 0$, $\dot{y} = 0$, $\ddot{y} = 0$, $\dddot{y} = 60\dot{x}^3$

At $\phi = \dfrac{\beta}{2}$: $x = \dfrac{1}{2}$, $\dot{y} = \dfrac{30}{4}\dot{x} = 7.5\dot{x}$, $\ddot{y} = (30 - 45 + 15)\dot{x}^2$,

$\dddot{y} = -30\dot{x}^3$

At $\phi = \beta$: $x = 1$, $\dot{y} = 0$, $\ddot{y} = 0$, $\dddot{y} = 60\dot{x}^3$

It remains to find the location and magnitude of $\dot{y}_{max}$ and $\ddot{y}_{max}$. Set

$$\ddot{y} = 0 = (60 - 360x + 360x^2) = x^2 - x + \tfrac{1}{6} = 0$$

This is a "quadratic form." Assuming that the roots are $a \pm ib$, our quadratic form can be written as

$$(x - a - ib)(x - a + ib) = 0$$

or

$$x^2 - 2ax + a^2 + b^2 = 0.$$

Equating coefficients, $-2a = -1$; *therefore,*

$$a = \frac{1}{2}; \qquad a^2 + b^2 = \frac{1}{6}, \qquad b^2 = \frac{1}{6} - \frac{1}{4} = \frac{4}{24} - \frac{6}{24} = -\frac{2}{24}$$

Thus

$$b = i\frac{1}{3.46}; \qquad x_1 = a + ib = \frac{1}{2} - \frac{1}{3.46} = 0.21$$

$$x_2 = a - ib = \frac{1}{2} + \frac{1}{3.46} = 0.79 \qquad \text{and} \qquad \ddot{y}_{max} = 5.77\dot{x}^2$$

$$\dot{y}_{max} = \dot{y}\left(x = \frac{1}{2}\right) = \left(\frac{30}{4} - \frac{60}{8} + \frac{30}{16}\right)\dot{x} = 1.88\dot{x}$$

Since

$$y = \frac{s}{L}, \qquad \dot{y} = \frac{1}{L}\dot{s}, \qquad \dot{s} = L\dot{y}$$

$$x = \frac{\phi}{\beta}, \qquad \dot{x} = \frac{1}{\beta}\dot{\phi} = \frac{\omega}{\beta}, \qquad \text{therefore } \dot{s}_{max} = 1.88\frac{L\omega}{\beta}$$

$$\ddot{y} = \frac{\ddot{s}}{L}, \qquad \ddot{s} = L\ddot{y}$$

$$\ddot{s}_{max} = L\ddot{y}_{max} = 5.77\frac{L\omega^2}{\beta^2}$$

$$\dddot{s}_{0,\beta} = L\dddot{y}_{0,\beta} = 60\frac{L\omega^3}{\beta^3}$$

$$\dddot{s}_{\beta/2} = L\dddot{y}_{\beta/2} = -30\frac{L\omega^3}{\beta^3}$$

Letting $L = 1$ in and ω (deg/sec) $= \beta°/$sec, the values shown in Fig. 6.19 will result.

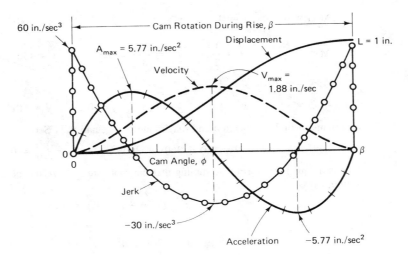

Figure 6.19 Displacement, velocity, acceleration, and jerk relations for the 3-4-5 polynomial motion. Extreme values of motion parameters are shown for $L = 1$ and ω (deg/sec) = $\beta°$/sec.

The results are comparable to the cycloidal profile, yet distinctly different. In general, this type of cam will begin and end its motion with a flatter displacement than the others. This presents difficulty in manufacturing the cam with sufficient accuracy to match the required displacement profile.

_____

Nonstandard Displacement Requirements

In some instances a required displacement profile will be other than the rise–dwell–fall type. In these cases a technique developed by Kloomok and Muffley [95, 110] is useful. They use three kinds of motions: (1) cycloidal and half-cycloidal, (2) harmonic and half-harmonic, and (3) eighth-degree polynomial. A plot and analytical functions for the displacement, velocity, and acceleration of the cycloidal and harmonic motions are shown in Figs. 6.20 and 6.21, together with their mathematical expressions while the same information for the polynomial was given in Fig. 6.18. The procedure is: Match as many derivatives as possible at the beginning and end of each segment. The better the match, the better the dynamic effects will be. For example, referring to Figs. 6.18, 6.19, 6.20 and 6.21, H-2 and P-2 can be designed to match in displacement, velocity, and acceleration, but H-2 and C-6 match only in position and velocity. Of course, polynomials of sufficiently high degree can be matched with one another for any arbitrary set of follower-motion programs with any desired order of junction matching between motion segments.

Example 6.2

The roller follower of a plate cam is required to start from a dwell, rise 1 inch, and immediately return to the original dwell. Figure 6.22 shows the displacement diagram with the given displacement curve in specified regions. (a) Select appropriate profile types to fit be-

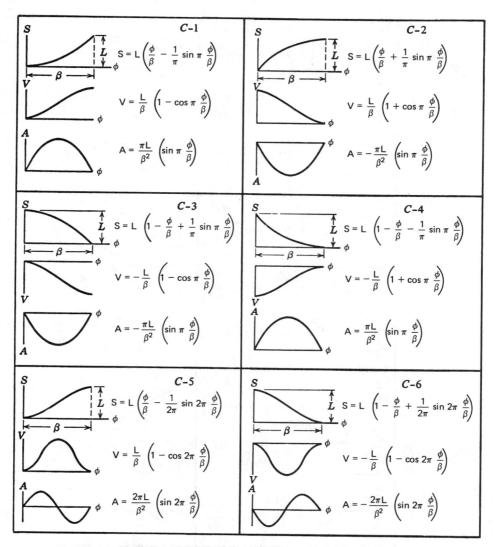

Figure 6.20 Cycloidal motion characteristics. S = displacement, inches; V = velocity, inches per degree; A = acceleration, inches per degree squared; ϕ = angle of cam rotation (degrees) measured from the start of the rise; β is the arc of total cam rotation for the rise as well as for the fall [95]. Note that these equations are kinematic. They do not include ω and (ω^2, α) in the velocity and acceleration terms — which would have to be included if a full dynamic analysis is the objective.

tween A-B, C-D, and E-F (refer to Figs. 6.18, 6.20, and 6.21). (b) Table 6.1 shows a partially completed chart of the instantaneous displacement, velocity, and acceleration values for the follower. Table 6.2 is to indicate the total angular travel of the cam required for each region. Fill in the missing values in both tables.

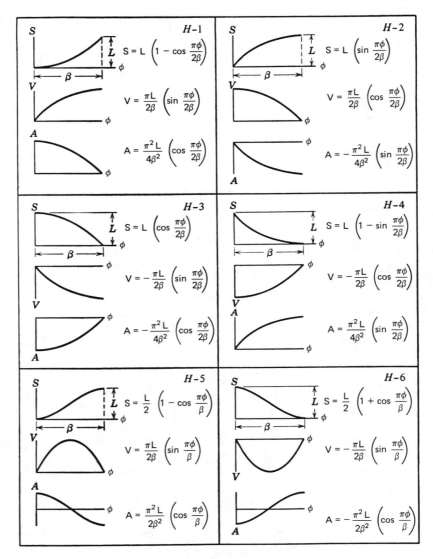

Figure 6.21 Harmonic motion characteristics. S = displacement, inches; V = velocity, inches per degree; A = acceleration, inches per degree squared. (*From* [95].) (See note in caption of Fig. (6.20).)

Solution (a) The cycloidal C-1 profile will match displacement, velocity, and acceleration conditions at both A and B. At A, all three are zero while the velocity levels off to a maximum at B.

Between C and D, harmonic H-2 profile matches boundary conditions. A cycloidal C-4 will fit the conditions at E and F. Figure 6.23 shows a sketch of the completed displacement diagram.

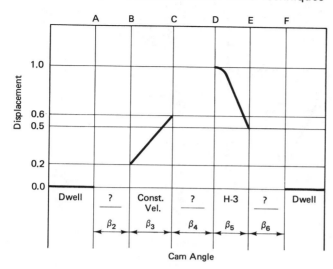

Figure 6.22 Partially specified displacement curve. See Example 6.2 about filling the gaps with segments, matching velocities and accelerations to avoid discontinuities.

TABLE 6.1 GIVEN VALUES OF DISPLACEMENT, VELOCITY, AND ACCELERATION FOR FIG. 6.22 (SEE EXAMPLE 6.2).

	A	B	C	D	E	F
S (in)	0	0.2	0.6	1.0	0.5	0
V (in/deg)	0	0.02	——	——	——	——
A (in/deg^2)	0	0	——	——	——	——

TABLE 6.2 20° CAM ROTATION IS GIVEN FOR THE CONSTANT VELOCITY SECTOR OF FIG. 6.22 (SEE EXAMPLE 6.2).

Section	β_2 A-B	β_3 B-C	β_4 C-D	β_5 D-E	β_6 E-F
β		20°			

(b) *Region A-B:* The unknown β_2 can be found from the C-1 velocity expression

$$V = \frac{L}{\beta_2}\left[1 - \cos\pi\left(\frac{\phi}{\beta_2}\right)\right] \text{in/deg}$$

At B, $V = 0.02$ in/deg, $L = 0.2$ in, and $\phi = \beta_2^\circ$; therefore,

$$\beta_2^\circ = 20°$$

Region C-D: By inspection at point C, $L_3 = L_C - L_B = 0.4$ in., $V_{BC} = 0.02$ in/deg and $A_C = 0$ in/deg^2. Then β_4 can be found from the H-2 velocity equation at $\phi_4/\beta_4 = 0$:

$$V_C = \frac{\pi L_3}{2\beta_4^\circ}\left[\cos\left(\frac{\pi}{2}\frac{\phi_4}{\beta_4}\right)\right] = \frac{\pi(0.4)}{2\beta_4^\circ} \text{ in/deg}, \quad \text{or} \quad \beta_4^\circ = \frac{\pi(0.4)}{2(0.02)} \text{ deg}$$

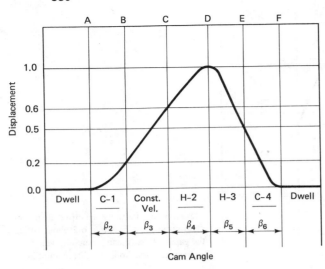

Figure 6.23 Sketch of the completed displacement diagram (see question a of Example 6.2).

and

$$\beta_4^\circ = 10\pi \text{ deg} \quad \text{or} \quad 31.4°$$

At D, $L_4 = L_D - L_C = 0.4$, $\dfrac{\phi_4}{\beta_4} = 1$, $V = 0$, and the acceleration equation for H-2 yields

$$A_D = -\frac{\pi^2 L_4}{4(\beta_4^\circ)^2}\left[\sin\left(\frac{\pi}{2}\frac{\phi_4}{\beta_4}\right)\right] \text{ in/deg}^2 = -0.001 \text{ in/deg}^2$$

Region D-E: By inspection, $A_E = 0$, $L_E = 0.5$ at E. Therefore the acceleration equation for H-3 may be applied at D to yield:

$$\beta_5^2 = \frac{-\pi^2 L}{4(-0.001)}$$

or

$$\beta_5 = 35.1°$$

Then the velocity equation for H-3 may be used at E: $V_E = \dfrac{-\pi(0.5)\text{in}}{2(35.1°)}$, from which

$$V_E = -0.022 \text{ in/deg}$$

Region E-F: By inspection, $A_E = A_F = V_F = 0$, $L = 0.5$ in between E and F. β_6 may be determined by use of the C-4 velocity equation at E: $\beta_6^\circ = -\dfrac{2L}{V_E} = \dfrac{-1}{-0.022}$

$$\beta_6 = 44.7°$$

The derived values for each of the regions are shown in Fig. 6.23 and Tables 6.3 and 6.4. This example demonstrates how to piece together various cam follower motions to custom design for a specific design situation.

TABLE 6.3   CALCULATED VALUES OF VELOCITIES AND ACCELERATIONS FOR STATIONS C, D, E AND F (SEE TABLE 6.1 AND EXAMPLE 6.2).

	A	B	C	D	E	F
S (in.)	0	0.2	0.6	1.0	0.5	0
V (in./deg)	0	0.02	0.02	0	−0.022	0
A (in./deg^2)	0	0	0	−.001	0	0

TABLE 6.4   CALCULATED VALUES OF CAM ROTATIONS β_2, β_4, β_5 AND β_6 (SEE TABLE 6.2 AND EXAMPLE 6.2).

Section	β_2 A-B	β_3 B-C	β_4 C-D	β_5 D-E	β_6 E-F
β	20°	20°	31.4°	35.1°	44.7°

6.7 GRAPHICAL CAM PROFILE SYNTHESIS

Now that we are aware of several follower motions, the task remains to generate the cam surface profile. As pointed out earlier, because of possible curvature mismatches between followers and the cam, which may cause interference or undercutting, procedures must be developed for cam profile synthesis which avoid such anomalies.

Disk Cam with Flat-Face Radially Translating Follower

A disk cam with a flat-face radially translating follower is shown in Fig. 6.24. Notice the numbered tic marks along the vertical centerline of the follower. These represent the specified rise at 30° intervals of cam rotation. For example, these could have come from Fig. 6.5, such that the follower travel for each station point number ($0 \rightarrow 12$) would be marked off on the follower (stations 0 and 12 are located at the base circle).

The following graphical procedure is independent of how the tic marks were generated for the follower travel.

To synthesize the cam contour, it is necessary to invert the mechanism so that the cam is held stationary while the follower moves around it *in the direction opposite to* the cam rotation. The procedure is as follows:

1. Move the follower about the center of the cam in the direction opposite to the cam rotation by an appropriate angle which matches the desired cam rotation during the follower travel (in this case 30° ccw).
2. Move the follower radially outward from the base circle of the cam by the distance indicated by the corresponding tic mark on the 0° radial.
3. The cam contour is then drawn tangent to the polygon that is formed by all the positions of the follower face.

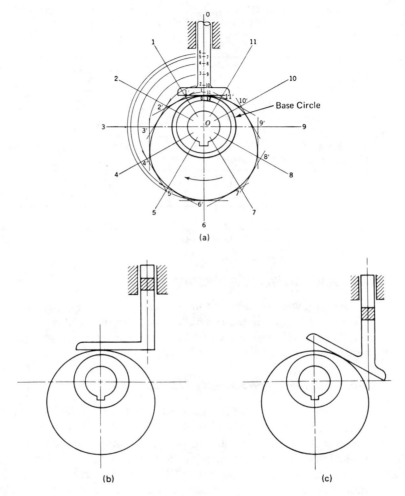

Figure 6.24 (a) Graphical generation of disc cam profile with radially translating right-angle flat-faced follower; (b) kinematic equivalent of the cam-and-follower system of Fig. 6.24(a); (c) non-right-angle flat-face follower.

For better accuracy, the cam cycle may be broken up into more divisions and/or the scale of the drawing may be increased. Alternatively, the mathematical equations that model this geometrical process may be programmed for computer-assisted synthesis (see Sec. 6.8).

In Fig. 6.24a and b the flat face of the follower is perpendicular to the direction of translation. In Fig. 6.24c it is not perpendicular. Design conditions will govern which of the three configurations to use.

Disk Cam with Radially Translating Roller Follower

The same synthesis procedure as described above is valid for the disk cam with the radially translating roller follower shown in Fig. 6.25a. The only difference is that the cam contour is tangent to the roller positions rather than those of the flat face.

Notice the resulting pressure angle (the angle between a radial line from the cam center and the normal to the cam surface through the center of the roller) at station 2 of the displacement diagram. If this angle becomes too large, it will tend to bind the follower stem due to excessive side pressure in the guide. If this angle is higher than acceptable, the follower could be offset (as shown in Fig. 6.25b) as one of the redesign options. The same cam contour design procedure is valid for Fig. 6.25b. Notice the reduced pressure angle. Again, there is no "free lunch" because the pressure angle on the fall portion of the displacement cycle will be larger; but this may very well be acceptable because in the rise the cam follower spring resists the motion, while in the descent the spring assists the motion.

Disk Cam with Flat-Face Oscillating Follower

Figure 6.26 shows a flat-face oscillating follower. Again the principle of inversion is employed and the fixed pivot of the follower arm is rotated about the cam center in the direction opposite the proposed cam rotation. At the same time the follower arm must be rotated about its own pivot by the specified displacement angle for each position. This is implemented by the intersection of two arcs: an arc centered at the center of the cam shaft and going through the appropriate station point tic mark; and another arc of a radius equal to the initial distance from the follower pivot, to the initial contact point on the base circle of the cam, centered at the new follower pivot position. Then a line representing the face of the follower is drawn through the intersection of the arcs, tangent to the face-offset circle. As with the translating flat-face follower, the cam contour is inscribed in the resulting polygon.

The disk cam with oscillating roller follower uses the same procedure except that the intersection of the arcs locates the center of the roller follower. The contour is completed by drawing the cam surface tangent to the follower roller contours in all relative positions of the roller.

6.8 ANALYTICAL CAM PROFILE SYNTHESIS*

Owing to the need for highly accurate cams in many situations, it is desirable to be able to determine the coordinates of points on the cam surface as well as the corresponding coordinates of the center of the milling cutter of arbitrary diameter to cut the cam profile.

---

* Draft and revision of this section is based on a contribution by Greg Vetter, Truth, SPX Corp., Owatonna, Minnesota.

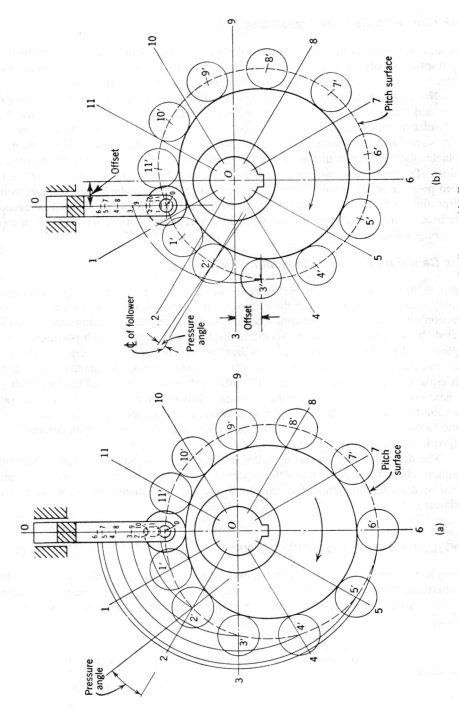

Figure 6.25 (a) Disc cam with radially translating roller follower; (b) disc cam with offset translating roller follower.

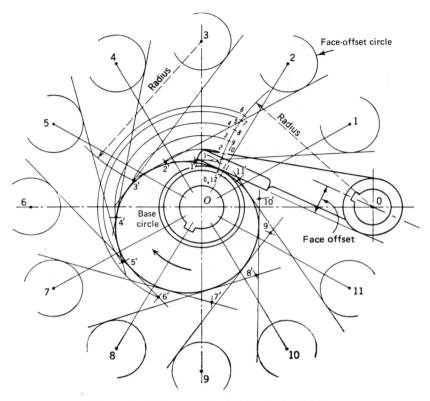

Figure 6.26 Disc cam with oscillating flat-faced follower.

The following subsections give the equations[†] for profile and cutter coordinate determination for disc cams with four different types of followers: translating flat face, oscillating flat face, translating roller, and oscillating roller.

Radially Translating Right-Angle Flat-Face Follower

In Fig. 6.27 the angle θ between the direction of radial translation of the follower and the point of contact between the cam and follower is

$$\theta = \arctan\left(\frac{1}{L}\frac{dL}{d\phi}\right) \qquad (6.36)$$

where ϕ = angle of cam rotation measured (against the direction of cam rotation) from an arbitrary reference radial on the cam to the centerline of follower translation

---

[†] Note that all angles are given in radians.

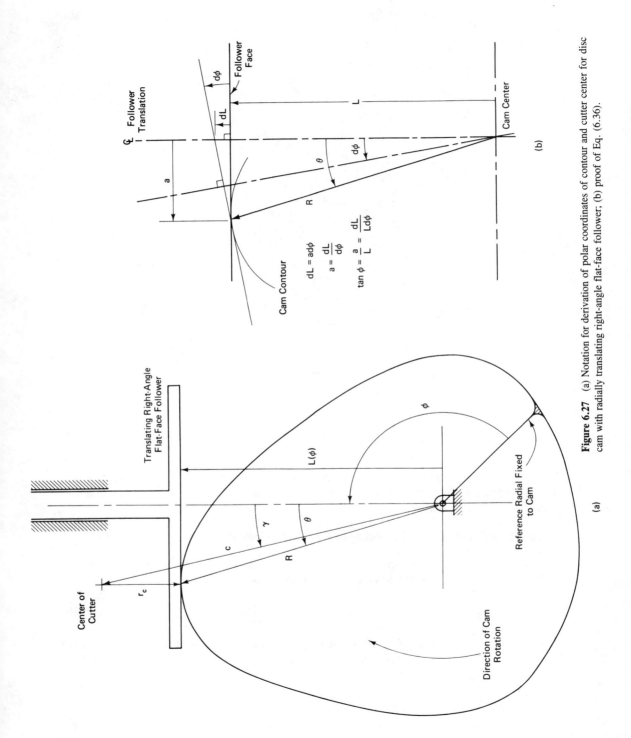

$dL = ad\phi$

$a = \dfrac{dL}{d\phi}$

$\tan\phi = \dfrac{a}{L} = \dfrac{dL}{Ld\phi}$

(b)

Figure 6.27 (a) Notation for derivation of polar coordinates of contour and cutter center for disc cam with radially translating right-angle flat-face follower; (b) proof of Eq. (6.36).

$L = L(\phi)$ = perpendicular distance of follower face from cam center as a function of ϕ

$\dfrac{dL}{d\phi}$ = current value of the rate of change of follower displacement with respect to cam angle

Proof of Eq. (6.36) appears in Fig. 6.27b. The radial distance from the cam center to the point of contact is

$$R = \frac{L}{\cos \theta} \tag{6.37}$$

The polar *profile coordinates* with respect to the reference radial on the cam are

$$[R, (\phi + \theta)]$$

The center of the milling cutter will be on the normal to the profile erected at the point of contact. The angle γ between the direction of radial translation of the follower and the cutter center is

$$\gamma = \arctan\!\left(\frac{R\,\sin\theta}{L + r_c}\right) \tag{6.38}$$

where r_c is the radius of the milling cutter, while the radial distance from the cam center to the cutter center is

$$c = \frac{L + r_c}{\cos \gamma} \tag{6.39}$$

The polar *cutter center coordinates* with respect to the reference radial on the cam are $[c, (\phi + \gamma)]$.

Oscillating Flat-Face Follower

In Fig. 6.28 the angle θ between the normal to the follower face and the point of contact between the cam and follower is [135]

$$\theta = \arctan\!\left[\left(\frac{+(d\zeta/d\phi)}{1 - (d\zeta/d\phi)}\right)\frac{m\,\cos\zeta}{f + m\,\sin\zeta}\right] \tag{6.40}$$

where ϕ = cam rotation angle measured (against direction of cam rotation) from an arbitrary reference radial on the cam to the line between the cam center and follower pivot point

$\zeta = \zeta(\phi)$ = angular position of follower face with respect to centerline between cam center and follower pivot

$\dfrac{d\zeta}{d\phi}$ = current value of the rate of change in follower angular position with respect to cam angle

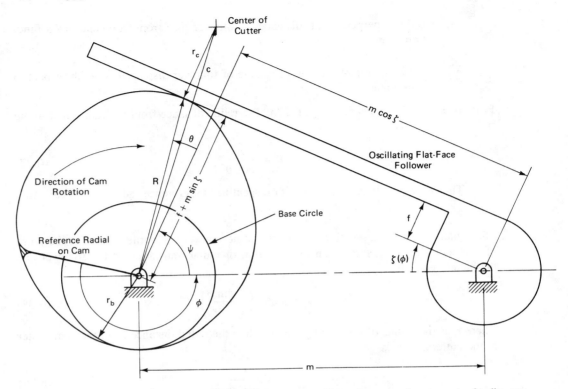

Figure 6.28 Notation for determining coordinates of cam contour and cutter center for disc cam with oscillating flat-faced follower.

f = follower face offset from follower pivot point: this quantity is negative if the follower face is offset from the pivot point toward the cam center

m = distance from cam center to follower pivot point

r_b = radius of the base circle of the cam

$\zeta_0 = \sin^{-1}[(r_b - f)/m]$ = the starting angle of the follower

Notice that

$$\psi = \frac{\pi}{2} - \zeta \qquad (6.41)$$

The radial distance from the center of the cam to the point of contact is

$$R = \frac{f + m \sin \zeta}{\cos \theta} \qquad (6.42)$$

The polar *profile coordinates* with respect to the reference radial on the cam are $[R, (\theta + \psi + \phi)]$. To determine the coordinates of the center of the milling cutter, we will temporarily use a Cartesian coordinate system. The coordinate system is chosen so that the origin is at the cam center, the positive x axis is along the radial to the point of contact, and the positive y axis (perpendicular to the x axis) points in the general direction of the follower pivot point.

The xy coordinates of the cutter center are

$$c_y = r_c \sin \theta$$
$$c_x = R + r_c \cos \theta \tag{6.43}$$

where r_c is the radius of milling cutter.

The polar *cutter center coordinates* with respect to the reference radial on the cam are

$$(c_x^2 + c_y^2)^{1/2}, \qquad [\theta + \psi + \phi - \arctan(c_y/c_x)]$$

Offset Translating Roller Follower

In Fig. 6.29 the angle α (seen from the follower center) between the cam-follower contact point and the cam center is [135]

$$\alpha = \arctan\left[\frac{L(dL/d\phi)}{m^2 + L^2 - m(dL/d\phi)}\right] \tag{6.44}$$

where m = distance the line of translation of the follower center is offset from the cam center (m is negative if the offset is in the direction the cam is turning); in the position shown it is positive

ϕ = cam rotation angle measured against direction of cam rotation from reference radial to cam centerline parallel to roller translation

$L = L(\phi)$ = distance of follower center from cam center measured parallel to roller translation

$\dfrac{dL}{d\phi}$ = current value of the rate of change of the follower displacement with respect to cam angle

The pressure angle δ is

$$\delta = \alpha - \psi \tag{6.45}$$

where

$$\psi = \arctan\left(\frac{m}{L}\right) \tag{6.46}$$

Also, the distance between cam and follower centers is

$$F = \sqrt{L^2 + m^2} \tag{6.47}$$

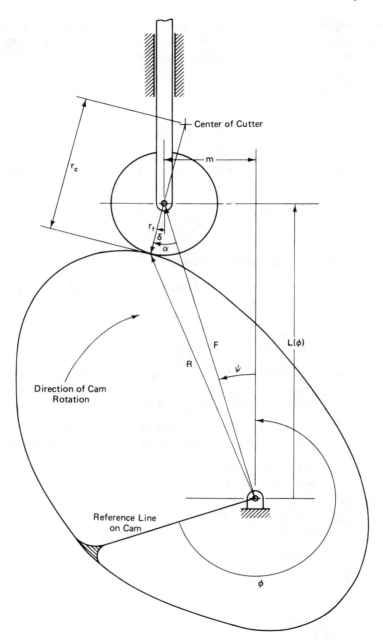

Figure 6.29 Contour and cutter center coordinates for disc cam with offset translating roller follower.

Using Cartesian coordinates with the cam center as the origin, the positive x axis along the radial to the follower center, and the positive y axis $90°$ ccw from the x axis, the Cartesian profile coordinates are

$$R_x = F - r_f \cos \alpha \qquad (6.48)$$

$$R_y = r_f \sin \alpha \qquad (6.49)$$

where r_f is the radius of the roller follower. This coordinate system will not be stationary. The polar *profile coordinates* with respect to the reference radial on the cam are $(R_x^2 + R_y^2)^{1/2}$, $[\phi + \psi + \arctan(R_y/R_x)]$.

The Cartesian cutter center coordinates are

$$c_x = F + (r_c - r_f)\cos \alpha$$
$$c_y = (r_f - r_c)\sin \alpha \qquad (6.50)$$

where r_c is the radius of the milling cutter.

The polar *cutter center coordinates* with respect to the reference radial on the cam are

$$(c_x^2 + c_y^2)^{1/2}, \qquad [\phi + \psi + \arctan(c_y/c_x)]$$

Oscillating Roller Follower

Let $\zeta(\phi)$ be the angular displacement of follower arm specified as a function of cam rotation. In Fig. 6.30 the angle α (seen from the follower center) between the cam center and cam-and-follower contact point is [135]

$$\alpha = \arctan\left[\frac{+(A \sin \gamma)(d\zeta/d\phi)}{L - (+A \cos \gamma)(d\zeta/d\phi)}\right] \qquad (6.51)$$

where A = arm length of follower from pivot point to follower center

ϕ = cam rotation angle measured against direction of cam rotation from reference radial on cam to centerline of cam follower-arm pivot

$\zeta = \zeta(\phi)$ = angular displacement of follower arm from the line drawn from the cam center to the follower arm pivot point

$\dfrac{d\zeta}{d\phi}$ = current value of the rate of change of follower displacement with respect to cam angle

L = distance of follower center from cam center

γ = angle (seen from follower center) between the cam center and follower pivot point

m = distance from cam center to follower pivot point

r_b = radius of the base circle of the cam

$\zeta_0 = \cos^{-1}((A^2 + m^2 - (r_b + r_f)^2)/(2mA))$ = the starting angle of the follower

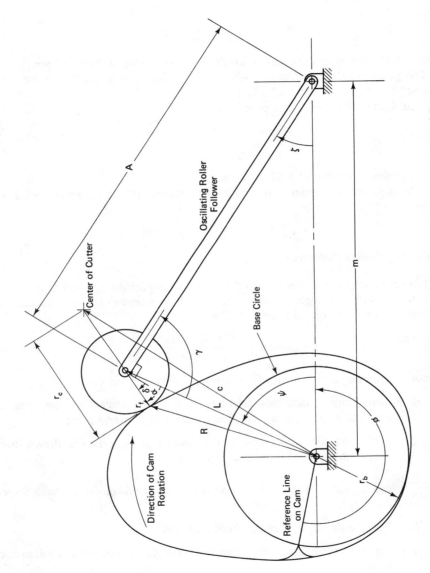

Figure 6.30 Contour and cutter center coordinates for disc cam with oscillating roller follower.

The pressure angle δ is

$$\delta = \gamma + \alpha - \frac{\pi}{2} \tag{6.55}$$

Choosing Cartesian coordinates with the cam center as the origin, the positive x axis along the radius to the follower center, and the positive y axis (90° ccw from the x axis) in the general direction of the follower pivot point, we get the Cartesian profile coordinates

$$R_x = L - r_f \cos \alpha$$
$$R_y = r_f \sin \alpha \tag{6.56}$$

where r_f is the radius of the roller follower.

The polar *profile coordinates* with respect to the reference radial on the cam are

$$(R_x^2 + R_y^2)^{1/2}, \qquad \phi + \left[\psi + \arctan\left(\frac{R_y}{R_x} \right) \right]$$

The Cartesian cutter coordinates are

$$c_x = L + (r_c - r_f)\cos \alpha$$
$$c_y = (r_f - r_c)\sin \alpha \tag{6.57}$$

where r_c is the milling cutter radius.

The polar *cutter center coordinates* with respect to the reference radial on the cam are

$$(c_x^2 + c_y^2)^{1/2}, \qquad \phi + \left[\psi + \arctan\left(\frac{c_y}{c_x} \right) \right]$$

In the foregoing derivation, the angle of oscillation of the roller follower was specified as a function of cam rotation. Alternatively, let $L(\phi)$ be the radial distance of roller center from cam center specified as a function of cam rotation. Referring again to Fig. 6.30, the angle α (as seen from the follower center) between the contact point and the cam center can be expressed this time as [135]

$$\alpha = \arctan\left[\frac{+dL/d\phi}{L - (+\cos \gamma)(dL/d\phi)} \right] \tag{6.58}$$

Here A, ϕ, γ, m, and $+$ are the same as for oscillating roller followers with angular displacement specified. Again

$$\gamma = \arccos\left(\frac{A^2 + L^2 - m^2}{2AL} \right) \tag{6.59}$$

but the expression for ψ is

$$\psi = \arccos\left(\frac{L^2 + m^2 - A^2}{2Lm} \right) \tag{6.60}$$

The rest of the expressions for the oscillating roller follower with angular displacements specified [Eq. (6.57)] are also appropriate here.

CAMSYN Program

The previous analytical development has been programmed by Chris Huber at the University of Minnesota and is part of a *Cam Synthesis* program: CAMSYN. The flowchart of the original, main frame program, CAMSYNG, is shown in Fig. 6.31. Example 6.3

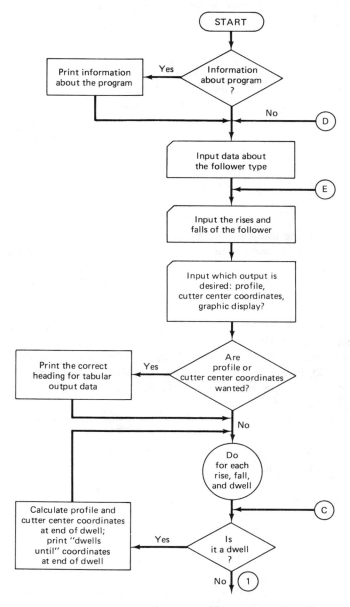

Figure 6.31 Flowchart of program CAMSYN. (Courtesy of Greg Vetter.)

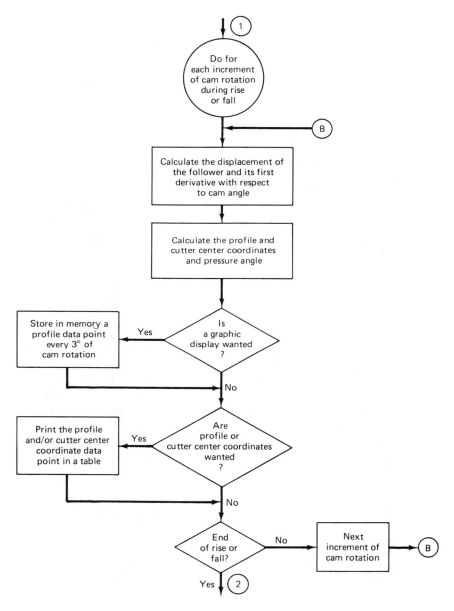

Figure 6.31 (cont.)

shows output from the original program which was run on a CDC main frame computer with a Tektronix terminal. Thanks to the dramatic evolution in computers and the hard work of Chris Huber, an updated, more versatile program called CAMSYN is available with this book along with other software modules for kinematic synthesis and analysis. The disc should run on an IBM PC (or compatible) with MS DOS 3.1 or later.

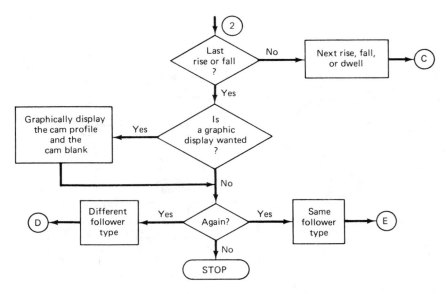

Figure 6.31 (cont.)

The computer-aided cam synthesis package has been developed for the purpose of enhancing the material covered in this chapter. CAMSYN allows the user to easily and efficiently develop a cam design. The pertinent characteristics of that design can then be displayed graphically. By introducing a computer in the design process, the student can more thoroughly explore some of the intricacies associated with cams and cam design.

CAMSYN is fairly easy to use and requires little explanation. On-line help will be included wherever possible to facilitate the use of the software. The development of the CAMSYN software package incorporated the analytical techniques discussed in Secs. 6.1 through 6.7. These sections should serve as a guide while using CAMSYN.

Example 6.3

A translating offset roller follower (Fig. 6.29) is to have a fall (from a cam blank diameter of 8 units) of 0.5 unit, a dwell, and a rise back up to the blank diameter. The fall should be in 30°, the dwell in 300°, and the rise in 30°. The offset is to be 0.75 unit, the follower roller diameter 2 units, and the cutter diameter is 3 units. Harmonic motion is recommended.

Table 6.5 shows the hard copy from the Tektronix screen, indicating the interactive input to the CAMSYNG program. Table 6.6 lists the cam profile coordinates and the cam cutter center coordinates for this task. Figure 6.32 shows the cam profile (solid line) with respect to the cam blank (dotted line).

6.9 CAM SYNTHESIS FOR REMOTE FOLLOWER

In the foregoing sections it was assumed that the motion of the follower in direct contact with the cam was specified, and that the cam rotation is uniform. This, however, is often not the case, because the ultimate end effector actuated by the cam is frequently

TABLE 6.5   HARD COPY OF INTERACTIVE SCREEN: INPUT FOR CAMSYNG PROGRAM

INPUT THE CAM FOLLOWER TYPE (1 = TRANSLATING ROLLER;
2 = OSCILLATING ROLLER; 3 = TRANSLATING FLAT FACE; 4 = OSCILLATING
FLAT FACE).
? 1
 INPUT THE OFFSET.
? .75
 INPUT THE FOLLOWER ROLLER DIAMETER.
? 2
INPUT THE CAM BLANK DIAMETER AND THE CAM ROTATION
ANGLE (DEG.) AT THE START OF THE FIRST FALL.
? 8,0
INPUT THE TOTAL NUMBER OF RISES + FALLS + DWELLS.
? 3
 INPUT THE RISE (FALL IS NEGATIVE) OF THE FOLLOWER,
THE CAM ROTATION ANGLE (DEG.) AT THE END OF THE RISE, AND
THE CAM FOLLOWER MOTION FOR EACH OF THE 3 RISES/FALLS/DWELLS.
FOLLOWER MOTIONS: 1 = CYCLOIDAL
                    2 = HARMONIC
                    3 = MODIFIED SINE
                    4 = MODIFIED TRAPEZOID
                    5 = 4–5–6–7 POLYNOMIAL
                    6 = DWELL
                    7 = ANY POLYNOMIAL YOU SPECIFY
? −.5,30,2
? 0,330,6
? .5,360,2
 WOULD YOU LIKE A PRINTOUT OF THE CAM PROFILE COORDINATES?
? Y
 WOULD YOU LIKE A PRINTOUT OF THE CAM CUTTER CENTER COORDINATES?
? Y
 INPUT THE CAM CUTTER DIAMETER.
? 3.0
 WOULD YOU LIKE A GRAPHICAL DISPLAY OF THE CAM PROFILE?
? Y
FOR WHAT INCREMENTS OF CAM ROTATION ANGLE (DEGREES) SHOULD DATA BE
OBTAINED?
? 3

---

driven by way of a kinematic chain starting with the cam follower. Thus, if the motion of the end effector is to be specified, the cam-follower motion must be determined by kinematic analysis of the intervening mechanism and the cam must then be designed to impart this motion to the follower. A typical example is an electric typewriter, where, for instance, a Watt II chain connects the end effector (type element) to the roller follower in contact with the cam on the camshaft of the typewriter (Fig. 8.8).

In other cases, the end effector, say a gripper, must be attached to a floating link of a motion-generator mechanism, and the gripper needs to be opened and closed at specified points in the motion cycle. Here the cam may be either rotating or stationary, with a moving follower system which actuates the end effector. In this case the body

TABLE 6.6 HARD COPY OF CAMSYNG PROGRAM OUTPUT (SEE EXAMPLE 6.3).
LAST TWO LINES OF OUTPUT WERE OFF THE SCREEN.

Cam profile coordinates		Cam cutter center coordinates	
Radius	Angle (deg)	Radius	Angle (deg)
4.000	8.627	5.500	8.627
3.993	10.353	5.486	12.119
3.971	12.272	5.446	15.598
3.932	14.460	5.386	19.029
3.875	16.958	5.313	22.384
3.805	19.790	5.234	25.637
3.726	22.981	5.159	28.761
3.646	26.565	5.095	31.727
3.575	30.562	5.046	34.514
3.525	34.946	5.016	37.120
3.506	39.581	5.006	39.581

The maximum pressure angle calculated is 25.613 deg.

The cam dwells until

3.506	339.581	5.006	339.581
3.506	339.581	5.006	339.581
3.525	344.219	5.016	341.970
3.578	348.582	5.045	344.381
3.652	352.481	5.092	346.893
3.736	355.863	5.156	349.551
3.816	358.764	5.230	352.372
3.884	1.253	5.309	355.356

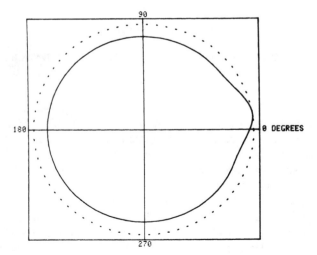

Figure 6.32 Hard copy of disc-cam contour plotted by CAMSYNG program (see Example 6.3).

motion and the opening/closing of the gripper must be combined and the intervening ac-tuating mechanism kinematically analyzed to determine the required cam-follower mo-tion and then synthesize the cam surface.

Combination of the linkage analysis and cam synthesis techniques presented in this text should enable the designer to deal with the kinds of problems described in this section. Two examples of such tasks will be given.

Example 6.4: Stamping Mechanism

Stamping platen (1) (Fig. 6.33),* supported on flexures (2) and maintained in its upper posi-tion against stop (3) by extension spring (4) is to be cyclically depressed against stationary anvil (5) according to the time-displacement curve (6). It is left to the reader to design the cam-follower linkage and the cam contour. The cam shaft location is fixed, but the follower linkage design is at the designer's discretion. Figure 6.33 shows the schematic sketch of one possible solution.

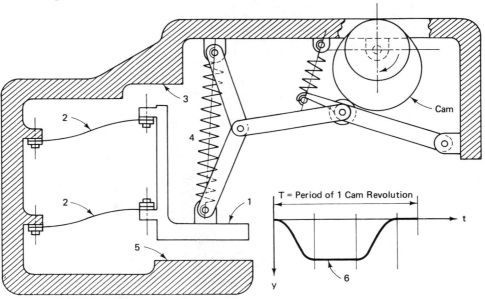

Figure 6.33   Cam design for remote end effector motion requirements (see Example 6.4). Courtesy Burroughs Corporation, Rochester, New York.

Example 6.5: Assembly Mechanism (Fig. 6.34)

Individual equally spaced products (*p*) are arriving on a continuously running upper con-veyor at a speed of V_1. They are to be collected in bins of the intermittently moving lower conveyor at a speed about one-eighth of V_1 to avoid damage to the product as it drops into the bin.

* Courtesy Burroughs Corporation, Rochester, New York.

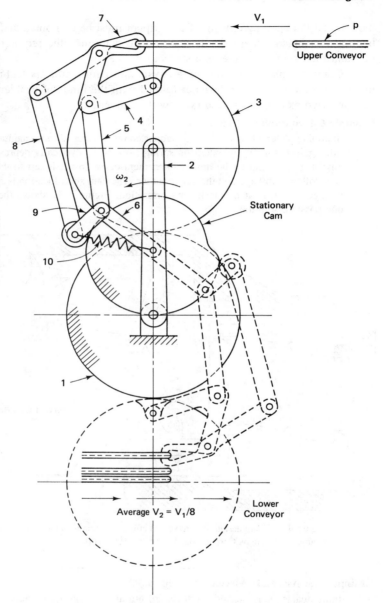

Figure 6.34 Stationary cam with moving follower and end effector system.

One possible solution is a cam-actuated cycloidal-crank nine-bar linkage, consisting of fixed sun gear (1), input crank (2), rotating ccw at constant angular velocity ω_2, planet gear (3), gripper jaw (4), coupler (5), follower crank (6) (pivoted at the center of the cam), gripper lip (7), gripper push-pull rod (8), and cam follower arm (9). The cam follower is spring-urged into contact with the stationary cam by spring (10).

As input crank (2) revolves, planet gear (3) orbits about the sun gear. The mechanism is shown in two positions: the upper or "take" position (full lines) and the lower or "release" position (dashed lines). It is easy to show by the method of instant centers that the moving pivot joining links (3) and (4) has a rightward linear speed in the lower position which is approximately one-eighth of its leftward speed in the upper position. The cam closes the gripper near the upper position, keeps it closed during the downward, decelerating motion, opens the gripper near the lower position, and keeps it open during the accelerating upward swing.

6.10 CAM-MODULATED LINKAGES [127]

In this section a method is presented for the number synthesis of cam-modulated linkages (CMLs). This method includes the classification and enumeration of CMLs having one degree of freedom and producing continuous output motion, which can be coordinated with a constant angular velocity input.

A cam-modulated linkage is a mechanism consisting of one or more cam-and-follower pairs in combination with a linkage. Whenever the problem to generate a motion is encountered, the first attempt at a solution is generally to use either a cam-and-follower mechanism or a linkage mechanism. The primary difference between CMLs and cam-and-follower mechanisms is that CMLs are capable of producing function, path and coplanar motion outputs, whereas cam-and-follower mechanisms can in general produce only function outputs. While a CML may be considered a more complicated device than a linkage because of the presence of at least one higher-pair connection, it provides for precise control over position and velocity of the output link throughout the motion cycle, including dwells during any part of the output cycle. Uses in printing, knitting, and packaging machines show clear evidence of the usefulness of CMLs in machines that require complete and precise control of the output motion.

The first step in the design of a CML for a specified motion requirement is the proper selection of its structural configuration (associated linkage; see Sec. 8.3). The designer's selection of a proper CML for a specific task is facilitated by the enumeration and classification of all CMLs having the same constructional type, namely, the same number of links and degrees of freedom. Moreover, the classification also depends on the intended kind of output, that is, function, path, or coplanar motion generation. Number synthesis of CMLs deals with these aspects of classification and enumeration of CMLs (see Sec. 8.3).

The Necessity Criteria of Conversion

The necessity criteria of conversion from linkages to CMLs determine the necessary number of binary links (other than the input, output, and fixed links) required in a linkage for converting it to a CML. Such binary links are referred to as convertible links. The number of convertible links required is equal to the number of resulting higher-pair connections, because each convertible link can only be replaced by one higher-pair connection, such as a cam-and-follower pair.

In the case of function generators, only one parametric relationship is required to determine the position and velocity of the output link assuming a constant angular velocity for the input link. This indicates that, to convert a linkage to a function generator CML, only one convertible link is necessary in the linkage.

For path generation, two parametric relationships are required to describe the position and velocity of the tracer point as it travels along the specified path. Here, again, it is assumed that the input link is rotating with a constant angular velocity. Therefore, in this case, two convertible links are necessary in a linkage so as to convert it to a path-generating CML. Such a CML will have two higher-pair connections.

For generation of coplanar motion of a link, three parametric relationships between the input variable (which is the parameter) and the three output variables are required, describing the two-dimensional path of a reference point of the moving plane and the rotation of a reference line embedded in that plane. If we assume a constant input velocity, the position and velocity of the moving plane can be determined by these three parametric relationships. Therefore, to convert a linkage to a motion generator CML, three convertible links are necessary in the linkage. The resulting CML will have three higher-pair connections.

The Existence Criteria of Conversion

The existence criteria deal with the requirements of the structural configuration of a given linkage for conversion to a function, path, or motion generator, one-degree-of-freedom CML. Specifically, the existence criteria determine which link(s), if any, in the linkage can be the output link for the desired type CML. Assuming that the input link and the fixed link have been preselected, these criteria are stated as follows (see Fig. 6.35):

1. *Function generator.* There must be two or more links between the input and output links. One of these must be a convertible link. In the case of standard cam-and-follower mechanisms, which are also used as function generators, there is only one link that is convertible and is also between the input and output links.

2. *Path generator.* The output link may be any link that is separated from both the input link and all fixed pivots by at least one link.

3. *Motion generator.* The output link may be any link that is separated from both the input link and all fixed pivots by a binary link or by at least two links. If a binary link intercedes between the output link and either the input link or a fixed pivot, it must be converted to a cam-and-follower pair.

Schematic diagrams illustrating the requirements of structural configurations of a linkage to satisfy these existence criteria are shown in Fig. 6.35.

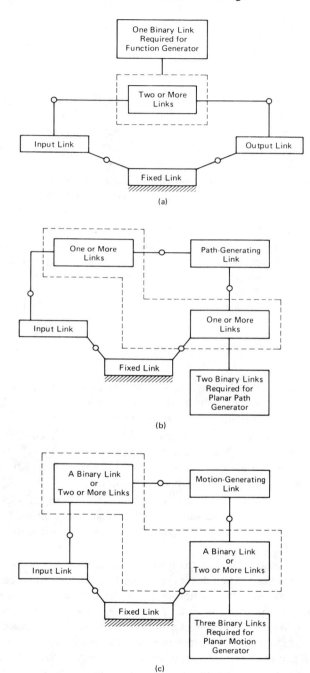

Figure 6.35 Diagram showing the linkage configurations satisfying the existence criteria for function, path, and motion generator CMLs: (a) function generator; (b) path generator; (c) coplanar motion generator.

Developing CMLs from Linkages

To derive a class of CMLs from a given linkage, the following 12-step procedure can be applied.

1. Number all links in the given linkage. Specify the fixed link.
2. Choose a link constrained to rotate about a fixed pivot as an input link.
3. Apply the necessity criterion of conversion for the desired class of CMLs. If the number of convertible links required exceeds the number of binary links remaining after the input link has been chosen, go to step 12.
4. Apply the existence criterion of conversion for the desired class of CML and determine which link(s), if any, can be used for the output. If no links can be used, go to step 12.
5. Choose one of the links as the output link from those found in step 4.
6. Let the total number of convertible links that remain (after the output has been chosen) be m, and let the number required by the necessity criterion be n. Subtract n from m. If the result is negative, go to step 12; if it is not, determine T, the total number of possible combinations of convertible links that may be used to convert the given linkage to the desired class of CML. T is given by

$$T = \frac{m!}{(m-n)!\,n!}$$

7. The number T obtained in step 6 indicates the number of possible combinations of binary links that may be converted to cam-and-follower pairs. List these combinations using the link numbers as assigned in step 1.
8. For each combination listed in step 7, redraw the linkage leaving out the binary links in that combination. These new configurations are referred to as test chains. The unconnected pivot joints in these test chains, where a binary link has been left out, are referred to as open joint pairs.
9. Check if there are duplications among the test chains. Duplicate test chains generally result when two convertible links connect the same two links. Converting either convertible link to a cam-and-follower pair in this case results in the same CML structure.
10. Assume an arbitrary output of the desired type. Consider the output link and the input link to be moving as though they were both input to the test chain and determine if this specifies the motion of every open joint pair. If these motions can thus be specified but cannot be analyzed by dyads, the test chain is not converted to CML. Also, if the motions of the open joints cannot be specified by means of the input and output motions, the test chain is not converted to a CML. When the motions can be specified by way of the input and output motions and can be analyzed by dyads, the test chain is converted to a CML by replacing the open joints with cam-and-follower pairs.

11. Repeat steps 6 to 10 for each of the remaining possible output links, if any, found in step 4.

12. Choose another possible input link and go back to step 3. If all possible inputs have been investigated, all possible CMLs have been found.

Example 6.6

Consider a linkage having 10 links as shown in Fig. 6.36. Among the ten links, five are binary, four are ternary, and one is quaternary. It is required to convert this linkage to a coplanar motion-guiding CML. Following the foregoing 12-step procedure, for possible input links 1, 7, and 5, it was found that one can obtain 13 test chains and five CMLs. These CMLs are shown in Fig. 6.37.

When link 1 was chosen as an input, six test chains were obtained. Out of these six test chains, it was found that only two could be converted to motion-guiding CMLs. These two CMLs are shown in Fig. 6.37a and b. These two CMLs were obtained by converting the binary links 2, 4, and 7 to higher pairs in the first case and by converting the binary links 2, 4, and 9 to higher pairs in the second case.

When link 7 was chosen as an input link, four test chains were obtained. Then, from step 10, only one CML, shown in Fig. 6.37c, was obtained by converting binary links 2, 4, and 9 to higher-pair connections.

When link 5 was chosen as an input link, three test chains were obtained. From these three test chains, one CML was obtained by converting the binary links 7, 4, and 10 to higher-pair connections and one CML was obtained by converting the binary links 9, 4, and 10 to higher-pair connections. These are shown in Fig. 6.37d and e.

Enumeration of CMLs for Path and Motion Generation

Using the necessity and existence criteria in conjunction with the 12-step procedure, six path-generating CMLs were found for two types of six-link chains. It was also found that, for 16 types of eight-link chains, 41 motion-generating CMLs can be obtained.

The two types of six-link chains selected for enumeration of path-generating CMLs are designated as I and II, respectively, and are shown in Fig. 6.38. The path-generating CMLs derived from I and II are shown in Fig. 6.39. Each CML is identified

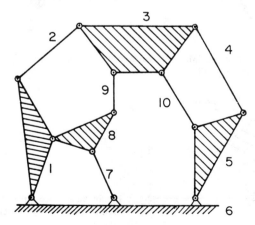

Figure 6.36 Ten-link chain of Example 6.6 to be converted to motion generator CMLs (Cam Modulated Linkages).

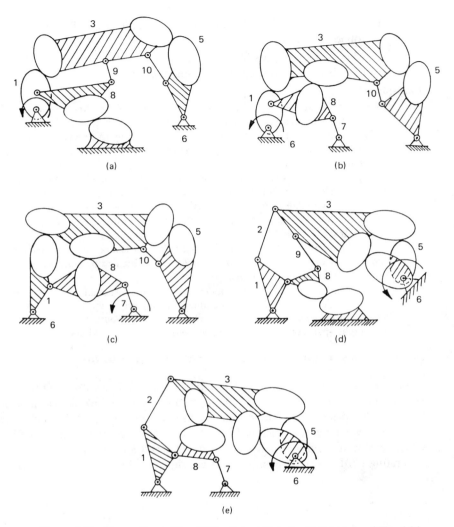

Figure 6.37 Five motion-guiding CMLs derived from the 10-link chain of Fig. 6.36. Input links are marked with rotation arrows.

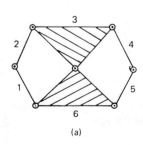

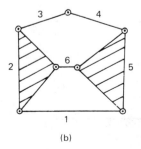

Figure 6.38 Two types of six-link chains used to enumerate path-generating CMLs: (a) type I; (b) type II.

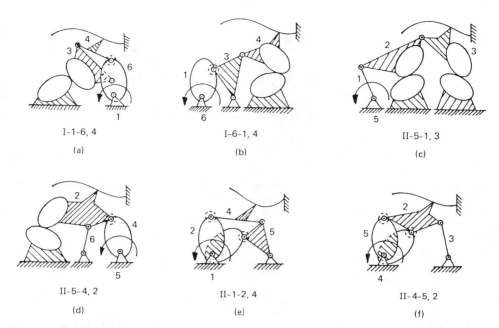

I-1-6, 4

(a)

I-6-1, 4

(b)

II-5-1, 3

(c)

II-5-4, 2

(d)

II-1-2, 4

(e)

II-4-5, 2

(f)

Figure 6.39 Six path-generating CMLs derived from the two six-link chains shown in Fig. 6.38. In the labels the roman number gives the type of 6-link mechanism, and the three arabic numbers designate the fixed, input, and output links in that order; (a) I-1-6,4; (b) I-6-1,4; (c) II-5-1,3; (d) II-5-4,2; (e) II-1-2,4; (f) II-4-5,2.

by the label describing the type of six-link chain, either I or II, followed by a set of three numbers. The first number indicates the fixed link, the second number denotes the input link, and the third number denotes the output link. When a cam is attached to an input link, as is the case in Fig. 6.39b, the follower contact is indicated as a circle penetrating the oval of the cam, so that the oval represents the pitch curve of the cam. When two or more cams are attached to the input link, as is the case in Fig. 6.39e, they are shown as overlapping ovals having hatch marks in the region common to them, as an indication that the cams are solidly attached to each other.

Applying the number synthesis techniques described in this section, 41 motion-generating CMLs have been derived from 16 types of eight-link chains. Figure 6.40

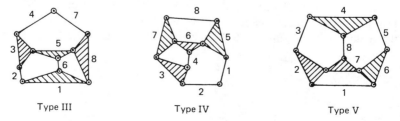

Type III

Type IV

Type V

Figure 6.40 Three of sixteen types of eight-link chains for the enumeration of coplanar motion-guiding CMLs.

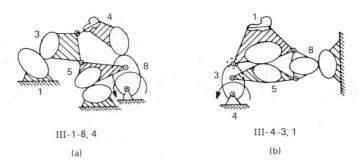

III-1-8, 4   III-4-3, 1

(a)   (b)

Figure 6.41 Coplanar motion-guiding CMLs derived from the three chains shown in Fig. 6.40. The labels have meanings similar to those in Fig. 6.39. For example, in Fig. 6.41(b), link 1 is the output.

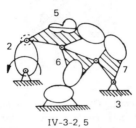

IV-3-2, 5

Figure 6.42 Coplanar motion-guiding CMLs derived from Type IV of the three chains shown in Fig. 6.40.

shows three of the 16 types of eight-link chains, designated as III, IV, and V, respectively. Figure 6.41 shows the two motion-generating CMLs derived from the eight-link chain of type III. Similarly, Figs. 6.42 and 6.43 show the one and seven motion-generating CMLs derived from types IV and V, respectively. In each CML, the link performing the prescribed guided motion is designated by the symbol of a nonsymmetric dumbbell with a straight line drawn along that symbol. The motion-generating CMLs shown in Figs. 6.41, 6.42, and 6.43 are labeled in the same way as the path-generating CMLs described earlier.

Cam-modulated linkages have a unique advantage over either cam-and-follower or linkage mechanisms, since CMLs can be synthesized not only at a limited number of points, but rather at every point over the entire range of the output motion. Practically all linkage dimensions can be chosen arbitrarily to suit design conditions. Then the follower types and their contours can be decided, and the cam contours synthesized with the method outlined in Sec. 6.9 for cams with remote end-effector follower systems.

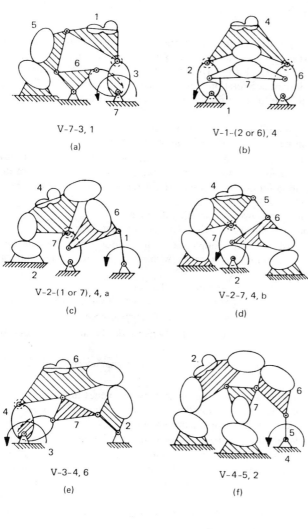

V-7-3, 1

(a)

V-1-(2 or 6), 4

(b)

V-2-(1 or 7), 4, a

(c)

V-2-7, 4, b

(d)

V-3-4, 6

(e)

V-4-5, 2

(f)

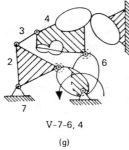

V-7-6, 4

(g)

Figure 6.43   Coplanar motion-guiding CMLs derived from Type V of the three chains shown in Fig. 6.40. (a) V-7-3,1; (b) V-1-(2 or 6),4; (c) V-2-(1 or 7),4a; (d) V-2-7,4, b; (e) V-3-4,6; (f) V-4-5,2; (g) V-7-6,4.

PROBLEMS

6.1. A follower is to move gradually outward 1 in with constant acceleration while the cam turns through 90°. During the next 90° of cam rotation the follower is to continue to move outward 1 in with constant deceleration and is to have zero velocity at the end of this time. The follower is to return with constant acceleration (for 70°) and deceleration (for 80°) during the next 150° of cam rotation. The follower dwells or is at rest during the remaining 30° of cam rotation.
  (a) Construct and dimension the follower acceleration diagram.
  (b) Construct and dimension the follower velocity and displacement diagrams.

6.2. A follower is to move outward 2 in. It is to have constant acceleration for 60°, then move with uniform velocity for 30°, then decelerate for 90° of the cam rotation. The return motion is to be constant acceleration for 60° and constant deceleration for 90°. The follower is to dwell for the remaining 30°.
  (a) Construct and dimension the follower acceleration diagram.
  (b) Construct and dimension the follower velocity and displacement diagrams.

6.3. A follower is to move outward 2 in with simple harmonic motion while the cam turns through one-half revolution. The follower is to return with simple harmonic motion during the next 150° and then dwell for 30°.
  (a) Construct and dimension the follower acceleration diagram.
  (b) Construct and dimension the follower velocity and displacement diagrams.

6.4. Same as Prob. 6.1 except use cycloidal motion instead of constant acceleration.

6.5. Same as Prob. 6.2 except use another motion with better jerk characteristics than the constant-acceleration motion.

6.6. Same as Prob. 6.3 except use another motion with better jerk characteristics than simple harmonic motion.

6.7. Draw the displacement-time curve of a cam follower that rises 3 inches in the first 180° of cam revolution, dwells for 45°, and then falls in the remaining rotation of the cam. Make the rise a parabolic motion and the fall a simple harmonic motion.

6.8. Shown in Fig. P6.1 is a roller follower and cam system. In each of the four quadrants is shown one position of the centerline of the follower guide relative to the cam. These are labeled A, B, C, and D. Utilizing the pressure-angle definition, check to see which position has the pressure angle greater than the others and then measure only the largest pressure angle.

6.9. Determine the maximum pressure angle for the positions of the reciprocating roller follower shown in Fig. P6.2.

6.10. A cam is to be designed using standard cam surfaces for discrete intervals of the cam's rotation. The displacement (S) versus cam rotation angle (ϕ) is shown in Fig. P6.3. The lift, velocities, and accelerations at points A, B, and C are specified as follows:

Point A	Point B	Point C
$S = 0$	$S = L$	$S = 3L/2$
$V = 0$	$V = +V_1$	$V = 0$
$A = 0$	$A = 0$	$A = 0$

Recommend standard curves to be used for the displacement graph and the ratio between β_1 and β_2 to match velocities at point B.

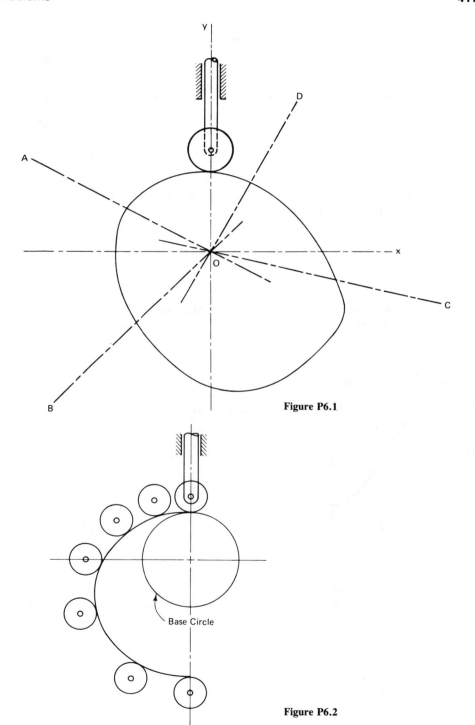

Figure P6.1

Figure P6.2

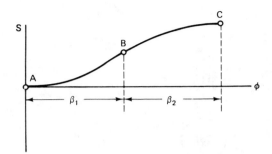

Figure P6.3

6.11. A follower is to have cyclical motion according to the displacement curve shown in Fig. P6.4. The displacement and velocity requirements are:

Point A	Point B	Point C
$S = L$	$S = 0$	$S = L$
$V = 0$	$V = 0$	$V = 0$

Recommend the displacement profile to be used and the ratio between β_1 and β_2 to match accelerations at points A, B, and C.

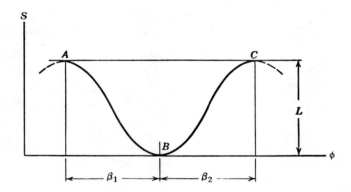

Figure P6.4

6.12. In Fig. P6.5, the displacement–time, velocity–time, and acceleration–time curves are sketched for a follower. The cam rotates at a constant angular velocity ω and the maximum (peak) value of the acceleration is 5 units. The equation for the acceleration–time curve is as follows:

$$A = \left(\frac{2\pi h\omega^2}{\beta^2}\right)\sin\left(\frac{2\pi}{\beta}\phi\right)$$

(a) Write the equations for the velocity–time and displacement–time curves.

(b) Determine the maximum values of displacement and velocity.

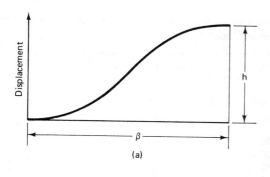

(a)

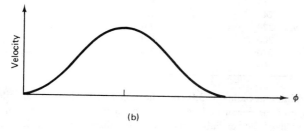

(b)

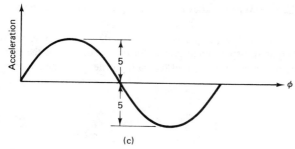

(c)

Figure P6.5

6.13. A roller follower is to move through a displacement and return with no dwells in the cycle. It is required, however, that a portion of the rise be covered at constant velocity. Figure P6.6 shows a sketch of the displacement curve with the constant velocity portion between B and C.
   (a) Determine the profiles to be used in the remaining segments (pick from among those presented in the chapter).
   (b) Sketch the resulting velocity and acceleration profiles.

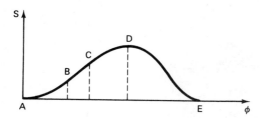

Figure P6.6

6.14. A disk cam rotating clockwise drives an oscillating flat-face follower through a total angle of 20° with the following displacement figures:

Cam angle (deg)	Follower angle (deg)
0	0.0
30	2.0
60	6.0
90	10.0
120	14.5
150	18.5
180	20.0
210	18.5
240	14.5
270	10.0
300	6.0
330	2.0
360	0.0

Lay out the cam using a minimum radius of 3.0 cm. The center of the hub of the follower is to be 9.0 cm to the right of the center and on the horizontal centerline of the cam. The distance from the center of the follower hub to the arc of the displacement scale is to be 8.0 cm. The extension of the follower face goes through the pivot. Determine the length of the follower face.

6.15. The base circle of a disk cam with a reciprocating knife-edge follower is shown in Fig. P6.7 together with a required displacement profile for one revolution of the cam. The cam is to rotate in a clockwise direction. Construct the cam profile to produce the required follower motion profile.

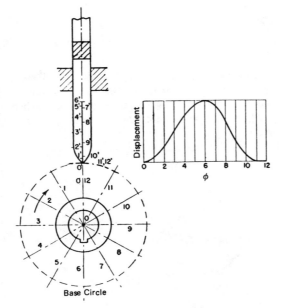

Figure P6.7

6.16. Figure P6.8a shows a disk cam with a reciprocating roller follower. The follower is to move radially according to the scale shown on the centerline. The lowest point on the scale, O' is at the center of the roller, and $O'O$ is the radius of the pitch circle. The cam is to rotate in a clockwise direction.

(a) Construct the cam contour that will produce the desired follower displacement.

(b) Determine the pressure angle at the 90° position (location 3).

(c) Assume that the pressure angle at the 90° position is too large (due to binding at the follower stem). An offset of the follower (see Fig. P6.8b) is suggested as a way to decrease this angle. Construct the new cam profile with the offset. Compare the old and new pressure angles at the 90° position. Also compare the required follower stem lengths between the two cases.

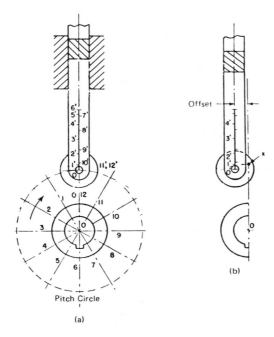

(a)

(b)

Figure P6.8

6.17. Figure P6.9 shows a disk cam with a radially reciprocating flat-face follower. The cam is to turn clockwise and move the follower according to the scale shown on its centerline. (Each interval of follower displacement represents 30° of cam rotation.) Construct the cam profile to produce the desired follower displacement. Determine the required length of the flat face follower on each side of the initial contact point (position O').

6.18. A disk cam with oscillating flat-face follower is to be constructed according to Fig. P6.10. This cam is to turn clockwise and the follower is to oscillate according to the scale shown. The flat face extended is tangent to a circle of radius r with center at O'''. Construct the cam profile and determine the required length of the flat-face follower on both sides of the intersection with the arc of radius R.

6.19. The base circle of a cam with primary and secondary followers is shown in Fig. P6.11. The offset is too large to use a single follower. The scale for the required motion of the secondary follower is shown on its centerline as $0'$, $1'$, $2'$, and so on. This scale is transferred to

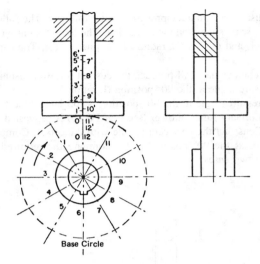

Figure P6.9

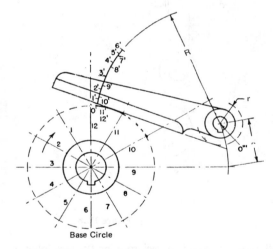

Figure P6.10

the primary follower on a convenient arc with O as the center. Construct the cam contour for the required follower motions.

6.20. Prove Eq. (6.21). Show all steps.

6.21. Prove Eq. (6.22). Show all steps.

6.22. Prove Eq. (6.25). Show all steps.

6.23. Prove Eq. (6.26). Show all steps.

6.24. Prove Eq. (6.27). Show all steps.

6.25. Derive the piecewise expressions for the acceleration, velocity, and displacement for the modified trapezoidal motion of Fig. 6.16.

6.26. Show that in Example 6.1 $\dot{y}_{max}$ occurs at $x = \frac{1}{2}$.

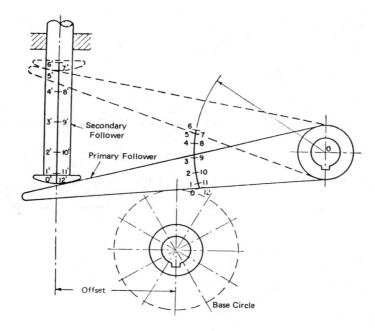

Figure P6.11

6.27. Design a polynomial profile to satisfy the following conditions:

At $\phi = 0$: $S = L$, $\dot{S} = 0$, $\ddot{S} = 0$

At $\phi = \beta$: $S = 0$, $\dot{S} = 0$, $\ddot{S} = 0$

6.28. Design a polynomial profile to satisfy the following conditions:

At $\phi = 0$: $S = 0$, $\dot{S} = 0$, $\ddot{S} = 0$

At $\phi = \beta$: $S = L$, $\dot{S} = V$, $\ddot{S} = 0$

6.29. Design a polynomial profile to satisfy the following conditions:

At $\phi = 0$: $S = L$, $\dot{S} = V$, $\ddot{S} = 0$

At $\phi = \beta$: $S = 0$, $\dot{S} = 0$, $\ddot{S} = -A$

6.30. A partial high-speed cam displacement profile is shown in Fig. P6.12 with $S(A) = .7$ in, $S(B) = 1.2$ in, $S(C) = .9$ in, $S(D) = 1.1$ in, and $S(E) = .9$ in.
   (a) What profiles should be used between A and B and between F and G?
   (b) Determine the values for β_1, β_2, β_3, β_4, β_5, β_7, β_8, L_6, and L_7 if $\beta_6 = 30°$ and the follower velocity at point A is .052 in/degree.
   (c) What is the maximum follower velocity and acceleration with respect to time if the angular velocity, ω, is constant at 10.472 rad/sec?

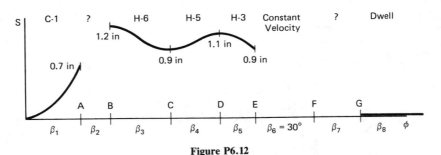

Figure P6.12

6.31. Fig. P6.13 shows a full-fall displacement curve made up of the H-3, constant velocity, and C-4 profiles. Suppose $\beta_1 = \beta_2 = \beta_3 = 30°$ and the absolute value of the maximum acceleration achieved during the fall cycle is $|.00163|$ in/degree^2. What is the total fall distance, L?

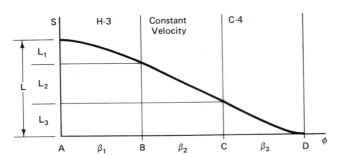

Figure P6.13

6.32. Most of a full rise trapezoidal acceleration profile curve is shown in Fig. P6.14. Unfortunately, β_6 is not known. If the full rise is 1.5 in, determine β_6 and complete and dimension the A, V, and S curves.

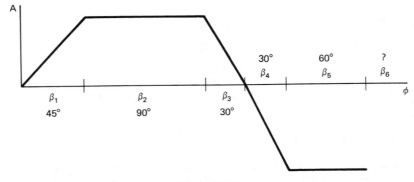

Figure P6.14

6.33. Describe the follower motion and velocity for the cam described in Prob. 6.32 if $\beta_6 = 10°$.

6.34. Both of the following acceleration profiles rise 1.5 in while $\beta_1 = 165°$ and $\beta_2 = 110°$. Sketch and dimension the acceleration curves and compare your results to those of Prob. 6.32.
   (a) Simple harmonic motion cam made up of H-1 and H-2.
   (b) Cycloidal motion cam made up of C-1 and C-2.

6.35. A rise-dwell-return cam is sketched in Fig. P6.15. The cam profiles listed in order are H-1, Constant Velocity, C-2, Dwell, C-3, and H-4. It is known that $L_3 = L_5 = L_6 = 1$ in, $\beta_3 = 120°$ and $\beta_6 = 60°$.
 (a) Determine β_1, β_2, β_4, β_5, L_1, and L_2. Sketch and dimension the S, V, and A diagram.
 (b) Calculate y, the follower acceleration with respect to time, for the positions $\phi = \beta_{3/2}$ and $\phi = \beta_{6/2}$ if $\omega = 120/\pi$ rad/sec and $\alpha = 120/\pi$ rad/sec^2.

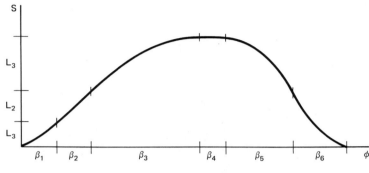

Figure P6.15

6.36. The S diagram of a rise-dwell-return-dwell cam using the uniform motion profile is shown in Fig. P6.16. The pressure angle for a nonoffset translating roller follower is given by:

$$\delta = \tan^{-1}\{V/(r_0 + S)\}$$

where V and S are both evaluated at the point of interest and r_0 is the base circle radius.
 (a) Sketch the pressure angle curve and determine at what value of ϕ the maximum δ will occur if $r_0 = 1$ in.
 (b) What must r_0 be if δ is everywhere to be less than 5°?
 (c) Repeat parts (a) and (b) using the full-rise and full-return cycloidal motion profiles and compare to the uniform motion cam.

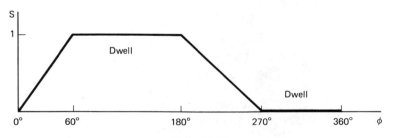

Figure P6.16

chapter seven
Gears and Gear Trains

7.1 INTRODUCTION

Basic principles and general nomenclature of gears are presented in this section. Since the emphasis of this text is on the use of various mechanisms for design tasks, the majority of this chapter is devoted to gear trains. The reader is referred to other texts for more detailed discussions on gear types and tooth profiles [e.g., 110].

Gears are machine elements that transmit motion by means of successively engaging teeth. Gears transmit motion from one rotating shaft to another, or to a *rack* (see Fig. 7.1) which translates in a straight line. Numerous applications exist in which a constant angular velocity ratio (or constant torque ratio) must be transmitted between shafts.

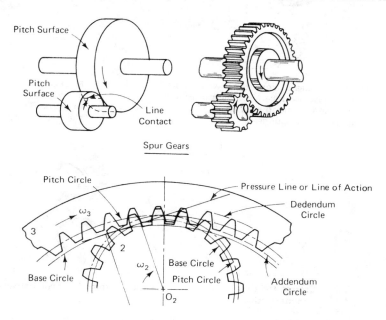

Spur Gears

Internal Ring Gear and Pinion: ring, internal, or annular gear all have the same meaning.

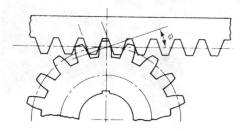

Involute Pinion and Rack: a rack is a spur gear having infinitely large pitch diameter.

Figure 7.1 Standard gear types.

Double Helical Gears: shown here as part of an open gear box. Loads along the shafts are minimal. (Courtesy of Kreiter Gear and Machine, Inc.)

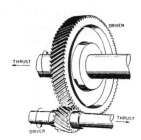

Stepped Gears: consists of two or more gears fastened together. These gears are quieter and smoother in action than conventional spur gears.

Parallel Helical Gears: can be thought of as stepped gears with an infinite number of parts. They run very quietly. Notice the loading along the shafts.

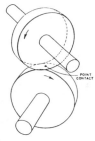

Crossed Helical Involute Gears: are used to transmit power between nonparallel, nonintersecting shafts.

Noncircular Gears: generate a nonconstant angular velocity ratio between input and output. Applications include printing presses, packaging machines, conveyors, and low-speed precision instruments. (Courtesy of Cunningham Industries, Inc.)

Figure 7.1 (cont.)

Straight Bevel Gears.

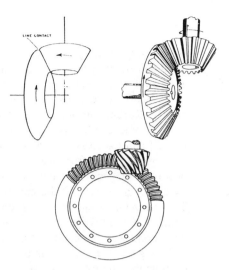

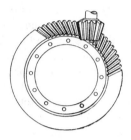

Skew Bevel Gears: are used to connect shafts whose axes do not intersect. The teeth are straight lines and there is sliding action along the tooth element as well as at right angles to the tooth element.

Hypoid Gears: are used to connect shafts whose axes do not intersect, allowing a certain amount of sliding action in the direction of the tooth elements. Although the correct pitch surfaces of these gears are hyperboloids, they are approximated by sections of cones in actual practice. They are used, for example, in rear-axial drive for passenger cars.

Worm and Worm Gear: the worm has one tooth sometimes called a thread. This gear set is used for nonparallel, nonintersecting shafts.

Herringbone Gear: shown here being generated on a gear generator. These gears are used to transfer large loads where loading along the shaft must be minimized. (Courtesy of Kreiter Gear and Machine, Inc.)

Figure 7.1 (cont.)

Based on the variety of gear types available (see Fig. 7.1), there is no restriction that the input and output shafts need be either in-line or parallel. Variable angular velocity ratios are also available by using noncircular gears (Fig. 7.1).

Gears are often used in applications in which power must be transmitted efficiently. Rolling surfaces are replaced by toothed gears so as to eliminate slipping. The first references [146, 88, 133, 35, 17] to the use of gears in mechanical devices appeared in the first or second century B.C. The Egyptians, the Greeks, and the Chinese used gears to transmit power from water. The design features that led to the use of gears were the positive engagement of the teeth (resulting in greater efficiency than could be achieved by earlier devices) and the capability for continuous rotation while transmitting power. The mechanism shown in the chapter heading on p. 420 depicts the Chinese south-pointing chariot some believe of the Han dynasty (200 B.C.–200 A.D.). This navigational instrument was designed for use in the Gobi desert. Based on pure rolling of the wheels and the gear train, the monk on the top of the chariot always points in the same direction even as the chariot changes direction.

In order to maintain a constant angular velocity (between an input and output gear) with teeth in contact, the individual tooth profile must obey the *fundamental law of gearing:* for a pair of gears to transmit a constant angular velocity ratio, the shape of their contacting profiles must be such that the common normal passes through a fixed point on the line of centers. This stems from Kennedy's Theorem (Sec. 3.7, Fig. 3.55), which states that three bodies in coplanar relative motion have their three instant centers in one straight line. Furthermore, their relative velocity ratios are equal to the ratios of the signed distances between their respective instant centers (I.C.-s).

Below we describe what this means for a spur gear pair, which is a three-link mechanism whose links are in coplanar relative motion (Fig. 7.2). The frame is link 1, the driver gear is link 2, and the driven gear is link 3. The instant centers are I.C. (1,2) and (1,3) at the gear centers, and (2,3) at the pitch point or the rolling contact of the pitch circles.

Figure 7.2 can be interpreted as showing two circular discs in true-rolling contact. Then the angular-velocity ratio between them is $\omega_3/\omega_2 = (1,2 - 2,3)/(1,3 - 2,3)$, or

$$\left|\frac{\omega_2}{\omega_3}\right| = \frac{O_3 P}{O_2 P} \tag{7.1}$$

The right side of the first equation is, in words, the ratio of the signed distance from 1,2 to 2,3 divided by the signed distance from 1,3 to 2,3. Since the I.C. distances point in

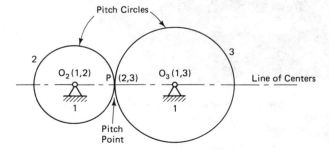

Figure 7.2 Two circular discs in true-rolling contact. This depicts the three-link mechanism.

opposite directions, their algebraic signs are opposite, and, therefore, the angular velocity ratio is negative: the discs rotate in opposite directions.

True-rolling discs are in frictional contact. To make the true-rolling contact positive and thus independent of friction, the discs are replaced by toothed gearing. The tooth profiles must be such as to maintain constant angular velocity ratio. This is accomplished if the instant center I.C. (2,3) remains stationary. This means that the normal to the tooth-profile contact must go through the pitch point in all positions while the gears are rotating. When this requirement is satisfied, the teeth of the two gears are said to have *conjugate profiles,* and the condition is called *conjugacy.* Figure 7.3 illustrates this. Thus, any two mating tooth profiles that satisfy the fundamental law of gearing are called *conjugate profiles.* Although there are many tooth shapes possible in which a mating tooth could be designed to satisfy the fundamental law, only two are in general use: the *cycloidal* and *involute* profiles. The involute has important advantages — it is easy to manufacture and the center distance between a pair of involute gears can be varied without changing the velocity ratio. Thus close tolerances between shaft locations are not required when utilizing the involute profile.

Figure 7.4 shows how an involute curve can be generated. A string is wrapped around the *base circle* (a cylinder). As the string is unwrapped from the surface, a point on the string (point *P*) traces an involute profile.

Figure 7.1 shows several gear types. For applications with parallel shafts, straight spur, stepped, helical, double-helical, or herringbone gears are usually used. In the case

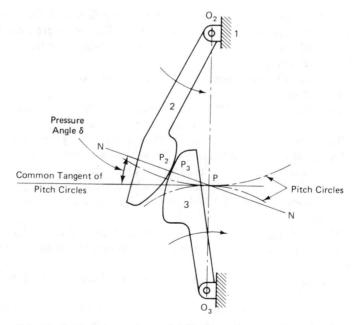

Figure 7.3 The fundamental law of gearing: for constant velocity ratio, the common normal of contacting tooth flanks must always pass through the *pitch point P*.

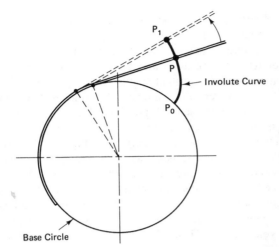

Base Circle

Figure 7.4 Involute generation by a point on a string unwrapped from a circular disk.

of intersecting shafts, straight bevel, spiral bevel, or face gears are employed. For non-intersecting shafts that are nonparallel, crossed helical, worm, face, skew bevel, or hypoid gears would be acceptable choices.

7.2 GEAR TOOTH NOMENCLATURE

Figure 7.5 shows part of a typical spur gear together with the standard nomenclature. Some of the expressions are explained below. Spur gears are cylindrical in form (have cylindric pitch surfaces) and operate on parallel axes. Their teeth are straight and parallel to the axes (Fig. 7.1).

The *pitch circles* of mating gears are tangent to each other. They roll on one another without sliding.

The *circular pitch* (*CP*) is the distance (in inches) along the arc of the pitch circle between corresponding profiles of neighboring teeth.

The *addendum* (*a*) is the height by which a tooth projects beyond the pitch circle (also the radial distance between the pitch circle and the addendum circle).

The *clearance* (*c*) is the amount by which the *dedendum* (tooth height below the pitch circle) in a given gear exceeds the addendum of its mating gear.

The *working depth* (h_k) is the depth of engagement of two gears (the sum of their addendums).

The *whole depth* (h_t) is the total depth of a tooth space (addendum plus dedendum).

The *tooth thickness* (*t*) is the distance across the tooth along the arc of the pitch circle.

The *tooth space* is the distance between adjacent teeth along the arc of the pitch circle.

The *backlash* (not shown) is the amount by which the width of the tooth space exceeds the thickness of the engaging tooth at the pitch circles.

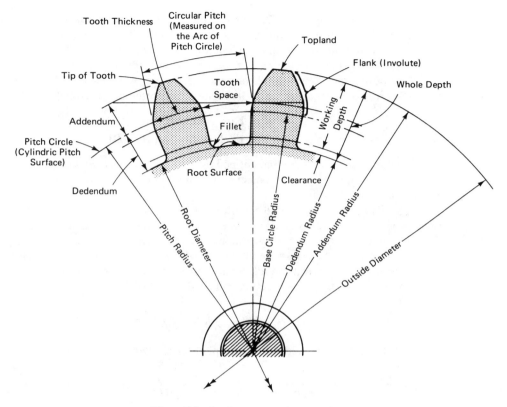

Figure 7.5   Involute gear tooth nomenclature.

The *diametral pitch* (*P*) (not shown) is the number of teeth on a gear per inch of pitch diameter. Thus

$$P = \frac{N}{D} \qquad (7.2)$$

where P = diametral pitch (teeth/in)

   N = number of teeth

   D = pitch diameter (in)

The circular pitch can be expressed as

$$CP = \frac{\pi D}{N} \qquad (7.3)$$

and $(P)(CP) = \pi$.

The *pressure angle* (Fig. 7.3) or δ is the angle that the line of action of the forces between mating teeth make with a perpendicular to the line of gear centers through the pitch point. It is equivalent to the deviation angle shown in Fig. 3.19.

The *flank* is the involute part of the side surface (or *profile*) of a tooth.

The *base pitch* (*BP*) is the distance along the arc of the base circle between corresponding profiles of adjacent teeth. Its relationship to the circular pitch is

$$BP = CP \frac{\text{base circle diameter } (D_b)}{\text{pitch circle diameter } (D)} \tag{7.4}$$

The *pressure line* is the line defined by the pressure angle and tangent to both base circles of a meshing gear set. Contact and, hence, motion and power transference is accomplished along this line.

The *line of action* (ℓ) is the portion of the pressure line along which gear tooth contact occurs. Contact is initiated when the tip of the driven tooth is pushed by the flank of the driver tooth or when the addendum circle of the driven gear crosses the pressure line. Contact ceases when the flank of the driven tooth is acted upon by the tip of the driver tooth or when the addendum circle of the driver gear crosses the pressure line. The length of the line of action can be determined by Eq. (7.5) with reference to Fig. 7.6 is

$$\ell = |\overline{AC} - \overline{AP}| + |\overline{DB} - \overline{DP}| \tag{7.5}$$

After substitution of the addendum and pitch circle radii, the length of the line of action is (see Problem 7.30)

$$\ell = |\sqrt{(r_2 + a_2)^2 - (r_2 \cos \delta)^2}| - r_2 \sin \delta + |\sqrt{(r_3 + a_3)^2 - (r_3 \cos \delta)^2}| - r_3 \sin \delta \tag{7.5a}$$

The *contact ratio* (*CR*) is the length of the line of action divided by the base pitch. It is the average number of teeth in contact at any one time as the gears rotate in mesh. It must be greater than 1 or else no teeth would mesh for some time interval.

$$CR = \frac{\ell}{BP} \tag{7.6}$$

where *BP* is the base pitch ($2\pi r_B/N$) and where *N* is the number of teeth in the gear.

The involute profile is defined from the base circle to the outside diameter at the tip of the tooth, the addendum circle. If the dedendum circle lies within the base circle, a noninvolute profile will result below the base circle. Therefore, nonconjugate curves will mesh if contact between two teeth occurs below the base circle. This violates the fundamental law of gearing, and such a condition leads to *interference*. This condition can be prevented by ensuring that the addendum circle of either gear does not intersect the line of action beyond the *interference points* defined to be where the pressure line is tangent to the base circle. The following discussion is derived with respect to Fig. 7.6. The addendum radius of gear 2 must be less than the interference point distance from O_2

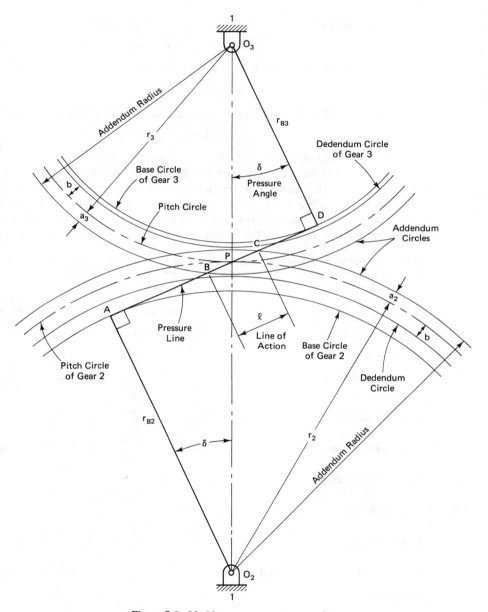

Figure 7.6   Meshing spur gear set nomenclature.

to D, while the addendum radius of gear 3 must be less than the interference point distance from O_3 to A. If c denotes the gear center distance, then the pressure line distance is

$$\overline{AD} = c \, \sin \delta = O_2 O_3 \sin \delta \qquad (7.7)$$

The interference point distance for gear 2 is $\overline{O_2D}$, given by

$$\overline{O_2D} = |\sqrt{r_{b2}^2 + (c \sin \delta)^2}| \tag{7.8}$$

Therefore, the noninterference condition for gear 2 is

$$r_2 + a_2 \leq |\sqrt{r_{b2}^2 + (c \sin \delta)^2}| \tag{7.9}$$

The noninterference condition for gear 3 is derived in a similar manner.

$$r_3 + a_3 \leq |\sqrt{r_{b3}^2 + (c \sin \delta)^2}| \tag{7.10}$$

The limiting condition for interference of standard gears is based upon the larger gear. Interference in the set will not occur if the addendum of the larger gear does not interfere with the smaller gear. If interference does occur, it can be eliminated by decreasing the addendum of the interfering gear, creating a *stub-tooth gear,* or by increasing the pressure angle. All of these methods decrease the contact ratio which causes noisier operation. Table 7.1 gives the AGMA tooth proportions for standard involute spur gears.

TABLE 7.1 STANDARD AGMA AND USASI TOOTH SYSTEMS FOR SPUR GEARS

Quantity	Coarse pitch[a] (up to 20P) full depth		Fine pitch (20P and up) full depth
Pressure angle δ (deg)	20°	25°	20°
Addendum a	$\dfrac{1.000}{P}$	$\dfrac{1.000}{P}$	$\dfrac{1.000}{P}$
Dedendum b	$\dfrac{1.250}{P}$	$\dfrac{1.250}{P}$	$\dfrac{1.200}{P} + 0.002$ in
Working depth h_k	$\dfrac{2.000}{P}$	$\dfrac{2.000}{P}$	$\dfrac{2.000}{P}$
Whole depth h_t, minimum	$\dfrac{2.250}{P}$	$\dfrac{2.250}{P}$	$\dfrac{2.200}{P} + 0.002$ in
Circular tooth thickness t_p	$\dfrac{\pi}{2P}$	$\dfrac{\pi}{2P}$	$\dfrac{1.5708}{P}$
Fillet radius of basic rack r_f	$\dfrac{0.300}{P}$	$\dfrac{0.300}{P}$	Not standardized
Basic clearance c, minimum	$\dfrac{0.250}{P}$	$\dfrac{0.250}{P}$	$\dfrac{0.200}{P} + 0.002$ in
Clearance c (shaved or ground teeth)	$\dfrac{0.350}{P}$	$\dfrac{0.350}{P}$	$\dfrac{0.3500}{P} + 0.002$ in
Minimum number of pinion teeth	18	12	18
Minimum width of top land t_0	$\dfrac{0.25}{P}$	$\dfrac{0.25}{P}$	Not standardized

[a] But not including 20P.

Source: Standardized by the American Gear Manufacturers' Association (AGMA) and the United States of America Standards Institute. The spur gear standards are the AGMA publications 201.02 and 201.02A, "Tooth Proportions for Coarse-Pitch Involute Spur Gears," and 207.04, "20-Degree Involute Fine-Pitch Systems for Spur and Helical Gears," and are available from the AGMA.

7.3 FORMING OF GEAR TEETH

There are many ways to form teeth of gears. Two distinct processes are used commercially: *casting* and *machining*. Popular methods for casting include sandcasting, investment casting, die casting, and centrifugal casting. Gears made of zinc, tin, aluminum, and copper alloys are usually die cast, resulting in good accuracy and surface finish.

Gears used in high-speed applications and/or those required to carry high loads in comparison to their size are usually made of steel and are cut with either *form cutters* or *generating cutters*. In the forming method, the teeth are cut by either a rotary milling cutter (see Figs. 7.7 and 7.8) or a reciprocating shaper cutter that has been formed *to the exact shape required*. Other forming methods use either a planing tool (which forms the outline required by following a previously shaped template) or broaching. Extrusion (drawing of pinion wire) is used for small, nonferrous and plastic-material gears. The latter are also produced by injection molding.

In the generating method a machine generates mathematically correct tooth profiles by virtue of the motions given to the cutter and the gear blank. The generating tool has a shape different from the tooth profile. The pitch surface of the cutter (cylindric for a *pinion cutter* resembling a small gear, or planar for a *rack cutter*) is constrained to roll with the pitch circle of the gear blank while the cutter also reciprocates to cut the tooth spaces on the blank. Instead of the reciprocating rack, a rotating toothed worm cutter may be used. In the generating method, interference results in *undercutting*. Unfortunately, undercutting along the dedendum will result in a weaker tooth near the base. If the minimum number of pinion teeth given in Table 7.1 is observed, such undercutting is avoided.

Figure 7.7 Large spur gear being form cut on a milling machine using a finger-like form cutter, one tooth space at a time. (*Courtesy of Kreiter Gear and Machine, Inc., Houston, TX.*)

Figure 7.8 Close-up of the same spur gear. Note chatter marks of form cutter. (*Courtesy of Kreiter Gear and Machine, Inc., Houston, TX.*)

Tools for generating gear teeth are of three types: a rotating worm-shaped cutter called a *hobbing tool* (see Figs. 7.9, 7.10, and 7.11), a reciprocating pinion-shaped cutter (see Fig. 7.12a), and a reciprocating rack-shaped cutter (see Fig. 7.12b).

The reader is referred to other texts on gears and gear forming for more detailed descriptions of these processes.

Figure 7.9 Single helical gear being completed on a Gould and Eberhardt gear hobber. (*Courtesy of Kreiter Gear and Machine, Inc., Houston, TX.*)

Figure 7.10 Spur gear being hobbed on a gear hobber similar to that in Fig. 7.8. Note partially cut teeth all around, indicating method of generation. (*Courtesy of Kreiter Gear and Machine, Inc., Houston, TX.*)

Figure 7.11 Double helical pinion just completed on a gear hobber. (*Courtesy of Kreiter Gear and Machine, Inc., Houston, TX.*)

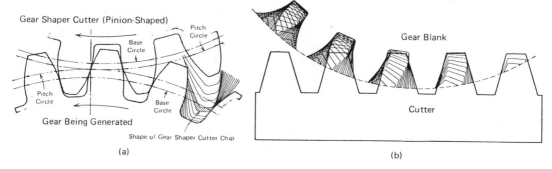

Figure 7.12 Generating gear teeth on a blank; (a) reciprocating pinion cutter; (b) rack cutter.

7.4 GEAR TRAINS

One of the major purposes for a mechanism is to transmit motion from one location to another, usually changing that motion in the process. In many instances one desires to transmit the rotation of one shaft into the rotation of another shaft. If these shafts are parallel and there is a nonlinear relationship between their rotations, either a planar function generator linkage or a cam with an oscillating follower would be considered (particularly if the output shaft is not required to rotate completely around through 360°).

In many instances, however, a constant relationship or *angular velocity ratio* is required between input and output shafts. For example, a design may call for a 2700-rpm counterclockwise input to a machine. Since most standard motors are 1800-rpm clockwise, a set of spur gears (an example of a *simple* gear train, one that has only one gear on each axis) between the parallel shafts of the motor and the machine (Fig. 7.13) would serve the purpose.

Since the pitch velocities of gears 2 and 3 ($\mathbf{V}_{P2}$, $\mathbf{V}_{P3}$) must be equal

$$\mathbf{V}_{P2} = \dot{\mathbf{z}}_2 = r_2\omega_2 ie^{i0°} = \mathbf{V}_{P3} = \dot{\mathbf{z}}_3 = r_3\omega_3 ie^{i\pi} \tag{7.11}$$

or

$$\frac{\omega_3}{\omega_2} = -\frac{r_2}{r_3} = -\frac{30 \text{ mm}}{20 \text{ mm}} \tag{7.12}$$

Thus if $\omega_2 = 1800$ rpm cw,

$$\omega_3 = -\tfrac{3}{2}\omega_2 = -\tfrac{3}{2}(-1800) = 2700 \text{ rpm ccw}$$

It can be seen that the angular velocity ratio ω_{in}/ω_{out} is inversely proportional to either the pitch-radius ratio, the pitch-diameter ratio, the pitch-circumference ratio, or the ratio of the number of teeth (since each gear must have the same circular pitch)

$$-\frac{\omega_{in}}{\omega_{out}} = \frac{r_{out}}{r_{in}} = \frac{d_{out}}{d_{in}} = \frac{c_{out}}{c_{in}} = \frac{N_{out}}{N_{in}} \tag{7.13}$$

where c_{out} and c_{in} are the respective pitch circumferences. For this case, the minus sign is present due to the opposite direction of rotation of the two gears where the centers of the gears are on opposite sides of the common tangent of the pitch circles.

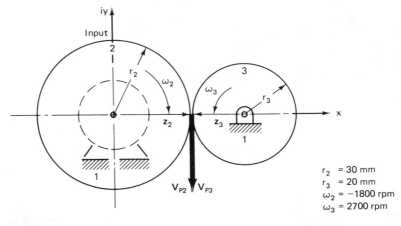

Figure 7.13 The angular speeds of a pair of spur gears in mesh are inversely proportional to their pitch-radii.

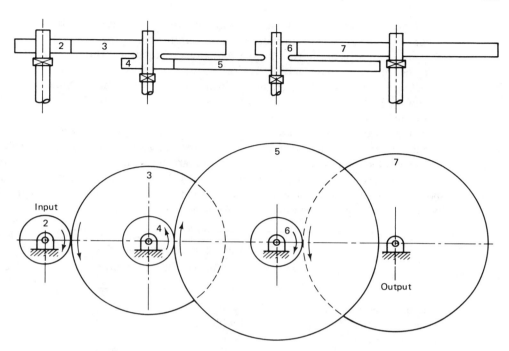

Figure 7.14 A compound gear train has one or more gears on each of several parallel axes.

Suppose that a design calls for an angular velocity ratio of 60:1. One set of gears with diameters of 60 units and 1 unit may very well be impractical due to space limitations (and perhaps cost). An *ordinary gear train** is a promising alternative (see Fig. 7.14). The overall angular velocity ratio ω_2/ω_7 can be determined as

$$\frac{\omega_2}{\omega_7} = \frac{\omega_2}{\omega_3}\left(\frac{\omega_3}{\omega_4}\right)\frac{\omega_4}{\omega_5}\left(\frac{\omega_5}{\omega_6}\right)\frac{\omega_6}{\omega_7} \tag{7.14}$$

where

$$\frac{\omega_2}{\omega_3} = -\frac{N_3}{N_2}, \qquad \frac{\omega_3}{\omega_4} = 1, \qquad \frac{\omega_4}{\omega_5} = -\frac{N_5}{N_4}$$

$$\frac{\omega_5}{\omega_6} = 1, \qquad \frac{\omega_6}{\omega_7} = -\frac{N_7}{N_6}$$

or

$$\frac{\omega_2}{\omega_7} = -\left(\frac{N_3}{N_2}\frac{N_5}{N_4}\frac{N_7}{N_6}\right) \tag{7.15}$$

* The centers of each gear are pinned through bearings to ground. Thus the centers of all gears are "fixed" to ground.

The minus sign in Eq. (7.15) can be confirmed visually by drawing arrows at the point of contact of each gear set to indicate direction of rotation. As Fig. 7.14 indicates, the input and output gears have opposite directions of rotation. This is a *compound gear train,* one that has two or more gears on one or more axes.

One can obtain a 60:1 angular velocity ratio with several combinations of N_2 and N_7. For example, tooth ratios[†] of $N_3/N_2 = 3$, $N_5/N_4 = 4$, and $N_7/N_6 = 5$ would provide the desired ratios.

This example provides us with a general rule for angular velocity ratios for compound gear trains. Notice in Eq. (7.15) that the numerator contains the product of the numbers of teeth on the gears that are driven gears; the denominator contains the product of the numbers of teeth on the drive gears: gear 2 drives gear 3, gear 4 drives gear 5, and so on. Also, ω_2 is the angular velocity of the driver gear and ω_7 is the angular velocity of the driven gear. Thus in general for gear sets with *fixed centers*

$$\left| \frac{\omega_{\text{driver}}}{\omega_{\text{driven}}} \right| = \frac{\text{product of numbers of teeth on driven gears}}{\text{product of numbers of teeth on driver gears}} \qquad (7.16)$$

This expression also holds for other types of gears (e.g., bevel, worm)[*] as long as the centers are fixed. The sign associated with Eq. (7.16) can be determined visually from a drawing like Fig. 7.14.

Example 7.1

Figure 7.15 is a typical runaway escapement mechanism described in [119]. The verge is an oscillating mass. It interferes with the motion of the escape wheel as the wheel attempts to rotate due to the torque, τ, applied to the ordinary gear train. The escape wheel cannot rotate freely because it continually collides with the oscillating verge. If τ is maintained at a constant level, the escape wheel will rotate at a constant average angular velocity and so will the gear train. Thus, the motion of the gears can be used as a mechanical analog of time.

The numbers of teeth on each gear are shown in the figure. According to [119], with an input torque τ of .022 N·M, the escape wheel angular velocity is 1000 rpm.

(a) Which direction will the escape wheel rotate?

(b) What will be the angular velocity (and direction) of gears A and C?

(c) If gear C is to be used to drive the second hand of a clock, what should be the rotational speed of the escape wheel?

Solution   Using Eq. (7.16)

$$\left| \frac{\omega_A}{\omega_F} \right| = \frac{N_F N_D N_B}{N_E N_C N_A}$$

Notice from inspection that the escape wheel rotates clockwise. (Use the methods of the arrows to determine angular direction as shown in Fig. 7.14.) Thus,

$$\omega_A = -\omega_F(.01209)$$

$$= 12.09 \text{ rpm ccw}$$

[†] "Tooth ratio" is a compact version of "ratio of numbers of teeth."

[*] Sprockets and pulleys are included also but, for the latter, instead of the number of teeth, the diameter or circumference would be used.

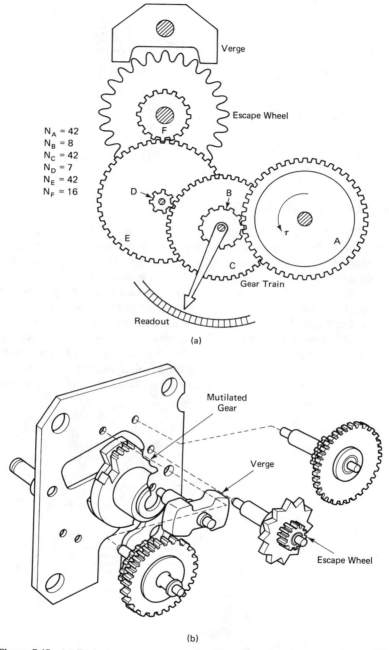

$N_A = 42$
$N_B = 8$
$N_C = 42$
$N_D = 7$
$N_E = 42$
$N_F = 16$

Figure 7.15   (a) Typical runaway escapement; (b) exploded view of the escapement assembly.

Also

$$\omega_C = +\omega_F \frac{N_F}{N_E} \frac{N_D}{N_C}$$

$$= \omega_F(.0635)$$

$$= 63.48 \text{ rpm cw}$$

If ω_C is to be 1 rpm cw,

$$\omega_F = 1/(.0635)$$

$$= 15.748 \text{ rpm cw}$$

7.5 PLANETARY GEAR TRAINS

An alternative way of using gears to transmit rotary motion is to detach one or more of the gear centers from ground and allow them to rotate about one or more fixed gears resulting in planetary motion. Figure 7.16 shows the simplest planetary (epicyclic) gear train where gear 2 (the *sun gear*) is fixed to ground and arm 4 carries the center of gear 3 (the *planet gear*) in a circle about gear 2. Notice that this figure is a kinematic inversion (Sec. 3.1) of Fig. 7.13. Arm 4 is equivalent to the fixed frame 1 in Fig. 7.13, whereas gear 2 is fixed here. Both inversions have a single degree of freedom.

If a known rotation speed is applied to arm 4 of Fig. 7.16, what is the *absolute* rotation of gear 3 (ω_{31})? Note that the relative rotation of gear 3 with respect to arm 4 does not change with inversion, but the absolute rotation with respect to ground (link 1) does change. The absolute rotation of the planetary gear ω_{31} may be derived from the following angular velocity expression:

$$\omega_{31} = \omega_{41} + \omega_{34} \tag{7.17}$$

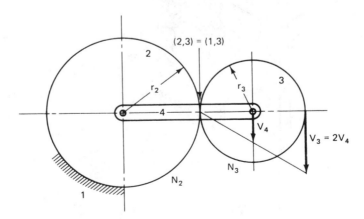

Figure 7.16 Epicyclic gear train.

Dividing through by ω_{41} and noticing that $\omega_{42} = \omega_{41}$, we obtain

$$\frac{\omega_{31}}{\omega_{41}} = 1 + \frac{\omega_{34}}{\omega_{42}}$$

or   (7.18)

$$\frac{\omega_{31}}{\omega_{41}} = 1 - \frac{\omega_{34}}{\omega_{24}}$$

Notice that the ratio ω_{34}/ω_{24} is the ratio of angular velocities with respect to the arm; this means that in this term we may consider the arm fixed and therefore treat the mechanism as an ordinary gear train in this kinematic inversion. Thus Eq. (7.16) applies:

$$\frac{\omega_{34}}{\omega_{24}} = -\frac{N_2}{N_3}$$

Thus

$$\omega_{31} = \omega_{41}(1 + N_2/N_3), \qquad \text{or } \omega_{31} = \omega_{41}(1 + r_2/r_3) \qquad (7.19)$$

Equation (7.19) can also be derived by observing the instant center between links 2 and 3 (2,3) (see Fig. 7.16):

$$V_4 = \omega_{41}(r_2 + r_3), \qquad V_3 = 2\omega_{41}(r_2 + r_3), \qquad \omega_{31} = \frac{V_3}{2r_3} = \frac{2\omega_{41}(r_2 + r_3)}{2r_3}$$

$$\omega_{31} = \omega_{41}\left(\frac{r_2}{r_3} + 1\right) = \omega_{41}\left(1 + \frac{N_2}{N_3}\right) \qquad\qquad (7.20)$$

Another way to verify Eq. (7.19) is to use a superposition of two independent motions. Temporarily fix gear 3 rigidly to arm 4, disregarding gear mesh. Then, as arm 4 makes one revolution, so will gear 3. However, in reality, gear 3 has been rolling on the pitch surface of gear 2. This causes it to rotate an additional N_2/N_3 times. Therefore, while arm 4 made 1 revolution, gear 3 had made $1 + N_2/N_3$ revolutions, which agrees with Eq. (7.20). This is the basis for the tabular method described later in this chapter. Figure 7.17 shows another simple planetary form in which gear 3 rotates inside gear 2 (a hypocyclic gear set). Here $\omega_{34}/\omega_{24} = +N_2/N_3$, so that

$$\omega_{31} = \omega_{41}\left(1 - \frac{N_2}{N_3}\right) \qquad\qquad (7.21)$$

Planetary gear trains are often employed to make a more compact gear reducer than an ordinary compound gear train. Another important application of the planetary gear train is to make use of the two degrees of freedom of this mechanism when the sun gear is released from ground. Figure 7.18 shows the same epicyclic train as Fig. 7.16, but the sun gear (link 2) is not fixed to ground. This two-degree-of-freedom mechanism requires two input conditions to determine completely the motion of the gear train. If rotations of the sun gear ω_{21} and the arm ω_{41} are given, an expression for the absolute angular veloc-

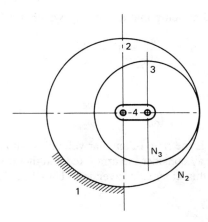

Figure 7.17 Hypocyclic gear train.

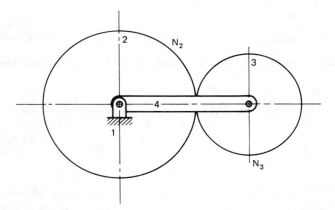

Figure 7.18 Two-degree-of-freedom epicyclic gear train.

ity of the planetary gear ω_{31} can be derived. Again, relative angular velocity relationships will be used.

$$\omega_{31} = \omega_{41} + \omega_{34} \tag{7.22}$$

$$\omega_{21} = \omega_{41} + \omega_{24} \tag{7.23}$$

or

$$\omega_{34} = \omega_{31} - \omega_{41} \tag{7.24}$$

$$\omega_{24} = \omega_{21} - \omega_{41} \tag{7.25}$$

Dividing Eq. (7.24) by Eq. (7.25) yields

$$\frac{\omega_{34}}{\omega_{24}} = \frac{\omega_{31} - \omega_{41}}{\omega_{21} - \omega_{41}} \tag{7.26}$$

Notice that the left side of Eq. (7.26) represents the ratio of the angular velocities of the planetary and the sun gear relative to the arm (a kinematic inversion which is an ordinary gear train) so that Eq. (7.16) may be used.

$$\frac{\omega_{31} - \omega_{41}}{\omega_{21} - \omega_{41}} = \frac{\omega_{34}}{\omega_{24}} = -\frac{N_2}{N_3} \tag{7.27}$$

Before solving for ω_{31}, a second general expression may be derived from Eq. (7.26). In general, if the planetary gear (link 3) is considered to be the *last* gear in the train, the sun gear (link 2) the *first* gear and link 4 the *arm*, then Eq. (7.26) may be written as

$$\boxed{\frac{\omega_{LA}}{\omega_{FA}} = \frac{\omega_L - \omega_A}{\omega_F - \omega_A}} \tag{7.28}$$

where* ω_{LA} = angular velocity of the last gear relative to the arm

ω_{FA} = angular velocity of the first gear relative to the arm

ω_L = absolute angular velocity of the last gear

ω_F = absolute angular velocity of the first gear

ω_A = absolute angular velocity of the arm

Equation (7.28) may only be employed when the first gear and the last gear are mounted on shafts whose bearings are fixed to ground. Also, the first and last gears must mesh with each other directly or through intermediate gears whose centers orbit with the arm (since more than one planetary gear set may be part of a gear system). This rule will be illustrated, for instance, in Example 7.2.

Returning to Eq. (7.27) and Fig. 7.18, the absolute velocity of the planetary gear may be obtained by solving for ω_{31}:

$$\omega_{31} = -\frac{N_2}{N_3}\omega_{21} + \left(1 + \frac{N_2}{N_3}\right)\omega_{41} \tag{7.29}$$

Equation (7.29) shows that the total absolute angular velocity of the planetary gear is the sum of two components that represent two degrees of freedom of the mechanism. The contribution from the rotation of the sun gear (the arm being fixed) is $(-N_2/N_3)\omega_{21}$, which agrees with Eq. (7.13). The contribution due to the rotation of the arm alone (the sun gear being fixed) is $(1 + N_2/N_3)\omega_{41}$, in agreement with Eq. (7.19).

Type Synthesis of Planetary Gear Trains

If a planetary gear train is suggested during the conceptual phase of mechanism design, numerous types of gear train topologies are possible. Graph theory has been demon-

---

* A single-subscripted angular velocity is an absolute angular velocity.

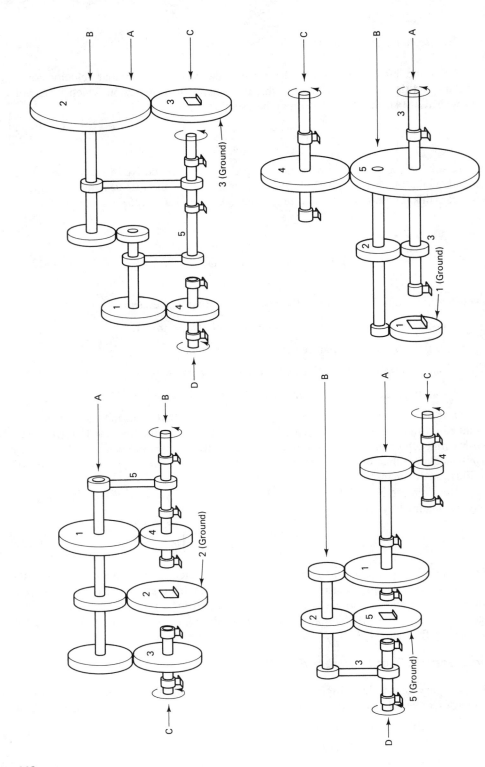

Figure 7.19 Four distinct planetary gear trains with 5 links. Courtesy of Dan Olson, Univ. of Rhode Island.

strated to be useful for enumerating both single- and two-degree-of-freedom planetary gear trains [116-118]. For example, all of the distinct planetary gear trains with five links are shown in Fig. 7.19 [118]. No dimensions are to be implied by these figures, and only epicyclic forms are shown. Also, the planet gears are typically doubled or tripled in a symmetric fashion around the sun gears for load sharing and to prevent imbalance.

Many options exist for type selection for a particular task. Once that choice is made, then three dimensional synthesis methods are presented in this chapter: *the formula, the tabular,* and *the instant center methods.*

7.6 THE FORMULA METHOD

Equations (7.16) and (7.28) can be used to solve any planetary gear problem. If one logically describes the first gear as the driver and the last gear as the driven gear, then

$$\left| \frac{\omega_{LA}}{\omega_{FA}} \right| = \left| \frac{\omega_L - \omega_A}{\omega_F - \omega_A} \right| = \frac{\text{product of numbers of teeth on driver gears}}{\text{product of numbers of teeth on driven gears}}$$

$$(7.30)$$

We will call this approach the *formula method* and illustrate it with the following examples.

Example 7.2

Figure 7.20 shows a planetary gear train that has two inputs: sun gear 2 rotates at 50 rad/sec and arm 6 rotates at 75 rad/sec, both clockwise as viewed from the left. Determine the magnitude and direction of the angular velocity of the sun gear, ω_{51}.

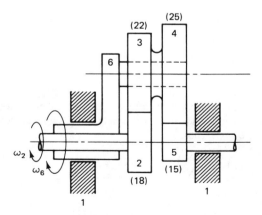

Figure 7.20 Compound planetary gear train with two inputs and one output (see Examples 7.2 and 7.5). See Fig. 7.25 for an end view sketch of this gear train.

Solution   Recall that Eqs. (7.28) and (7.30) are valid for the planetary gear train:

$$\frac{\omega_{LA}}{\omega_{FA}} = \frac{\omega_L - \omega_A}{\omega_F - \omega_A}$$

If gear 2 is the first (driver) gear, 5 the last (driven) gear, and 6 the arm, looking at the gear system from the right side so that ω_2 and ω_6 are ccw and therefore positive, we obtain

$$\frac{\omega_{56}}{\omega_{26}} = \frac{\omega_5 - \omega_6}{\omega_2 - \omega_6} = \frac{\omega_5 - 75}{50 - 75}$$

Equation (7.16) may be utilized to replace the left-hand side of this equation (since the ratios of angular velocities are with respect to the arm):

$$\frac{\omega_{\text{driven}}}{\omega_{\text{driver}}} = \frac{\omega_{56}}{\omega_{26}} = \frac{N_2 N_4}{N_3 N_5}$$

Therefore,

$$\frac{\omega_{56}}{\omega_{26}} = \frac{(18)(25)}{(22)(15)} = \frac{\omega_5 - 75}{-25}$$

and

$$\omega_5 = \frac{-(25)(25)(18)}{(22)(15)} + 75 = -34.09 + 75$$

$$\omega_5 = 40.91 \text{ rad/sec}$$

which is positive, and therefore has the same sense as the inputs.

Example 7.3

In the gear train of Fig. 7.21 the inputs are the sun gear 5 and the ring gear 2. For given angular velocities of $\omega_{51} = 300$ rpm and $\omega_{21} = 500$ rpm (both counterclockwise as seen from the right), find the resulting rotation of arm 6.

Solution If gear 2 is considered the first and the driver, gear 5 the last and the driven, then from Eq. (7.30)

$$\frac{\omega_{26}}{\omega_{56}} = \frac{\omega_2 - \omega_6}{\omega_5 - \omega_6} = -\frac{N_3 N_5}{N_2 N_4}$$

$$\frac{500 - \omega_6}{300 - \omega_6} = -\frac{(45)(48)}{(120)(27)}$$

Notice that there is a minus sign in front of the product of the numbers of teeth because, if arm 6 is held fixed, gear 2, gear 3, and gear 4 will rotate in the same direction, while gear 5 rotates in the opposite direction. Therefore,

$$\omega_6\left(1 + \frac{N_3 N_5}{N_2 N_4}\right) = \omega_5 \frac{N_3 N_5}{N_2 N_4} + \omega_2$$

$$\omega_6\left(1 + \frac{(45)(48)}{(120)(27)}\right) = 300\left(\frac{(45)(48)}{(120)(27)}\right) + 500$$

and $\omega_6 = 420$ rpm ccw as viewed from the right.

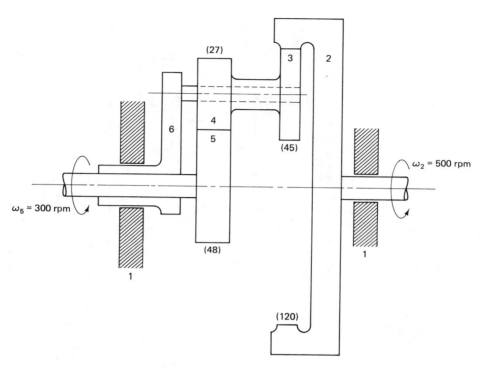

Figure 7.21   Two-degree-of-freedom planetary gear train; 2 and 5 are inputs, 6 is output (see Examples 7.3, 7.6, and 7.8). See Fig. 7.26 for an end view sketch of this gear train.

Example 7.4

This example is similar to Example 7.3 but the number of degrees of freedom of the gear train is reduced to one by adding a fixed ring gear in mesh with planetary gear 4 (Fig. 7.22). If $\omega_{21} = 500$ rpm ccw as seen from the right, what is the magnitude and direction of ω_{51}?

Solution   We can begin similarly to Ex. 7.3 with gear 2 the first gear and gear 5 the last gear,

$$\frac{\omega_{26}}{\omega_{56}} = \frac{\omega_2 - \omega_6}{\omega_5 - \omega_6} = -\frac{N_3 N_5}{N_2 N_4} = -\frac{(45)(48)}{(120)(27)} = -\frac{2}{3} = -0.67$$

We seem to be stuck here with one equation and two unknowns, but we have not included any information regarding gear 7. Therefore, another equation is possible. If gear 7 is specified as the last gear and gear 2 the first, then

$$\frac{\omega_{26}}{\omega_{76}} = \frac{\omega_2 - \omega_6}{\omega_7 - \omega_6} = \frac{N_3 N_7}{N_2 N_4}, \qquad \text{where } N_7 = N_5 + 2N_4 = 48 + 54 = 102$$

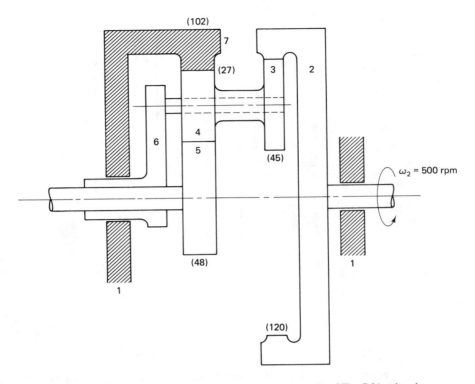

Figure 7.22 The degrees of freedom of the planetary gear train of Fig. 7.21 reduced to one by the addition of fixed ring gear 7 (see Examples 7.4 and 7.7). See Fig. 7.27 for an end view sketch of this gear train.

and

$$\frac{500 - \omega_6}{0 - \omega_6} = \frac{(45)(102)}{(120)(27)}, \qquad \omega_6 = \omega_6 \frac{(45)(102)}{(120)(27)} - 500$$

$$\omega_6 = \frac{-500}{1 - [(45)(102)/(120)(27)]}$$

$$= -1200.0 \text{ rpm (cw as viewed from the right, because we entered the counterclockwise } \omega_2 \text{ as } 500)$$

Thus

$$\frac{500 - (-1200)}{\omega_5 - (-1200)} = -0.67, \qquad 1700 = -(0.67)\omega_5 - 800, \qquad \omega_5 = \frac{1700 + 800}{-0.67}$$

and $\omega_5 = -3750$ rpm in the same sense as ω_6.

Notice that Exs. 7.2, 7.3, and 7.4 were solved essentially the same way, by combined use of Eqs. (7.16), (7.28), or (7.30). Rather than dealing with the planetary gear train all at once (which is difficult to visualize), the total motion has been separated so that the left side of Eq. (7.28) consists of a ratio of the product of the number of teeth on the driven gears to that of the drive gears in an ordinary compound gear train, while the right side of the equation is a ratio of the difference of the absolute angular velocity of the first gear and the arm to that of the last gear and of the arm. Thus Eq. (7.30) summarizes the *formula method*, applicable to any planetary gear train with either one or two degrees of freedom.

Quite often a gear system may contain combinations of one or more compound gear trains and one or more planetary gear trains. To avoid confusion, each component gear train should be analyzed by itself. One must be able to visualize how the gear train works so as to recognize gear trains with separate output components. The way to start is to identify the shafts whose centers move (carried around by arms about fixed shafts). The gears that are carried by the moving shaft then make up that planetary unit of the gear train. Figure 7.23 can be used as a guide in gear train analysis. Notice that this procedure

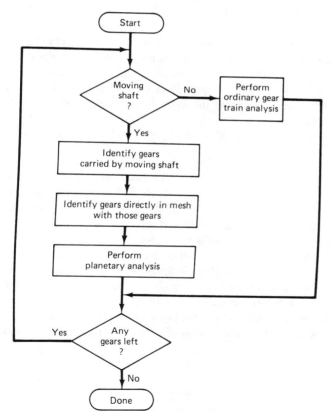

Figure 7.23 Flowchart of general computer program for kinematic analysis of gear trains. (*Suggested by Don Riley, University of Minnesota.*)

is independent of which technique one chooses to use for planetary-gear-train analysis. A second method of analysis, the tabular method, follows.

7.7 THE TABULAR METHOD

A second method of analysis of gear trains, the *tabular method,* is based on kinematic inversion—two easily describable parts of the total motion are analyzed separately, then added together: (1) motion with all components rigidly fixed to the rotating arm, and (2) motion of all components relative to the arm. This method, which is equivalent to the formula approach, formalizes the superposition of the two components by the following steps:

1. Disconnect any gears from ground (if there are any fixed to ground) and fix all gears rigidly to the rotating arm.
2. *Motion with arm.* Rotate the arm with the rigidly attached gears by a number of revolutions proportional to the angular velocity of the arm.* (If the angular speed is unknown, rotate the arm by x revolutions; x will be determined later in the analysis.)
3. *Motion relative to arm.* Now the arm is assumed to be in its "final" orientation but some components of the rest of the drive are not. Therefore, unlock the gears from the arm and while holding the arm fixed, rotate the rest of the drive back so that the total rotation (step 2 + step 3) of one or more[†] of the gears matches their given rotations (again use the same linear relationship between angular velocity and number of rotations). The total number of rotations of each gear may be found by algebraically adding its numbers of rotations in steps 2 and 3.

The epicyclic train of Fig. 7.24, which may be a subsystem of a more complex gear train, will be used to illustrate this method (refer to Table 7.2). Let us say that the arm rotates at 20 rpm cw and the number of teeth of the gears are $N_2 = 40$ and $N_3 = 20$. For each gear and arm in the planetary (sub) system a column appears in the table. The rows represent the superposition of motions.

Steps 1 and 2 are entered into Table 7.2 as motion with arm. Thus the arm as well as gears 2 and 3 rotate by 20 (revolutions). Note that clockwise rotation is taken as positive here. The arm is required to have a total motion of 20 revolutions; so no further motion is necessary and zero is entered in the second row (motion relative to arm) and 20 in the third row (total motion) under arm 4. At this point in this analysis, any other total motion information must be used. The total motion of gear 2 is zero; so -20 should

---

* This relies on a proportional relationship between the angle of rotation θ and the angular velocity ω: $\theta = \omega t$, i.e., the angular velocity is assumed constant for the purpose of kinematic analysis.

† Two if the motion of the arm is not known (x was used in step 2).

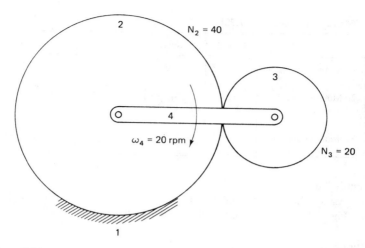

Figure 7.24   Epicyclic planetary gear set kinematically analyzed by the tabular method (see Table 7.2).

TABLE 7.2   (SEE FIG. 7.24)

	Gear 2	Gear 3	Arm 4
1. Motion with arm	20	20	20
2. Motion relative to arm	-20	$+20(N_2/N_3)$	0
3. Total motion	0	$20(1 + N_2/N_3)$	20

be inserted as the value of the motion relative to arm under gear 2. If gear 2 has -20 revolutions with the arm considered fixed, gear 3 has $+20 \times N_2/N_3$ or $+40$ revolutions (relative to the arm) in this intermediate step. Thus the total motion of gear 3 is 60 rev cw. This result agrees with the formula method [Eq. (7.19)].

The following examples will show that this technique may be applied to the same gear trains as Examples 7.2 to 7.4. Notice that the same tabular procedure is repeated but involving a greater number of components than those in Table 7.2.

Example 7.5

Referring again to Fig. 7.20, determine the magnitude and direction of the angular velocity of the output shaft connected to sun gear 5 with the same input data as Example 7.2.

Solution   Table 7.3 shows the tabular solution of this example. In the first row, motion with arm, we enter 75 revolutions for each gear and the arm. The total motion of gear 2 is 50, so in the motion-relative-to-arm step, the second row of Table 7.3, we enter -25 for gear 2. The relative angular velocities of gears 3 to 5 are determined from gear 2 in consecutive steps, so that the total motion of gear 5 is $\omega_5 = 75 - 25(N_2/N_3)(N_4/N_5) = 40.91$ rad/sec cw (as seen from the left side).

TABLE 7.3 (SEE FIG. 7.20)

	Gear 2	Gear 3	Gear 4	Gear 5	Arm 6
1. Motion with arm	75	75	75	75	75
2. Motion relative to arm	−25	$-25\left(\dfrac{-N_2}{N_3}\right)$	$25\left(\dfrac{N_2}{N_3}\right)$	$25\left(\dfrac{N_2}{N_3}\right)\left(\dfrac{-N_4}{N_5}\right)$	0
3. Total motion	50			$75 - 25\left(\dfrac{N_2}{N_3}\right)\left(\dfrac{N_4}{N_5}\right)$	75

Example 7.6

For angular velocities of the sun gear 5 of 300 rpm and the ring gear 2 of 500 rpm (see Fig. 7.21 and Ex. 7.3), find the rotation of the arm 6.

TABLE 7.4 (SEE FIG. 7.21)

	Gear 2	Gear 3	Gear 4	Gear 5	Arm 6
1. Motion with arm	x	x	x	x	x
2. Motion relative to arm	$\left(\dfrac{N_5}{N_4}\right)\left(\dfrac{N_3}{N_2}\right)(x-300)$	$\left(\dfrac{N_5}{N_4}\right)(x-300)$	$\left(\dfrac{N_5}{N_4}\right)(x-300)$	$300 - x$	0
3. Total motion	$x + \left(\dfrac{N_5}{N_4}\right)\left(\dfrac{N_3}{N_2}\right)(x-300) = 500$			300	x

Solution   In Table 7.4 the rotation of the arm is yet unknown, so x is inserted across the first row. In the total motion row, arm 6 has x revolutions but the number of revolutions of gear 2 and 5 are both known (500 and 300, respectively). With these two knowns, x can be determined by working toward gear 2 from gear 5 (or vice versa) to establish one expression in x. The table shows that the total rotation of gear 2 (500) is equal to the expression obtained by starting with gear 5.

Thus $x + (N_5/N_4)(N_3/N_2)(x - 300) = 500$ and $x = \omega_6 = 420$ rpm ccw as viewed from the right.

Example 7.7

Referring to Fig. 7.22 (and Example 7.4), determine ω_5 for an input of $\omega_2 = 500$ rpm ccw observed from the right.

Solution   Table 7.5 shows that the rotation of the arm is unknown; so x is placed everywhere in the first row. The two known total motions of gear 2 and gear 7 are 500 and 0 revolutions, respectively.

Here we work from gear 7 to gear 2 to determine x. The following equation results: $x[1 - (N_7/N_4)(N_3/N_2)] = 500$, so that $\omega_6 = x = -1200$ rpm, cw as seen from the right. Table 7.6 completes the analysis, yielding $\omega_5 = -1200 - 1700(N_2 N_4/N_3 N_5) = -3750$, cw as seen from the right.

TABLE 7.5 (SEE FIG. 7.22)

	Gear 2	Gear 3	Gear 4	Arm 6	Gear 7
1. Motion with arm	x	x	x	x	x
2. Motion relative to arm	$-x\left(\dfrac{N_7}{N_4}\right)\left(\dfrac{N_3}{N_2}\right)$	$-x\left(\dfrac{N_7}{N_4}\right)$	$-x\left(\dfrac{N_7}{N_4}\right)$	0	$-x$
3. Total motion	$x\left(1 - \left(\dfrac{N_7}{N_4}\right)\left(\dfrac{N_3}{N_2}\right)\right) = 500$			x	0

TABLE 7.6 (SEE FIG. 7.22)

	Gear 2	Gear 3	Gear 4	Gear 5	Arm 6
1. Motion with arm	-1200	-1200	-1200	-1200	-1200
2. Motion relative to arm	1700	$1700\left(\dfrac{N_2}{N_3}\right)$	$1700\left(\dfrac{N_2}{N_3}\right)$	$-1700\left(\dfrac{N_2}{N_3}\right)\left(\dfrac{N_4}{N_5}\right)$	0
3. Total motion	500			$-1200 - 1700\left(\dfrac{N_2}{N_3}\right)\left(\dfrac{N_4}{N_5}\right)$	-1200

7.8 THE INSTANT CENTER METHOD (OR TANGENTIAL VELOCITY METHOD)

This method has been introduced briefly in connection with Fig. 7.16. Now it will be applied to the gear train of Fig. 7.20. First we draw an end view of the system, say, from the right. This is shown in Fig. 7.25. Next, we label the instant centers between inputs and ground I_{21} and I_{61}, which coincide at the main bearing. The peripheral velocities of the input links are

$$V_2 = \omega_2 r_2 = (50)(18) = 900, \qquad V_6 = \omega_6 r_6 = (75)(40) = 3000$$

where we took r_2 proportional to N_2 and, for the sake of simplicity, numerically equal to it. Then $r_6 = N_2 + N_3 = 40$. We draw these velocities to scale in Fig. 7.25.

Note that I_{26} coincides with I_{21} and I_{61}. I_{36} is at the center of the planetary shaft and I_{23} is the pitch point at the mesh of gears 2 and 3. I_{31} is found by intersecting the line of $I_{23}I_{36}$ with a line through the tips of the absolute velocity vectors $\mathbf{V}_2$ and $\mathbf{V}_6$. This is also I_{41}, because gears 3 and 4 are rigidly connected.

Observe that $\mathbf{V}_6$ is the velocity of the center of gear 4. Therefore, $V_5 = V_{I45}$ can be constructed as shown in Fig. 7.25. By similar triangles, we see that

$$\frac{V_6 - V_2}{V_6 - V_5} = \frac{r_3}{r_4} \quad \text{or} \quad \frac{3000 - 900}{3000 - V_5} = \frac{22}{25}$$

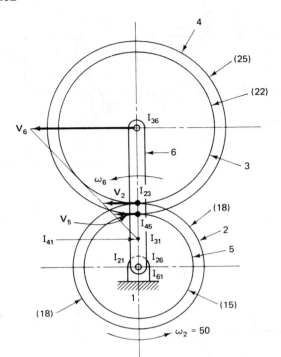

Figure 7.25 Kinematic analysis of the planetary gear train of Fig. 7.20 by way of the instant-center (or tangential-velocity) method.

and $2100 = -(22/25)V_5 + 3000(22/25)$, so that

$$V_5 = 613.64; \qquad \omega_5 = \frac{V_5}{r_5} = \frac{613.64}{15} = 40.19 \text{ rpm ccw}$$

which checks with the results obtained by the formula and tabular methods.

An advantage of the instant center method is its power in aiding the visualization of the motion of the planetary system and that it yields the direction of output rotation without the risk of a sign error. It is a truly kinematic method, essentially graphical. Furthermore, by combining it with torque and tangential force analysis, this method can be readily extended to power-flow analysis (determination of the percentage of power in each section of the gear train) in complex, multiloop planetary gear systems.

Example 7.8

In analyzing the gear train of Fig. 7.21, we start with the inputs (see Fig. 7.26). The scale of gear radii is arbitrary, because only proportions are important. We choose to make r_2 and r_3 numerically equal to the respective numbers of teeth. Then $r_2 = 120$, $r_3 = 45$, $r_6 = r_2 - r_3 = 75$, $r_4 = 27$, and $r_5 = 48$. Also, since we are interested only in the proportions of tangential velocities, the dimensional units and drawing scale of velocities are arbitrary. Therefore, we need not convert angular velocities to radians per second, but may work directly with rpm. Accordingly,

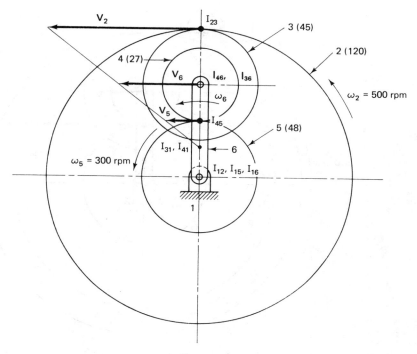

Figure 7.26 Right-end view of planetary gear train in Fig. 7.21 which illustrates its kinematic analysis by the instant-center (or tangential-velocity) method.

$$V_2 = r_2\omega_2 = (120)(500) = 60,000$$

$$V_5 = r_5\omega_5 = (48)(300) = 14,400$$

With these, the instant center $I_{31} \equiv I_{41}$ and $\mathbf{V}_6$ can be constructed. From similar triangles,

$$\frac{V_2 - V_6}{45} = \frac{V_6 - V_5}{27}, \qquad 27V_2 - 27V_6 = 45V_6 - 45V_5$$

$$72V_6 = 27V_2 + 45V_5 = (27)(60,000) + (45)(14,400)$$

$$V_6 = 31,500, \qquad \omega_6 = \frac{V_6}{r_6} = 420 \text{ rpm ccw}$$

which checks with the other two methods.

_____

Example 7.9

Let us apply the instant center method to the gear train of Fig. 7.22. Since the gear proportions are the same as those shown in Fig. 7.26, in the diagram of this example (Fig. 7.27)

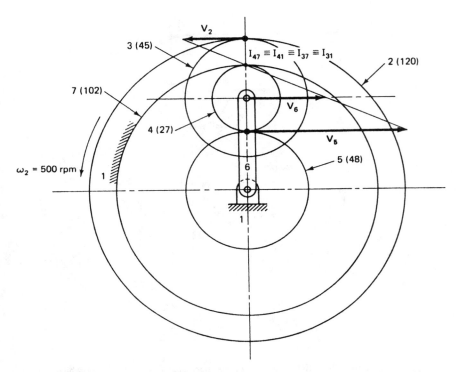

Figure 7.27 Link numbers and numbers of teeth are the same as in Fig. 7.26. In this endview of the planetary gear train of Fig. 7.22, the kinematic analysis is illustrated by the instant-center (or tangential-velocity) method.

we may leave off much of the labeling. We use a shorter scale for velocity vectors to keep the construction within the drawing. From similar triangles,

$$V_6 = V_2\left(\frac{r_4}{r_3 - r_4}\right) = 60,000\left(\frac{27}{45 - 27}\right) = 90,000,$$

$$\omega_6 = \frac{V_6}{r_6} = \frac{90,000}{75} = 1,200 \text{ rpm cw}$$

$$V_5 = V_2\left(\frac{2r_4}{r_3 - r_4}\right) = 2V_6 = 180,000, \qquad \omega_5 = \frac{V_5}{r_5} = \frac{180,000}{48} = 3750 \text{ rpm cw}$$

It is quite remarkable to compare the simplicity of calculations of this method compared to the formula or tabular methods. This leads to greater economy and accuracy, because the fewer the steps in the computation, the smaller the accumulated rounding-off error. Also, the geometric construction is a priceless aid in visualizing the kinematics of the gear train and plainly reveals the unknown directions of rotation, which avoids the risk of a sign error that can easily happen in the formula and tabular methods.

Example 7.10

Note that the power flow in the previous examples was directed from the input to the output. Let us now apply the instant center method to a more complex planetary gear system, one with internal power circulation. Figure 7.28a is a partial cross section showing the upper half of an axisymmetric arrangement. Axisymmetry makes high-speed planetary gears easier to balance and reduces individual gear tooth loads. However, it requires greater manufacturing precision to avoid binding and provide uniform load sharing.

Referring to Fig. 7.28a, link 2 is the input and link 4 the output. Observe that the power flow follows two parallel paths: $2 \rightarrow 3 \rightarrow 4$ and $2 \rightarrow 3 \rightarrow 5 \rightarrow 6 \rightarrow 4$. Furthermore, links 2, 5, 6, and 4 form a closed loop. This is the loop in which so-called power circulation may occur.

View A-A (Fig. 7.28b) pictures the input gear side: gear 2 rotates at input rpm ω_2, drives planet gear 3 orbiting with arm 5 and driving output ring gear 4. The tangential velocity of gear 2 is $V_2 = \omega_2 r$, where r is the radius of the gear and ω_2 is assumed ccw viewed from the right. This is the only known information. Therefore, to draw a velocity diagram, we must assume a temporary value for another velocity, say, the peripheral speed of the arm 5. Now we can draw the velocity diagram (not to scale) from which, since the gears are equal size,

$$V_5 = \frac{V_2 + V_4}{2} \quad \text{or} \quad 2V_5 - V_4 = V_2 \tag{7.31}$$

Equation (7.31) has two unknowns, V_4 and V_5. We derive another equation from view B-B (Fig. 7.28c) as follows. Here the sun gear is 5, with an idler 6 rotating about a fixed center and connecting gear 5 with output gear 4, which is equal in size to gear 4 of view A-A. Gears 5 and 6 are equal in size. Because of the rigidity and proportions of link 5 (refer to Fig. 7.28a), the tangential velocity of gear 5 is $\frac{1}{2}V_5$. The top of gear 5, in mesh with gear 4, must move at V_4. But the center of gear 5 is fixed, therefore (see not-to-scale sketch in Fig. 7.28c):

$$V_4 = -\tfrac{1}{2}V_5 \tag{7.32}$$

This is an apparent mismatch with the diagram in Fig. 7.28b. However, the only purpose of these not-to-scale sketches was to derive Eqs. (7.31) and (7.32). Solving these simultaneously for V_4 yields

$$-4V_4 - V_4 = V_2, \qquad V_4 = -\frac{V_2}{5}, \qquad V_5 = \frac{2}{5}V_2$$

With this we can combine the velocity diagrams of Figs. 7.28b and c to scale as shown in Fig. 7.28d. From this

$$\omega_4 = \frac{V_4}{3r} = -\frac{V_2}{15r} = -\frac{\omega_2 r}{15r} = -\frac{1}{15}\omega_2 \text{ (cw)}$$

Thus the overall velocity ratio from input to output is $-\frac{1}{15}$, a 15:1 reduction with reverse output.

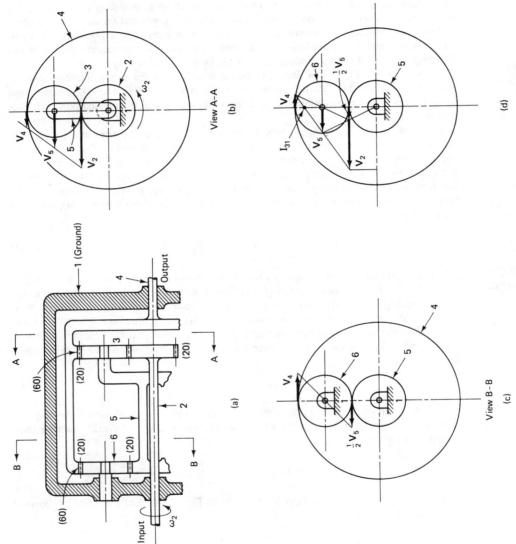

Figure 7.28 Kinematic analysis of compound planetary gear train with branching power flow by means of the instant-center (or tangential velocity) method.

The intermediate angular velocities are

$$\omega_3 = \frac{V_4 - V_2}{2r} = \frac{-\frac{6}{5}V_2}{2r} = \frac{-6V_2}{10r} = -\frac{3}{5}\omega_2 \text{ (cw)}$$

$$\omega_5 = \frac{V_5}{2r} = \frac{2V_2}{10r} = \frac{1}{5}\omega_2 \text{ (ccw)}$$

$$\omega_6 = \frac{V_4}{r} = \frac{-\frac{1}{5}V_2}{r} = -\frac{1}{5}\omega_2 \text{ (cw)}$$

Although the direction of rotation of each gear is clear from the diagram, note that in each case, to obtain the correct sign for the angular velocity, we took the tangential velocity *on top* and divided it by its distance from the instant center of the link with respect to the ground, positive downward. This stems from our decision to show the input tangential velocity V_2 on top of the input gear 2, above the instant center I_{21}. This is obvious in all cases except for gear 3. However, if we observe the location of I_{31} in Fig. 7.28d, from similar triangles we have

$$\omega_3 = \frac{V_4}{\frac{1}{3}r} = \frac{-\frac{1}{5}V_2}{\frac{1}{3}r} = -\frac{3}{5}\omega_2 \text{ (cw)}$$

as before.

7.9 TOOTH LOADS AND POWER FLOW IN BRANCHING PLANETARY GEAR SYSTEMS

The instant center method lends itself to an easily visualized study of the tooth loads, power flow, power branching, and power circulation in planetary gear systems.

Having determined angular velocities and peripheral speeds of all gears and gear-carrying links (or arms) of the system, one can readily set up a system of linear equations for all gear forces on the basis of static equilibrium applied to free-body diagrams. We will introduce this method by way of an example.

Example 7.11

Refer to the system of Fig. 7.28. The schematic shown there was quite adequate for purely kinematic analysis. However, for force, torque, and power analysis, we need a somewhat more complete schematic; one that shows bearings and their locations. Figure 7.29 was prepared with this in mind. It also shows an *OXYZ* coordinate system, to which all force, torque, and location vectors will be referred.

As indicated earlier, in addition to the one planet gear shown, there will be additional ones located axisymmetrically with respect to the one on Fig. 7.29a for rotor balance. Nevertheless, the load is often assumed to be carried by only one. This leads to conservative design, allows for unequal load sharing due to manufacturing errors, and provides for statically determinate analysis. We will, therefore, apply this procedure.

To locate the bearings in the X direction, we choose the radius r of gears 2, 5, 5, and 6 as the unit of length. The dimensions are shown in Fig. 7.29a. Input torque $\mathbf{T}_{in}$ is shown as a double-arrowheaded vector applied to input shaft 2. Assume all straight spur gears with involute profile and pressure angle δ.

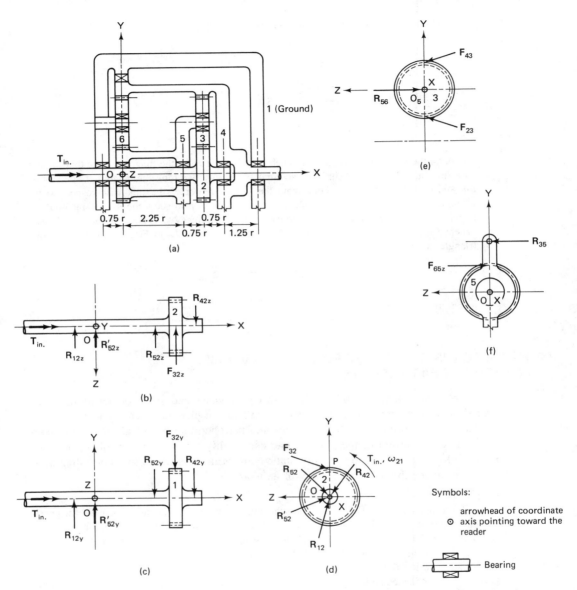

Figure 7.29 Compound planetary gear train with branching power flow; (a) cross section of gear train; (b), (c), and (d) orthographic views of free body diagram (FBD) of input shaft and gear; (e) FBD of planet gear 5; (f) FBD of planet carrier arm and the gear integral with it.

The free-body diagram (FBD) of link 2, the input shaft and sun gear, is shown in three orthographic views in Fig. 7.29b, c, and d. All forces and torques acting on this link are also shown. For example, the force exerted on the teeth of gear 2 by gear 3 is shown in true shape in view d and is labeled $\mathbf{F}_{32}$. Its y and z components, $\mathbf{F}_{32y}$ and $\mathbf{F}_{32z}$, are shown in views c and b, respectively. The bearing reactions, $\mathbf{R}_{42}$, $\mathbf{R}_{52}$, $\mathbf{R}'_{52}$, and $\mathbf{R}_{12}$, of unknown direction and magnitude, are similarly shown. Since we are dealing with straight spur gears, there are no thrust loads and, therefore, none of the forces have axial or x-directional components. The vector equations of static equilibrium for link 2 are

$$\sum \mathbf{M} = 0, \qquad \sum \mathbf{F} = 0 \tag{7.33}$$

Writing the moment equation about origin, O, we have

$$\mathbf{T}_{\text{in}} + \mathbf{r}_{12} \times \mathbf{R}_{12} + \mathbf{r}'_{52} \times \mathbf{R}'_{52} + \mathbf{r}_{52} \times \mathbf{R}_{52} + \mathbf{r}_{32} \times \mathbf{F}_{32} + \mathbf{r}_{42} \times \mathbf{R}_{42} = 0 \tag{7.34}$$

The radius vectors $\mathbf{r}_{ij}$ run from the origin O to any point on the line of action of the forces $\mathbf{R}_{ij}$ and $\mathbf{F}_{ij}$, respectively. For example, $\mathbf{r}_{32}$ runs from O to the pitch point P, where the tooth-load force $\mathbf{F}_{32}$ is acting.

We now write each of the vector products in determinant form:

$$\mathbf{r}_{12} \times \mathbf{R}_{12} = \begin{vmatrix} \mathbf{i} & \mathbf{j} & \mathbf{k} \\ r_{12x} & r_{12y} & r_{12z} \\ R_{12x} & R_{12y} & R_{12z} \end{vmatrix} = (r_{12y}R_{12z} - r_{12z}R_{12y})\mathbf{i} \\ + (r_{12z}R_{12x} - r_{12x}R_{12z})\mathbf{j} \\ + (r_{12x}R_{12y} - r_{12y}R_{12x})\mathbf{k} \tag{7.35}$$

Noting that in Eq. (7.35) only r_{12x} is nonzero, this equation reduces to

$$\mathbf{r}_{12} \times \mathbf{R}_{12} = -r_{12x}R_{12z}\mathbf{j} + r_{12x}R_{12y}\mathbf{k} \tag{7.36}$$

Since $\mathbf{r}'_{52}$ is zero,

$$\mathbf{r}'_{52} \times \mathbf{R}'_{52} = 0 \tag{7.37}$$

For the remaining terms, we have

$$\mathbf{r}_{52} \times \mathbf{R}_{52} = -r_{52x}R_{52z}\mathbf{j} + r_{52x}R_{52y}\mathbf{k} \tag{7.38}$$

We know the direction of $\mathbf{F}_{32}$. Therefore,

$$F_{32x} = 0, \qquad F_{32y} = -F_{32}\cos\delta\mathbf{j}, \qquad F_{32z} = -F_{32}\sin\delta\mathbf{k} \tag{7.39}$$

Furthermore, $r_{32z} = 0$. With these

$$\mathbf{r}_{32} \times \mathbf{F}_{32} = -r_{32y}\cos\delta F_{32}\mathbf{i} + r_{32x}\cos\delta F_{32}\mathbf{j} - r_{32x}\sin\delta F_{32}\mathbf{k} \tag{7.40}$$

Finally,

$$\mathbf{r}_{42} \times \mathbf{R}_{42} = -r_{42x}R_{42z}\mathbf{j} + r_{42x}R_{42y}\mathbf{k} \tag{7.41}$$

The force-balance equation is

$$(R_{12y} + R'_{52y} + R_{52y} - \sin\delta F_{32} + R_{42y})\mathbf{j} + (R_{12z} + R'_{52z} + R_{52z} - \cos\delta F_{32} + R_{42z})\mathbf{k} = 0 \tag{7.42}$$

Since

$$\mathbf{T}_{\text{in}} = T_{\text{in}x}\mathbf{i} \tag{7.43}$$

combining Eqs. (7.34) through (7.43), separating components and collecting terms, we obtain the following five equations from the FBD of link 1:

$$T_{\text{inx}} - r_{32y} \cos \delta F_{32} = 0 \tag{7.44}$$

$$-r_{12x}R_{12z} - r_{52x}R_{52z} + r_{32x} \cos \delta F_{32} - r_{42x}R_{42z} = 0 \tag{7.45}$$

$$r_{12x}R_{12y} + r_{52x}R_{52y} - r_{32x} \sin \delta F_{32} + r_{42x}R_{42y} = 0 \tag{7.46}$$

$$R_{12y} + R'_{52y} + R_{52y} - \sin \delta F_{32} + R_{42y} = 0 \tag{7.47}$$

$$R_{12z} + R'_{12z} + R_{52z} - \cos \delta F_{32} + R_{42z} = 0 \tag{7.48}$$

Equations (7.44) to (7.48) represent a set of five equations linear in the nine unknowns F_{32}, $\mathbf{R}_{12}$, $\mathbf{R}'_{52}$, $\mathbf{R}_{52}$, and $\mathbf{R}_{42}$. We need additional equations. They are derived similarly from FBDs of the other links. Table 7.7 shows how the set of equations is built up so that the number of equations and that of unknowns is equal. Thus the set can be solved for all tooth loads and bearing reactions.

In this section, however, we wish to deal only with power flow. Therefore, in addition to the angular velocities already known from previous sections, we only need the tangential tooth loads. Let us, therefore, concentrate on only those equations that contain these unknowns.

From the FBD of link 2, we have Eq. (7.34):

$$\mathbf{T}_{\text{in}} = T_{\text{inx}} = r \cos \delta F_{32} \quad \text{or} \quad F_{32} = \frac{T_{\text{inx}}}{r \cos \delta} \tag{7.49}$$

TABLE 7.7

Link number	Equations	Number of equations	Unknowns	Number of new unknowns	Total number of:	
					Equations	Unknowns
2	$\Sigma \mathbf{M} = 0$	3	$F_{32}, R_{12y}, R_{12z}, R'_{52y}, R'_{52z},$	9	5	9
	$\Sigma F_y = 0$	1	$R_{52y}, R_{52z}, R_{42y}, R_{42z}$			
	$\Sigma F_z = 0$	1				
4	$\Sigma \mathbf{M} = 0$	3	$R_{14y}, R_{14z}, R_{24y}, R_{24z},$	8	10	17
	$\Sigma F_y = 0$	1	$F_{34}, F_{64}, R'_{14y}, R'_{14z}$			
	$\Sigma F_z = 0$	1				
5	$\Sigma \mathbf{M} = 0$	3	$R_{35y}, R_{35z}, (R_{25y}, R_{25z})^{\text{a}}$	3	15	20
	$\Sigma F_y = 0$	1	$(R'_{52y}, R'_{52z}), F_{65}$			
	$\Sigma F_z = 0$	1				
1 (ground)						
3	$\Sigma M_x = 0$	1	$(F_{23}), (F_{43}), (R_{53y}), (R_{53z})$	0	18	20
	$\Sigma F_y = 0$	1				
	$\Sigma F_z = 0$	1				
6	$\Sigma M_x = 0$	1	$(F_{56}), (F_{46}), R_{26}$	1	21	21
	$\Sigma F_y = 0$	1	Direction of $\mathbf{R}_{26}$ is known to			
	$\Sigma F_z = 0$	1	be in the Z direction due to symmetry			

ᵃ "Unknowns" in parentheses are not new, because, for example, $R_{25y} = -R_{25y}$.

From the FBD of link 3 (Fig. 7.29e):

$$F_{43} = F_{23} \quad \text{and} \quad \mathbf{R}_{52} = R_{52z} = -2 \cos \delta F_{23} = \frac{-2T_{inx}}{r} \tag{7.50}$$

From the FBD of link 5 (Fig. 7.29f),

$$F_{65z} = 2R_{53z} = \frac{-4T_{inx}}{r} \tag{7.51}$$

From the FBD of link 6 (not shown),

$$F_{42z} = F_{26z} = \frac{4T_{inx}}{r} \tag{7.52}$$

Now, with the use of the known angular velocities of each link, we can find the absolute peripheral velocity of each gear mesh. Since all such velocities are parallel with the z axis, we need only the z components of the tooth forces. Let us use the following notation:
Power flow from link i to link j is

$$P_{ij} = \mathbf{F}_{ij} \cdot \mathbf{V}_{ij} = F_{ijz} V_{ijz} \tag{7.53}$$

where $\mathbf{V}_{ij}$ is the linear velocity of the pitch point between gears i and j.
We start with the power input to link 2:

$$P_{02} = \mathbf{T}_{in} \cdot \omega_{in} = T_{inx} \omega_1 \tag{7.54}$$

For the sake of clarity, let us use some numerical values. Let $\mathbf{T}_{in} = (10 \text{ in·lbf})\mathbf{i}$, or $T_{inx} = 10$ in·lbf, $r = 10$ in, and $\omega_1 = 15$ rad/sec. With these, referring to Figs. 7.28 and 7.29, we have

$$P_{02} = (10)(15) = 150 \text{ in·lbf/sec}$$

$$V_{23z} = (10)(15) = 150 \text{ in/sec}$$

$$V_{35z} = 2r\omega_5 = \frac{2r\omega_2}{5} = (20)(3) = 60 \text{ in/sec}$$

$$V_{34z} = 3r\omega_4 = \frac{-3r\omega_2}{15} = -(30)(1) = -30 \text{ in/sec}$$

$$V_{56z} = r\omega_5 = 30 \text{ in/sec}$$

$$V_{64z} = V_{32z} = -30 \text{ in/sec}$$

Now, for the tangential gear forces

$$F_{23z} = \frac{T_{inx}}{r} = 1 \text{ lbf}$$

$$R_{35z} = 2 \text{ lbf}$$

$$F_{34z} = -1 \text{ lbf}$$

$$F_{56z} = 4 \text{ lbf}$$

$$F_{64z} = -4 \text{ lbf}$$

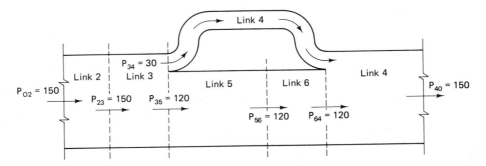

Figure 7.30 Stream diagram of branching power flow in the planetary gear train of Figs. 7.28 and 7.29. Power flow from link i to link j, P_{ij}, are shown in in·lbf/sec.

With these, we calculate the power flow:

$$P_{23} = F_{23z} V_{23z} = (1)(150) = 150 \text{ in·lbf/sec}$$

$$P_{35} = R_{35z} V_{35z} = (2)(60) = 120 \text{ in·lbf/sec}$$

$$P_{34} = F_{34z} V_{34z} = (-1)(-30) = 30 \text{ in·lbf/sec}$$

$$P_5 = F_{5z} V_{5z} = (4)(30) = 120 \text{ in·lbf/sec}$$

$$P_4 = F_{4z} V_{4z} = (-4)(-30) = 120 \text{ in·lbf/sec}$$

The stream diagram of the power flow is shown in Fig. 7.30, without losses. If losses are taken into account, say, by assigning an efficiency of $\eta = 0.9$ to each gear mesh, then at each transit there would be a 10% loss branching off in the stream diagram. It is easy to see that, in that case, the output would become

$$P_{24} = (0.9)(0.9)(0.9)(120) + (0.9)(0.9)30 = 103 \text{ in·lbf/sec}$$

or an overall efficiency of

$$\eta = \frac{103}{150} = 0.69$$

Example 7.12

Figure 7.31 shows a coupled planetary gear drive in which the spiders of both planet systems are rigidly connected to rotate together. To find the angular velocity ratios, we apply the method of instant centers.

Note that, if sun gear 2 is the input, the velocity analysis has to work its way through the unknown velocities of links 3, 4, 5, and 6 before reaching a point of known velocity: the stationary mesh point or pitch point of links 1 and 5. On the other hand, if we regard link 6 as the input, only link 5 intervenes before reaching the same zero-velocity point. The results then obtained can easily be inverted to give angular velocity ratios, tangential velocities and forces, and thus the power flow through the system from input link 2 to output link 6.

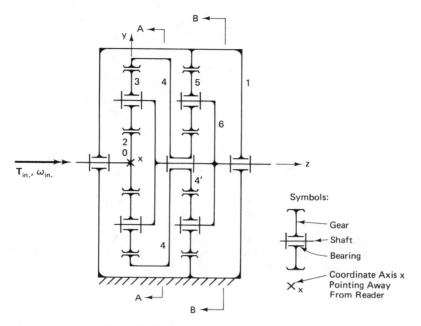

Figure 7.31   Coupled planetary gear train.

   To take advantage of this observation, we start with link 6 as the input. As before, we will "lump" all forces and power flows onto one planet gear in each set. Keep in mind, though, that, owing to axisymmetry, in actuality these forces and power flows are shared among all (two, three, or more) planets in each set more or less equally, depending on the precision of manufacture.

   Figure 7.32 is view B-B (Fig. 7.31), showing link 6. Since we seek only velocity ratios and proportions of power flow, we might as well let the input torque, input angular velocity, and the radius of all sun and planet gears be unity. Thus

$$\mathbf{T}_{in} = 1\mathbf{k} \text{ in·lbf,} \qquad \omega_6 = 1\mathbf{k} \text{ rad/sec,} \qquad \mathbf{V}_6 = -2\mathbf{i} \text{ in/sec}$$

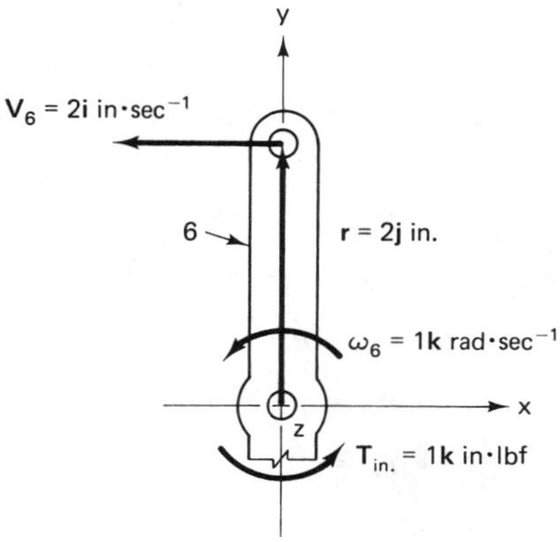

$$\mathbf{V}_6 = 2\mathbf{i} \text{ in·sec}^{-1}$$

$$r = 2\mathbf{j} \text{ in.}$$

$$\omega_6 = 1\mathbf{k} \text{ rad·sec}^{-1}$$

$$\mathbf{T}_{in.} = 1\mathbf{k} \text{ in·lbf}$$

Figure 7.32   View B-B of link 6 of Fig. 7.31.

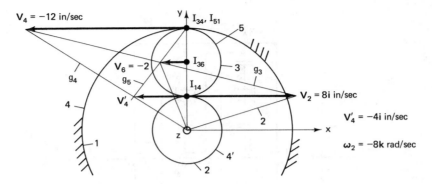

Figure 7.33 Views A-A and B-B of Fig. 7.31 superposed on one another, showing instant centers and peripheral velocities.

Next, we draw Fig. 7.33, showing gears 4' and 5, as well as stationary internal gear 1 in mesh with gear 5. Drawing in the known velocity $\mathbf{V}_6$ and the gauge line g_5 of gear 5 going through I_{51} and the tip of $\mathbf{V}_6$, we see that

$$\mathbf{V}'_4 = -4\mathbf{i} \text{ in/sec}$$

Now we observe that Fig. 7.33 also represents view A-A of Fig. 7.31 and label sun gear 2, planet gear 3, and internal ring gear 4, keeping in mind that these are in the layer behind the gears labeled previously. Noting that gears 4 and 4' are both on link 4 and drawing gauge line g_4 for this link, we see that

$$\mathbf{V}_4 = -12\mathbf{i} \text{ in/sec}$$

Then, by drawing the gauge line g_3 for link 3 through the tips of $\mathbf{V}_4$ and $\mathbf{V}_6$, we obtain

$$\mathbf{V}_2 = 8\mathbf{i} \text{ in/sec}$$

from which the output angular velocity (that of link 2) becomes

$$\omega_2 = -8\mathbf{k} \text{ rad/sec}$$

We are now ready to consider the power flow through the system. The input is received by link 6.

$$P_{\text{in}} = \mathbf{T}_{\text{in}} \cdot \omega_6 = 1\mathbf{k} \cdot 1\mathbf{k} = 1 \text{ in·lbf/sec}$$

Referring to Fig. 7.31, we see that this power is split in two and passed on from link 6 to links 3 and 5. Therefore,

$$P_{\text{in}} = P_{63} + P_{65} \tag{a}$$

Similarly, following the flow of power P_{ij} from link i to link j, we find that

$$P_{54} = P_{65} \tag{b}$$

$$P_{43} = P_{54'} = P_{54} \tag{c}$$

$$P_{43} + P_{63} = P_{32} = P_{\text{out}} = P_{\text{in}} \tag{d}$$

where we neglected all friction losses.

For tangential pitch-line forces and bearing reactions ($\mathbf{F}_{ij}$ or $\mathbf{R}_{ij}$) exerted by link i on link j, we can write equations of equilibrium from free-body diagrams. Thus, from the free-body diagram of link 6 (FBD6, not shown), we have

Equations: Total number of real equations and unknowns:

$$(\mathbf{R}_{65} + \mathbf{R}_{63})\cdot\mathbf{V}_6 = P_{\text{in}} = 1 \qquad \text{(1 eq., 2 unknowns)} \quad (1)$$

FBD5 yields: $\mathbf{F}_{54} = \tfrac{1}{2}\mathbf{R}_{65}$ (2 eqs., 3 unknowns)   (2)

From FBD4: $\mathbf{F}_{43} = \tfrac{1}{3}\mathbf{F}_{54}$ (3 eqs., 4 unknowns)   (3)

From FBD2: $\mathbf{F}_{32} = \tfrac{1}{2}\mathbf{R}_{63} - \mathbf{F}_{43}$ (4 eqs., 5 unknowns)   (4)

From FBD1: $\mathbf{F}_{32}\cdot\mathbf{V}_2 = P_{\text{out}} = 1$ (5 eqs., 5 unknowns)   (5)

Observing that all vectors are parallel with the x axis and solving Eqs. (1) to (5) simultaneously, we get

$$R_{63x} = \tfrac{1}{16} \text{ lbf}, \qquad F_{43x} = -\tfrac{3}{32} \text{ lbf}, \qquad F_{54x} = -\tfrac{9}{32} \text{ lbf}, \qquad R_{65x} = -\tfrac{9}{16} \text{ lbf},$$
$$F_{32x} = \tfrac{1}{8} \text{ lbf}$$

Forming the dot product of appropriate forces and velocities yields the power flows (all given in lbf·in/sec):

$$P_{65} = R_{65x} V_{6x} = \left(-\tfrac{9}{16}\right)(-2) = \tfrac{9}{8}$$
$$P_{63} = R_{63x} V_{6x} = \left(\tfrac{1}{16}\right)(-2) = -\tfrac{1}{8}$$
$$P_{54} = F_{54x} V'_{4x} = \left(-\tfrac{9}{32}\right)(-4) = \tfrac{9}{8}$$
$$P_{43} = F_{43x} V_{4x} = \left(-\tfrac{3}{32}\right)(-12) = \tfrac{9}{8}$$
$$P_{32} = F_{32x} V_{2x} = \left(\tfrac{1}{8}\right)8 = 1$$

Substituting these values into power-balance equations (a), (b), (c), and (d) will verify these results. The power flow is diagrammed in Fig. 7.34. Observe the negative power feedback or power circulation from link 3 to link 6.

Although they were not needed for power-flow calculations, the angular velocities of intermediate links and their angular velocity ratios with respect to that of the input link are

POWER CIRCULATION

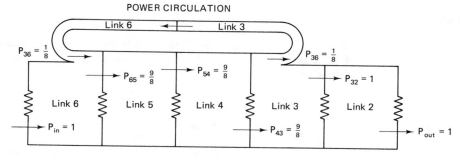

Figure 7.34 Stream diagram of inverted power flow through the compound planetary gear set of Fig. 7.31. Observe the power circulation branch of power feedback from link 2 to link 6.

easily determined from Fig. 7.33. Inspection of appropriate gauge lines and associated velocities readily yields the following values:

$$\omega_2 = 10\mathbf{k}, \qquad \omega_5 = -2\mathbf{k}, \qquad \omega_4 = 4\mathbf{k}$$

Since all angular velocity vectors are parallel to the z axis, and since the input velocity is
$\omega_6 = 1\mathbf{k}$, the velocity ratios are

$$\frac{\omega_5}{\omega_6} = -2, \qquad \frac{\omega_4}{\omega_6} = 4, \qquad \frac{\omega_3}{\omega_6} = 10, \qquad \frac{\omega_2}{\omega_6} = -8$$

Thus we see that with link 6 as the input and link 2 the output, the planetary system
of Fig. 7.31 is a "speed amplifier." If we now revert to link 2 as input, the power flow of
Fig. 7.34 is reversed. Power feedback or circulation will now occur from link 6 back to link
3. The angular velocity ratios will be (see Fig. 7.31);

$$\frac{\omega_3}{\omega_2} = \frac{10}{-8} = \frac{-5}{4}$$

$$\frac{\omega_4}{\omega_2} = \frac{4}{-8} = \frac{-1}{2}$$

$$\frac{\omega_5}{\omega_2} = \frac{-2}{-8} = \frac{1}{4}$$

$$\frac{\omega_6}{\omega_2} = \frac{1}{-8} = \frac{-1}{8}$$

Conclusion

The foregoing derivations and examples demonstrate how rather complex planetary gear
trains can be analyzed for velocity ratios, static forces, and power flow, all with straightforward readily visualized means which make an excellent tool for the practicing designer. Further practice in the application of these techniques is provided by Probs. 7.25,
7.26, and 7.27.

PROBLEMS

7.1. Figure P7.1 shows a simple gear train. If the input is provided by gear A, and the output
taken off the shaft of gear E, what is the ratio and the sign of ω_{out}/ω_{in} in terms of the number of teeth on the gears?

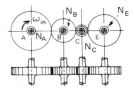

Figure P7.1

7.2. The compound gear train shown in Fig. P7.2 is attached to a motor that drives gear A at ω_{in} clockwise as viewed from below. What is the expression for the angular velocity of gear H in terms of the number of teeth on the gears? What is the direction of rotation of gear H as viewed from below?

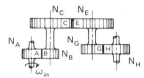

Figure P7.2

7.3. Gear A of the gear train in Fig. P7.3 is driven at 200 rpm clockwise as viewed from below. The shafts of all gears are located in grounded bearings. Determine ω_G, ω_K, and ω_I given the number of teeth on each gear as indicated in the figure. Determine the direction of rotation of these three gears as viewed from the right.

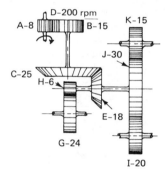

Figure P7.3

7.4. The gear train shown in Fig. P7.4 is arranged to feed lumber between the two 6-in-diameter rollers (R) into a cutting blade for the purpose of ripping the lumber. The blade is being directly driven at 500 rpm. A power takeoff which consists of the belt system shown drives the rollers for the purpose of feeding the stock. Determine the speed of the stock in ft/min and the direction of stock movement.

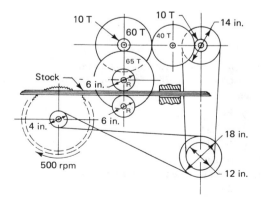

Figure P7.4

7.5. Design a simple gear train to couple the two shafts in Fig. P7.5 so that the driver will rotate at 30 rpm and the follower at 21,000 rpm in the directions indicated. Design must be based on catalog gears of from 12 to 25 teeth listed every integer, and from 26 to 60 teeth listed every other integer. Prepare a drawing of the train arrangement and label each gear.

ω_D = 30 rpm

ω_F = 21,000 rpm   **Figure P7.5**

7.6. Figure P7.6 shows a three-speed transmission with the drive shaft rotating in place at 450 rpm carrying the cluster of gears A, B, and C. The follower shaft is splined and may be shifted along its axis by means of the shift link to bring gears E, G, and H into contact with the gears on the drive shaft. Design the transmission to produce speeds of 150, 350, and 550 rpm, and describe how motion is transmitted in these three speeds. Gear teeth are to be kept within the limits of from 10 to 80 teeth.

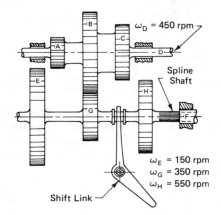

ω_E = 150 rpm
ω_G = 350 rpm
ω_H = 550 rpm

Shift Link

Figure P7.6

7.7. A gear train is required to have a train value of $(1/\sqrt{3}) \pm 0.00001$. One constraint on feasible designs is that the least number of teeth any gear can have is 15. Competing designs will be judged on the basis of cost of parts. The following unit costs can be used to estimate the total cost for your design.

Gear blanks	$ 5.00 per gear
Cutting teeth	$ 0.25 per tooth
Shafts (with bearings)	$20.00 per shaft

The following information is requested:

(a) A sketch showing the train arrangement for your design.

(b) Tooth numbers for each gear.

(c) A cost estimate.

7.8. For each planetary gear train in Fig. P7.7, determine the ratio indicated below (a) by the formula method, (b) by the tabular method, and (c) by the instant center method:

Fig. P7.7a: ω_2/ω_3 Fig. P7.7d: ω_2/ω_4
Fig. P7.7b: ω_3/ω_4 Fig. P7.7e: ω_3/ω_4
Fig. P7.7c: ω_2/ω_3 Fig. P7.7f: ω_4/ω_5

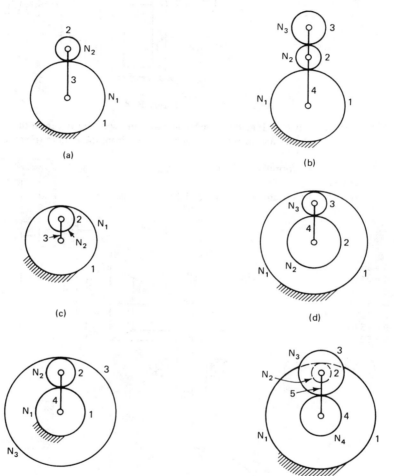

Figure P7.7

7.9. **(a)** Determine the number of teeth on gear E of the gear train hoist in Fig. P7.8 which yields a gear reduction between B and A of $\omega_B/\omega_A = 25$, given $N_B = 20$, $N_C = 80$, and $N_D = 30$.

(b) Determine the numerical value of the mechanical advantage of this gear train hoist, W/F_{in} assuming no losses.

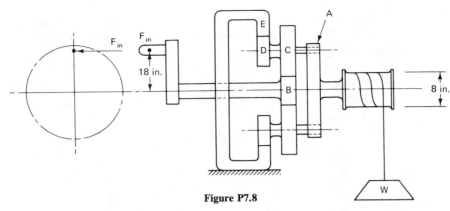

Figure P7.8

7.10. The sun gear B of Fig. P7.9 rotates at 100 rpm cw as viewed from the right. Determine the angular velocity and direction of ω_G as viewed from the bottom (note: an idler gear may be required).

  (a) Use the formula method.

  (b) Use the tabular method.

  (c) Use the instant center/tangential velocity method.

A = arm
B = 24 teeth
C = 60 teeth
D = 18 teeth
E = 102 teeth (fixed)
F = 25 teeth
G = 50 teeth

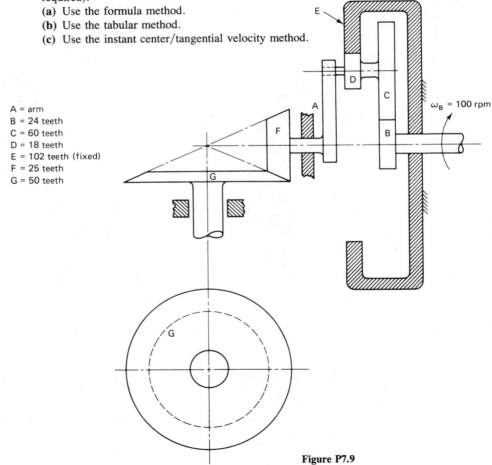

Figure P7.9

7.11. For the gear train shown in Fig. P7.10, you must choose suitable gears (F and G) to connect the shaft of gear E to the output shaft. Notice that $\omega_A = 1000$ rpm, $\omega_G = 20$ rpm in the same sense as specified.

(a) Find ω_F.

(b) What are suitable tooth numbers for gears F and G? Be careful.

(c) Should F and G be a pair of external gears or an external-internal pair?

(d) For 10 pitch gears, what would the distance between the two shafts have to be?

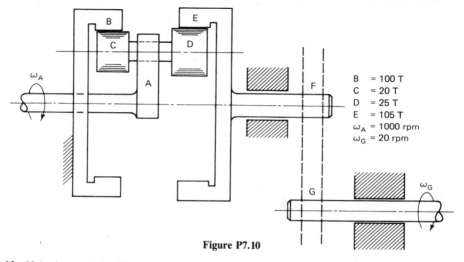

B	= 100 T
C	= 20 T
D	= 25 T
E	= 105 T
ω_A	= 1000 rpm
ω_G	= 20 rpm

Figure P7.10

7.12. If the input shaft of the planetary gear train in Fig. P7.11 is rotating at 150 rpm clockwise as viewed from the right, determine ω_E by:

(a) The formula method.

(b) The tabular method.

(c) The instant center/tangential velocity method.

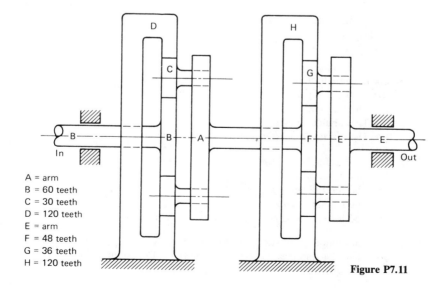

A = arm
B = 60 teeth
C = 30 teeth
D = 120 teeth
E = arm
F = 48 teeth
G = 36 teeth
H = 120 teeth

Figure P7.11

7.13. Figure P7.12 shows a gear train with two degrees of freedom. If gear 3 rotates at 100 rpm clockwise and gear 1 rotates by 200 rpm counterclockwise (both viewed from the right), find ω_6 by:
   (a) The formula method.
   (b) The tabular method.
   (c) The instant center/tangential velocity method.

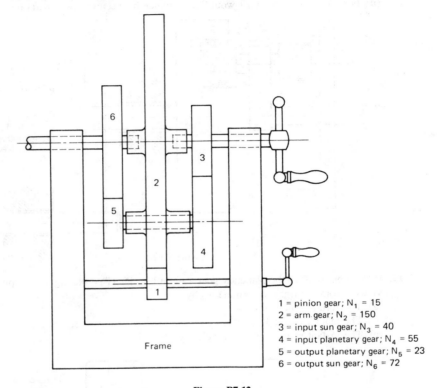

1 = pinion gear; N_1 = 15
2 = arm gear; N_2 = 150
3 = input sun gear; N_3 = 40
4 = input planetary gear; N_4 = 55
5 = output planetary gear; N_5 = 23
6 = output sun gear; N_6 = 72

Figure P7.12

7.14. For the gear train in Fig. P7.13, if $\omega_1 = 1$ rpm cw and $\omega_5 = 1$ rpm ccw as viewed from the left, find the angular velocity of the arm by:
   (a) The formula method.
   (b) The tabular method.
   (c) The instant center/tangential velocity method.

7.15. If gear A in Fig. P7.14 rotates by 72 rpm clockwise as viewed from the right, find the angular velocity of gear E by:
   (a) The formula method.
   (b) The tabular method.
   (c) The instant center/tangential velocity method.

7.16. The differential gear system in Fig. P7.15 requires angular velocities of $\omega_A = 10$ rpm cw and $\omega_B = 24$ rpm ccw as seen from the right. Determine the input angular velocity ω_D.

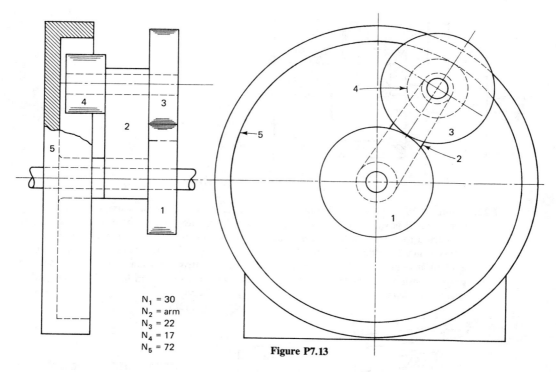

$N_1 = 30$
$N_2 = $ arm
$N_3 = 22$
$N_4 = 17$
$N_5 = 72$

Figure P7.13

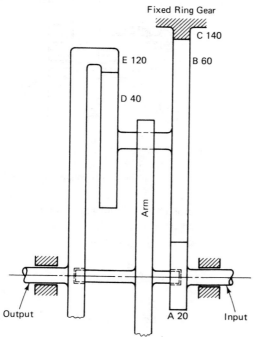

Fixed Ring Gear

C 140

E 120   B 60

D 40

Arm

Output   A 20   Input

Figure P7.14

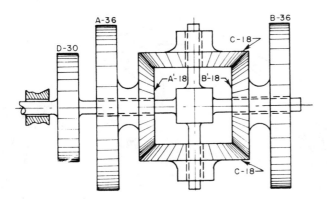

Figure P7.15

7.17. Figure P7.16 shows a variation of a planetary gear system known as Humpage's gear reduction. Gear D is stationary and acts as a bearing for the arm shaft. The arm shaft is directly connected to gear 3, while the arm itself acts as a bearing for the gear cluster B, C, E. Gears 1 and 2 are directly connected to gears F and A, respectively. A single input can be applied at either gear 1, 2, or 3 with an output occurring at each of the two remaining gears. Consider that an input of 800 rpm clockwise (as seen from the right) is applied at gear 2 and determine the resulting outputs at gears 1 and 3.

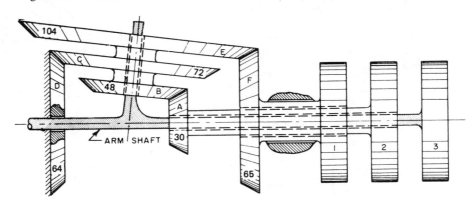

Figure P7.16

7.18. Figure P7.17 shows a planetary bevel gear train. If the driver rotates at 10 rpm cw as viewed from the right, what is ω_E?

7.19. By determinant expansion of the vector products and reference to Fig. 7.29, verify Eqs. (7.38), (7.40), and (7.41). Show all steps.

7.20. Using FBDs (free-body diagrams), verify the entries in Table 7.7 (see Fig. 7.29) for
  (a) Link 4.
  (b) Link 5.
  (c) Link 3.
  (d) Link 6.

7.21. With the help of the applicable FBDs, verify Eqs. (7.49) to (7.52) (see Fig. 7.29).

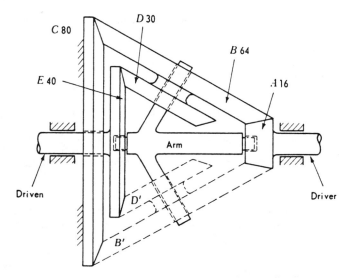

7.22. Using the data given following Eq. (7.54), and by reference to Figs. 7.28 and 7.29, verify the values of the absolute velocities V_{23z}, V_{35z}, V_{34z}, V_{56z}, and V_{54z}.

7.23. Similar to Prob. 7.22, but the task is to verify the forces R_{35z}, F_{34z}, F_{56z}, and F_{54z}.

7.24. Using the techniques of Sec. 7.9, for the system of Fig. P7.18, determine:
(a) The absolute velocities of all pitch points.
(b) All gear forces and bearing reactions.

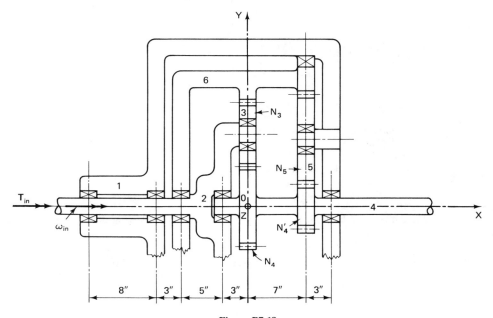

Figure P7.18

(c) The power flow; draw the power-flow stream diagram with the following data:

Number of teeth	(A)	(B)	(C)	(D)
N_3	40	48	26	30
N_4	50	60	36	40
N_4'	30	20	16	18
N_5	55	40	32	36
T_{in}	10 in·lbf i	12 in·lbf i	−22 in·lbf i	30 in·lbf i
ω_{in} (rad/sec)	10	6	15	4

In each case, link 1 is ground.

7.25. Using the method in Fig. 7.31, determine the velocity ratios and power flow for the coupled planetary gear of Fig. P7.19 with gears 1 and 3 at 5 in pitch diameter and 5 and 6 at 6 in pitch diameter.

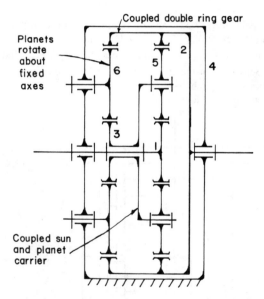

Coupled double ring gear

Planets rotate about fixed axes

Coupled sun and planet carrier

Figure P7.19

7.26. Same as Prob. 7.25, but use Fig. P7.20, with the following dimensions:

Gears	Pitch Diameter (in)
1	$8\frac{1}{2}$
2, 3'	13
4	11

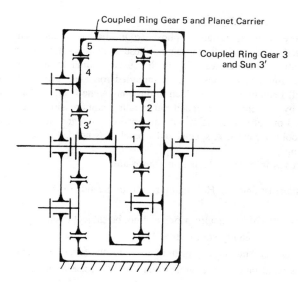

Figure P7.20

7.27. Same as Prob. 7.25, but use Fig. P7.21. Let

Gears	Pitch Diameter (in)
1, 3	9
5, 6	$11\frac{1}{2}$

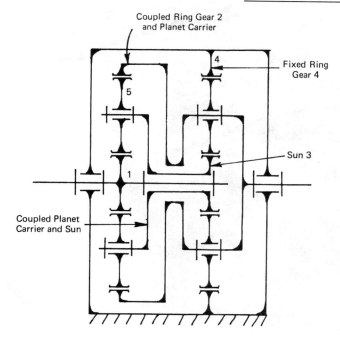

Figure P7.21

7.28. Starting with Eq. (7.27) prove Eq. (7.29).

7.29. Scoliosis (lateral curvature of the spine) and kyphosis (round back) are common spinal deformities. A standard method of surgically correcting these spinal curvatures is by inserting compression instrumentation on the convex side of the curve (similar to jacking up a car). The Harrington compression system consists of a threaded rod and various metal hooks which slide over this rod. Each hook is embedded in the bony spine from the back and each hook slides on the rod by tightening a threaded nut creating compression (Figure P7.22). The conventional method of nut tightening is with an open-end wrench, a time-consuming process when 6 to 24 hooks are used. A specialized wrench was needed to reduce the time involved in nut tightening.

The design objectives in order of their importance are:

1). The wrench must fit onto the Harrington rod from the side of the rod — not "over the top."

2). The wrench must be mechanically strong and not be subject to damage during use.

3). The wrench must be readily sterilizable.

4). The wrench must be "human engineered" to allow the surgeon a sensitive feel.

5). The surgeon's hand input to the wrench must be at least 8.8 cm from the center line of the Harrington rod.

6). The "opened" wrench (during sterilization) must have a minimum number of "loose" parts which would require re-assembly.

7). Wrench thickness at the driving tip must not exceed 6 mm to allow access between hooks. Wrench width at the driving tip must not exceed 2 cm.

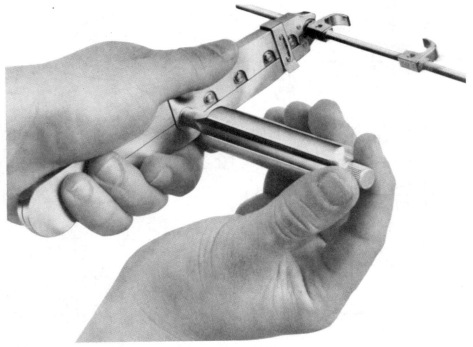

Figure P7.22

Figure P7.23

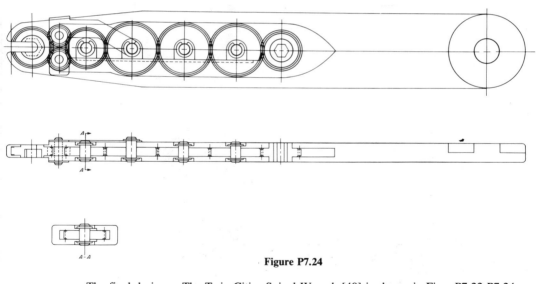

Figure P7.24

The final design—The Twin Cities Spinal Wrench [40] is shown in Figs. P7.22-P7.24.

(a) Does the nut turn in the same direction as the input?

(b) What is ω_{out}/ω_{in}?

(c) What is T_{out}/T_{in}?

(d) What is the relationship between the diameter of the Harrington rod (d) and the radius of the smallest gears?

(e) Can you design your own spinal wrench?

7.30 Prove Equation (7.5a) by referring to Figure 7.6.

chapter eight

Introduction to Kinematic Synthesis—Graphical and Linear Analytical Methods

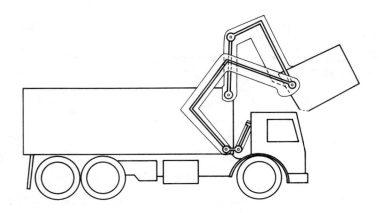

8.1 INTRODUCTION

Ampère defined kinematics as "the study of the motion of mechanisms and methods of creating them." The first part of this definition deals with kinematic *analysis*. Given a certain mechanism, the motion characteristics of its components will be determined by kinematic analysis (as described in Chap. 3). The statement of the task of analysis contains all principal dimensions of the mechanism, interconnections of its links, and the specification of the input motion or method of actuation. The objective is to find the displacements, velocities, accelerations, shock or jerk (second acceleration), and perhaps higher accelerations of the various members, as well as the paths described and motions performed by certain elements. In short, *in kinematic analysis we determine the performance of a given mechanism.* The second part of Ampère's definition may be paraphrased in two ways:

1. The study of methods of creating a given motion by means of mechanisms.
2. The study of methods of creating mechanisms having a given motion.

In either version, the *motion* is given and the mechanism is to be found. This is the essence of *kinematic synthesis*. Thus kinematic synthesis deals with the *systematic design of mechanisms for a given performance.*

The areas of synthesis may be grouped into two categories.

1. *Type synthesis.* Given the required performance, what type of mechanism will be suitable? (Gear trains? Linkages? Cam mechanisms?) Also: How many links should the mechanism have? How many degrees of freedom are required? What configuration is desirable? And so on. Deliberations involving the number of links and degrees of freedom are often referred to as the province of a subcategory of type synthesis called *number synthesis,* pioneered by Gruebler (see Chap. 1). One of the techniques of type synthesis which utilizes the "associated linkage" concept is described in Sec. 8.3.

2. *Dimensional synthesis.* The second major category of kinematic synthesis is best defined by way of its objective:

Dimensional synthesis seeks to determine the significant dimensions and the starting position of a mechanism of preconceived type for a specified task and prescribed performance.

Principal or *significant dimensions* mean link lengths or pivot-to-pivot distances on binary, ternary, and so on, links, angle between bell-crank levers, cam-contour dimensions and cam-follower diameters, eccentricities, gear ratios, and so forth (Fig. 8.1). Configuration or *starting position* is usually specified by way of an angular position of an input link (such as a driving crank) with respect to the fixed link or frame of reference, or the linear distance of a slider block from a point on its guiding link (Fig. 8.2).

A *mechanism of preconceived type* may be a slider-crank, a four-bar linkage, a cam with flat follower, or a more complex linkage of a certain configuration defined topologically but not dimensionally (geared five-bar, Stevenson or Watt six-bar linkage, etc.), as depicted in Fig. 8.3.

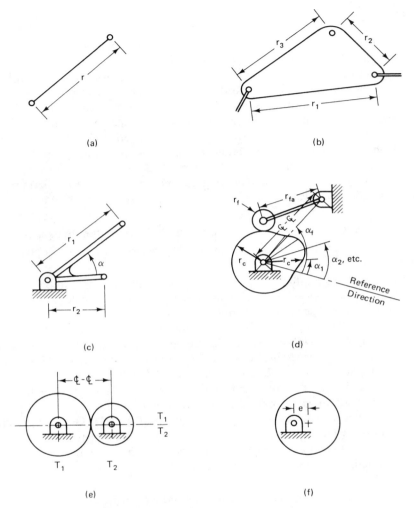

Figure 8.1 Significant dimensions; (a) binary link: has one length only; (b) ternary link: 3 lengths, 2 lengths and one angle, or 1 length and two angles; (c) bell crank: same as for ternary link; (d) cam and roller follower: center line distance, follower arm length r_{fa}, follower radius r_f, and an infinite number of radial distances to the cam surface, r_c, at angles α_1, α_2, etc., specified from a reference direction; (e) gear pair: center line distance and gear tooth ratio; (f) eccentric: eccentricity only (this is a binary link).

8.2 TASKS OF KINEMATIC SYNTHESIS

Recall from Chap. 1 that there are three customary *tasks* for kinematic synthesis: *function*, *path*, and *motion* generation.

In *function generation* rotation or sliding motion of input and output links must be correlated. Figure 8.4a is a graph of an arbitrary function $y = f(x)$. The kinematic synthesis task may be to design a linkage to correlate input and output such that as the input

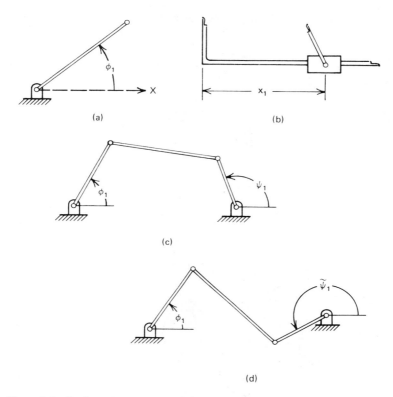

Figure 8.2   Configuration or starting position: (a) starting position of a crank; (b) starting position of a slider; (c) starting position of a four-bar linkage requires two crank angles, because one crank angle leaves two possibilities for the other crank, as shown in Fig. 8.2(d).

moves by x, the output moves by $y = f(x)$ for the range $x_0 \leq x \leq x_{n+1}$. Values of the independent parameter, $x_1, x_2, \ldots, x_n$ correspond to prescribed *precision points* $P_1, P_2, \ldots, P_n$ on the function $y = f(x)$ in a range of x between x_0 and x_{n+1}. In the case of rotary input and output, the angles of rotation ϕ and ψ (Fig. 8.5a) are the linear analogs of x and y, respectively. When the input is rotated to a value of the independent parameter x, the mechanism in the "black box" causes the output link to turn to the corresponding value of the dependent variable $y = f(x)$. This may be regarded as a simple case of a mechanical analog computer.

The subscript j indicates the jth prescribed position of the mechanism; the subscript 1 refers to the *first* or *starting* prescribed position of the mechanism, and $\Delta\phi$, Δx, $\Delta\psi$, and Δy, are the desired *ranges* of the respective variables ϕ, x, ψ, and y (e.g., $\Delta x \equiv |x_{n+1} - x_0|$, $\Delta\phi \equiv |\phi_{n+1} - \phi_0|$, etc.). Since there is a linear relationship between the angular and linear changes,

$$\frac{\phi_j - \phi_1^0}{x_j - x_1} = \frac{\Delta\phi}{\Delta x} \tag{8.1}$$

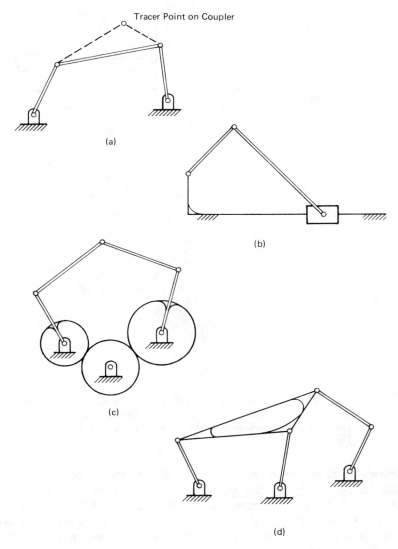

Figure 8.3 Some mechanisms of preconceived type: (a) four-bar linkage; (b) slider-crank; (c) geared five-bar linkage; (d) Stephenson III six-link mechanism.

where ϕ_1 is the datum for ϕ_j, and therefore $\phi_1 = 0$. It follows that

$$\phi_j = \frac{\Delta\phi}{\Delta x}(x_j - x_1)$$

$$\psi_j = \frac{\Delta\psi}{\Delta y}(y_j - y_1)$$

(8.2)

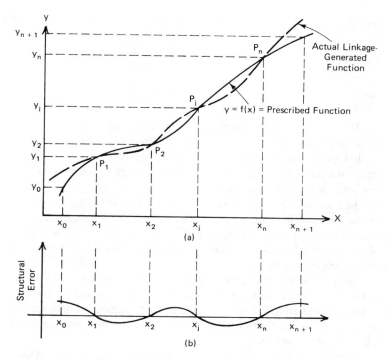

Figure 8.4   Function-generation synthesis: (a) ideal function and generated function; (b) structural error.

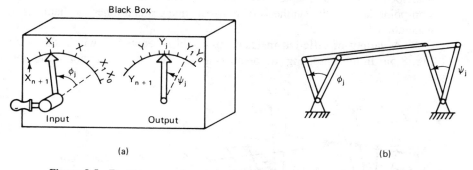

Figure 8.5   Function-generator mechanism; (a) exterior view; (b) schematic of the mechanism inside.

These relationships may also be written as

$$\phi_j = R_\phi(x_j - x_1) \tag{8.3}$$

$$\psi_j = R_\psi(y_j - y_1) \tag{8.4}$$

Where R_ϕ and R_ψ are the *scale factors* in degrees per unit variable defined by

$$R_\phi = \frac{\Delta\phi}{\Delta x} \tag{8.5}$$

$$R_\psi = \frac{\Delta\psi}{\Delta y} \tag{8.6}$$

The four-bar linkage is not capable of error-free generation of an arbitrary function and can match the function at only a limited number of precision points (see Fig. 8.4a). It is, however, widely used in industry in applications where high precision at many points is not required because the four-bar is simple to construct and maintain. The number of precision points that are used in the dimensional synthesis of the four-bar linkage varies in general between two and five.* It is often desirable to space the precision points over the range of the function in such a way as to minimize the *structural error* of the linkage. Structural error is defined as the difference between the generated function (what the linkage actually produces) and the prescribed function for a certain value of the input variable (see Fig. 8.4b).

Notice that the first precision point ($j = 1$) is not at the beginning of the range (see Fig. 8.4). The reason for this is to reduce the extreme values of the structural error. It is also evident from Eq. (8.1) that angles of rotation are measured from the first position (e.g., $\phi_1 = 0$). Section 8.10 will discuss optimal spacing of precision points for minimizing structural error.

Figure 8.6 shows a not-to-scale schematic of the input and output links of a four-bar function generator mechanism in four precision positions, illustrating the relationship between x_j and ϕ_j as well as y_j and ψ_j. The dimensional synthesis techniques described later in this chapter and Chap. 3 of Vol. 2 will show us how to use such precision-point data for the synthesis of four-bar linkages and other mechanisms for function generation.

A variety of different mechanisms could be contained within the "black box" of Fig. 8.5a. In this case, Fig. 8.5b shows a four-bar linkage function generator. A typical

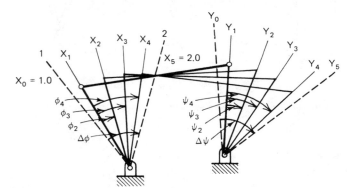

Figure 8.6 Not-to-scale schematic of a function-generator four-bar mechanism with four precision positions of the input and output links x_i and y_i, $i = 1, 2, 3, 4$, within the range $\Delta x = x_5 - x_0$ and $\Delta y = y_5 - y_0$. Input rotations ϕ and output rotations ψ are the analogs of independent and dependent variables x and y, respectively.

* Function generation synthesis up to 7 and path generation synthesis up to 9 precision points are possible, but they generally require numerical rather than the preferable closed-form methods of synthesis.

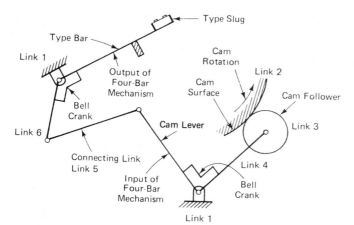

Figure 8.7 Four-bar mechanism used as the impact printing mechanism in an electric typewriter.

example of a function generator is shown schematically in Fig. 8.7. A four-bar linkage connects a cam follower, driven by the cam, to a type bar of a typewriter printing mechanism. Here the type must be moved, first by smaller then by larger angles per increment of input rotation, in order to throw the type against the platen roller with an impact. Another application of function generation would be an engine where the mixing ratios of fuel to oxidant might vary as the function $y = y(x)$. Here ϕ might control the fuel valve while ψ would control the oxidant valve. Flow characteristics of the valves and the required ratio at various fuel rates would dictate the functional relationship to be generated. Yet another example is a linkage to correlate steering positions of the front wheels of an all-terrain vehicle with the relative speed at which each individually driven wheel should rotate to avoid scuffing. Here the input crank is connected to the steering arm, while the output adjusts a potentiometer controlling the relative speed of the two drive wheels.

Mechanical function generators may also be of the type shown in Fig. 8.8 in which a *rectilinear* displacement may be the linear analog of one variable and the crank rotation may be the linear analog of another, a functionally related variable. As illustrated in Fig. 8.9, a function generator may have more degrees of freedom than one; an output variable may be a function of two or more inputs. For example, such a linkage might be used to simulate the addition, multiplication, or any other algebraic or transcendental functional correlation of several variables. Figure 8.10 shows a six-link single-degree-of-freedom function generator mechanism in which two four-link mechanisms

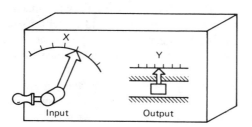

Figure 8.8 Function generator with rotary input and translational output, analogs of the independent and dependent variables of the function $y = f(x)$.

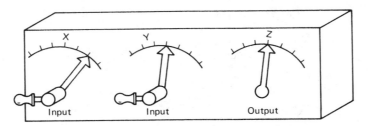

Figure 8.9 Two-degree-of-freedom function generator for generating the function $z = f(x, y)$.

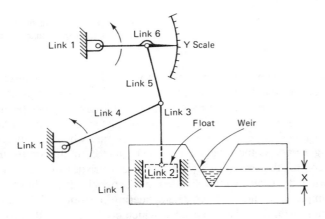

Figure 8.10 Flow-rate-indicator mechanism, $y = K_1 x K_2$, where K_1 and K_2 are constants.

are joined in a series. The objective in this linkage is to provide a measure of flow rate through the weir where the input is the vertical translation x of the water level.

In *path generation* a point on a "floating link" (not directly connected to the fixed link) is to trace a path defined with respect to the fixed frame of reference. If the path points are to be correlated with either time or input-link positions, the task is called *path generation with prescribed timing*. An example of path generation is a four-bar linkage designed to pitch a baseball or tennis ball. In this case the trajectory of point P would be such as to pick up a ball at a prescribed location and to deliver the ball along a prescribed path with prescribed timing for reaching a suitable throw-velocity and direction.

In Fig. 8.11, a linkage whose floating link will contain point P is desired such that point P will trace $y = f(x)$ as the input crank turns. Typical examples are where $y = f(x)$ is the path desired for a thread-guiding eye on a sewing machine (Fig. 8.12) or the path to advance the film in a movie camera (Fig. 8.13). Various straight-line mechanisms, such as Watt's and Robert's linkages, are examples of a special kind of path generator (see Fig. 8.14) in which geometric relationships assure the generation of straight-line segments within the cycle of the linkage's motion.

Motion-generation or *rigid-body guidance* requires that an entire body be guided through a prescribed motion sequence. The body to be guided usually is a part of a floating link. In Fig. 8.15 not only is the path of point P prescribed, but also the rotations α_j

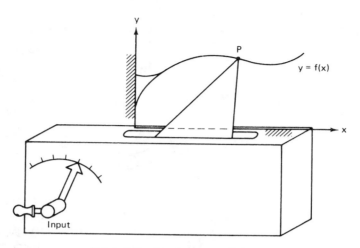

Figure 8.11   A path generator linkage.

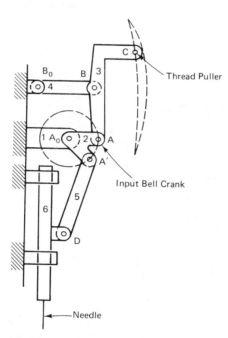

Figure 8.12   In a sewing machine, one input (bell crank 2) drives a path generator (four-bar mechanism 1, 2, 3, 4) and a function generator slider-crank (1, 2, 5, 6). The first generates the path of thread-guide C and the second generates the straight-line motion of the needle, whose position is a function of crank rotation.

of vector **Z** embedded in the moving body. The corresponding input rotations may or may not be prescribed. For instance, vector **Z** might represent a carrier link in automatic machinery where a point located on the carrier link (the tip of **Z**) has a prescribed path while the carrier has a prescribed angular orientation (see Fig. 8.16). Prescribing the movement of the bucket for a bucket loader is another example of motion generation.

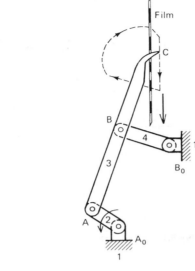

Figure 8.13 Film-advance mechanism of a movie camera or projector generates the path of point C as a function of the angle of rotation of crank 2.

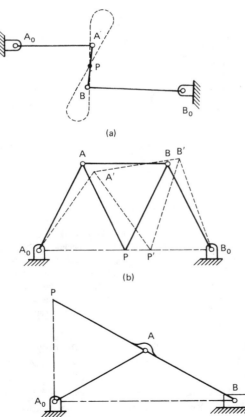

(a)

(b)

(c)

Figure 8.14 Straight-line mechanisms (a) Watt's mechanism — approximate straight-line motion traced by point P; $AP/PB = BB_0/AA_0$; (b) Robert's mechanism — approximate straight-line motion traced by point P; $A_0A = AP = PB = BB_0$, $A_0B_0 = 2AB$; (c) Scott-Russele mechanism gives exact straight-line motion traced by point P. Note the equivalence to Cardan motion (see chapt. 3, vol. 2); $A_0A = AB = AP$.

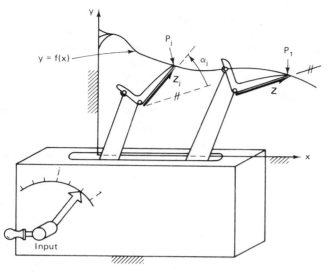

Figure 8.15   Motion-generator mechanism.

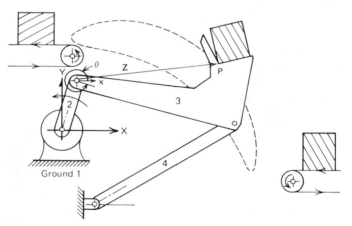

Figure 8.16   Carrier mechanism in an assembly machine.

The path of the tip of the bucket is critical since the tip must perform a scooping trajectory followed by a lifting and a dumping trajectory. The rotations of the bucket are equally important to ensure that the load is dumped from the correct position.

Since a linkage has only a finite number of significant dimensions, the designer may only prescribe a finite number of *precision conditions;* that is, we may only *prescribe* the *performance* of a linkage at a finite number of *precision points.* There are three methods of specifying the *prescribed performance* of a mechanism: *first-order* or *point approximation, higher-order approximation,* and combined *point-order approximation.**

---

* Approximate (rather than precise) generation of greater numbers of prescribed conditions are possible by the use of least squares or non-linear programming methods. These, however, are numerical procedures rather than closed-form solutions.

In *first-order approximation* for function and path generation, discrete points on the prescribed (or ideal) function or path are specified. Recall that Fig. 8.4a showed precision points P_1 to P_n of the ideal function. The synthesized mechanism will generate a function that will coincide with the ideal function at the precision points but will generally deviate from the ideal function between these points (Fig. 8.4b).

Structural error for path generation may be defined as the vector from the ideal to the generated path perpendicular to the ideal path or it may be defined as the vector between corresponding points on an ideal and a generated path taken at the same value of the independent variable. The latter definition is used when there is prescribed timing. In motion generation there will be both a path and an angular-structural-error curve to analyze.

In some cases a mechanism is desired to generate not only a position but also the velocity, acceleration, shock, and so on, at one or more positions (see Fig. 8.17). For example, the blade of a cutter that must slice a web of paper into sheets while the web is in motion would not only be required to match the correct position at the instant of the cut, but also several derivatives at that position in order to cut straight across and to preserve the sharpness of the blade. For *higher-order approximation,* the first derivative, dy/dx, prescribes the slope of function (or path) at that point; the second derivative, d^2y/dx^2, implies prescribing the radius of curvature; the third derivative, d^3y/dx^3, prescribes the rate of change of curvature; and so on (see Sec. 8.24).

The combination of both point and order approximations is called *point-order approximation* or approximation by *multiply separated precision points* [157]. For example, one might desire to prescribe a position and a velocity at one precision point, only a position at a second precision point, and a position and velocity at a third point.

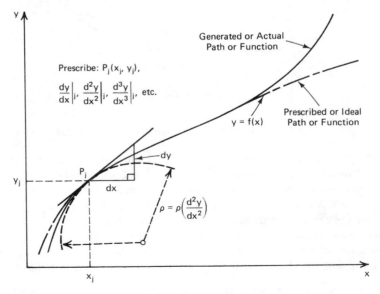

Figure 8.17   Higher-order approximation of function or path.

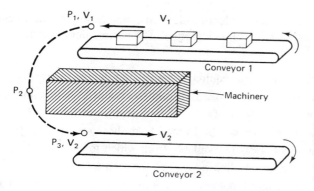

Figure 8.18 Point-order approximation for path generation. There are five prescribed conditions: three-path points and velocities at two of these, tantamount to two infinitesimally close prescribed position at P_1 and P_3.

Figure 8.18 shows such an application where a mechanism is desired to pick up an item from conveyor belt 1 traveling at velocity V_1 and deposit it on a conveyor belt 2 traveling at V_2, having traversed the intervening space in such a way as to avoid some machinery components. Typical application of this occurs in bookbinding, where signatures (32- or 64-page sections) of a book from conveyor 1 are to be stacked on conveyor 2 to form the complete book (see Fig. 6.34).

Kinematic synthesis has been defined here as a combination of type and dimensional synthesis. Most of the rest of this chapter and Chap. 3 of Vol. 2 are devoted to dimensional synthesis. Before moving on to dimensional synthesis, however, one of the methods to creatively discover suitable types of linkages for a prescribed task will be introduced. The method is based on structural models or associated linkages. A case study of type synthesis using another method can be found in the appendix to this chapter.

8.3 TYPE SYNTHESIS [160]

Type synthesis strives to predict which combination of linkage topology and type of joints may be best suited to solve a particular task. Frequently, a novice designer may settle for a solution which merely satisfies the requirements, since there appears to be no method to find a "best" solution. Many experienced designers perform a rudimentary form of type synthesis, sometimes without being aware they are doing so. These experts have an innate "feel" for which type of linkage will work and which will not. This ability is developed only after designing linkages for many years and is difficult to pass on to younger engineers. Many times, type synthesis is skipped due to ignorance or because the designer was not aware of the required relations between the form and function of the linkage. When this happens, a linkage may be chosen which is not capable of meeting the problem requirements. An example would be to choose a single-degree-of-freedom-linkage topology for a two-degree-of-freedom task. This is an expensive mistake, since no choice of dimensions or joint types will yield a viable solution. Besides being relatively unknown, type synthesis is difficult to apply because the principles are not as well defined as those for dimensional synthesis, and so the technique is usually not utilized to its full potential usefulness.

Type synthesis consists of many identifiable steps. For example, the questions listed below may be considered in the following order. The desired degree of freedom is known from the problem.

1. How many links and joints are required for a desired degree of freedom?
2. What are the link types and how many of each are needed for this link set?
3. How many different link sets satisfy the desired degrees of freedom?
4. How many linkage topologies can be formed from these sets of links?
5. How many unique topologies are available from which to choose?
6. How many ways can a ground link be chosen for each topology?
7. How can one predict if any topologic inversions are inherently better than all others for the task at hand?
8. How many ways can the particular types of joints, required to satisfy the task, be distributed throughout the linkage?
9. How many different links could serve as the input driver?

Type synthesis can be subdivided into topological synthesis, topological analysis, and number synthesis [117]. Questions 1 through 3 constitute number synthesis, questions 4 through 6 are topological synthesis, and questions 7 through 9 are typical of topological analysis. Figure 8.19 lists the divisions within type synthesis and shows the place of the field within kinematics.

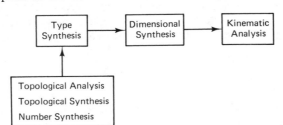

Figure 8.19   The field of type synthesis within kinematics.

The first step in type synthesis is to determine the number and type of links needed to form linkages with the correct degree of freedom. This can be done by using a modified form of the Gruebler equation (see Chap. 1) listed below as Eq. (8.7). Solution of this equation determines all the sets of *higher-order links* (those larger than binaries) which satisfy the desired degree of freedom.

$$n - (F + 3) = T + 2Q + 3P + \ldots \qquad (8.7)$$

Where n = the total number of links in a mechanism

B = the number of binary links*

T = the number of ternary links

---

* Not included in Eq. (8.7), because links are not *higher-order links*.

$Q =$ the number of quarternary links

$P =$ the number of pentagonal links

$F =$ the degree of freedom required to perform the desired task

Each higher order link set is combined with the necessary number of binary links to total the number of links required by Gruebler's equation for the mechanism. Each set of n links is known as a *kinematic link set solution* (KLSS). There are methods to generate these solutions exhaustively and to determine a priori what the final count should be for any combination of number of links and degrees of freedom [160].

The collection of links comprising each KLSS are assembled into figures using pin joints at all link connection points. These figures define the topologic structure of the linkages formed from this set and are called *isomers*. These isomers are guaranteed to have the desired overall degree of freedom. Each isomer obtained from all kinematic link set solutions for a desired degree of freedom and number of links is called a *basic kinematic chain* (BKC). It is important to have a complete set of BKCs. The urge to accomplish this has attracted much attention over the years [32, 162–164]. Care must be taken when forming these topologic structures to exclude "bad" BKCs, those which fail the *degree-of-freedom-distribution* criterion. This criterion demands that a kinematic chain not have an imbedded zero-freedom subchain. Such a chain would be an unnecessarily complex version of a simpler one, and it should be eliminated before continuing. All KLSS, except those which describe the binary chain mechanisms, have some bad isomers. It is predictable which KLSS will yield only bad isomers, but discovering the useful isomers in the remaining KLSS can be tedious. For example, the one degree of freedom six bar chains have two kinematic link-set solutions. All isomers of one set fail the degree of freedom distribution criterion and degenerate to other linkages, while only ⅔ of the isomers from the second KLSS fail. These isomers must be individually checked.

The next step is to generate all *topologic inversions* of a given BKC. These are formed by grounding each link in a BKC, one at a time, and determining which of the resulting mechanisms are topologically unique. For example a Watt I is topologically different than a Watt II and may be capable of performing different tasks as was pointed out in Chap. 1.

Few methods exist to determine which topologic inversion is best suited for a particular task. One successful technique is called the *associated linkage concept*. It is presented below.

The final three steps determine how drivers and different types of joints can be chosen and distributed throughout the mechanism. Fig. 8.20 outlines the entire procedure described above for the case of a six-bar one-freedom linkage. At this point the topology and joint pairs for a mechanism have been determined and all that remains is to perform a dimensional synthesis.

The Associated Linkage Concept

The *associated linkage concept* was developed by R. C. Johnson and K. Towligh [91, 92] to act as a spur to creativity. An engineer armed with this technique should be able to

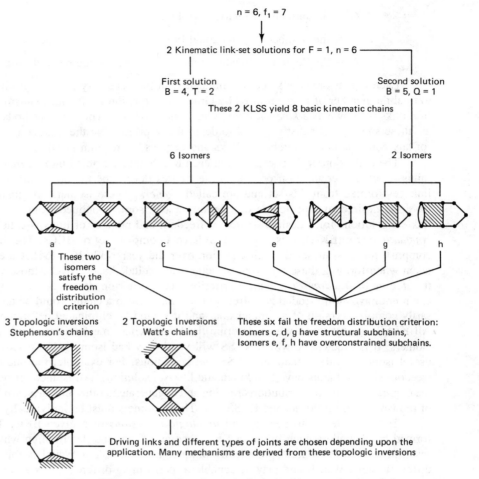

Figure 8.20 Demonstration of type synthesis for 6-bar, one-freedom chains.

generate many mechanisms for a specific task. Design rules are translated into their topologic equivalents (steps 6, 8 and 9 from above) and suitable BKCs (step 5 above) are chosen. The method consists of the following procedure:

1. The determination of rules that must be satisfied for the selection of a suitable "associated linkage." These rules are derived by observing the specific design application.

2. The application of suitable asociated linkages to the synthesis of different types of devices. (See Table 1.2 for equivalent lower-pair joints for velocity matching of higher-pair connections.)

This technique of applying number synthesis to the creative design of practical devices will be illustrated by several examples.

Synthesis of Some Slider Mechanisms

Suppose that it is desired to derive types of mechanisms for driving a slider with rectilinear translation along a fixed path in a machine. Assume that the drive shaft will be fixed against translation and that it must rotate with unidirectional rotation. Also, assume that the slider must move with a reciprocating motion.

A basic rule for this example is that a suitable associated linkage must have a single degree of freedom ($F = +1$) when one link is fixed. Let us start with the least complicated associated linkage chain (which is the four-bar) since simplicity is an obvious design objective (Fig. 8.21a). The four-bar associated linkage has four revolute joints. If one of the revolutes (joint *c-d*) is replaced by a slider, the slider-crank mechanism is derived as shown in Fig. 8.21b.

Increasing the degree of complexity, a Stephenson six-bar chain (in which ternary links are not directly connected) is considered next as a suitable associated linkage (Fig. 8.22a). By varying the location of the slider one creates the slider mechanisms of Fig. 8.22b–f, different from the slider-crank of Fig. 8.21. Finally, in Fig. 8.23, from a Watt six-bar chain (in which the ternary links are direct connected) we derive only one new mechanism (Fig. 8.23b), which is of the same degree of complexity as those in Fig. 8.22; Fig. 8.23c, d, and e are merely slider-cranks, with an added passive dyad. Thus five different six-link mechanisms, each having only a single slider joint, can be derived for this problem.

This general procedure could be extended to other suitable linkages of greater complexity, including those containing *higher pairs*.* Thus cams and sliding pivots may be incorporated in the derivations of different types of mechanisms, such as those illustrated in Fig. 8.24, derived from the four-bar chain as the associated linkage.

Synthesis of Some Gear-Cam Mechanisms

A typical meshing gear set is shown in Fig. 8.25 with two typical teeth in contact. At the instant of observation the meshing gear set is equivalent to a hypothetical quadric

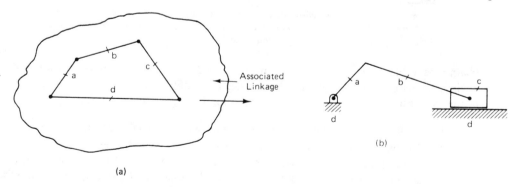

(a)

(b)

Figure 8.21   Slider-crank mechanism and its associated linkage; (a) four-bar chain; (b) slider-crank mechanism.

* Sec. 6.10 describes this technique applied to cam-modulated linkages.

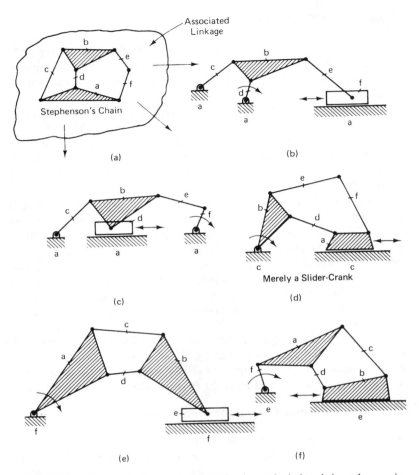

Figure 8.22 Slider mechanisms derived from Stephenson's six-bar chain as the associated linkage. Note that (d) shows merely a slider crank with redundant (superfluous) links, the passive dyad consisting of links e and f.

chain (see Table 1.2). Hence, as shown in Fig. 8.25, a meshing gear set has a four-bar chain as an associated linkage. The basic rules for a suitable associated linkage involved in the synthesis of a mechanism containing a meshing gear set are as follows:

1. The number of degrees of freedom with one link fixed must be $F = +1$.
2. The linkage must contain at least one four-sided closed loop. This is true since the meshing gear set corresponds to a four-sided closed loop containing two centers of rotation, $R_{p/f}$ and $R_{g/f}$, and two base points, B_p and B_g, which are the instantaneous centers between gear p and the fictitious coupler C and between gear g and C, respectively. In the gear set, coupler C is replaced by the higher-pair contact between the tooth profiles. Hence B_p and B_g coincide with the cen-

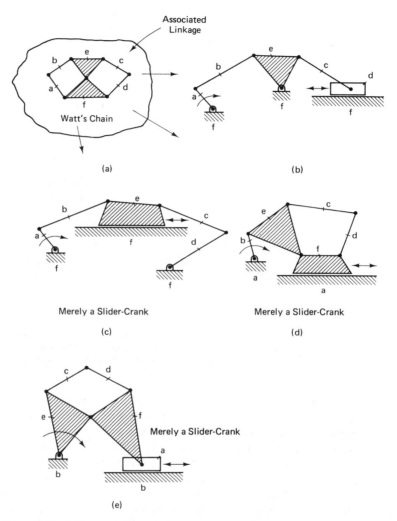

Figure 8.23   Slider mechanisms derived from Watt's chain six-bar as the associated linkage.

ters of curvature of the respective involute tooth profiles at their point of contact. In traversing this four-sided closed loop, the two centers of rotation must be encountered in succession, such as *RRBB* rather than *RBRB*.

3. The four-sided closed loop must contain at least one binary link. This is true because in the four-sided closed loop the link connecting the two base points must be a binary link. This is evident since the base points on the meshing gears are instantaneous and they are joined by a hypothetical connecting rod in the equivalent quadric chain.

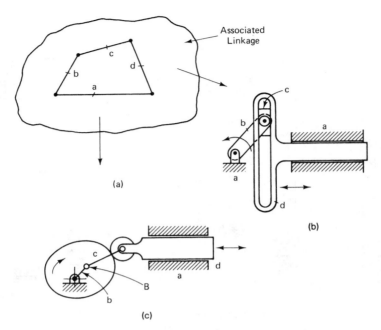

Figure 8.24 Derivation of some slider mechanisms containing cams and sliding pivots from the four-bar chain as the associated linkage. Notice that point B is the center of curvature of the cam contour at the point of contact of the cam; (a) four-bar chain; (b) Scotch yoke; (c) disk cam with translating follower.

Suppose that it is required to design a gear mechanism for driving a slider with arbitrary motion along fixed ways in a machine. Assume that the driving shaft must have unidirectional rotation and the slider must have a reciprocating motion. One possible design would be the mechanism shown in Fig. 8.26, where the driving cam provides arbitrary motion and a gear and rack drive the slider. In Fig. 8.27 an equivalent linkage for this mechanism is shown together with its associated linkage. Incidentally, a gear and rack is a special gear type with one base point and one center of rotation at infinity.

Simplicity in design is a practical goal worth striving for. Suppose that we wish to explore different, simpler mechanism types for the basic problem described in the preceding paragraph (assuming that a cam, follower, gear, and rack are to be employed for driving the slider). The simplest suitable associated linkage for this application would be either Watt's chain or Stephenson's chain. From these chains three different mechanism types are derived (Figs. 8.28 and 8.29), where Fig. 8.29c would require a flexible shaft for driving the cam.

Synthesis of Some Internal-Force-Exerting Devices

Kurt Hain [83] has applied number synthesis to the design of differential brakes and differential clamping mechanisms by recognizing the analogy with preloaded structures. This analogy shows that, for the synthesis of internal-force-exerting devices in general,

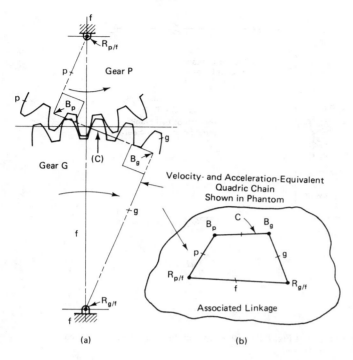

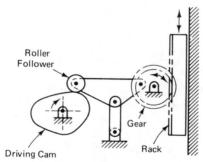

Figure 8.25 Meshing gear set with its associated linkage. B_g and B_p are the centers of curvature of the involutes at the contact point of gear G and gear P, respectively; (a) gear pair; (b) associated linkage.

Figure 8.26 Slider mechanism with cam and gear.

a suitable associated linkage must have $F = -1$ for the number of degrees of freedom with one link fixed. Also, *forces* exerted by the device on the work piece correspond to *binary links* in the associated linkage, recognizing that a binary link is a two-force member. Let us apply this technique to the synthesis of two practical devices. First, different types of compound-lever snips are explored, followed by several types of yoke riveters.

Synthesis of Compound-lever Snips.   Simply constructed compound-lever snips are to be designed for cutting through tough materials with a relatively small amount of effort. The actuating force is designated by P and the resisting force by F_r. We will assume that the compound-lever snips should be hand-operated and mobile.

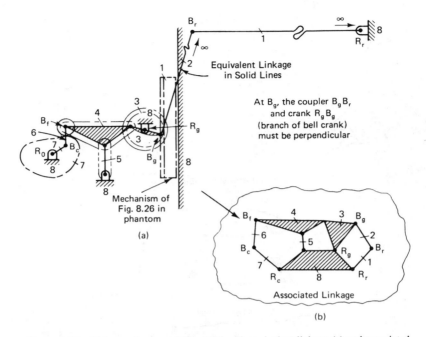

Figure 8.27 Slider mechanism of Fig. 8.26 with equivalent linkage (a) and associated linkage (b) from which it was derived.

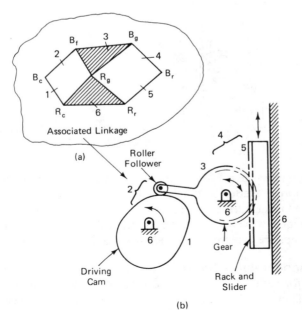

Figure 8.28 Cam-gear-slider mechanism derived from Watt's chain.

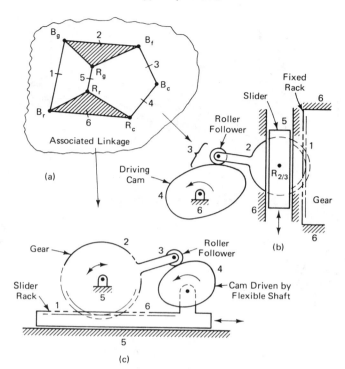

Figure 8.29 Cam-gear-slider mechanisms derived from Stephenson's chain.

Hence there will be no ground link in the construction. However, a high amplification of force is required in the device. Therefore, in the associated linkage, binary links P and F_r must not be connected by a single link; otherwise, a simple lever type of construction will result in relatively low force amplification.

In summary, for application to the synthesis of compound-lever snips, the rules or requirements for a suitable associated linkage are as follows:

1. $F = -1$.
2. There must be at least two binary links because of the two forces P and F_r.
3. Two binary links P and F_r must not connect the same link, because in that case the snips will be simple instead of compound.

The associated linkages in Figs. 8.30, 8.31, and 8.32 satisfy the requirements. Each suitable associated linkage yields a different mechanism for compound-lever snips.

Synthesis of yoke riveters.   The configuration for an existing yoke-riveter design [91] is shown in Fig. 8.33. Let us apply number synthesis in the creation of other types of yoke-riveter designs.

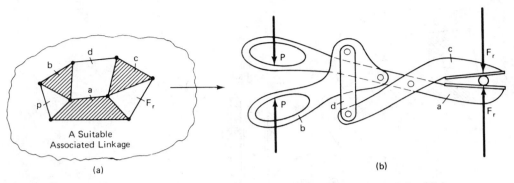

Figure 8.30 Synthesis of compound lever snips from a suitable associated linkage.

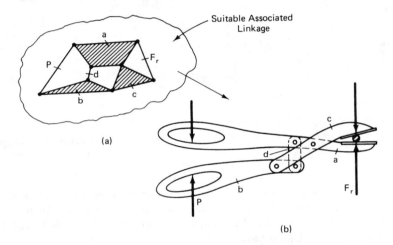

Figure 8.31 Synthesis of compound-lever snips from a suitable associated linkage.

The following characteristics are assumed to be requirements for a suitable yoke riveter in our particular application:

1. Simple features of construction.
2. Self-contained, portable unit.
3. High force amplification between power piston and rivet die.
4. One part of the two-part rivet die and the relatively large pneumatic power cylinder are fixed to the frame link.
5. Another part of the rivet die and the power piston are to slide relative to the frame link.

From Fig. 8.33 of the existing yoke-riveter design the associated plane linkage with single pin joints is derived as shown in Fig. 8.34. Applying Gruebler's equation (Chap. 1) to the linkage in Fig. 8.34, we obtain $F = -1$, which is expected, since this

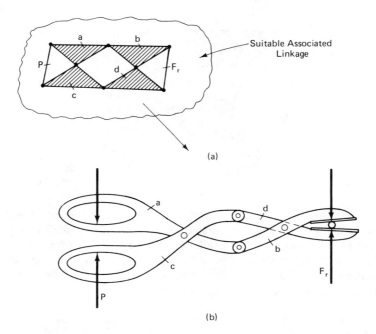

(a)

(b)

Figure 8.32 Different design derived from another suitable associated linkage for compound-lever snips.

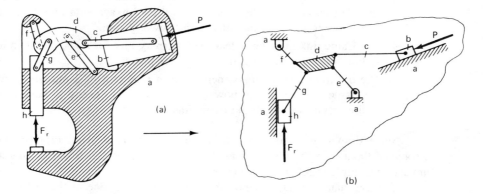

(a)

(b)

Figure 8.33 Existing yoke riveter (a) and the equivalent toggle linkage in inset (b).

value of F is characteristic of the associated linkage for any internal-force-exerting device: $F = 3(n - 1) - 2f_1$. Note that $n = 10$, including the binary links representing P and F_r, connecting a with b and a with h, respectively. Note also that the number of pin joints, f_1, is 14. Therefore,

$$F = 3(10 - 1) - 2(14) = -1$$

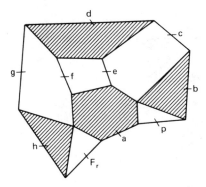

Figure 8.34 Associated linkage for the existing yoke riveter of Fig. 8.33.

In the synthesis of new configurations of yoke riveters, it will be necessary to reverse the procedure just illustrated in going from Fig. 8.33 to Fig. 8.34. Thus first it will be necessary to select a suitable associated linkage for a new yoke-riveter design. From a careful study of Figs. 8.33 and 8.34, and from a consideration of the desired features of a suitable yoke riveter listed previously, the following rules or requirements for a suitable associated linkage are obtained.

1. $F = -1$.
2. There must be at least two binary links (for P and F_r).
3. The binary links corresponding to P and F_r must be connected to the same link at one end, which is the frame link, and to different *ternary* links at their other end. This assures simple construction of the linkage with high force amplification between the rivet die set and the power piston.
4. The frame link must be at least a quaternary link for P, F_r, and two lower-pair sliding joints for the rivet die and power piston.
5. The different *ternary* links mentioned in requirement 3 must be connected to the frame link, since the power piston and rivet die are to have a lower-pair sliding connection with the frame link.

Since simplicity of construction is a feature of practical importance, the simpler associated linkage in the inset of Fig. 8.35a is a suitable choice. From this associated linkage the simple toggle-type riveter is derived.

The associated linkage method for type synthesis is one of the useful techniques used for synthesizing mechanism *types*. Similar methods of analysis are sometimes employed in patent cases in determining whether a device is of the same or different type than others. Another type-synthesis method is described in the appendix of this chapter by way of a case study.

Observe that nothing yet has been said regarding actual dimensions of these type-synthesized mechanisms. The specific dimensions will control the relative motions and the force transmission characteristics of the examples given above.

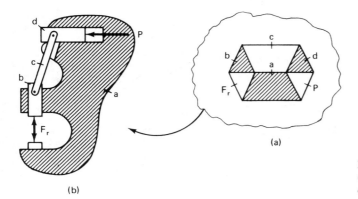

(a)

(b)

Figure 8.35 Simple toggle-type riveter; (a) associated linkage; (b) the mechanism derived from (a).

8.4 TOOLS OF DIMENSIONAL SYNTHESIS

The two basic tools of dimensional synthesis are geometric contruction and analytical (mathematical) calculation.

Geometric or graphical methods of synthesis provide the designer with a fairly quick, straightforward method of design. Graphical techniques do have limitations of accuracy due to drawing error, which can be very critical and because of complexity of solution to achieve suitable results, the geometric construction may have to be repeated many times.

Analytical methods of synthesis are suitable for automatic computation and have the advantages of accuracy and repeatability. Once a mechanism is modeled mathematically and coded for a computer, mechanism parameters are easily manipulated to create new solutions without further programming. Although this text emphasizes analytical synthesis, it is important to have experience in graphical techniques for use in the initial phases of kinematic synthesis. The next several sections present a review of useful geometric approaches before moving on to analytical synthesis.

8.5 GRAPHICAL SYNTHESIS—MOTION GENERATION: TWO PRESCRIBED POSITIONS [139]

Suppose that we wish to guide a link in a mechanism in such a way that it will assume several arbitrarily prescribed distinct (finitely separated) positions. For two positions of motion generation, this can be accomplished by a simple rotation (Fig. 8.36) about a suitable center of rotation. This *pole* (see Sec. 4.2 of Vol. 2), P_{12}, is found graphically by way of the *midnormals* a_{12} and b_{12} to the connecting line segments of two *corresponding* positions each of points A and B, namely A_1, A_2 and B_1, B_2.

If pole P_{12} happens to fall off the frame of the machine, we may use a four-bar linkage to guide link AB from position 1 to position 2 (Fig. 8.37). Two fixed pivots, one each anywhere along the two midnormals, will accomplish this task. The construction is as follows.

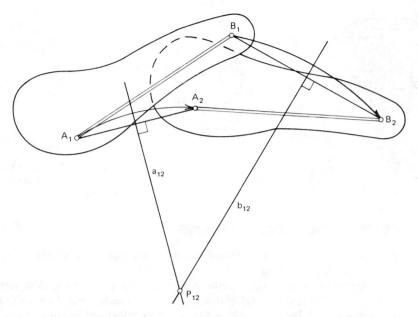

Figure 8.36 Two prescribed coplanar positions of a rigid body can be reached from one another by rotation about pole P_{12}.

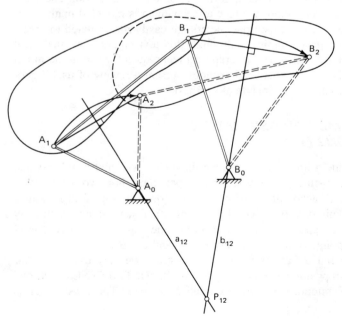

Figure 8.37 Two-position graphical synthesis of a four-bar motion generator mechanism. Fixed pivots A_0 and B_0 can be located anywhere along the midnormals between A_1A_2 and B_1B_2, respectively.

Draw the perpendicular bisector (or midnormal) to $A_1 A_2$, the first and second positions of the *circle point A* —so named because a circular arc can be drawn through its corresponding positions. Any point along this midnormal, say A_0, is a possible fixed pivot or *center point,* conjugate to circle point A. A link between a center and circle point will guide A from A_1 to A_2. This construction is now repeated for another circle point, B, to yield B_0.

Figure 8.37 shows one of the possible four bar linkages that will act as a motion generator for two positions. Notice that the construction of each circle point–center point pair involved *three free choices*. For two prescribed positions, a circle point A may be chosen anywhere in the plane or its extension, located by two independent coordinates along the x and y axes of a Cartesian system fixed in the moving body, and the conjugate center point may be selected anywhere along the midnormal of its corresponding positions. Thus there are ideally *three* infinities of solutions (for each pair of center point and circle point) to build a four-bar linkage. For instance, if the entire midnormal a_{12} represent undesirable locations for fixed pivots, we can rigidly attach point C to A and B by means of a triangle in the plane of the moving (or "floating") link and use C as a crank pin. Figure 8.38 shows the construction yielding an alternative linkage replacing the $A_1 A_0$ link of Figure 8.37 with $C_1 C_0$.

8.6 GRAPHICAL SYNTHESIS—MOTION GENERATION: THREE PRESCRIBED POSITIONS

Let us now consider three arbitrary positions of a plane, $A_1 B_1$, $A_2 B_2$, and $A_3 B_3$ (Fig. 8.39). There will be three poles associated with these positions, P_{12}, P_{23}, P_{31} (note that

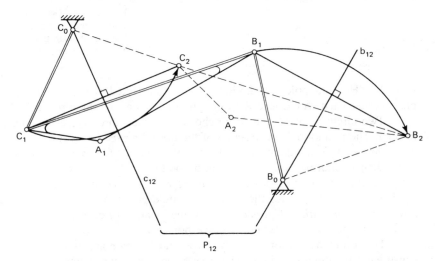

Figure 8.38 If the midnormal of $A_1 A_2$ does not contain suitable locations for ground pivot A_0, another point C can be located in the moving body. Then the midnormal of $C_1 C_2$ may yield a suitable ground pivot C_0.

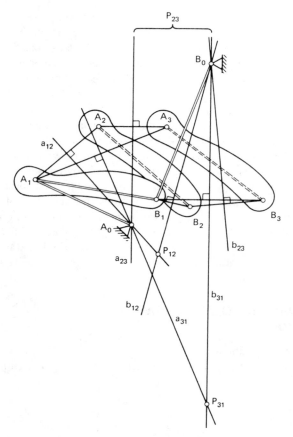

Figure 8.39 Geometrically (graphically) synthesized four-bar mechanism guides body *AB* through three prescribed positions A_1B_1, A_2B_2, and A_3B_3.

$P_{ij} \equiv P_{ji}$). Here the poles can no longer be used as fixed pivots even if they are accessible, because each would lead *AB* through only two of the three prescribed positions.

Two circle points *A* and *B* are chosen and their three corresponding positions are located. The midnormal construction of the preceding section is repeated twice for point *A* (a_{12} and a_{23}). Since the center point for each pair of two positions may lie anywhere along their midnormal, the intersection of the two midnormals locates the common center point A_0 for all three positions. Figure 8.39 shows the resulting unique four-bar mechanism synthesized for the choices of circle points *A* and *B*. Notice that there are, however, *two infinities* of possibilities for the location of each circle point (x_a and y_a for *A*), and thus for each center point–circle point pair.

The following sections illustrate how four-bar path and function generators can be constructed for three positions. The very same technique of intersection of the perpendicular bisectors is used, but only after a kinematic inversion is performed. The following sections describe these procedures.

8.7 GRAPHICAL SYNTHESIS FOR PATH GENERATION: THREE PRESCRIBED POSITIONS [105]

A very similar construction is involved for graphical synthesis of a four-bar path generator for three positions. Let us design a four-bar mechanism so that a path point P on the coupler link will pass through three selected positions, P_1, P_2, and P_3 (Fig. 8.40).

In designing for three prescribed positions, the positions of A_0 and B_0 (defining the length and inclination of the fixed link) are free choices. Also, the length of the input crank and the distance between A and P are arbitrary. (As the number of design positions is increased, restrictions are imposed on some of these free choices.) The construction is as follows (Fig. 8.40):

1. After selecting the prescribed path points, P_1, P_2, and P_3, select locations for fixed pivots, center pivots A_0 and B_0, establishing the fixed link.
2. Choose a length for the crank and draw in the path of A (a circle). Pick a point for A_1 (position of A for position P_1).
3. With the length of AP established, locate A_2 and A_3. A, P, and B are all points on the coupler and thus remain the same distance apart in all positions.
4. The position of B is found by means of a kinematic inversion (see Sec. 3.1.) This is accomplished by fixing the coupler in position 1. The rest of the mechanism, including the frame, must move so that the same relative motion exists between all links in this inversion as well as the original arrangement. The relative positions of B_0 with respect to position 1 of the coupler are obtained by the construction shown in Fig. 8.41 as follows (see Figs. 8.40 and 8.41). Rotate A_0 about A_1 by $(\alpha_2 - \alpha_1)$ (where $\alpha_2 = \sphericalangle A_0 A_2 P_2$ and $\alpha_1 = \sphericalangle A_0 A_1 P_1$ of Fig. 8.40) to A_0'. Draw an arc about A_0' with radius $\overline{A_0 B_0}$. Draw an arc about P_1 with radius $\overline{P_2 B_0}$ measured in Fig. 8.40. The intersection of these two arcs locates B_0'. The construction of B_0'' (not shown) follows the same procedure with A_0'' [rotated about A_1 from A_0 by $(\alpha_3 - \alpha_1)$] as the center of arc with radius $\overline{A_0 B_0}$, and with $\overline{P_3 B_0}$ as the radius of a second arc from center P_1.

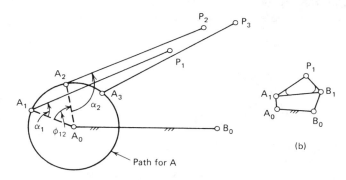

(a)

(b)

Figure 8.40   Three-position design of a path generator mechanism (a) initial layout indicating design parameters; (b) schematic of the desired mechanism.

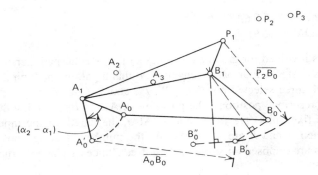

Figure 8.41   Three-position path-generator design. Inversion to locate B_1.

5. Erect perpendicular bisectors to lines $B_0 B_0'$ and $B_0' B_0''$ (not shown). The point of intersection locates B_1 as the center of the circle that will pass through the three relative positions of B_0: B_0, B_0', and B_0''.

6. Draw the mechanism in all three positions to check the design (Fig. 8.42). If the design is not satisfactory, these steps can be repeated with different choices for A_0, B_0, and A_1.

Notice that there are ideally six infinities of four-bar linkages that will accomplish this path-generation task, since location of A_0 (x, y coordinates) and the vectors $\overrightarrow{A_0 B_0}$ and $\overrightarrow{A_0 A_1}$ were arbitrarily chosen in the fixed plane of reference. This is tantamount to *three infinities* of solutions for each side of the linkage for path generation, compared with two infinities of solutions for motion generation. If path generation with prescribed timing (i.e., prescribed rotations of the input link ϕ_{12} and ϕ_{13} correlated with the path points) is the objective, there are two infinities of solutions for each side, or a total of four infinities for the four-bar linkages, as shown in the following section.

An important point should be made here that has relevance to all the graphical techniques. In step 5 of this section, the intersection of the perpendicular bisectors located B_1. Slight error in locating B_0, B_0', or B_0'' will result in a magnified error in the location of B_1. In fact, as lines $B_0 B_0'$ and $B_0' B_0''$ become close to being parallel, the error magnification is very large. The designer must be aware of these inherent drawbacks of graphical construction.

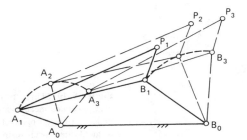

Figure 8.42   Three-position path-generator design. Checking the completed mechanism.

8.8 PATH GENERATION WITH PRESCRIBED TIMING: THREE PRESCRIBED POSITIONS

The preceding construction must be modified in order to prescribe input crank rotations which are to correspond with the prescribed path positions. The same example will be used as in Fig. 8.40, except that input crank rotations are prescribed: 58° cw corresponding to the movement of point P from P_1 to P_2 and 108° cw from P_1 to P_3 (see Fig. 8.43). The construction, shown in Fig. 8.44, is as follows:

1. Pick the fixed pivot of the input link (A_0) with respect to the prescribed path precision points $P_1 P_2 P_3$ (two infinities of choices, one for x and one for y of A_0).
2. Draw line $\overline{P_2 A_0}$ and $\overline{P_3 A_0}$.
3. Inverting the motion (by fixing the yet unknown input link $A_0 A$), rotate $\overline{P_2 A_0}$ 58° ccw about A_0 and $\overline{P_3 A_0}$ 108° ccw around A_0 locating P_2' and P_3'.
4. Draw lines $\overline{P_2' P_1}$ and $\overline{P_3' P_1}$.
5. The intersection of the perpendicular bisectors p_{12}' and p_{13}' locates A_1, the first position of A.

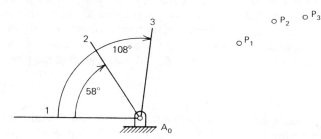

Figure 8.43 Prescribed path points and crank rotations for path generation with prescribed timing with three finitely separated precision points.

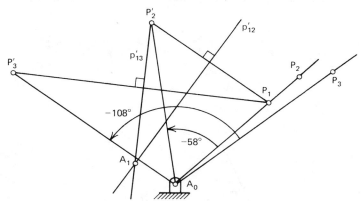

Figure 8.44 Graphical construction of the starting position of crank $A_0 A_1$ for the path generator with the prescribed data of Fig. 8.43. Completion of the geometric synthesis of the four-bar mechanism proceeds according to Figs. 8.40 and 8.41. $P_0 A_0$ does not necessarily go through P_1.

6. The rest of the construction proceeds as illustrated in the preceding section. Thus path generation with prescribed timing involves two free choices for the left side of the four-bar (the x and y location of A_0 with respect to P_1) and therefore ideally yields *two infinities* of solutions.

8.9 GRAPHICAL SYNTHESIS FOR PATH GENERATION (WITHOUT PRESCRIBED TIMING): FOUR POSITIONS

A design procedure similar to that of Fig. 8.41 may be employed for path generation (without prescribed timing) for four precision points using the *point-position reduction method* [83, 105].

The point-position reduction method is based on the fact that a circle can be drawn through three points. Three different relative positions for a point on a link are determined, then a circle is drawn through the points. The center and radius of the circle determine the position and lengths of the remaining links of the mechanism. Up to six precision points [105] can be satisfied in this method. However, the design parameters are chosen so that some corresponding positions of a design point, usually a pin joint, coincide and thereby the total number of distinct positions is reduced to three. This is demonstrated in designs 1 and 2, in which the number of distinct positions is reduced from four to three. This is accomplished by locating either point B_0 or B at one of the poles of the coupler. Designs will be presented first with B_0 and then with B at the pole.

Design 1

The task.   Design a four-bar mechanism such that the coupler point P will pass through four arbitrarily selected positions in the order P_1, P_2, P_3, and P_4 (Fig. 8.45). Lo-

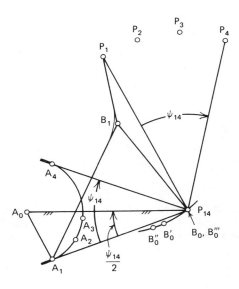

Figure 8.45 Four-position design. Layout showing parameters and design procedure. Pivot at pole.

cate the fixed pivot B_0 at one of the poles of the coupler motion. The procedure is as follows.

1. Choose two positions to make coincident in the inversion. Positions 1 and 4 were picked so that B_0 is positioned at pole P_{14}. The pole is located on the perpendicular bisector of the line $P_1 P_4$ (any convenient point on this line will do). This determines the angle ψ_{14}, rotation of the follower link from position 1 to 4.

2. Since B_0 is at the pole P_{14}, the coupler can be rotated about B_0 from position 1 to position 4. This means that A and B, both points on the coupler, must also rotate the same angle ψ_{14} about B_0 from position 1 to 4.

3. Select some direction for $A_0 B_0$ and draw two lines through B_0 at angle $\pm \psi_{14}/2$ from $B_0 A_0$ (Fig. 8.45). A_1 and A_4 must lie on these lines equidistant from B_0.

4. Choose positions for A_1 and A_0. This establishes A_0 and the lengths of the fixed and input links and the distance AP.

5. Locate A_2 and A_3 on the arc about A_0 with radius $A_0 A_1 = A_0 A_4$, such that $P_2 A_2 = P_3 A_3 = P_1 A_1$.

6. B_0 and B_0''' are located at P_{14}. Fix the coupler (a kinematic inversion) and locate the relative position of B_0 for positions 2 and 3 (B_0', B_0'') by constructing $\Delta A_1 P_1 B_0' = \Delta A_2 P_2 B_0$ and $\Delta A_1 P_1 B_0'' = \Delta A_3 P_3 B_0$. The center of the circle that passes through B_0, B_0', and B_0'' is B_1. This establishes the lengths of coupler and output links and completes the design.

7. Figure 8.46 shows the mechanism in all four positions as a check on the design.

It is to be noted that, between positions 3 and 4, the input crank rotates beyond B_4, and then rotates back to B_4, until the path point P finally comes into coincidence with the prescribed position P_4. During this forward and backward rotation of the input crank, point P goes off the prescribed path. This behavior is characteristic of designs resulting from point-position reduction methods and may be objectionable in some applications of path generators.

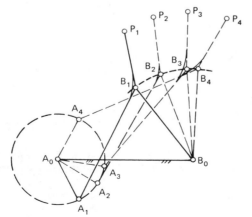

Figure 8.46 Four-position design. Check of completed mechanisms.

Design 2

The task.   Design a four-bar mechanism such that the coupler point P will pass through the prescribed positions P_1, P_2, P_3, and P_4 in that order (Fig. 8.47). Locate the coupler point B at one coupler pole. The procedure is as follows:

1. Locate the pole P_{14} on the perpendicular bisector of the line $P_1 P_4$ arbitrarily. Let B_1 and B_4 be collocated with P_{14}. Angle $P_1 P_{14} P_4$ is ψ_{14}.
2. Since the coupler triangle ABP is rigid, the angle $A_1 B_1 P_1$ must equal angle $A_4 B_4 P_4$. With B_1 and P_1 located, a line can be drawn from B_1 in an arbitrary direction to establish a locus for A_1. The distance $B_1 A_1$ is arbitrary.
3. Locate A_4 so that angle $A_1 B_1 A_4 = \psi_{14}$ in magnitude and sense and $\overline{A_4 B_1} = \overline{A_1 B_1}$.
4. Select the pivot A_0 for the input link on the bisector of angle $A_1 B_1 A_4$. Thus $\overline{A_0 A_1} = \overline{A_0 A_4}$. Draw the circular arc path of A from A_1 to A_4.
5. Locate A_2 so that $\overline{A_2 P_2} = \overline{A_1 P_1}$ and A_3 so that $\overline{A_3 P_3} = \overline{A_1 P_1}$.
6. $\Delta A_1 B_1 P_1 = \Delta A_2 B_2 P_2 = \Delta A_3 B_3 P_3 = \Delta A_4 B_4 P_4$. Use this information to locate B_2 and B_3.
7. Since B_1 and B_4 are collocated, a circle can be drawn that passes through B_1, B_4, B_2, and B_3. The center of this circle is the fixed pivot B_0. The radius is the length of the output link $\overline{B_0 B}$. This establishes the mechanism.

These two designs show how the pole is used in reducing the number of four-point positions to three. The graphical procedure is somewhat simpler when the coupler point B is at the pole than when the pivot B_0 is at the pole. The design situation may dictate which to use.

Notice that each of these designs involved choosing four parameters (e.g., in design 1 we picked arbitrarily the position of B_0 along the perpendicular bisector of $P_1 P_4$,

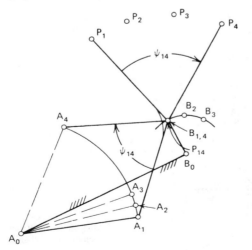

Figure 8.47 Four-position geometric synthesis of four-link path generator mechanism. First position of coupler point, B_1, is chosen at pole P_{14}. Point-position reduction method.

the x and y coordinate of A_0, and the radius $\overline{A_0 A}$). Thus there are *two infinities* of solutions per side for path generation for four prescribed positions. If path generation with prescribed timing (i.e., rotations of $\overline{A_0 A_1}$, $\overline{A_0 A_2}$, $\overline{A_0 A_3}$, and $\overline{A_0 A_4}$) were the objective, there would be *one infinity* of solutions per side. Lindholm [104, 105] has also presented the graphical procedures for five and six prescribed path positions using point-position reduction procedures.

8.10 FUNCTION GENERATOR: THREE PRECISION POINTS

Before describing the geometric construction method for function generation synthesis, optimal spacing of prescribed precision points will be presented.

Chebyshev determined that the best linkage approximation to a function occurs when the absolute value of the maximum structural error between precision points and at both ends of the range are equalized. *Chebyshev spacing* [86] of precision points is employed to minimize the structural error. This technique, based on *Chebyshev polynomials* [28, 86], is often used as a "first guess," although it is applicable only in special cases (such as symmetric functions). After the synthesis is completed, the resultant structural error of the mechanism can be determined, followed by assessment and alteration of the placement of precision points to improve the mechanism accuracy. Two techniques for locating precision points for minimized structural error are the *Freudenstein respacing formula* [72] and the *Rose-Sandor direct optimal spacing technique* [134]. Both are based on the fact that reducing the space between adjacent precision points reduces the extreme error between them, and vice-versa.

A simple construction is available for determining Chebyshev spacing as an initial guess (see Fig. 8.48). Precision points may be located graphically; a circle is drawn whose diameter is proportional to the range of the independent parameter (Δx). A regular equilateral polygon having $2n$ sides (where n = the number of prescribed precision points) is then inscribed in the circle such that two sides of the polygon are vertical.

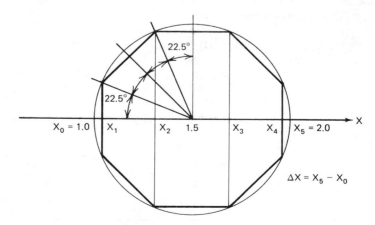

Figure 8.48   Chebyshev spacing of four precision points.

Lines drawn perpendicular to the horizontal diameter through each corner of the polygon intersect the diameter at points spaced at distances proportional to Chebyshev spacing of precision points. This procedure is now explained by way of examples.

Example 8.1

Determine the Chebyshev spacing for a four-bar linkage generating the function $y = 2x^2 - 1$, in the range $1 \le x \le 2$, where four precision points are to be prescribed ($n = 4$).

Solution   The first step is to draw a circle with diameter $\Delta x = x_{n+1} - x_0 = 2.0 - 1.0 = 1.0$. Next, construct a polygon of $2n = 8$ sides, with two sides vertical, as shown in Fig. 8.48. The corners of the polygon projected vertically onto the horizontal axis are the prescribed precision points. Measurements from this geometric construction yield

$$x_0 = 1.00, \qquad x_3 = 1.69$$
$$x_1 = 1.04, \qquad x_4 = 1.96$$
$$x_2 = 1.31, \qquad x_5 = 2.00$$

The foregoing construction for Chebyshev spacing is tantamount to the following formulas:

$$\Delta x_j = x_j - x_0 = \tfrac{1}{2}\Delta x\left[1 - \cos\left(\frac{\pi(2j - 1)}{2n}\right)\right], \qquad j = 1, 2, \ldots, n$$

and

$$x_j = x_0 + \Delta x_j, \qquad j = 1, 2, \ldots, n$$

where Δx_j is the distance from the beginning of the x range to the jth precision point

$$\Delta x = x_{n+1} - x_0 = \text{range in } x$$
$$j = \text{precision point number}, \qquad j = 1, 2, \ldots, n$$
$$n = \text{total number of precision points}$$

Thus, in this example

$$\Delta x_1 = \tfrac{1}{2}(1)\left[1 - \cos\left(\frac{\pi}{8}\right)\right] = 0.038$$
$$x_1 = 1.04$$

and

$$\Delta x_2 = 0.309, \qquad \Delta x_3 = 0.691, \qquad \Delta x_4 = 0.962$$

so that

$$x_2 = 1.31$$
$$x_3 = 1.69$$
$$x_4 = 1.96$$

Example 8.2

Given the Chebyshev precision points derived in Example 8.1 and the ranges in the input and output link rotations $\Delta\phi = 60°$, $\Delta\psi = 90°$, find ϕ_2, ϕ_3, ϕ_4, ψ_2, ψ_3, and ψ_4.

Solution y_j is found by substituting the values of x_j into the function $y = 2x^2 - 1$:

$$y_0 = 1.00, \qquad y_3 = 4.71$$
$$y_1 = 1.16, \qquad y_4 = 6.68$$
$$y_2 = 2.43, \qquad y_5 = 7.00$$

Using Eqs. (8.1) and (8.2), where $\Delta x = 1$, $\Delta y = 6$, $\Delta\phi = 60°$, and $\Delta\psi = 90°$, we have

$$\phi_2 = 16.2°, \qquad \psi_2 = 19.1°$$
$$\phi_3 = 39.0°, \qquad \psi_3 = 54.3°$$
$$\phi_4 = 55.2°, \qquad \psi_4 = 82.8°$$

The graphical procedure for three-precision-point function generation is very similar to that of motion and path generation for the same number of precision points. Again, kinematic inversion and the intersection of midnormals are used. An illustrative example [39] will be employed to demonstrate the method.

A mechanism will be synthesized to generate the function $y = \sin(x)$ for $0° \leq x \leq 90°$. The input range is chosen arbitrarily to be $\Delta\phi = 120°$ and the output range is similarly chosen to be $\Delta\psi = 60°$. For this case the scale factors R_ϕ and R_ψ are found to be

$$R_\phi = \frac{\Delta\phi}{\Delta x} = \frac{120°}{90°} = \frac{4}{3}$$

$$R_\psi = \frac{\Delta\psi}{\Delta y} = \frac{60°}{1} = 60° \tag{8.8}$$

The next task is to pick three precision points, x_1, x_2, and x_3. Chebyshev spacing will be used for these precision points [28]. Referring to Fig. 8.49, we find that

$$x_0 = 0°, \qquad x_3 = 84°$$
$$x_1 = 6°, \qquad x_4 = 90°$$
$$x_2 = 45°$$

Eqs. (8.3) and (8.4) that

$$\phi_j = R_\phi(x_j - x_1)$$
$$\psi_j = R_\psi(y_j - y_1)$$

so that

$$\phi_2 = (4/3)(45 - 6) = 52°, \qquad \psi_2 = 60(0.7071 - 0.1045) = 36.15°$$
$$\phi_3 = (4/3)(84 - 6) = 104°, \qquad \psi_3 = 60(0.9945 - 0.1045) = 53.40°$$

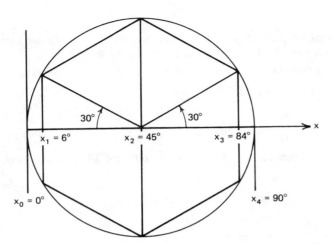

Figure 8.49 Graphical determination of three precision points with Chebyshev spacing.

See Fig. 8.50 for a geometric interpretation of the function generation synthesis task. The graphical construction procedure is as follows (refer to Figs. 8.50 and 8.51).

1. Pick the position of the ground pivots (A_0, B_0) and the output link $(B_0 B)$. Here the ground pivots are along the x axis, the length of the fixed link $\overline{A_0 B_0} = 1$ unit, the length of the output link $\overline{B_0 B} = 0.75$ unit, and $\psi_0 = 60°$. Notice that the initial position of the output link is therefore $\psi_0 + (60°)(0.1045 - 0) = 66.27°$ (see Fig. 8.50).

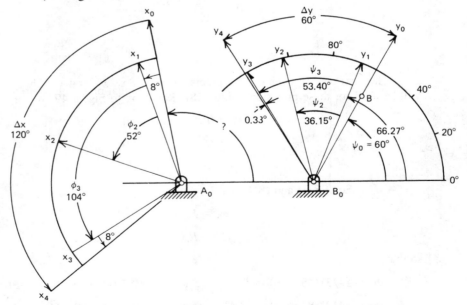

Figure 8.50 Prescribed values of input and output rotations for four-bar function generator synthesis with three finitely separated precision points.

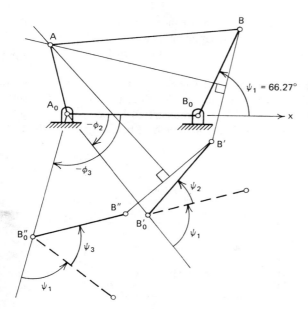

$\psi_1 = 66.27°$

Figure 8.51 Kinematic inversion applied to the three-point function generation synthesis of a four-bar mechanism.

2. Using inversion, fix the input link (although the position of the input link is unknown). The mechanism will now be moved through the specified precision points preserving the same relative motion between links. Therefore, in the second precison position the fixed link rotates by $-\phi_2 = -52°$ about A_0, locating B_0', while the output rotates by $\psi_2 = 36.15°$ about B_0' locating B'. The third precision-point position may be generated by rotations (from the first position) of $-104°$ for the fixed link about A_0, locating B_0'', and $53.40°$ for the output link about B_0'', locating B''.

3. Lines $B_0 B$, $B_0' B'$, and $B_0'' B''$ represent the actual precision positions of the output link relative to the input link. The center of the circular arc $B - B' - B''$ will locate A, found by intersection of the perpendicular bisectors of BB' and $B''B'$ (see Fig. 8.51).

Two infinities of solutions are available for each side of the four-bar for function generation for three prescribed finitely separated positions, since both the positions of A_0 and B relative to B_0 (four parameters) and thus the ground and output links, were picked arbitrarily in the construction.

Before moving on to analytical methods, another popular function-generation technique, the overlay method will be described.

8.11 THE OVERLAY METHOD

Another graphical method often used for kinematic synthesis (primarily for function generation) is the overlay technique. It consists of constructing a part of the solution to a problem on transparent paper and another part of the solution on a separate sheet. The transparency (overlay) is placed over the separate sheet and a search is made by moving the transparency until precision points are matched between the transparency and the separate sheet.

The technique can be used for the synthesis of mechanisms involving two to five positions although the solution procedure is more difficult as the number of required precision points increases. The method will be demonstrated by way of a five-precision-point design [39, 104]. A four-bar function generator is to be designed for the following precision points:

Precision point number	Crank rotation from starting positions (deg)	
	Input (cw)	Output (cw)
1	0	0
2	$\phi_2 = 15°$	$\psi_2 = 20°$
3	$\phi_3 = 30$	$\psi_3 = 35$
4	$\phi_4 = 45$	$\psi_4 = 50$
5	$\phi_5 = 60$	$\psi_5 = 60$

Method

1. On tracing paper lay out the input crank positions and select lengths for the input and coupler links (see Fig. 8.52). Draw a family of circular arcs with centers at successive crank pin positions with a radius equal to the arbitrarily chosen coupler length.

2. On a second piece of paper (Fig. 8.53) lay out the output crank positions and add several arcs, indicating possible lengths of link 4.

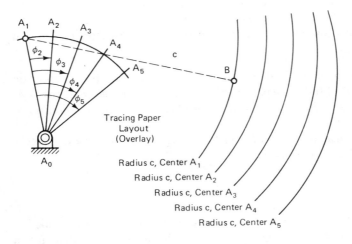

Tracing Paper
Layout
(Overlay)

Radius c, Center A_1
Radius c, Center A_2
Radius c, Center A_3
Radius c, Center A_4
Radius c, Center A_5

Figure 8.52 Overlay technique. A five-position design. Input crank and connecting rod side.

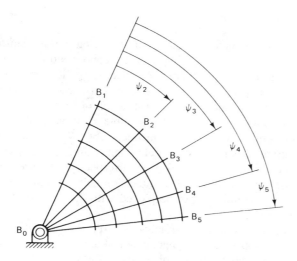

Figure 8.53   Five-position design. Layout of output crank possibilities.

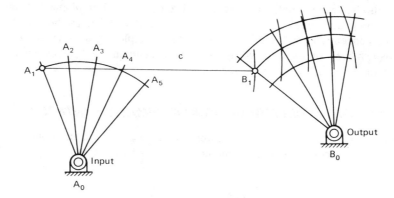

Figure 8.54   Five-position overlay technique design. Fitting of the overlay and the resulting mechanism.

3. Place the first layout on the second and move until the family of arcs of Fig. 8.52 falls on the respective positions of the output crank as shown in Fig. 8.54. This establishes the lengths of the ground link and the output link.

It may be necessary to try different lengths for the coupler link in order to achieve a match between the overlay of Fig. 8.52 and the layout of Fig. 8.53. With practice, this method should be accurate within 1°.

Notice that the scale factors ($\Delta\phi$, $\Delta\psi$) and the coupler length are free choices here. There is no guaranteed solution, however. This is not a closed-form solution; it is a trial-and-error technique.

8.12 ANALYTICAL SYNTHESIS TECHNIQUES

Figures 8.37 to 8.39 show that the geometric construction of four-bar motion generators for two and three prescribed positions is a fairly simple task. Suppose, however, that we wish to find an "optimal" four-bar motion generator for a specific application — perhaps a case that has constraints on ground and moving pivot locations, transmission angle, link-length ratio, and/or mechanical advantage. The construction of Fig. 8.39, although simple, may be too time consuming to repeat until a suitable solution is obtained. A graphical search through two infinities of solutions is inconceivable. What other alternatives are available? By choosing the position of the circle point A_1 in Fig. 8.39, we have arbitrarily picked two free choices — those free choices in turn specify the corresponding center point A_0. These two free choices for the three-precision-point motion-generation synthesis of one side of the four-bar linkage can be picked with different strategies in mind toward various design objectives.

In order to obtain a handle on the design variables and free choices, an analytical model of the linkage must be developed. Several mathematical techniques for modeling linkages have been utilized for planar synthesis objectives. These include algebraic methods, matrix methods, and complex numbers. For planar linkages, the complex numbers technique is the simplest, yet the most versatile method. In this text we therefore concentrate on the latter method. Before exploring the question of free choices versus synthesis options, the complex-number technique will be reviewed,* especially as it relates to modeling linkages for synthesis.

8.13 COMPLEX-NUMBER MODELING IN KINEMATIC SYNTHESIS

Any planar mechanism can be represented by a general chain, consisting of one or more loops of successive bar-slider members (Fig. 8.55). For example, the offset slider-crank

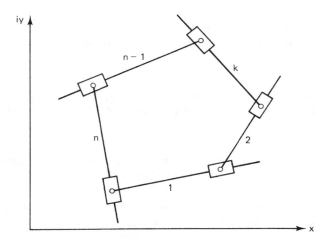

Figure 8.55 General planar chain.

---

* A more complete review of complex numbers is given in the appendix to Chap. 3.

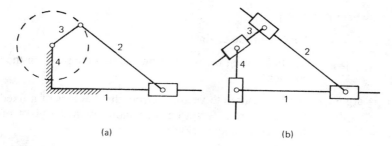

Figure 8.56   (a) offset slider-crank mechanism; (b) its equivalent general chain.

mechanism of Fig. 8.56a may be derived from the general chain (Fig. 8.56b) by fixing the sliders to their respective bars between members 1 and 4, 4 and 3, and 3 and 2 as well as fixing bars 1 and 4 to ground.

Complex numbers readily lend themselves as an ideal tool for modeling linkage members as parts of planar chains. For each bar-slider member of Fig. 8.55, the position of the pivot on the slider with respect to the pivot of the bar can be defined by the relative position vector $\mathbf{Z}_k$ (Fig. 8.57a) expressible as a complex number. The first or starting position of the kth bar can be written as

$$\mathbf{Z}_k = Z_k e^{i\theta_1} = Z_k(\cos \theta_1 + i \sin \theta_1) \tag{8.9}$$

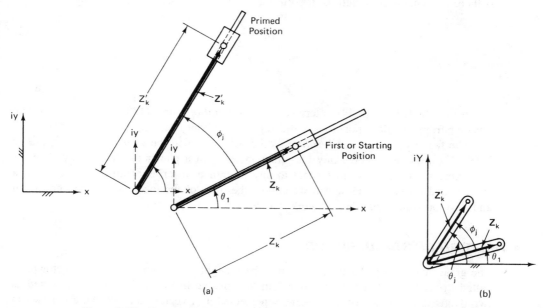

Figure 8.57   Complex-vector representation of a bar-slider pair; (a) stretch rotation; (b) pure rotation.

where $i \equiv \sqrt{-1}$

$k = k$th bar of the chain

$Z_k = |\mathbf{Z}_k| =$ length between the pivot of the bar and the pivot on the slider in the first position

$\theta_1 = \arg \mathbf{Z}_k =$ angle measured to vector $\mathbf{Z}_k$ from the real axis of a fixedly oriented rectangular coordinate system translating with the pivot of the bar (angles measured counterclockwise are positive)

If there is no change in the length of the kth bar in the chain from the first to the primed (jth) position as shown in Fig. 8.57b, $\mathbf{Z}_k'$ is expressible as

$$\mathbf{Z}_k' = Z_k e^{i(\theta_1 + \phi_j)} = Z_k e^{i\theta_1} e^{i\phi_j} \tag{8.10}$$

where

$$\phi_j = \theta_j - \theta_1 \tag{8.11}$$

Notice that as a link moves in the plane, a coordinate system is pinned to the base of the link (Fig. 8.57a). This coordinate system remains parallel to a fixed set of coordinates so that θ_j and θ_1 are arguments of $\mathbf{Z}$ in the jth and first positions respectively while ϕ_j is the angle of rotation from position 1 to j. Using Eq. (8.9) yields

$$\mathbf{Z}_k' = \mathbf{Z}_k e^{i\phi_j} \tag{8.12}$$

If there is a change in length of the kth bar, and if this change is defined by

$$\rho_j \equiv \frac{Z_k'}{Z_k} \tag{8.13}$$

then

$$\mathbf{Z}_k' = \mathbf{Z}_k \rho_j e^{i\phi_j} \tag{8.14}$$

$e^{i\phi_j}$ in Eqs. (8.12) and (8.14) is termed the *rotational operator* [138, 140] and will rotate a vector from its initial position by the angle ϕ_j without changing the length of the vector. The factor ρ_j is the *stretch ratio,** while $\rho_j e^{i\phi_j}$ is called the *stretch rotation operator* [138]. We may now model any bar-slider member in a planar mechanism by a vector and express its motion with respect to any reference in terms of an initial position, a stretch, and a rotation. How can we collect the links of the mechanism into one model and develop some equations to work with?

8.14 THE DYAD OR STANDARD FORM

The great majority of planar linkages may be thought of as combinations of vector pairs called *dyads* [145]. For example, the four-bar linkage in Fig. 8.58 can be perceived as two dyads: the left side of the linkage represented as a vector pair ($\mathbf{W}$ and $\mathbf{Z}$) shown in

* See, for example, the slider-crank of Fig. 8.82 and Eq. (8.83).

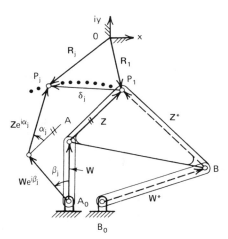

Figure 8.58 Notation associated with a dyad shown as it would model the left half of a four-bar linkage. The dyad (**W** and **Z**) is drawn in its first and jth positions.

solid lines, and the right side represented by the dashed dyad (**W*** and **Z***). The vectors that represent the coupler $\overrightarrow{AB}$ and the ground link $\overrightarrow{A_0B_0}$ are easily determined by vector addition when these dyads are synthesized [see Eqs. (8.25) and (8.26)]. The path point of the coupler link moves along a path from position P_1 to P_j defined in an arbitrary complex coordinate system by $\mathbf{R}_1$ and $\mathbf{R}_j$.

All vector rotations are measured from the starting position, positive counterclockwise (Figs. 8.58 and 8.59). Angle β_2 is the rotation of vector **W** from the first to the second position, while β_3 is the rotation from the first to the third position. Similarly, angles α_j are rotations of the vector **Z** from its first to its jth position (see Fig. 8.59).

Suppose that we specify two positions for an unknown dyad by prescribing the values of $\mathbf{R}_1$, $\mathbf{R}_j$, α_j, and β_j (Fig. 8.58). To find the unknown starting position vectors of the dyad, **W** and **Z**, a loop closure equation may be derived by summing the vectors clockwise around the loop containing $\mathbf{W}e^{i\beta_j}$, $\mathbf{Z}e^{i\alpha_j}$, $\mathbf{R}_j$, $\mathbf{R}_1$, **Z**, and **W**:

$$\mathbf{W}e^{i\beta_j} + \mathbf{Z}e^{i\alpha_j} - \mathbf{R}_j + \mathbf{R}_1 - \mathbf{Z} - \mathbf{W} = 0 \tag{8.15}$$

or

$$\boxed{\mathbf{W}(e^{i\beta_j} - 1) + \mathbf{Z}(e^{i\alpha_j} - 1) = \boldsymbol{\delta}_j} \tag{8.16}$$

where the displacement vector along the prescribed trajectory from P_1 to P_j is

$$\boldsymbol{\delta}_j \equiv \mathbf{R}_j - \mathbf{R}_1 \tag{8.17}$$

Equation (8.16) is the *standard-form* equation. This equation is simply the vector sum around the loop containing the first and jth positions of the dyad forming the left side of the four-bar linkage. As we will see, Eq. (8.16) is called the standard form if $\boldsymbol{\delta}_j$

* See, for example, the slider-crank of Fig. 8.82 and Eq. (8.83).

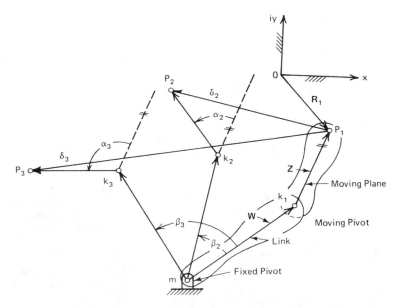

Figure 8.59   Schematic of the **W**, **Z** dyad shown in three positions. The precision points P_1, P_2, P_3 are located by $\mathbf{R}_1$, $\mathbf{R}_2$, $\mathbf{R}_3$, while all rotations are expressed from the first dyad position.

and either α_j or β_j are prescribed or known. This requirement is consistent with the definitions of the usual tasks of kinematic synthesis: motion generation, path generation with prescribed timing, and function generation.

8.15 NUMBER OF PRESCRIBED POSITIONS VERSUS NUMBER OF FREE CHOICES

For how many positions can we synthesize a four-bar linkage for motion, path, or function generation? A finite number of parameters (the two components of each vector) completely describe this linkage in its starting position. Therefore, there are only a finite number of prescribable parameters which can be imposed in a synthesis effort. The four-bar motion generator will be used to determine how many positions may actually be prescribed. In Fig. 8.58, the path displacement vectors δ_j and coupler rotations α_j will be prescribed in a *motion-generation task*.

Table 8.1 illustrates how to determine the maximum number of prescribable positions for the synthesis of a four-bar motion generator. Although Table 8.1 is based on the left side of the linkage of Fig. 8.58 [Eq. (8.16)], the right side of the linkage will yield the same results [see Eq. (8.24)]. The table shows that, for two positions there are two independent scalar equations contained in the vector equation Eq. (8.16): the summation of x components and the summation of the y components of the vectors. These are called the *real* and *imaginary* parts of the equation, each a scalar equation in itself. This system of two scalar equations contains five scalar unknowns: two coordinates each

TABLE 8.1 MAXIMUM NUMBER OF SOLUTIONS FOR THE UNKNOWN DYAD **W, Z** WHEN δ_j AND α_j ARE PRESCRIBED IN THE EQUATION:

$$\mathbf{W}(e^{i\beta_j} - 1) + \mathbf{Z}(e^{i\alpha_j} - 1) = \delta_j \tag{8.16}$$

Number of positions (n): $j = 2, 3, \ldots, n$	Number of scalar equations	Number of scalar unknowns	Number of free choices (scalars)	Number of solutions
2	2	5($\mathbf{W}$, $\mathbf{Z}$, β_2)	3	$O(\infty^3)$
3	4	6(above + β_3)	2	$O(\infty^2)$
4	6	7(above + β_4)	1	$O(\infty^1)$
5	8	8(above + β_5)	0	Finite

of the vectors **W** and **Z** (W_x, W_y, Z_x, and Z_y) and the input rotation β_2. If three of the five unknowns are chosen arbitrarily, the equations can be solved for the remaining two unknowns. Since in general there is an infinite number of choices for each of the three free choices, the number of possible solutions for the two-position synthesis problem is on the order of *infinity cubed*, symbolized by $O(\infty^3)$.

In the case of three-prescribed positions of the moving plane, specified by three precision points P_1, P_2 and P_3 and two angles of rotation, α_2 and α_3, there are two more real equations but only one more scalar unknown (β_3). Thus two free choices can be made and $O(\infty)^2$ solutions are available. Each additional prescribed position in Table 8.1 adds two scalar equations and one scalar unknown. Thus, for four positions, there is one free choice and a single infinity of solutions. For five prescribed positions there are no free choices available, and at best a finite number of solutions will exist (see Chap. 3 of Vol. 2). Five prescribed positions is therefore the maximum number of precision points possible for the standard-form solutions for the motion-generation dyad of Fig. 8.58.

Table 8.1 correlates the number of prescribed positions, the number of free choices, and the number of closed-form solutions expected for the standard form. However, Table 8.1 does not say anything directly about the difficulty in solving the sets of standard-form equations in closed form. An important question is this: can a linear equation-solver technique be applied for two, three, four, and five prescribed positions?

The answer lies in the form of the respective sets of equations of synthesis: are they linear or nonlinear in the unknown reals? A nonlinearity test will be applied to Eq. (8.16) for each row of Table 8.1.

Two positions.   There are three free choices to be made in Eq. (8.16). For example, if δ_2 and α_2 were prescribed, **Z** and β_2 could be chosen arbitrarily, yielding a simple linear solution for the remaining unknown, **W**:

$$\mathbf{W} = \frac{\delta_2 - \mathbf{Z}(e^{i\alpha_2} - 1)}{e^{i\beta_2} - 1} \tag{8.18}$$

This case of motion generation for two positions for a dyad is analogous to the graphical technique described in Sec. 8.5. *In both cases there are three infinities of solutions.* Design methods can be developed for two prescribed positions with the ability to optimize other indices of performance *a priori*, such as the transmission angle.

Three positions. Here, according to Table 8.1, two free choices must be made, so one may expect *two infinities of solutions,* as in the graphical method (Sec. 8.6). The system of equations for three positions is

$$\mathbf{W}(e^{i\beta_2} - 1) + \mathbf{Z}(e^{i\alpha_2} - 1) = \boldsymbol{\delta}_2$$
$$\mathbf{W}(e^{i\beta_3} - 1) + \mathbf{Z}(e^{i\alpha_3} - 1) = \boldsymbol{\delta}_3 \tag{8.19}$$

If $\boldsymbol{\delta}_2$, $\boldsymbol{\delta}_3$, α_2, and α_3 are prescribed, β_2 and β_3 can be picked arbitrarily. Thus system (8.19) is a set of complex equations, *linear* in the complex unknowns $\mathbf{W}$ and $\mathbf{Z}$ (the vectors representing the dyad in its first position) with known coefficients. The system can be solved by Cramer's rule:

$$\mathbf{W} = \frac{\begin{vmatrix} \boldsymbol{\delta}_2 & e^{i\alpha_2} - 1 \\ \boldsymbol{\delta}_3 & e^{i\alpha_3} - 1 \end{vmatrix}}{\begin{vmatrix} e^{i\beta_2} - 1 & e^{i\alpha_2} - 1 \\ e^{i\beta_3} - 1 & e^{i\alpha_3} - 1 \end{vmatrix}} \tag{8.20}$$

$$\mathbf{Z} = \frac{\begin{vmatrix} e^{i\beta_2} - 1 & \boldsymbol{\delta}_2 \\ e^{i\beta_3} - 1 & \boldsymbol{\delta}_3 \end{vmatrix}}{\begin{vmatrix} e^{i\beta_2} - 1 & e^{i\alpha_2} - 1 \\ e^{i\beta_3} - 1 & e^{i\alpha_3} - 1 \end{vmatrix}} \tag{8.21}$$

Equations (8.20) and (8.21) are readily programmed on a hand calculator or microcomputer.

The three-position motion-synthesis case yields a linear solution if β_2 and β_3 are free choices. The two free choices in Eq. system (8.16) may be made with different strategies (as will be explored in Sec. 8.20 and 8.21), but will always involve two infinities of solutions [43, 44, 58, 61, 64, 108, 124, 125].

Four positions. The system of equations for four prescribed positions of the moving plane is as follows:

$$\mathbf{W}(e^{i\beta_2} - 1) + \mathbf{Z}(e^{i\alpha_2} - 1) = \boldsymbol{\delta}_2$$
$$\mathbf{W}(e^{i\beta_3} - 1) + \mathbf{Z}(e^{i\alpha_3} - 1) = \boldsymbol{\delta}_3 \tag{8.22}$$
$$\mathbf{W}(e^{i\beta_4} - 1) + \mathbf{Z}(e^{i\alpha_4} - 1) = \boldsymbol{\delta}_4$$

Table 8.1 allows only one free choice from among the seven real unknowns: coordinates of $\mathbf{W}$, $\mathbf{Z}$ and angles β_2, β_3, and β_4. Recall that $\boldsymbol{\delta}_j$ and α_j, $j = 2, 3, 4$, are prescribed. Thus only one of the rotations or one coordinate of a link vector can be picked arbitrarily. System (8.22) contains three unknown angles β_j in transcendental expressions. Even if we pick one β_j as a free choice, Eq. system (8.22) requires a nonlinear equation-solving technique. Thus three precision points comprise the maximum number which may be prescribed and yet obtain a linear solution. For the four-position problem, Chap. 3 of Vol. 2 presents a closed-form nonlinear solution for Eq. (8.22), yielding up to an infinity of solutions. The LINCAGES and KINSYN software packages are built around the three and four prescribed position cases [6, 50, 51, 57, 60, 62, 93, 114, 124, 125, 136, 171]. Refer to color inserts in this book for LINCAGES example output.

Five positions.   The system of equations for five positions, which adds one equation added to Eq. system (8.22) with $j = 5$, is also nonlinear in the unknowns β_j, and there are no free choices available. This case is also solved in closed form in Chap. 3 of Vol. 2.

8.16 THREE PRESCRIBED POSITIONS FOR MOTION, PATH, AND FUNCTION GENERATION

This chapter concentrates on kinematic synthesis objectives that yield linear solutions — those easily solved graphically, on a hand calculator, or by a simple computer program. In the preceding section we discovered that, for motion generation of a dyad, three positions were the limit for a linear solution. Table 8.1 shows that there are two free choices to be made amongst the variables **W**, **Z**, β_2 and β_3. Although there is good logic in choosing β_2 and β_3, as was done in Eqs. (8.20) and (8.21), other free choices may be made to satisfy strategies other than one that has a simple equation set. This and the next section will continue with the standard form solution procedure, followed by other design strategies for three prescribed positions in subsequent sections.

Notice that the balancing of the number of equations and the number of unknowns in Table 8.1 was based on motion synthesis. Equation (8.16) has been termed the *standard form* with the understanding that both δ_j and α_j or β_j were prescribed. The numbers in Table 8.1 will be the same if β_j were prescribed instead of α_j, which, as we know from Sec. 8.2, is the case for path generation with prescribed timing.

We will look at the four-bar mechanism of Fig. 8.60, as well as a six-bar linkage in Sec. 8.20, and attempt to express the synthesis of these linkages in the standard form for motion, path, and function generation. If this can be accomplished, only one computer program will be needed to synthesize these linkages for either of these tasks. (This generality of the standard form also extends to the nonlinear solutions of Chap. 3 of Vol. 2.)

Synthesis of a Four-Bar Motion Generator for Three Precision Points

The four-bar linkage of Fig. 8.60 is to be synthesized for motion generation. As suggested in Sec. 8.14, there are two independent dyads in the four-bar linkage, which will be called the *left-hand side* and the *right-hand side*. Each dyad connects a ground pivot (a center point) to the path point on the coupler by way of joint A or B of the coupler (the circle point). The equations describing the displacements of the left-hand side have already been derived, but in the notation of Fig. 8.60 the standard form is

$$\mathbf{Z}_2(e^{i\phi_j} - 1) + \mathbf{Z}_5(e^{i\gamma_j} - 1) = \delta_j, \qquad j = 2, 3 \tag{8.23}$$

where δ_j and γ_j are prescribed.

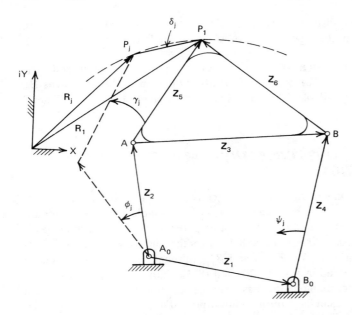

Figure 8.60 Four-bar motion- and path-generator mechanism.

The displacement equations for the right-hand side of the linkage may be written as

$$\mathbf{Z}_4(e^{i\psi_j} - 1) + \mathbf{Z}_6(e^{i\gamma_j} - 1) = \boldsymbol{\delta}_j, \qquad j = 2, 3 \tag{8.24}$$

where $\boldsymbol{\delta}_j$ and γ_j are prescribed.

If we assume ϕ_j and ψ_j arbitrarily, Eqs. (8.23) and (8.24) can be solved by Cramer's rule for $\mathbf{Z}_2$, $\mathbf{Z}_5$, $\mathbf{Z}_4$, and $\mathbf{Z}_6$. See the form of solution in Eqs. (8.20) and (8.21). The other two linkage vectors are simply

$$\mathbf{Z}_3 = \mathbf{Z}_5 - \mathbf{Z}_6 \tag{8.25}$$

and

$$\mathbf{Z}_1 = \mathbf{Z}_2 + \mathbf{Z}_3 - \mathbf{Z}_4 \tag{8.26}$$

Example 8.3

This example* will help demonstrate the correlation between the graphical and complex number methods for three-precision-point standard-form motion synthesis. The graphical solution will be demonstrated first. Figure 8.61a shows a rigid body in three desired positions. The angular orientation and precision positions of the body in three positions are known. Suppose points A and B are chosen as proposed circle-point locations. Choosing these locations constitutes making four free choices, the x and y coordinates of both points. These choices dictate the location of the ground pivots A_0 and B_0 shown in Fig. 8.61b, found by the intersection of the perpendicular bisectors as described in Sec. 8.6 (refer to Fig. 8.39). The resulting starting positions of the input and output dyads, $\mathbf{W}_A \mathbf{Z}_A$ and $\mathbf{W}_B \mathbf{Z}_B$, and the arguments of the input and output links at the three precision positions are

* Contributed by Ray Giese and John Titus.

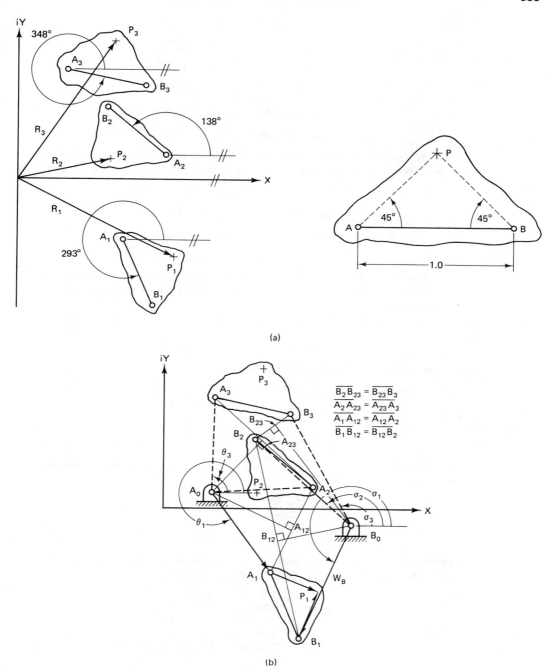

(a)

(b)

Figure 8.61   Three position motion synthesis of a four-bar linkage: (a) coupler link *AB* shown in three prescribed coplanar positions as a rigid body; (b) graphical construction for locating A_0 and B_0.

$$\mathbf{W}_A = .72 - 1.06i \qquad \mathbf{W}_B = -.66 - 1.55i$$

$$\theta_1 = 304.2°, \qquad\qquad \sigma_1 = 247.1°$$

$$\theta_2 = 2.6°, \qquad\qquad \sigma_2 = 136.7°$$

$$\theta_3 = 88.4°, \qquad\qquad \sigma_3 = 116.6°$$

$$\mathbf{Z}_A = .656 - .265i \qquad \mathbf{Z}_B = .265 + .656i$$

($\mathbf{Z}_A$ and $\mathbf{Z}_B$ are known from the choices for points A and B)

Now we shall try to match the graphically generated solution with the standard-form method. Solving for the input side first, using path tracer point P, we compute from Fig. 8.61.

$$\delta_2 = \mathbf{R}_2 - \mathbf{R}_1 = (1.2 + .25i) - (2.0 - 1.1i) = -.8 + 1.35i$$

$$\delta_3 = \mathbf{R}_3 - \mathbf{R}_1 = (1.25 + 1.9i) - (2.0 - 1.1i) = -.75 + 3.0i$$

$$\alpha_2 = 138° - 293° = -155° = 205°$$

$$\alpha_3 = 348° - 293° = 55°$$

The input link rotation angles are free choices in the motion generation solution. We choose the same angles as found in the graphical solution to demonstrate the correlation between the two solutions:

$$\beta_2 = \theta_2 - \theta_1 = 2.6° - 304.2° = -301.6° = 58.4°$$

$$\beta_3 = \theta_3 - \theta_1 = 88.4° - 304.2° = -215.8° = 144.2°$$

The input link solution as given by Eq. (8.20) is

$$\mathbf{W}_A = \frac{\delta_2(e^{i\alpha_3} - 1) - \delta_3(e^{i\alpha_2} - 1)}{(e^{i\beta_2} - 1)(e^{i\alpha_3} - 1) - (e^{i\beta_3} - 1)(e^{i\alpha_2} - 1)}$$

$$\mathbf{W}_A = \frac{\delta_2 e^{i\alpha_3} - \delta_3 e^{i\alpha_2} + \delta_3 - \delta_2}{e^{i(\beta_2+\alpha_3)} - e^{i(\beta_3+\alpha_2)} - e^{i\beta_2} - e^{i\alpha_3} + e^{i\beta_3} + e^{i\alpha_2}}$$

Using Euler's equation ($e^{i\theta} = \cos\theta + i\sin\theta$) and substituting the values for the variables, we obtain

$$\mathbf{W}_A = \frac{-3.462 + 4.171i}{-4.194 - .403i}$$

$$\mathbf{W}_A = .723 - 1.064i = 1.287 \angle 304.21°$$

which, when compared to the graphical solution, shows less than 1% difference. This is well within graphical accuracy. Vector $\mathbf{Z}_A$ is calculated from Eq. (8.21).

$$\mathbf{Z}_A = \frac{\delta_3(e^{i\beta_2} - 1) - \delta_2(e^{i\beta_3} - 1)}{(e^{i\beta_2} - 1)(e^{i\alpha_3} - 1) - (e^{i\beta_3} - 1)(e^{i\alpha_2} - 1)}$$

$$\mathbf{Z}_A = \frac{\delta_3 e^{i\beta_2} - \delta_2 e^{i\beta_3} - \delta_3 + \delta_2}{e^{i(\beta_2+\alpha_3)} - e^{i(\beta_3+\alpha_2)} - e^{i\beta_2} - e^{i\alpha_3} + e^{i\beta_3} + e^{i\alpha_2}}$$

$$\mathbf{Z}_A = \frac{-2.856 + .847i}{-4.1941 - .4025i}$$

$$\mathbf{Z}_A = .656 - .265i = .707 \angle 338°$$

This is exactly the value that was chosen for $\mathbf{Z}_A$ in the graphical solution. The output side dyad is synthesized using the same values for δ_2, δ_3, α_2 and α_3 but requires two free choices for the output link rotation angles. Again, we choose the values determined in the graphical synthesis.

$$\beta_2 = \sigma_2 - \sigma_1 = 136.7° - 247.1° = -110.4° = 249.6°$$

$$\beta_3 = \sigma_3 - \sigma_1 = 116.6° - 247.1° = -130.5° = 229.5°$$

The same procedure is followed for the output side as for the input side, resulting in

$$\mathbf{W}_B = \frac{-3.462 + 4.171i}{-1.481 - 2.852i}$$

$$\mathbf{W}_B = -.655 - 1.554i = 1.686 \; \measuredangle \; 247.15°$$

This is exactly the graphical solution for this link. Finally, the output coupler side is found to be:

$$\mathbf{Z}_B = \frac{-1.477 - 1.727i}{-1.481 - 2.852i}$$

$$\mathbf{Z}_B = .265 + .656i = .707 \; \measuredangle \; 68°$$

Again, this is the same as that found in the graphical solution.

Synthesis of a Four-Bar Path Generator with Prescribed Timing

Suppose that the four-bar linkage of Fig. 8.60 is to be synthesized for path generation with prescribed timing. The very same equations as derived for motion generation, Eqs. (8.23) to (8.26) will apply in this case, but the prescribed angles will be different. Instead of γ_j in Eq. (8.32), ϕ_j will be prescribed and γ_j, $j = 2, 3$, are free choices. Thus Eq. (8.23) will still be in the standard form. As for Eq. (8.24), in order to connect the right-hand side with the left side, vector $\mathbf{Z}_6$ must rotate by the same rotations (γ_j) as $\mathbf{Z}_5$. Thus the same γ_j, $j = 2, 3$, that were picked as free choices for Eq. (8.23), are prescribed in Eq. (8.24). Therefore, the four-bar path generator with prescribed timing has the same solution procedure as the four-bar motion generator.

Synthesis of a Four-Bar Function Generator

The standard form for a four-bar function generator can be derived from Fig. 8.60 as follows. Recall that in function generation we wish to correlate the prescribed rotations of the input link (ϕ_j) and the output link (ψ_j). Therefore, the upper portion of the coupler link ($\mathbf{Z}_5$ and $\mathbf{Z}_6$) is of no concern for this task. Figure 8.62 shows the basic four-bar of Fig. 8.60 in the first and jth position. The vector loop containing $\mathbf{Z}_2$, $\mathbf{Z}_3$, and $\mathbf{Z}_4$ is

$$\mathbf{Z}_2(e^{i\phi_j} - 1) + \mathbf{Z}_3(e^{i\gamma_j} - 1) - \mathbf{Z}_4(e^{i\psi_j} - 1) = 0 \qquad (8.27)$$

Since this vector equation is not in the standard form, Table 8.2 is formulated to help correlate the number of free choices and the number of prescribed positions. The

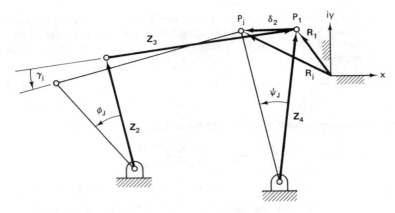

Figure 8.62 Four-bar function-generator mechanism.

TABLE 8.2 NUMBER OF AVAILABLE SOLUTIONS IN THE SYNTHESIS OF FOUR-BAR FUNCTION GENERATORS (FIG. 8.62) ACCORDING TO THE EQUATION

$$\mathbf{Z}_2(e^{i\phi_j} - 1) + \mathbf{Z}_3(e^{i\gamma_j} - 1) - \mathbf{Z}_4(e^{i\psi_j} - 1) = 0 \qquad (8.27)$$

Number of positions (n): $j = 2, 3, \ldots, n$	Number of scalar equations	Number of scalar unknowns	Number of free choices (scalars)	Number of solutions
2	2	7($\mathbf{Z}_2$, $\mathbf{Z}_3$, $\mathbf{Z}_4$, γ_2)	5	$O(\infty)^5$
3	4	8(above + γ_3)	4	$O(\infty)^4$
4	6	9(above + γ_4)	3	$O(\infty)^3$
5	8	10(above + γ_5)	2	$O(\infty)^2$
6	10	11(above + γ_6)	1	$O(\infty)^1$
7	12	12(above + γ_7)	0	Finite

same development as that done in connection with Table 8.1 is repeated here. Notice that the maximum number of prescribed positions is seven when a triad (three links) is used and when two of the three rotations are prescribed, as they must be for function generation.

Picking $\mathbf{Z}_4$ as an arbitrary choice ($\mathbf{Z}_2$ could be picked instead) will convert Eq. (8.27) to the standard form:

$$\mathbf{Z}_2(e^{i\phi_j} - 1) + \mathbf{Z}_3(e^{i\gamma_j} - 1) = \boldsymbol{\delta}_j = \mathbf{Z}_4(e^{i\psi_j} - 1) \qquad (8.28)$$

The justification for picking $\mathbf{Z}_4$ is twofold: first, by comparing Tables 8.1 and 8.2, the latter becomes equivalent to the first if two of the original seven real unknowns of Table 8.2 are picked arbitrarily; second, by choosing $\mathbf{Z}_4$, we are actually specifying the scale and orientation of the function generator. In fact, once a four-bar is synthesized for function generation, the entire linkage may be scaled up or down and oriented in any direction without changing the functional relationship between input and output link rotations $[\psi_j = f(\phi_j)]$. Therefore, new function-generation solutions do not result from allowing $\mathbf{Z}_4$ to be an unknown. (Path and motion generators *do change* their prescribed path with a change in scale; therefore, only in function generation synthesis of four-bar linkages do we pick one of the link vectors arbitrarily).

Equation (8.28) is now in the standard form. In fact, function generation can be thought of as a special case of path generation with prescribed timing, the path of $\mathbf{Z}_4$ being along a circular arc. Note also that the only dyad that needs to be synthesized for function generation is that of Eq. (8.28) for $j = 2, 3$.

Sections 8.21 and 8.22 show other techniques for generating design equations for function generation: Freudenstein's equation and the loop-closure-equation technique. The number of free choices here does coincide with the latter method. These other techniques do not necessarily yield the standard form, although they can also be so formulated (see Chap. 3 of Vol. 2).

8.17 THREE-PRECISION-POINT SYNTHESIS PROGRAM FOR FOUR-BAR LINKAGES*

A program can be written to synthesize a four-bar motion, path, or function generator mechanism for three finitely separated precision points utilizing the notation of Fig. 8.60 (Fig. 8.63 is a flowchart for this program). The system of equations, Eq. (8.23), $j = 2,3$ for the left side of the four-bar and the equations for the right side, Eq. (8.24), $j = 2,3$ are solved by Cramer's rule as suggested in Eqs. (8.20) and (8.21). The input data required are the rotations of the input, output, and coupler links: PHI2, PHI3, GAM2, GAM3, PSI2, PSI3, (ϕ_2, ϕ_3, γ_2, γ_3, ψ_2, ψ_3) and the path displacements: XDEL2, YDEL2, XDEL3, YDEL3 (the x and y coordinates of $\boldsymbol{\delta}_2$ and $\boldsymbol{\delta}_3$). As can be seen by the examples below, the output of the program can include a repeat of the input data, link vectors in the starting position of the synthesized linkage in both Cartesian and polar form, as well as the coordinates of the coupler points: A, B, and P with respect to A_0. Figures 8.64, 8.65, and 8.66 show linkages that have been synthesized for motion, path, and function generation, respectively.

Notice that arbitrary choices must be made for all three examples according to Table 8.1. Thus for motion generation, ϕ_2, ϕ_3, ψ_2, and ψ_3 are free choices. For path generation with prescribed timing γ_2, γ_3, ψ_2, and ψ_3 are free choices. The procedure of Sec. 8.16 is not used here in the function-generation case. Rather than expanding the program of Fig. 8.63, the function generator is synthesized by prescribing ϕ_2, ϕ_3, ψ_2, ψ_3. With this method $\boldsymbol{\delta}_2$, $\boldsymbol{\delta}_3$, and γ_2, γ_3 are free choices. The only portion of the output of interest would be $\mathbf{Z}_2$, $\mathbf{Z}_3$, $\mathbf{Z}_4$, and $\mathbf{Z}_1$ (see Fig. 8.60 and the example of Fig. 8.66).

Example 8.4: Motion Generation

Completion of an assembly line requires the synthesis of a motion generator linkage to transfer boxes from one conveyor belt to another as depicted in Fig. 8.64a. A pickup and release position plus an intermediate location are specified. For simplicity, a four-bar link-

---

* The computer disc which accompanies this text includes a three position synthesis program that parallels this section.

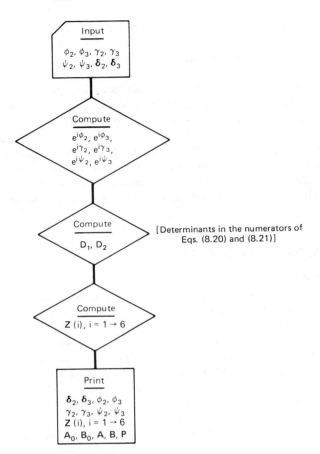

Input

$\phi_2, \phi_3, \gamma_2, \gamma_3$
$\psi_2, \psi_3, \delta_2, \delta_3$

Compute

$e^{i\phi_2}, e^{i\phi_3},$
$e^{i\gamma_2}, e^{i\gamma_3},$
$e^{i\psi_2}, e^{i\psi_3}$

Compute

D_1, D_2 [Determinants in the numerators of
                Eqs. (8.20) and (8.21)]

Compute

$Z(i), i = 1 \rightarrow 6$

Print

$\delta_2, \delta_3, \phi_2, \phi_3$
$\gamma_2, \gamma_3, \psi_2, \psi_3$
$Z(i), i = 1 \rightarrow 6$
A_0, B_0, A, B, P

Figure 8.63 Flowchart of three-precision-point four-bar synthesis program (see Fig. 8.60).

age (Fig. 8.60) is the type chosen for the task. From Fig. 8.64a, prescribed quantities for the motion generation are

$$\delta_2 = -6 + 11i, \qquad \gamma_2 = 22°$$
$$\delta_3 = -17 + 13i, \qquad \gamma_3 = 68°$$

The free choices are arbitrarily set* as

$$\phi_2 = 90°, \qquad \psi_2 = 40°$$
$$\phi_3 = 198°, \qquad \psi_3 = 73°$$

Table 8.3 shows a copy of the computer printout for this example, while Fig. 8.64b shows the solution drawn in three positions. Section 3.9 analyzes this linkage throughout its cycle of motion (see Example 3.12).

---

* By specifying input rotations about twice as large as output rotations, these choices are meant to bring about a crank-rocker type of four-bar solution. The ability to bring this about is a valuable attribute of the three-position standard-form method.

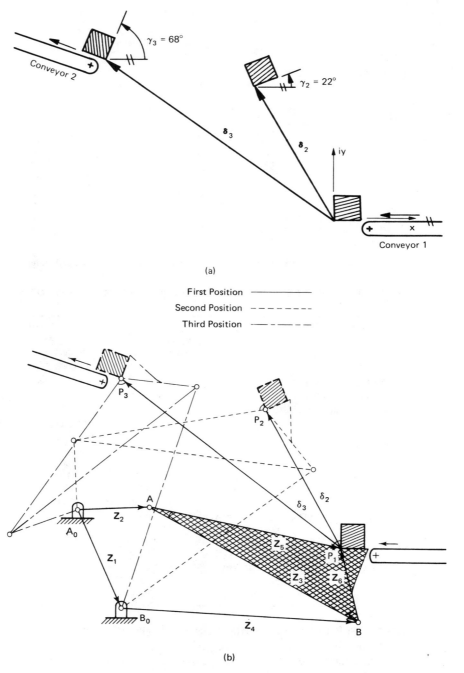

Figure 8.64 (a) three prescribed positions for four-bar motion synthesis; (b) synthesized conveyor linkage of Example 8.3 using the program of Fig. 8.63.

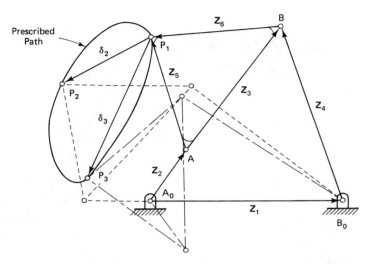

Figure 8.65 Four-bar path generator with three prescribed path points.

TABLE 8.3 COMPUTER PRINTOUT OF 3-POSITION MOTION-GENERATOR SYNTHESIS OF FOUR-BAR LINKAGE

INPUT DATA

	X COMPONENT	Y COMPONENT
DELTA 2 =	−6.0000	11.0000
DELTA 3 =	−17.0000	13.0000

PHI 2 = 90.000   GAMMA 2 = 22.000   PSI 2 = 40.000
PHI 3 = 198.000   GAMMA 3 = 68.000   PSI 3 = 73.000

COMPUTED VECTORS

	X COMPONENT	Y COMPONENT	LENGTH	DIRECTION (DEG)
Z(2) =	5.7550	.4809	5.7751	4.777
Z(5) =	14.6106	−3.4698	15.0169	−13.359
Z(4) =	18.3746	−.6611	18.3864	−2.061
Z(6) =	−1.4207	5.9518	6.1190	103.426
Z(3) =	16.0313	−9.4215	18.5948	−30.443
Z(1) =	3.4118	−8.2796	8.9550	−67.605

LINKAGE PIVOT AND COUPLER LOCATIONS

	X COMPONENT	Y COMPONENT
A_0 =	0	0
B_0 =	3.4118	−8.2796
A =	5.7550	.4809
B =	21.7863	−8.9407
P =	20.3656	−2.9889

Example 8.5: Path Generation with Prescribed Timing (Fig. 8.65)

A stirring operation requires the generation of an elliptical path. A four-bar linkage is picked for this task. Since a crank rocker is required, the input rotations are also to be prescribed. Specified quantities are

$$\delta_2 = -1.4 - 0.76i, \qquad \phi_2 = 126°$$
$$\delta_3 = -1.0 - 2.3i, \qquad \phi_3 = 252°$$

TABLE 8.4   COMPUTER PRINTOUT OF PATH GENERATION EXAMPLE

INPUT DATA

		X COMPONENT	Y COMPONENT	
	DELTA 2 =	−1.4000	−.7600	
	DELTA 3 =	−1.0000	−2.3000	

PHI 2 = 126.000	GAMMA 2 = −6.000	PSI 2 = 33.000
PHI 3 = 252.000	GAMMA 3 = 37.000	PSI 3 = 37.000

COMPUTED VECTORS

	X COMPONENT	Y COMPONENT	LENGTH	DIRECTION (DEG)
Z(2) =	.5919	.8081	1.0017	53.777
Z(5) =	−.5182	1.8246	1.8967	105.856
Z(4) =	−.9412	2.8331	2.9854	108.376
Z(6) =	−1.9958	−.1888	2.0047	−174.596
Z(3) =	1.4776	2.0134	2.4974	53.725
Z(1) =	3.0107	−.0117	3.0107	−.223

LINKAGE PIVOT AND COUPLER LOCATIONS

	X COMPONENT	Y COMPONENT
A_0 =	0	0
B_0 =	3.0107	−.0117
A =	.5919	.8081
B =	2.0695	2.8214
P =	.0737	2.6326

Arbitrarily chosen variables are

$$\gamma_2 = -6°, \qquad \psi_2 = 33°$$
$$\gamma_3 = 37°, \qquad \psi_3 = 37°$$

Table 8.4 is a copy of the computer-generated output for this example. Figure 8.65 illustrates the computer-generated linkage solution in its three prescribed positions.

Example 8.6: Function Generation

Figure 8.66a shows a barber's chair in which a single control arm is to actuate both the foot rest and the head rest. Notice the nonlinear relationship between the angles of rotation of the three members in the three specified positions. The type of linkage chosen for this task is a Watt II six-bar, which is simply two four-bars in series (usually connected through a bell crank). Specified quantities for the *first four-bar* function generator (between the head rest and the control arm) are

$$\phi_2 = 50°, \qquad \psi_2 = 22.5°$$
$$\phi_3 = 75°, \qquad \psi_3 = 45°$$

Arbitrarily chosen are

$$\delta_2 = -0.07 + 0.4i, \qquad \gamma_2 = 7°$$
$$\delta_3 = -0.3 + 0.7i, \qquad \gamma_3 = 12°$$

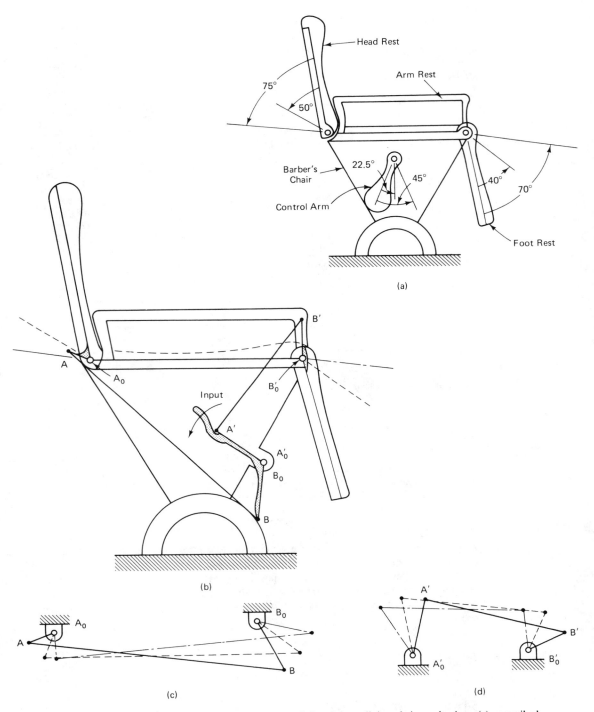

Figure 8.66 Four-bar function-generator linkages in reclining chair mechanism; (a) prescribed corresponding angular positions of foot rest, back rest, and control arm; (b) schematic of the completed mechanism; (c) head rest linkage; (d) foot rest linkage. The two four-bars in series constitute the Watt II six-bar.

The *second four-bar* function generator (between the control arm and the foot rest) has specified variables of

$$\phi_2 = 22.5°, \qquad \psi_2 = 40°$$
$$\phi_3 = 45°, \qquad \psi_3 = 70°$$

Arbitrarily chosen are

$$\delta_2 = -0.07 + 0.4i, \qquad \gamma_2 = 8°$$
$$\delta_3 = -0.3 + 0.7i, \qquad \gamma_3 = 13°$$

Table 8.5 is a printout for both sides of the six-bar linkage. Figure 8.66b illustrates how both these solutions are put together by appropriate rescaling and reorientation as one of the numerous possible Watt's II six-bar solutions to this problem, while Fig. 8.66c and d show the two four-bar halves in their three design positions.

TABLE 8.5 COMPUTER OUTPUT OF FOUR-BAR SYNTHESES OF RECLINER MECHANISM

HEAD REST LINKAGE

INPUT DATA

	X COMPONENT	Y COMPONENT		
DELTA 2 =	−.0700	.4000		
DELTA 3 =	−.3000	.7000		
PHI 2 = 50.000	GAMMA 2 = 7.000		PSI 2 = 22.500	
PHI 3 = 75.000	GAMMA 3 = 12.000		PSI 3 = 45.000	

COMPUTED VECTORS

	X COMPONENT	Y COMPONENT	LENGTH	DIRECTION (DEG)
Z(2) =	.0404	−.4640	.4657	−85.022
Z(5) =	1.8676	3.2580	3.7554	60.178
Z(4) =	1.0009	.2777	1.0388	15.506
Z(6) =	.2552	−.9384	.9725	−74.788
Z(3) =	1.6124	4.1965	4.4956	68.982
Z(1) =	.6518	3.4548	3.5158	79.315

LINKAGE PIVOT AND COUPLER LOCATIONS

	X COMPONENT	Y COMPONENT
A_0 =	0	0
B_0 =	.6518	3.4548
A =	.0404	−.4640
B =	1.6528	3.7325
P =	1.9080	2.7941

FOOT REST LINKAGE

INPUT DATA

	X COMPONENT	Y COMPONENT		
DELTA 2 =	−.0700	.4000		
DELTA 3 =	−.3000	.7000		
PHI 2 = 22.500	GAMMA 2 = 8.000		PSI 2 = 40.000	
PHI 3 = 45.000	GAMMA 3 = 13.000		PSI 3 = 70.000	

TABLE 8.5 (CONT.)

COMPUTED VECTORS

	X COMPONENT	Y COMPONENT	LENGTH	DIRECTION (DEG)
$Z(2) =$	.9642	.2270	.9906	13.247
$Z(5) =$	.3001	−.6696	.7338	−65.859
$Z(4) =$	.5189	−.4332	.6759	−39.857
$Z(6) =$	−.1359	1.6410	1.6466	94.733
$Z(3) =$	.4360	−2.3105	2.3513	−79.315
$Z(1) =$	.8813	−1.6503	1.8709	−61.897

LINKAGE PIVOT AND COUPLER LOCATIONS

	X COMPONENT	Y COMPONENT
$A_0 =$	0	0
$B_0 =$	.8813	−1.6503
$A =$	.9642	.2270
$B =$	1.4002	−2.0835
$P =$	1.2643	−.4426

8.18 CIRCLE-POINT AND CENTER-POINT CIRCLES

This section describes an alternative approach to choosing the two free choices indicated in Table 8.1. The angular unknowns will be considered as candidates for parameters on which the locations of the fixed and moving pivots of the solution dyads will depend. Loerch [108] discovered that, if an arbitrary value is chosen for one unprescribed angular parameter while the other angular parameter is allowed to assume all possible values, the resulting loci of corresponding fixed pivots m and moving pivots k_1 are found to be pairs of circles. For example, in Fig. 8.59, if δ_2, δ_3, α_2, β_2, and β_3 are chosen to have fixed values, the points m and k_1 trace circular loci as α_3 ranges between 0 and 2π. These will be referred to as M and K_1 circles, respectively. A complex-number formulation will be used to generate these circles analytically.

Dyad Equations

The vectors of the dyad m, k_1, P_1, are defined in Fig. 8.67. The loop-closure equations for the dyad in three finitely separated positions are

First position:

$$\mathbf{R} + \mathbf{W} + \mathbf{Z} = 0 \tag{8.29}$$

Second position:

$$\mathbf{R} + \mathbf{W}e^{i\beta_2} + \mathbf{Z}e^{i\alpha_2} = \boldsymbol{\delta}_2 \tag{8.30}$$

Third position:

$$\mathbf{R} + \mathbf{W}e^{i\beta_3} + \mathbf{Z}e^{i\alpha_3} = \boldsymbol{\delta}_3 \tag{8.31}$$

The unknown location of the moving pivot k_1 is defined by the vector $-\mathbf{Z}$ with respect to P_1, the origin of the fixed-coordinate system (as shown in Fig. 8.67), which coincides with the given initial position P_1 of the tracer point of the moving plane. The yet

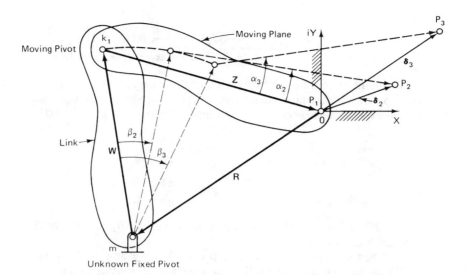

Figure 8.67 When **R**, **W**, and **Z** are unknown, δ_2 and δ_3 are prescribed and three of the four angles of rotation, α_2, α_3, β_2, and β_3 are chosen arbitrarily and held constant, varying the fourth angle of rotation $(\alpha_2, \alpha_3, \beta_2, \beta_3)$ through 0 to 360° will generate circular loci for m and k_1, the *centerpoint-* and *circlepoint-circles*. The first precision point P_1 is located at $(0, 0)$.

unknown fixed pivot m is located by vector **R**. Synthesis problems can be formulated by specifying δ_2 and δ_3, plus the appropriate angular parameters. Vectors **R** and $-\mathbf{Z}$ may be obtained from Eqs. (8.29) to (8.31) (using Cramer's rule). With $e^{i\alpha_j} = \boldsymbol{\alpha}_j$ and $e^{i\beta_j} = \boldsymbol{\beta}_j$, these equations yield

$$\mathbf{R} = \frac{\begin{vmatrix} 0 & 1 & 1 \\ \delta_2 & \boldsymbol{\beta}_2 & \boldsymbol{\alpha}_2 \\ \delta_3 & \boldsymbol{\beta}_3 & \boldsymbol{\alpha}_3 \end{vmatrix}}{\begin{vmatrix} 1 & 1 & 1 \\ 1 & \boldsymbol{\beta}_2 & \boldsymbol{\alpha}_2 \\ 1 & \boldsymbol{\beta}_3 & \boldsymbol{\alpha}_3 \end{vmatrix}} \tag{8.32}$$

or

$$\mathbf{R} = \frac{\delta_2(\boldsymbol{\beta}_3 - \boldsymbol{\alpha}_3) - \delta_3(\boldsymbol{\beta}_2 - \boldsymbol{\alpha}_2)}{\boldsymbol{\alpha}_2 - \boldsymbol{\alpha}_3 + \boldsymbol{\beta}_3 - \boldsymbol{\beta}_2 + \boldsymbol{\beta}_2\boldsymbol{\alpha}_3 - \boldsymbol{\alpha}_2\boldsymbol{\beta}_3} \tag{8.33}$$

and

$$-\mathbf{Z} = \frac{\begin{vmatrix} 1 & 1 & 0 \\ 1 & \boldsymbol{\beta}_2 & \delta_2 \\ 1 & \boldsymbol{\beta}_3 & \delta_3 \end{vmatrix}}{\begin{vmatrix} 1 & 1 & -1 \\ 1 & \boldsymbol{\beta}_2 & -\boldsymbol{\alpha}_2 \\ 1 & \boldsymbol{\beta}_3 & -\boldsymbol{\alpha}_3 \end{vmatrix}} \tag{8.34}$$

or

$$-\mathbf{Z} = \frac{-\delta_2(\beta_3 - 1) + \delta_3(\beta_2 - 1)}{\beta_2 - \beta_3 + \alpha_3 - \alpha_2 - \beta_2\alpha_3 + \alpha_2\beta_3} \tag{8.35}$$

If all parameters on the right-hand side of these expressions are fixed except for an angular parameter θ, which ranges over all possible values, the equations for $\mathbf{R}$ and $-\mathbf{Z}$ can be expressed as functions of θ forming "bilinear mappings" [108]

$$\mathbf{R}(\theta) = \frac{\mathbf{a}\theta + \mathbf{b}}{\mathbf{c}\theta + \mathbf{d}} \tag{8.36}$$

$$-\mathbf{Z}(\theta) = \frac{\mathbf{e}\theta + \mathbf{f}}{\mathbf{g}\theta + \mathbf{h}} \tag{8.37}$$

where $\theta = e^{i\theta}$ [θ stands for the angle to be varied (β_2, α_2, α_3, or β_3)], and where $\mathbf{a}$ through $\mathbf{h}$ are known.

When θ varies from 0 to 2π, θ describes the unit circle. Equations (8.36) and (8.37) are tantamount to the following sequence of transformations:

$\mathbf{p}(\theta) = \mathbf{a}\theta$,	a stretch rotation,	(8.38)
$\mathbf{q}(\theta) = \mathbf{a}\theta + \mathbf{b}$,	the same plus a change of origin,	(8.39)
$\mathbf{r}(\theta) = \mathbf{c}\theta$,	another stretch rotation,	(8.40)
$\mathbf{s}(\theta) = \mathbf{c}\theta + \mathbf{d}$,	another stretch rotation plus another change in origin,	(8.41)
$\mathbf{t}(\theta) = \dfrac{\mathbf{q}(\theta)}{\mathbf{s}(\theta)}$	a "bilinear mapping"	(8.42)

Since both $\mathbf{q}(\theta)$ and $\mathbf{s}(\theta)$ are circles, it can be shown that $\mathbf{t}(\theta)$ is also a circle [108]. Thus it is seen that the loci of $\mathbf{R}(\theta)$ and $-\mathbf{Z}(\theta)$ are circles, which in the limit can become straight lines. The complex constants $\mathbf{a}$ through $\mathbf{h}$ are found by appropriately rearranging Eqs. (8.33) and (8.35) in the form of Eqs. (8.36) and (8.37). The centers of the circular loci C_M and C_K can be found directly from the constants $\mathbf{a}$ through $\mathbf{h}$, or more simply by evaluating $\mathbf{R}$ and $-\mathbf{Z}$ at three θ values, yielding three points each, which define the circles. Either way the solutions are within the realm of programmable hand calculators. Computer programs have been written to display the M and K_1 circles on a computer graphics terminal in order to examine their properties. This leads to the manual graphical constructions presented below.

When the circles have been drawn, the associated fixed and moving pivots on a pair of circles remain to be coordinated. This is done using the pole relationship presented in Fig. 8.68. For example, rays emanating from P_{12} defining an angle $\frac{1}{2}\alpha_2$ will intersect the M and K_1 circles at fixed and moving pivot pairs m and k_1. An angle meter (adjustable protractor) rotated about the pole serves as a convenient tool for constructing such pivot pairs.

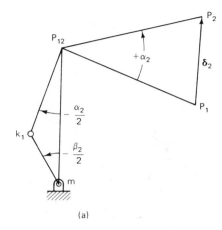

(a)

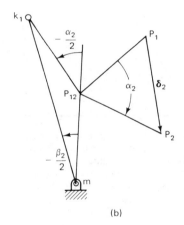

(b)

Figure 8.68 Pole, center-, and circle-point configuration for α_2 and β_2 having (a) opposite signs, and (b) equal signs (k_1 on K_1 circle, m on M circle).

Graphical Constructions: Motion and Path Generation with Prescribed Timing

Examples of motion and path generation with prescribed timing are presented here to illustrate how ground- and moving-pivot circles may be generated on a graphics terminal.*

Sample computer plots have been generated for each problem type (Figs. 8.69 and 8.70). Points m on the M circles (with solid arcs) are labeled with the values of the angle held fixed while the circle was generated. A short line segment is directed from a sample fixed pivot m on each M circle toward the moving pivot k_1 on the conjugate K_1 circle (with dotted arcs), generated concurrently. The poles used in finding conjugate $m-k_1$ pairs are represented by small rings.

* More examples may be found in Ref. 108.

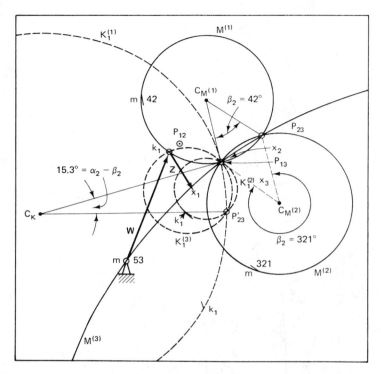

Figure 8.69 M and K_1 circles for motion generation (see Example 8.7). Prescribed are δ_2, δ_3, α_2, and α_3; $\mathbf{R}$, $\mathbf{M}$ and $\mathbf{K}_1$ are the unknowns (see Fig. 8.67). Each pair of conjugate M and K_1 circles (say, $M^{(1)}$ and $K_1^{(1)}$) is generated by arbitrarily choosing a value for β_2, holding this constant, and varying β_3 from 0 to 2π. For example, for $M^{(1)}$ and $K_1^{(1)}$, $\beta_2 = 42°$. A sample solution $\mathbf{W}$, $\mathbf{Z}$ dyad is shown. Short line segments on the circles point to associated pivot location on the conjugate circle.

In path generation with prescribed timing (Fig. 8.70), the pole P_{12} to be used for finding conjugate pivots on each MK_1 circle pair will be different for each such pair, since α_2 and α_3 are not the same for different pairs. A shorter line segment is then directed from the sample fixed pivot of each M circle toward the associated pole P_{12}.

Example 8.7: Motion Generation

Figure 8.69 illustrates a motion-generation example. The intersections of the symbol x in x_1, x_2, and x_3 mark the prescribed positions of the tracer point P, with $\delta_2 = 1 + i$, $\delta_3 = 2 + 0.5i$, $\alpha_2 = 1$ rad, and $\alpha_3 = 2$ rad. The M and K_1 circles are generated for three values of the angle β_2 (42°, 321°, and 353°), with β_3 ranging from 0 to 2π. These three circle-point circles and center-point circles are sufficient to display the properties of the diagram. It is interesting to note that the M and K_1 circles intersect at points which can be shown to be the poles P_{13} and P_{23}' for the M circles and P_{13}, P_{23}', for the K_1 circles, where P_{23}' is the *image pole* of P_{23} obtained by reflecting P_{23} about the line $P_{12}P_{13}$. Similarly notable are the angles subtended at the circle centers by the line connecting the intersection (see Fig. 8.69). The

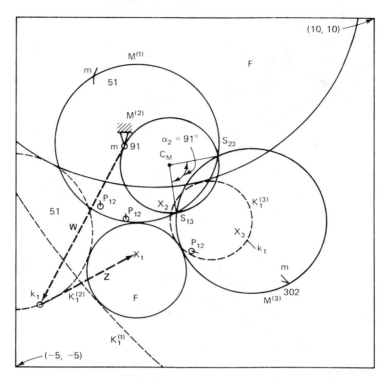

Figure 8.70 M and K_1 circles for path generation with prescribed timing (see Example 8.8). Referring to Fig. 8.67, $\mathbf{R}$, $\mathbf{W}$ and $\mathbf{Z}$ are unknown, δ_2, δ_3, β_2, and β_3 are prescribed. To generate a conjugate pair of M and K_1 circles, choose arbitrarily a value for α_2 and let α_3 vary from 0 to 2π. A sample solution dyad is shown: $\mathbf{W}$, $\mathbf{Z}$.

steps for obtaining a set of circular M and K_1 loci for a motion generation problem are therefore:

1. Find the circle-intersection poles: P_{13} and P_{23} for the M circles; P_{13} and P'_{23} for the K_1 circles.

2. Bisect the lines between the intersection pole pairs to find the lines of centers for the M and K_1 circles.

3. For each value of the angle β_2, lay off the circle centers so that $\sphericalangle P_{13} C_M P_{23} = \beta_2$ and $\sphericalangle P'_{23} C_K P_{13} = \alpha_2 - \beta_2$.

4. Draw the circle pairs through the intersection poles with centers C_M and C_K.

It can be shown (Fig. 8.71a) that the complex-number expressions for the poles are

$$\mathbf{P}_{12} = \frac{\delta_2}{1 - \alpha_2} \tag{8.43}$$

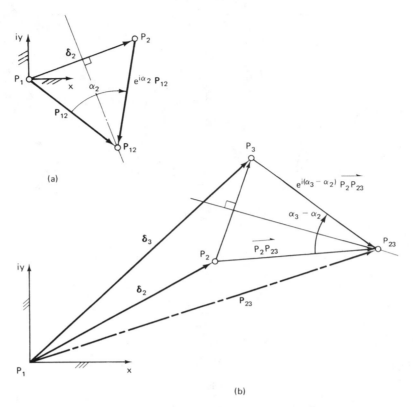

Figure 8.71 Derivation of Eqs. (8.43) and (8.45): (a) note that $\mathbf{P}_{12}(1 - e^{i\alpha_2}) = \boldsymbol{\delta}_2$, which leads to Eq. (8.43); (b) here $(\mathbf{P}_{23} - \boldsymbol{\delta}_2)(1 - e^{i(\alpha_3-\alpha_2)}) = \boldsymbol{\delta}_3 - \boldsymbol{\delta}_2$, which leads to Eq. (8.45).

$$\mathbf{P}_{13} = \frac{\boldsymbol{\delta}_3}{1 - \alpha_3} \tag{8.44}$$

$$\mathbf{P}_{23} = \frac{\boldsymbol{\delta}_3\alpha_2 - \boldsymbol{\delta}_2\alpha_3}{\alpha_2 - \alpha_3} \quad \text{(see Fig. 8.71b)} \tag{8.45}$$

where $\mathbf{P}_{ij}$ is the vector from the origin of $\mathbf{R}_i$ (Fig. 8.67) to the pole P_{ij} and where $\alpha_j = e^{i\alpha_j}$.

Example 8.8: Path Generation with Prescribed Timing

M and K_1 circles for an example of path generation with prescribed timing are shown in Fig. 8.70, with $\boldsymbol{\delta}_2 = 1 + 2i$, $\boldsymbol{\delta}_3 = 4 + i$, $\beta_2 = 1$ rad, and $\beta_3 = 2$ rad. The M and K_1 circles are generated for $\alpha_2 = 51°$, $91°$, and $302°$, with α_3 ranging from 0 to $360°$. The M circles all have common intersections at the pseudopoles* S_{13} and S_{23}, and exhibit properties that allow M-circle construction with the steps used in motion generation, except that for

---

* The pseudopoles are defined as poles that would be obtained if the moving plane rotations were those of the grounded link β, rather than α (see Fig. 8.67).

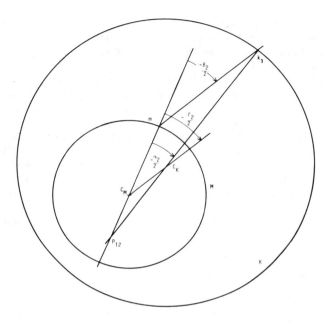

Figure 8.72 Graphical construction of a K_1 circle conjugate to a known M circle for path generation with prescribed timing (see Example 8.8, steps 1, 2, and 3).

each conjugate pair of M and K_1 circles the angle α_2 is chosen arbitrarily and then held constant, rather than β_2.

The K_1 circles have no intersections, but a useful property exists that permits easy construction: the M- and K_1-circle centers are coordinated about the poles, as are the m and k_1 pivots (see Fig. 8.70). Referring to Fig. 8.72, C_K and k_1 (the center of the K_1 circle and the moving pivot conjugate to either fixed pivot on the M circle diameter through P_{12}) are found accordingly, allowing the K_1 circle to be drawn about C_k with radius $C_K k_1$. Thus the M and K_1 circles of path generation with prescribed timing can be found as follows:

1. Construct the pseudopoles S_{13} and S_{23}.
2. Choose arbitrarily a value of α_2 and find C_M on the perpendicular bisector of $S_{13} S_{23}$ such that $\sphericalangle S_{13} C_M S_{23} = \alpha_2$.
3. For each M circle (see Fig. 8.72):
   a. Construct the pole P_{12}.
   b. Draw the diameter of the M circle through P_{12} and extend it.
   c. From this line, lay off the angle $-\alpha_2/2$ with P_{12} as the apex and the angle $-\beta_2/2$ with C_M as the apex. The intersection of these two lines is C_K. Choose one of the intersections of the $C_M P_{12}$ line with the M circle to be the fixed pivot m. With this point as the apex, lay off the angle $-\beta_2/2$ from the $P_{12} C_M m$ line. The intersection of this line with the $-\alpha_2/2$ line is k_1, and the K_1 circle can now be drawn.

Example 8.9: Intersections of Circles for Motion Generation*

Figure 1 of the color insert to this text illustrates sets of m and k circles based on the following motion generation problem. Precision points are located at $(12, 2)$, $(-5, 0)$, and $(-15, 6)$

* Contributed by Lisa Logan, University of Minnesota.

and are marked on the figure by the numbers 1, 2, and 3 respectively, with each number surrounded by a circle (in the case of the second and third precision points) or a square (in the case of the first). Coupler angles at each precision point are shown by angle markers. The three chosen coupler angles are 210°, 0°, and −57°.

The precision points can be translated into δ's and α's by subtracting the first position: $\delta_2 = -17 - 2i$, $\delta_3 = -27 + 4i$, $\alpha_2 = -210°$, and $\alpha_3 = 93°$. From this information, the poles P_{12}, P_{13}, and P_{23}, and the image pole P'_{23}, can be calculated in a manner analogous to the procedure given in the previous example. If a value is now chosen for either β_2 or β_3, m and k circles may be calculated using the analytical equivalent of the construction technique shown in Fig. 8.68. Figure 1 of the insert shows m (green) and k (blue) circles for values of β_3 ranging from 0° to 360° in 45° increments, with each successive increment shaded darker. Note that all m circles pass through the poles P_{12} and P_{23} and that all k circles pass through P_{12} and P'_{23}. If the circles had been based on β_2 instead of β_3, the intersection poles would have been P_{13}, P_{23}, and P'_{23}, as in the previous example.

Figure 2 shows a single pair of m and k circles for the same example. In this case, $\beta_3 = 265°$. Also shown are two dyads corresponding to β_2 values of $-12°$ and 280°.

Dyad Moving-Pivot Existence: Three Precision Points

When computer plots of the M and K_1 circles were made, it was found that moving pivots cannot exist within certain regions of the plane in path generation with prescribed timing: Two circles exist within which no moving pivots will be found. One of the circles surrounds the first path point position x_1. It is possible to define analytically a distance bound from this point, within which no moving pivots can exist [108]. All K_1 circles are tangent to both nonexistence circles and are arranged so that two k_1 pivot solutions occur for each point outside the nonexistence circles. Figures 8.70 and 8.73 show the nonexistence circles (labeled F, for "forbidden regions"). Of these, Fig. 8.73 is drawn for the example.

$$\delta_2 = 2 + 2i; \qquad \beta_2 = 0.5 \text{ rad}$$
$$\delta_3 = 4 + i; \qquad \beta_3 = 1 \text{ rad}$$

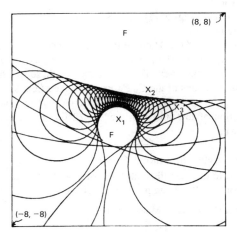

Figure 8.73 Circular non-existence regions F for path generation with prescribed timing, within which K_1 circles cannot exist. All existing K_1 circles are tangent to these "forbidden" F regions.

The circles shown in Fig. 8.73 are the K_1 circles which are tangent to the two nonexistence F circles. The smaller F circle surrounds the initial precision point, $\mathbf{x}_1$.

8.19 GROUND-PIVOT SPECIFICATION [109]

There is yet another useful strategy for choosing the two free choices in the system of equations for three finitely separated precision positions of a dyad (Table 8.1). Recall that in each case in Sec. 8.16 two rotation angles were chosen arbitrarily, yielding a simple set of linear equations. In the preceding section, it was observed that varying one of the free-choice angles as a parameter produces circular loci of center and circle points. With two free choices available, however, one of the link vectors ($\mathbf{W}$ or $\mathbf{Z}$) can be assumed arbitrarily instead. In fact, by writing the dyad equations in a different form, a ground- or moving-pivot location may be specified directly.

Figure 8.74 shows a dyad in three finitely separated positions. The synthesis equations can be written as*

$$\mathbf{W} + \mathbf{Z} = \mathbf{R}_1$$
$$\mathbf{W}e^{i\beta_2} + \mathbf{Z}e^{i\alpha_2} = \mathbf{R}_2 \qquad (8.46)$$
$$\mathbf{W}e^{i\beta_3} + \mathbf{Z}e^{i\alpha_3} = \mathbf{R}_3$$

Suppose that we wish to synthesize the dyad in Fig. 8.74 for motion generation. According to Sec. 8.14, this requires that $\boldsymbol{\delta}_2 = (\mathbf{R}_2 - \mathbf{R}_1)$, $\boldsymbol{\delta}_3 = (\mathbf{R}_3 - \mathbf{R}_1)$, α_2, and α_3 be specified. Subtracting the first equation from the second and third in Eq. (8.46) will in fact yield the standard form Eq. (8.16). Table 8.1 requires two additional free choices for this system of equations. Let us specify $\mathbf{R}_1$ (which locates the ground pivot). Thus the coefficients of $\mathbf{Z}$ and $\mathbf{R}_j$ are known.

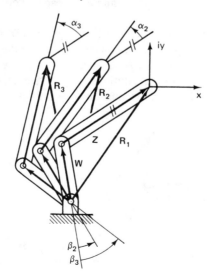

Figure 8.74 Three discrete positions of the unknown dyad $\mathbf{W}$, $\mathbf{Z}$. For synthesis for motion generation with specified ground pivot, $\mathbf{R}_j$, $j = 1, 2, 3$, α_2 and α_3 are prescribed and β_j, $j = 2, 3$, are to be found before the system of Eq. 8.46 can be solved for $\mathbf{W}$ and $\mathbf{Z}$.

* Note that in Fig. 8.74, $\mathbf{R}_1 = -\mathbf{R}$, as defined in Fig. 8.67.

If we view Eq. (8.46) temporarily as three complex equations linear and nonhomogeneous in the two complex unknowns $\mathbf{W}$ and $\mathbf{Z}$, this set has a solution for $\mathbf{W}$ and $\mathbf{Z}$ only if the determinant of the augmented matrix of the coefficients is identically zero:

$$\begin{vmatrix} 1 & 1 & \mathbf{R}_1 \\ e^{\beta_2} & e^{i\alpha_2} & \mathbf{R}_2 \\ e^{i\beta_3} & e^{i\alpha_3} & \mathbf{R}_3 \end{vmatrix} = 0 \tag{8.47}$$

Equation (8.47) represents a complex equation with two unknowns, β_2 and β_3. Since the unknowns are in the first column, the determinant is expanded about this column:

$$(\mathbf{R}_3 e^{i\alpha_2} - \mathbf{R}_2 e^{i\alpha_3}) + e^{i\beta_2}(-\mathbf{R}_3 + \mathbf{R}_1 e^{i\alpha_3}) + e^{i\beta_3}(\mathbf{R}_2 - \mathbf{R}_1 e^{i\alpha_2}) = 0 \tag{8.48}$$

or

$$\mathbf{D}_1 + \mathbf{D}_2 e^{i\beta_2} + \mathbf{D}_3 e^{i\beta_3} = 0 \tag{8.49}$$

which is transcendental in the unknowns β_2 and β_3, and where

$$\mathbf{D}_1 = \mathbf{R}_3 e^{i\alpha_2} - \mathbf{R}_2 e^{i\alpha_3}$$
$$\mathbf{D}_2 = \mathbf{R}_1 e^{i\alpha_3} - \mathbf{R}_3 \tag{8.50}$$
$$\mathbf{D}_3 = \mathbf{R}_2 - \mathbf{R}_1 e^{i\alpha_2}$$

are known from prescribed data.

A simple graphical construction aids in solving Eq. (8.49) for β_2 and β_3. Figure 8.75 shows a geometric solution where the knowns $\mathbf{D}_1$, $\mathbf{D}_2$, and $\mathbf{D}_3$ are represented as vectors. Notice that $\mathbf{D}_3$ and $\mathbf{D}_2$ are pinned to $\mathbf{D}_1$ but vector $\mathbf{D}_1$ is fixed. Note in Eq. (8.49) that vectors $\mathbf{D}_2$ and $\mathbf{D}_3$ are multiplied by $e^{i\beta_2}$ and $e^{i\beta_3}$, respectively. These quantities are regarded as *rotation operators*.

In Fig. 8.75, if the vectors form a closed loop, Eq. (8.49) will be satisfied. Thus $\mathbf{D}_2$ and $\mathbf{D}_3$ are rotated about their pin connections with $\mathbf{D}_1$ until they meet. The rotations required to close the loop are then β_2 and β_3. Notice that there are two solutions for the triangle: β_2, β_3 and $\tilde{\beta}_2$, $\tilde{\beta}_3$. One set of β solutions will be "trivial," however. This solution is $\tilde{\beta}_2 = \alpha_2$ and $\tilde{\beta}_3 = \alpha_3$ (these are turning-block solutions — see Vol. 2 Chap. 3).

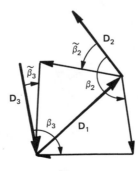

Figure 8.75 Graphical solution of Eq. (8.49) for β_2 and β_3. Note $\tilde{\beta}_2 = \alpha_2$ and $\tilde{\beta}_3 = \alpha_3$.

This can be verified by plugging the trivial roots back into Eq. (8.49). Based on this observation, it can be shown* that

$$\beta_2 = 2 \arg(-\mathbf{D}_1) - \arg(\mathbf{D}_2) - \arg(\mathbf{D}_2 e^{i\alpha_2}) \qquad (8.51)$$

$$\beta_3 = 2 \arg(-\mathbf{D}_1) - \arg(\mathbf{D}_3) - \arg(\mathbf{D}_3 e^{i\alpha_3}) \qquad (8.52)$$

Knowing values for β_2 and β_3, one can substitute these values along with the specified values for α_2, α_3, δ_2, and δ_3 into any two of Eqs. (8.46). Using Cramer's rule, values for $\mathbf{W}$ and $\mathbf{Z}$ may be calculated, thus determining the dyad for the specified ground pivot location.

Example 8.10

An engineering student who had recently purchased a tape unit for her sports car was concerned with possible theft of her investment. Therefore, the student envisioned synthesis of a four-bar linkage to hide the tape player behind the glove compartment when not in use. Figure 8.76 shows a cross section of the area of interest, including the glove compartment and heating duct as well as the three prescribed positions for the tape unit. Since there is a small acceptable area for possible ground pivots, the method of Sec. 8.19 is used as a synthesis tool.

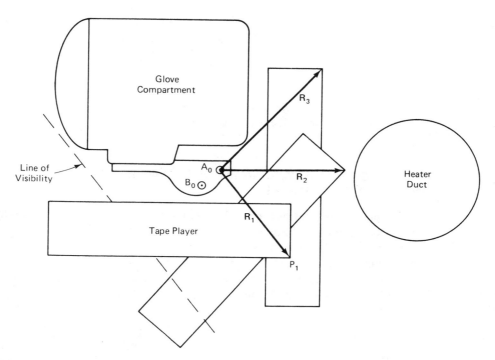

Figure 8.76   Four-bar motion-generator synthesis with prescribed ground pivots.

* Derived with the help of Tom Chase, Univ. of Minnesota.

Two ground-pivot locations are chosen: A_0 and B_0, as shown in Fig. 8.76. The three position vectors for the first dyad of the four-bar are

$$\mathbf{R}_1 = 2.14 - 3.68i$$

$$\mathbf{R}_2 = 4.46 - 0.63i$$

$$\mathbf{R}_3 = 4.10 + 3.22i$$

while the rotations of the coupler (the tape player) are

$$\alpha_2 = 50.7°$$

$$\alpha_3 = 91.9°$$

Using Eqs. (8.50) yields

$$\mathbf{D}_1 = -0.377 + 0.734i$$

$$\mathbf{D}_2 = -0.493 - 0.959i$$

$$\mathbf{D}_3 = 0.257 + 0.0448i$$

and with these, Eqs. (8.51) and (8.52) give

$$\beta_2 = 58.09°$$

$$\beta_3 = 122.70° \qquad \text{(see Fig. 8.77a)}$$

with which the results of simultaneous solution of any two equations in Eq. (8.46) will be

$$\mathbf{W} = -1.42 - 1.45i$$

$$\mathbf{Z} = 3.56 - 2.23i$$

Figure 8.77 shows the synthesized linkage in its starting, final, and two intermediate positions. The other side of the linkage is designed the same way.

8.20 EXTENSION OF THREE-PRECISION-POINT SYNTHESIS TO MULTILOOP MECHANISMS

Multiloop planar mechanisms can also be synthesized by recognizing key dyads that yield equations of the *same form as Eq. (8.16)*. In addition to the Watt six-bar of Fig. 8.66 and Example 8.6, the Stephenson III linkage of Fig. 8.78a will be used to demonstrate the extension of the dyad approach beyond four-bar linkages.

Inspection of Fig. 8.78a will yield three independent loops; one is a dyad loop and two are triad loops (see Fig. 8.78b).

Loop 1:

$$\mathbf{Z}_2(e^{i\phi_j} - 1) + \mathbf{Z}_3(e^{i\gamma_j} - 1) = \boldsymbol{\delta}_j \tag{8.53}$$

Loop 2:

$$\mathbf{Z}_5(e^{i\psi_j} - 1) + \mathbf{Z}_4(e^{i\beta_j} - 1) - \mathbf{Z}_9(e^{i\gamma_j} - 1) = \boldsymbol{\delta}_j \tag{8.54}$$

Loop 3:

$$\mathbf{Z}_6(e^{i\theta_j} - 1) + \mathbf{Z}_7(e^{i\beta_j} - 1) - \mathbf{Z}_9(e^{i\gamma_j} - 1) = \boldsymbol{\delta}_j \tag{8.55}$$

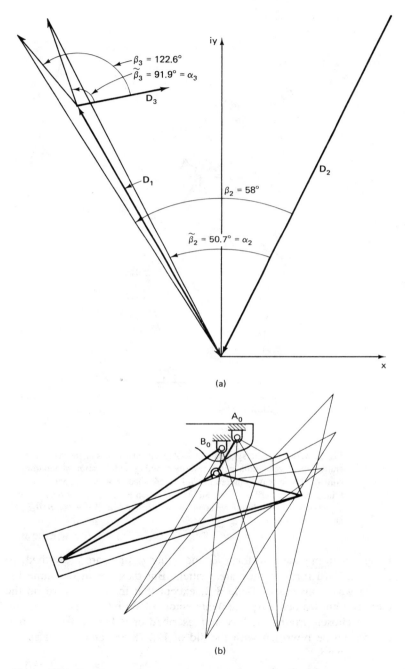

Figure 8.77 Four-bar motion generator synthesized to the requirements shown in Fig. 8.76. Note the specified ground-pivot locations: (a) graphical check of calculated values of β_2 and β_3 for the first dyad pivoted at A_0 (Eq. (8.49) and Fig. 8.75); (b) the synthesized four-bar mechanism in four intermediate positions.

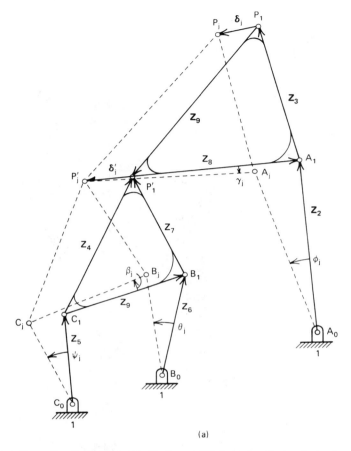

Figure 8.78 Dyadic synthesis of a Stephenson III six-bar mechanism for motion generation, path generation with prescribed timing and function generation. $\overline{P_1 P_j}$ is path cord δ_j. For motion generation, δ_j is prescribed. For path generation with prescribed timing, ϕ_j is prescribed. For additional function generation, not only ϕ_j but also ψ_j or θ_j is prescribed. Loop closure equations are written for $A_0 A_j P_j P_1 A_1 A_0$, $B_0 B_j P_j' P_j P_1 P_1' B_1 B_0$, and $C_0 C_j P_j' P_j P_1 P_1' C_1 C_0$.

Loop 1 is in the standard form: δ_j and either ϕ_j or γ_j are prescribed. Neither loop 2 or 3 is in standard form as they are written. But they are in the same form as Eq. (8.27), which was analyzed in Table 8.2, except that instead of zero on the right-hand side, there is a known path displacement vector δ_j. With two free choices available, vector $\mathbf{Z}_9$ can be chosen arbitrarily. If γ_j is prescribed or is known from loop 1, Eqs. (8.54) and (8.55) can be rewritten with the aid of Eq. (8.56) and solved as standard-form equations. If we let

$$\delta_j' = \delta_j + \mathbf{Z}_9(e^{i\gamma_j} - 1) \tag{8.56}$$

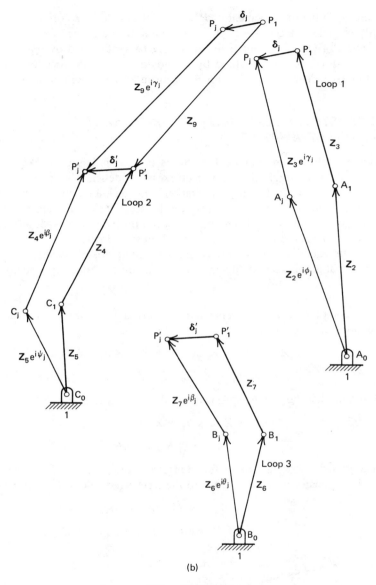

(b)

Figure 8.78 (cont.)

then, for loop 2,

$$\mathbf{Z}_5(e^{i\psi_j} - 1) + \mathbf{Z}_4(e^{i\beta_j} - 1) = \delta_j' \qquad (8.57)$$

and for loop 3,

$$\mathbf{Z}_6(e^{i\theta_j} - 1) + \mathbf{Z}_7(e^{i\beta_j} - 1) = \delta_j' \qquad (8.58)$$

where δ_j' is the displacement of P'. The free choice of $\mathbf{Z}_9$ offers the designer the possibility of picking the shape of the coupler link or the generation of different solutions by varying $\mathbf{Z}_9$. Other multiloop linkages may be synthesized by repeated use of the same standard-form solution method by employing a similar procedure as described in this section (see Chap. 3 of Vol. 2 for Stephenson III and other multiloop examples) [48, 49].

8.21 FREUDENSTEIN'S EQUATION FOR THREE-POINT FUNCTION GENERATION

Another well-known analytical synthesis method is based on *Freudenstein's Equation* [71, 148]. This algebraic method utilizes Freudenstein's displacement equations for three-precision-point function generation. This technique has been extended to four and five precision points and, by regarding the scale factors R_ϕ and R_ψ of input and output rotations as unknowns, also to six and seven precision points. It has also been extended to other linkages, but these cases are not presented here.

The equation can be derived from the loop-closure equation written for Fig. 8.79 (notice that in this section all angles are the arguments of the link vectors in all positions):

$$\mathbf{Z}_1 + \mathbf{Z}_2 + \mathbf{Z}_3 - \mathbf{Z}_4 = 0 \qquad (8.59)$$

If this complex equation is separated into real and imaginary components, two algebraic equations are produced:

$$Z_1 \cos \theta_1 + Z_2 \cos \theta_2 + Z_3 \cos \theta_3 - Z_4 \cos \theta_4 = 0 \qquad (8.60)$$

$$Z_1 \sin \theta_1 + Z_2 \sin \theta_2 + Z_3 \sin \theta_3 - Z_4 \sin \theta_4 = 0 \qquad (8.61)$$

Assuming that the ground link is along the x axis (as in Fig. 8.80), $\theta_1 = 180°$ and

$$-Z_1 + Z_2 \cos \theta_2 + Z_3 \cos \theta_3 - Z_4 \cos \theta_4 = 0 \qquad (8.62)$$

$$Z_2 \sin \theta_2 + Z_3 \sin \theta_3 - Z_4 \sin \theta_4 = 0 \qquad (8.63)$$

Since we wish to synthesize a function generator, θ_3 is not of interest and will be eliminated by transferring the Z_3 terms to the right-hand side of Eqs. (8.62) and (8.63).

$$-Z_1 + Z_2 \cos \theta_2 - Z_4 \cos \theta_4 = -Z_3 \cos \theta_3 \qquad (8.64)$$

$$Z_2 \sin \theta_2 - Z_4 \sin \theta_4 = -Z_3 \sin \theta_3 \qquad (8.65)$$

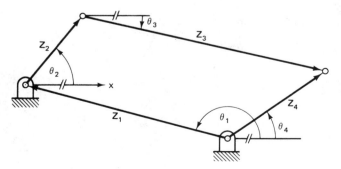

Figure 8.79 Freudenstein's equation is based on the closure of the four-bar loop.

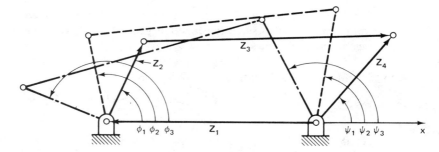

Figure 8.80 Notation for the four-bar mechanism for writing Freudenstein's equation, Eq. (8.70).

Next, Eqs. (8.64) and (8.65) are squared and added together to eliminate θ_3. The resulting equation is

$$Z_3^2 = Z_1^2 + Z_2^2 + Z_4^2 - 2Z_1 Z_2 \cos \theta_2 + 2Z_1 Z_4 \cos \theta_4$$
$$- 2Z_2 Z_4 (\cos \theta_2 \cos \theta_4 + \sin \theta_2 \sin \theta_4) \tag{8.66}$$

Since $\cos \theta_2 \cos \theta_4 + \sin \theta_2 \sin \theta_4 = \cos(\theta_2 - \theta_4)$, Eq. (8.66) can be rearranged as

$$\frac{Z_3^2 - Z_1^2 - Z_2^2 - Z_4^2}{2Z_2 Z_4} + \frac{Z_1}{Z_4} \cos \theta_2 - \frac{Z_1}{Z_2} \cos \theta_4 = -\cos(\theta_2 - \theta_4) \tag{8.67}$$

In a more compact form, *Freudenstein's equation* reads

$$K_1 \cos \theta_2 + K_2 \cos \theta_4 + K_3 = -\cos(\theta_2 - \theta_4) \tag{8.68}$$

where

$$K_1 = \frac{Z_1}{Z_4}$$

$$K_2 = \frac{-Z_1}{Z_2} \tag{8.69}$$

$$K_3 = \frac{Z_3^2 - Z_1^2 - Z_2^2 - Z_4^2}{2Z_2 Z_4}$$

Notice that the Ks are three independent algebraic expressions containing the three unknown lengths of the links. Freudenstein's equation is a displacement equation for the four-bar linkage which holds true for each position of the linkage. Thus, for three prescribed positions, the equation can be written for each position. The notation will be changed at this point to avoid double subscripts: the three angles for the three prescribed positions of Z_2 with respect to the fixed x axis will be ϕ_1, ϕ_2, and ϕ_3, while those of Z_4 will be ψ_1, ψ_2, and ψ_3, as in Fig. 8.80. Thus Freudenstein's equation for three prescribed positions is

$$K_1 \cos \phi_1 + K_2 \cos \psi_1 + K_3 = -\cos(\phi_1 - \psi_1)$$
$$K_1 \cos \phi_2 + K_2 \cos \psi_2 + K_3 = -\cos(\phi_2 - \psi_2) \tag{8.70}$$
$$K_1 \cos \phi_3 + K_2 \cos \psi_3 + K_3 = -\cos(\phi_3 - \psi_3)$$

Cramer's rule may be used to solve system Eq. (8.70). To find the Zs, one length, say Z_1, is arbitrarily picked to scale the function generator (as was done in the previous function-generator techniques).

Dealing with third-order determinants may be avoided by first subtracting the second and third equations from the first, eliminating K_3

$$K_1(\cos \phi_1 - \cos \phi_2) + K_2(\cos \psi_1 - \cos \psi_2) = -\cos(\phi_1 - \psi_1) + \cos(\phi_2 - \psi_2)$$
(8.71)

$$K_1(\cos \phi_1 - \cos \phi_3) + K_2(\cos \psi_1 - \cos \psi_3) = -\cos(\phi_1 - \psi_1) + \cos(\phi_3 - \psi_3)$$
(8.72)

and solving the resulting system of two equations for K_1 and K_2

$$K_1 = \frac{\omega_3\omega_5 - \omega_2\omega_6}{\omega_1\omega_5 - \omega_2\omega_4} \qquad K_2 = \frac{\omega_1\omega_6 - \omega_3\omega_4}{\omega_1\omega_5 - \omega_2\omega_4}$$
(8.73)

in which

$$\omega_1 = \cos \phi_1 - \cos \phi_2, \qquad\qquad \omega_4 = \cos \phi_1 - \cos \phi_3$$
$$\omega_2 = \cos \psi_1 - \cos \psi_2, \qquad\qquad \omega_5 = \cos \psi_1 - \cos \psi_3 \qquad (8.74)$$
$$\omega_3 = -\cos(\phi_1 - \psi_1) + \cos(\phi_2 - \psi_2), \qquad \omega_6 = -\cos(\phi_1 - \psi_1) + \cos(\phi_3 - \psi_3)$$

Substituting values of K_1 and K_2 into any part of Eq. (8.70) yields

$$K_3 = -\cos(\phi_i - \psi_i) - K_1 \cos \phi_i - K_2 \cos \psi_i, \qquad i = 1, 2, \text{ or } 3$$
(8.75)

The link lengths may be expressed in terms of the known Ks by using Eq. (8.69) (having chosen the length of Z_1):

$$Z_4 = \frac{Z_1}{K_1}$$

$$Z_2 = -\frac{Z_1}{K_2}$$
(8.76)

$$Z_3 = \sqrt{2K_3 Z_2 Z_4 + Z_1^2 + Z_2^2 + Z_4^2}$$

By using Freudenstein's equation, as in the other synthesis methods, one may obtain *two infinities* of solutions for the same set of precision points. All that is required is to shift the precision points so that the starting position input and output angles, ϕ_1 and ψ_1 vary between 0 and 360°. Each new ϕ_1 or ψ_1 will yield a new solution.

Example 8.11 [39]

Let us synthesize the same function generator as in Sec. 8.10 using the Freudenstein equation. Recall that the function to be synthesized was $y = \sin(x)$ for $0° \le x \le 90°$. The range in ϕ is 120° and the range in ψ is 60° (see Fig. 8.50). Chebyshev spacing yielded the following precision points

$$\phi_2 - \phi_1 = 52°, \qquad \psi_2 - \psi_1 = 36.15°$$

$$\phi_3 - \phi_1 = 104°, \qquad \psi_3 - \psi_1 = 53.40°$$

In order to obtain the same four-bar solution as Fig. 8.51, we set $\phi_1 = 105°$ and $\psi_1 = 66.27°$. Thus the absolute precision points (required for this method) are

$$\phi_1 = 105°, \qquad \psi_1 = 66.27°$$

$$\phi_2 = 157°, \qquad \psi_2 = 102.42°$$

$$\phi_3 = 209°, \qquad \psi_3 = 119.67°$$

Thus

$$\omega_1 = \cos(105.0) - \cos(157.0) = 0.662$$

$$\omega_2 = \cos(66.27) - \cos(102.42) = 0.618$$

$$\omega_3 = -\cos(105.0 - 66.27) + \cos(157.0 - 102.42) = -0.201$$

$$\omega_4 = \cos(105.0) - \cos(209.0) = 0.616$$

$$\omega_5 = \cos(66.27) - \cos(119.67) = 0.897$$

$$\omega_6 = -\cos(105.0 - 66.27) + \cos(209.0 - 119.67) = -0.768$$

Solving for K_1, K_2, and K_3 from Eqs. (8.73) and (8.75) yields

$$K_1 = \frac{-0.180 + 0.475}{0.594 - 0.381} = 1.385$$

$$K_2 = \frac{-0.508 + 0.124}{0.594 - 0.381} = -1.803$$

$$K_3 = 0.304$$

If $Z_1 = 52.5$ mm, then from Eq. (8.76),

$$Z_2 = 29.0 \text{ mm}$$

$$Z_4 = 38.0 \text{ mm}$$

$$Z_3 = 75.6 \text{ mm}$$

which agrees with the graphical solution of Fig. 8.51.

8.22 LOOP-CLOSURE-EQUATION TECHNIQUE

An alternative method for synthesizing function generators is the loop-closure-equation technique. In the case of function generation, the bar-slider members of the general chain of Fig. 8.55 form one or more closed polygons. Therefore, using the notation of Fig. 8.81 with the link vectors $\mathbf{Z}_k$ identifying the mechanism loop, one can write the *equation of closure* (or *loop-closure equation*) for the jth position of the four-bar chain as follows:

$$\sum_{k=1}^{4} \mathbf{Z}_k = 0 \qquad (8.77)$$

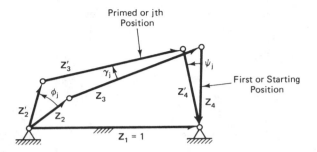

Figure 8.81 Notation for the four-bar mechanism for writing its loop-closure equation [Eq. (8.81)].

Specifically, the equation of closure for the four-bar linkage (Fig. 8.81) in its first position will be

$$\mathbf{Z}_2 + \mathbf{Z}_3 + \mathbf{Z}_4 - \mathbf{Z}_1 = 0 \qquad (8.78)$$

In case of function generation, only angular relationships of the four-bar linkage are of interest; doubling the size of the mechanism, or even a stretch-rotation does not alter the rotation between the links of that mechanism. Therefore, we may set $\mathbf{Z}_1 = 1$. Then

$$\mathbf{Z}_2 + \mathbf{Z}_3 + \mathbf{Z}_4 - 1.0 = 0 \qquad (8.79)$$

The jth position of the four-bar linkage of Fig. 8.81 can be expressed as

$$\mathbf{Z}_2' + \mathbf{Z}_3' + \mathbf{Z}_4' - 1.0 = 0 \qquad (8.80)$$

or, using Eq. (8.12),

$$\mathbf{Z}_2 e^{i\phi_j} + \mathbf{Z}_3 e^{i\gamma_j} + \mathbf{Z}_4 e^{i\psi_j} - 1.0 = 0 \qquad (8.81)$$

Equation (8.81) is an example of a *displacement equation* which is nonhomogeneous and linear in the complex unknowns $\mathbf{Z}_k$, $k = 2, 3, 4$, and has complex coefficients $e^{i\theta_j}$, $e^{i\psi_j}$, and $e^{i\gamma_j}$.

    In some cases the stretch rotation operator is useful in writing a displacement equation. The displacement equations for the first and jth position for the offset slider-crank of Fig. 8.82 are

$$\mathbf{Z}_4 + \mathbf{Z}_3 + \mathbf{Z}_2 - \mathbf{Z}_1 = 0 \qquad (8.82)$$

and

$$\mathbf{Z}_4 + \mathbf{Z}_3 e^{i\phi_j} + \mathbf{Z}_2 e^{i\gamma_j} - \rho_j \mathbf{Z}_1 = 0 \qquad (8.83)$$

where

$$\rho_j \equiv \frac{Z_1 + x_j}{Z_1}$$

is a stretch ratio.

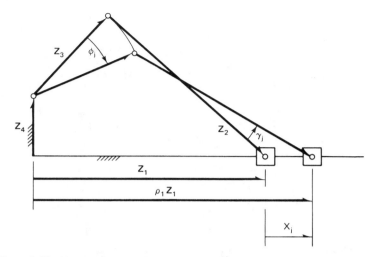

Figure 8.82   Notation for the offset slider-crank for writing its loop-closure equation, Eq. (8.83).

Applications of the Loop-Closure Method for Function Generation (Three Precision Points)

Suppose that we wish to synthesize a four-bar linkage for function generation where three precision points are to be prescribed (three-point approximation). From Eq. (8.81), the three equations that represent the loop equations of Fig. 8.81 for an initial position and two displacements are

$$\mathbf{Z}_2 + \mathbf{Z}_3 + \mathbf{Z}_4 = 1.0$$
$$\mathbf{Z}_2 e^{i\phi_2} + \mathbf{Z}_3 e^{i\gamma_2} + \mathbf{Z}_4 e^{i\psi_2} = 1.0 \qquad (8.84)$$
$$\mathbf{Z}_2 e^{i\phi_3} + \mathbf{Z}_3 e^{i\gamma_3} + \mathbf{Z}_4 e^{i\psi_3} = 1.0$$

Since we are synthesizing this linkage for function generation where the rotations of $\mathbf{Z}_2$ and $\mathbf{Z}_4$ are to be prescribed according to a functional relationship, ϕ_2, ϕ_3, ψ_2, and ψ_3 are prescribed. The unknowns in the system Eq. (8.84) are the vectors $\mathbf{Z}_2$, $\mathbf{Z}_3$, and $\mathbf{Z}_4$ (which represent the four-bar linkage in its first position) and the two rotations γ_2 and γ_3 of the coupler link. There are six independent equations (real and imaginary parts of each complex equation) and eight unknown reals. If we choose some arbitrary values for γ_2 and γ_3, the system Eq. (8.84) is nonhomogeneous and linear in the unknown link vectors $\mathbf{Z}_2$, $\mathbf{Z}_3$, and $\mathbf{Z}_4$. The complex coefficients of the link vectors are now known and the number of unknowns equals the number of equations. Cramer's rule may be used to solve these equations:

$$\mathbf{Z}_2 = \frac{\begin{vmatrix} 1 & 1 & 1 \\ 1 & e^{i\gamma_2} & e^{i\psi_2} \\ 1 & e^{i\gamma_3} & e^{i\psi_3} \end{vmatrix}}{\mathbf{D}} \qquad (8.85)$$

$$\mathbf{Z}_3 = \frac{\begin{vmatrix} 1 & 1 & 1 \\ e^{i\phi_2} & 1 & e^{i\psi_2} \\ e^{i\phi_3} & 1 & e^{i\psi_3} \end{vmatrix}}{\mathbf{D}} \tag{8.86}$$

$$\mathbf{Z}_4 = \frac{\begin{vmatrix} 1 & 1 & 1 \\ e^{i\phi_2} & e^{i\gamma_2} & 1 \\ e^{i\phi_3} & e^{i\gamma_3} & 1 \end{vmatrix}}{\mathbf{D}} \tag{8.87}$$

where

$$\mathbf{D} = \begin{vmatrix} 1 & 1 & 1 \\ e^{i\phi_2} & e^{i\gamma_2} & e^{i\psi_2} \\ e^{i\phi_3} & e^{i\gamma_3} & e^{i\psi_3} \end{vmatrix} \tag{8.88}$$

Notice that in Sec. 8.15, the standard form resulted in 2×2 determinants rather than the 3×3 determinants here. The same four-bar solution will result from either method, except that the scale and orientation of the synthesized linkages will differ. The standard form may be derived easily from the loop-closure technique by subtracting Eq. (8.79) from Eq. (8.81) and choosing one of the link vectors arbitrarily.

8.23 ORDER SYNTHESIS: FOUR-BAR FUNCTION GENERATION

In many instances a kinematic synthesis objective involves specifying not only finitely separated positions but velocities, accelerations, or higher derivatives as well. This is called *order synthesis,* and it may be accomplished using the complex-number loop-closure method by taking derivatives of the position equations.

Figure 8.79 shows a four-bar function generator in which we wish to specify the relative angular velocity, angular acceleration, and so on, of the output link with respect to those of the input link. Recall from the preceding section that the loop-closure (or position) equation is written for the starting position as follows:

$$\mathbf{Z}_2 + \mathbf{Z}_3 - \mathbf{Z}_4 = -\mathbf{Z}_1 \tag{8.89}$$

Since it is more convenient to take derivatives of vectors in the polar form, Eq. (8.89) is expressed as

Position equation: $Z_2 e^{i\theta_2} + Z_3 e^{i\theta_3} - Z_4 e^{i\theta_4} = -Z_1 e^{i\theta_1}$ \tag{8.90}

The velocity equation is formed by taking the derivative of Eq. (8.90) with respect to time ($\omega_i = d\theta_i/dt$):

Velocity equation: $Z_2 \omega_2 i e^{i\theta_2} + Z_3 \omega_3 i e^{i\theta_3} - Z_4 \omega_4 i e^{i\theta_4} = 0$ \tag{8.91}

or

$$\mathbf{Z}_2 \omega_2 + \mathbf{Z}_3 \omega_3 - \mathbf{Z}_4 \omega_4 = 0 \tag{8.92}$$

where the fixed link $\mathbf{Z}_1$ has dropped out because $\omega_1 = 0$.

Notice that Eq. (8.91) is equivalent to the velocity polygon technique of Chap. 3, in which all position vectors are rotated by 90°. The acceleration equation requires the second derivative of Eq. (8.90) obtained by taking the time derivative of Eq. (8.91) ($\alpha_i \equiv d\omega_i/dt$):

Acceleration equation: $Z_2(i\alpha_2 - \omega_2^2)e^{i\theta_2} + Z_3(i\alpha_3 - \omega_3^2)e^{i\theta_3} - Z_4(i\alpha_4 - \omega_4^2)e^{i\theta_4} = 0$

$$(8.93)$$

or

$$\mathbf{Z}_2(i\alpha_2 - \omega_2^2) + \mathbf{Z}_3(i\alpha_3 - \omega_3^2) - \mathbf{Z}_4(i\alpha_4 - \omega_4^2) = 0 \qquad (8.94)$$

Further derivatives may be obtained following the same procedure. The question arises: For how many derivatives can we synthesize the four-bar function generator? Following the logic laid out in Sec. 8.15, only a finite number of derivatives may be specified. Table 8.6 shows some not so surprising results for the four-bar function generator. With $\mathbf{Z}_1$ specified as $(-1.0 + 0.0i)$ (which establishes the linkage scale and orientation) and prescribing the derivatives of θ_2 and θ_4, the maximum number of positions plus derivatives is five.* However, if the scale factors R_{θ_2} and R_{θ_4} are also regarded as unknowns and the system of equations is written in terms of the independent and dependent variables x and y[†] [where $y = f(x)$ is the function to be generated], the number of infinitesi-

TABLE 8.6 NUMBER OF POSSIBLE SOLUTIONS USING EQ. (8.90), $\mathbf{Z}_2e^{i\theta_2} + \mathbf{Z}_3e^{i\theta_3} - \mathbf{Z}_4e^{i\theta_4} = -\mathbf{Z}_1$, AND ITS DERIVATIVES IN HIGHER-ORDER APPROXIMATE SYNTHESIS OF A FOUR-BAR MECHANISM. NOTE THAT FOR FUNCTION GENERATION θ_2, θ_4, ω_2, ω_4, α_2, α_4, ETC. ARE PRESCRIBED AND THAT $\mathbf{Z}_1 = -1$.

Number of positions plus derivatives	Number of scalar equations	Number of scalar unknowns	Number of free choices (scalars)	Number of solutions
1	2	$6 = (\mathbf{Z}_2, \mathbf{Z}_3, \mathbf{Z}_4)$	4	$O(\infty)^4$
2	4	$7 = $ (above) $+ \omega_3$	3	$O(\infty)^3$
3	6	$8 = $ (above) $+ \alpha_3$	2	$O(\infty)^2$
4	8	$9 = $ (above) $+ \dot{\alpha}_3$	1	$O(\infty)^1$
5	10	$10 = $ (above) $+ \ddot{\alpha}_3$	0	Finite

  * Note that each added derivative adds an "infinitesimally separated" position.

  [†] This is accomplished by solving the defining equations of R_{θ_2} and R_{θ_4} for the input and output angles, respectively, as follows:

$$R_{\theta_2} = \frac{\Delta\theta_2}{\Delta x} = \frac{\theta_2 - (\theta_2)_0}{x - x_0}$$

$$R_{\theta_2}(x - x_0) = \theta_2 - (\theta_2)_0, \qquad \theta_2 = (\theta_2)_0 + R_{\theta_2}(x - x_0)$$

$$\dot{\theta}_2 = R_{\theta_2}\dot{x}, \qquad \ddot{\theta}_2 = R_{\theta_2}\ddot{x}, \qquad \ldots, \qquad \overset{(j)}{\theta}_2 = R_{\theta_2}\overset{(j)}{x}$$

Similarly,

$$\theta_4 = (\theta_4)_0 + R_{\theta_4}(y - y_0)$$

$$\dot{\theta}_4 = R_{\theta_4}\dot{y}, \qquad \ddot{\theta}_4 = R_{\theta_4}\ddot{y}, \qquad \ldots, \qquad \overset{(j)}{\theta}_4 = R_{\theta_4}\overset{(j)}{y}$$

Substituting these in the position equation and its derivatives will yield a set of equations in which R_{θ_2} and R_{θ_4} are present and can be regarded as unknowns.

mally separated prescribed positions can be increased to 7. In other words, seventh-order approximation of function generation is possible with the four-bar linkage. This is the same result as for the synthesis of the four-bar function generators with finitely separated precision points. In fact, one would find that any mixture of point- and order-precision synthesis equations (i.e., any case of "multiply separated" precision points) will yield a similar table as Table 8.1. Also, the maximum number of positions, finitely or infinitesimally separated, that yields a set of linear equations is the same in each case. The maximum number of infinitesimally separated prescribed positions for order synthesis of path and motion generators will be found to be the same as in Table 8.1 for finitely separated positions.

Position, Velocity, and Acceleration Specification for the Four-Bar Function Generator

As was the case with the finite theory of Sec. 8.16, three multiply separated positions is the maximum number of positions available that still yields a set of linear equations in the starting position vectors of the movable links, $\mathbf{Z}_j$, $j = 2, 3, 4$. The position, velocity, and acceleration equations for the four-bar function generator (Fig. 8.79) from Eqs. (8.89), (8.92), and (8.94) are

$$\mathbf{Z}_2 + \mathbf{Z}_3 - \mathbf{Z}_4 = 1.0$$

$$\mathbf{Z}_2\omega_2 + \mathbf{Z}_3\omega_3 - \mathbf{Z}_4\omega_4 = 0.0 \tag{8.95}$$

$$\mathbf{Z}_2(i\alpha_2 - \omega_2^2) + \mathbf{Z}_3(i\alpha_3 - \omega_3^2) - \mathbf{Z}_4(i\alpha_4 - \omega_4^2) = 0.0$$

where $\mathbf{Z}_1$ was specified as $(-1.0 + 0.0i)$. Using Cramer's rule, we obtain

$$\mathbf{Z}_2 = \frac{-\omega_3(i\alpha_4 - \omega_4^2) + \omega_4(i\alpha_3 - \omega_3^2)}{D}$$

$$\mathbf{Z}_3 = \frac{\omega_2(i\alpha_4 - \omega_4^2) - \omega_4(i\alpha_2 - \omega_2^2)}{D} \tag{8.96}$$

$$\mathbf{Z}_4 = \frac{\omega_2(i\alpha_3 - \omega_3^2) - \omega_3(i\alpha_2 - \omega_2^2)}{D}$$

where

$$D = \omega_2[-\omega_4^2 + \omega_3^2 + i(-\alpha_3 + \alpha_4)] + \omega_3[-\omega_2^2 + \omega_4^2 + i(-\alpha_4 + \alpha_2)]$$
$$- \omega_4[\omega_3^2 - \omega_2^2 + i(\alpha_2 - \alpha_3)] \tag{8.97}$$

Example 8.12

Synthesize a four-bar linkage for function generation where the angular velocity and acceleration of the moving links are prescribed according to

$$\omega_2 = 2 \text{ rad/sec}, \qquad \alpha_2 = 0 \text{ rad/sec}^2$$

$$\omega_3 = 3.5 \text{ rad/sec}, \qquad \alpha_3 = 2 \text{ rad/sec}^2$$

$$\omega_4 = 5 \text{ rad/sec}, \qquad \alpha_4 = 4 \text{ rad/sec}^2$$

Using Eqs. (8.96) and (8.97), we have

$$\mathbf{D} = +6.75 + 0.0i$$

$$\mathbf{Z}_2 = 3.89 - 0.59i, \qquad \mathbf{Z}_3 = -4.44 + 1.19i, \qquad \mathbf{Z}_4 = -1.56 + 0.59i$$

8.24 THREE-PRECISION-POINT SYNTHESIS: ANALYTICAL VERSUS GRAPHICAL

Thus far, both graphical and analytical approaches have been presented for three finitely separated positions of motion-, path with timing-, and function-generation synthesis of a four-bar linkage. Both techniques are straightforward. Which is better? The answer: both are equally important. Graphical techniques are extremely useful in the initial stages of synthesis. If a graphical construction does not yield an "optimal" solution in a reasonable amount of time or if the error sensitivity is high (e.g., the need to locate the intersection of lines that form an acute angle), then the analytical standard-form method is very attractive. In such cases the preliminary graphical solution will yield reasonable values for arbitrarily assumed (free-choice) quantities, which will help obtain workable computer solutions. The Cramer's rule solution described above is easily programmed for digital computation (the flowchart of a three-precision-point program is shown in Fig. 8.63) and numerous accurate solutions can be obtained in a fraction of the time required for a graphical construction. (Section 8.18 shows an alternative computer graphics technique for three precision points, which is a combined graphical and analytical method.)

A notable correlation between the graphical and analytical methods should be emphasized at this point. In both techniques, for the three-position synthesis of each dyad, there are two infinities of solutions for motion-, path-generation with prescribed timing, and function generation. As pointed out above, a function generator four-bar linkage actually appears to require two additional scalars as free choices: the two components of the starting position vector of one of the links. However, picking that link specifies only the scale and orientation of the linkage. No new function generators are obtained by varying this link, because the functional relationship of the input and output rotations is not affected by this choice.

A very useful reference for linkage design is an atlas of four-bar coupler curves by *Hrones and Nelson* [89]. Approximately 7300 coupler curves of crank-rocker four-bar linkages are displayed (e.g., see Fig. 8.83). The black dots represent coupler points whose coupler curves are plotted. These can be used as "tracer points" in a path or motion generator linkage. Each dash on the coupler curves represents 10° of input crank rotation to provide an indication of coupler point velocity. The crank always has length 1 while the lengths of the coupler A, follower B and the fixed link C vary from page to page, yielding a variety of families of coupler curves. This atlas is extremely useful in the initial stages of a path- or in some cases motion-generation synthesis effort. A designer may be able to find several coupler curve forms that nearly accomplish the task at hand and then use these linkages to come up with proper "free-choices" (see Sec. 8.15) to help find a more nearly optimal linkage in a shorter time. Also, it may turn out that

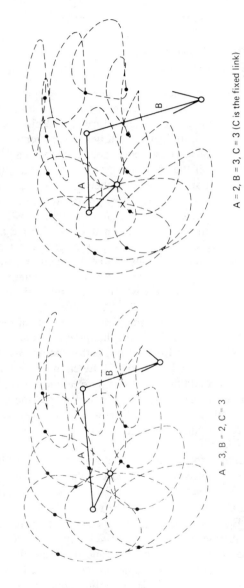

A = 3, B = 2, C = 3

A = 2, B = 3, C = 3 (C is the fixed link)

Figure 8.83 Sample pages from the atlas of four-bar coupler curves by Hrones and Nelson [89]. In [89], lengths of dashes of the curves indicate 10° increments of crank rotations. Here the lengths of dashes are not to scale.

there are no crank-rocker four-bars that satisfy the design requirements and the *type* synthesis step may have to be reconsidered.

APPENDIX: CASE STUDY—TYPE SYNTHESIS OF CASEMENT WINDOW MECHANISMS [54]

A powerful alternative method of type synthesis to the associated linkage approach presented in this chapter is applied here in an industrial application.

Structure Phase of Type Synthesis

Freudenstein and Maki [76] suggest separation of *structure* and *function* in the conceptual phase of mechanism synthesis. They point out that the degree of freedom of a mechanism imposes constraint on the structure of the mechanism. Rather than using Gruebler's equation (see Chap. 1) and Eq. (8.7) for degrees of freedom of mechanisms, they suggest the following forms:

$$F = \lambda(l - j - 1) + \sum_{i=1}^{j} f_i \tag{8.98}$$

and

$$L_{\text{IND}} = j - l + 1 \tag{8.99}$$

where F = number of degrees of freedom of mechanism

l = number of links of mechanism (including the fixed link; all links are considered as rigid bodies having at least two joints)

j = number of joints of mechanism; each joint is assumed as binary (i.e., connecting two links); if a joint connects more than two links, the number of joints $j = N - 1$, where N = the number of links at the common joint

f_i = degree of freedom of ith joint; this is the freedom of the relative motion between the connected links

λ = degree of freedom of the space within which the mechanism operates; for plane motion and motion on a surface $\lambda = 3$ and for spatial motions $\lambda = 6$

L_{IND} = number of independent circuits or closed loops in the mechanism

Combining Eqs. (8.98) and (8.99), we obtain

$$\sum f_i = F + \lambda L_{\text{IND}} \tag{8.100}$$

Since we are dealing with planar motion and a single degree of freedom,

$$F = 1, \qquad \lambda = 3$$

For the case of one closed loop,

$$L_{IND} = 1$$

From Eq. (8.100),

$$\sum f_i = 4$$

For example, if we investigate the four-link chain using Eq. (8.99),

$$j = L_{IND} + l - 1 = 4 \tag{8.101}$$

Thus the number of joints is four (as would be expected). Equation (8.98) shows that a five-bar chain with five joints (with pin and slider joints) yields a mechanism with two degrees of freedom:

$$F = 3(5 - 5 - 1) + 5 = +2$$

If four joints allow a single degree of freedom between connecting links and one joint allows two degrees of freedom of relative motion (e.g., gear connection), a five-bar chain has a single degree of freedom. In the case of a six-bar with two loops ($L_{IND} = 2$), from Eq. (8.101),

$$j = 2 + 6 - 1 = 7$$

and from Eq. (8.98) (assuming pin and slider joints),

$$F = 3(6 - 7 - 1) + 7 = 1$$

Design Objective

Casement-type windows (see Fig. 1.4) are generally defined as vertically pivoted, outward-swinging, ventilating windows. Screens, where used, are placed to the inside of the room. Casement windows were initially operated without screens and were merely pushed open and closed. Development of linkage operators for this window brought various improvements, such as

1. A method to lock the window at various open and closed positions.
2. Screens through which an opener could function.
3. Concealed hinges.
4. Weather stripping.
5. Geared operators to control window position.

These developments were achieved by about 1906. Since the year 1906, at least 44 U.S. patents have been issued (as discovered in a prior-art search) which make further improvements to casement window mechanisms. These improvements are necessary because over 5,000,000 operating casement-type windows are sold annually in the United States and Canada.

Consumer demands have brought about better insulation by (1) double- and triple-glazed sashes, and (2) multiple weather stripping. These changes result in significantly increased loading on the casement window operator and its associated linkages, causing objectionable operating characteristics. Existing casement window mechanisms function satisfactorily on windows up to about 50 lbf weight but require excessive operating torque on larger windows.

Some popular operators have good pull-in (closing) characteristics but they lose mechanical advantage as the window approaches the 90° open position. The transmission angle (also a critical factor in mechanism design) is also poor near the 90° position. Other operators found in use today have good (low) torque requirement at open positions but have poor pull-in characteristics. A low, uniform torque from fully open to the fully closed position is desirable. Low torque gives user satisfaction and long operator life. The prior art search disclosed many window mechanism concepts, but no satisfactory scheme was available to evaluate and compare the various designs.

Consequently, a plan was developed to analyze and categorize past mechanism concepts and new designs which would hopefully lead to an improved casement window mechanism.

Design Constraints

The casement window operator linkage design has many challenging constraints. The most important considerations are

1. The sash (window) must open 90° from the sill.
2. The end of the sash must slide at least 10.16 cm (4 in) in order to allow washing on both sides of the window from the inside.
3. An open sash must leave 50.8 cm (20 in) for egress codes (exit in case of emergency). Some local codes require even greater opening.
4. The operator linkage must support the weight of window with minimal sag of the sash.
5. The operator linkage must have single actuator arm (the mechanism should have one degree of freedom).
6. A new operator linkage must have improved transmission angle and better mechanical advantage than the present mechanism.
7. When the sash is in the closed position, all portions of the mechanism must be below the sill cover, not extend beyond the plane of the sash toward the outside, and have minimum extension into the room.
8. During deployment, all parts of the operator linkage must be between the sill and the sash (so as not to interfere with the weather stripping) and cannot extend further into the sill.
9. The casement operator must be as simple as possible due to economic considerations (e.g., pin and slider joints are preferred over gear and cam connections for both initial cost and maintenance considerations).
10. A maximum number of parts must be interchangeable for both left- and right-handed operators. (A sash may be hung at either the left or right side.)

Note that the need for self locking was not mentioned. This caused problems later, as will be seen.

Analysis of the Current Operator

Before looking for new casement window linkages, the performance of the current operator will be investigated.

Mechanical advantage analysis.   The instant center technique (Chap. 3) is used to perform a static force analysis (or mechanical advantage analysis) which is useful in determining possible improvements in the current operator. Figure 8.84 shows the current casement linkage with the operator arm in an intermediate position. The kinematic diagram (unscaled) of the mechanism in Fig. 1.4b is shown in Fig. 1.5. Notice that this mechanism is a six-bar chain. The pertinent instant centers have been located in Fig. 8.85 (operator arm at 60° to the horizontal of the drawing) and also in Fig. 8.86 (fully open position).

Consider the operator arm (link 2) as the input and the window (link 4) as the output. If one assumes that the energy losses in a linkage as it moves are small, then (as is described in Chap. 3) power in should be equal to power out:

$$P_{in} = P_{out}$$

or

$$\omega_{in} T_{in} = \omega_{out} T_{out}$$

where ω = angular velocity (rad/sec) and T = torque (in·lb)
In this case

$$|\omega_2 T_2| = |\omega_4 T_4|$$

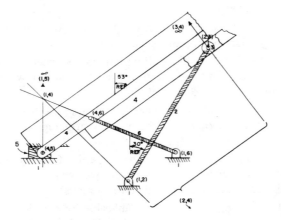

Figure 8.84   Current casement window at 60° operator arm position showing locations of instant centers.

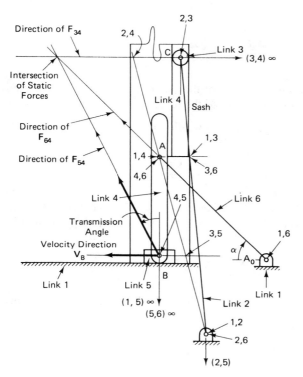

Figure 8.85 Current casement window mechanism at fully open position showing locations of instant centers.

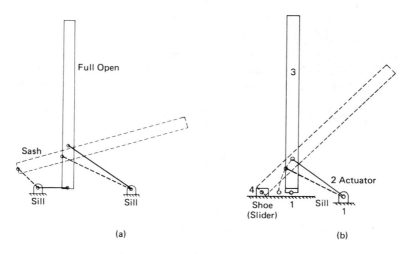

(a)   (b)

Figure 8.86 (a) four-bar operator; (b) slider-crank operator.

or

$$|\omega_2| F_2 R_2 = |\omega_4| F_4 R_4$$

where F = a force (lbf)

  R = a radius from the instant center of the link (with respect to the ground) perpendicular to the line of action of the force (in)

Using instant centers, the angular velocity ratio may be expressed in terms of linkage geometry (see Chap. 3)

$$\left|\frac{\omega_2}{\omega_4}\right| = \frac{(2,4 - 1,4)}{(2,4 - 1,2)}$$

where 2,4, 1,4, and 1,2 are the locations of the instant centers between links 2 and 4, 1 and 4, and 1 and 2, respectively. The right-hand side of the foregoing equation is negative if center 2,4 lies between 1,4 and 1,2. From Fig. 8.84,

$$\frac{\omega_2}{\omega_4} \simeq +1.1$$

The mechanical advantage (M.A.) is defined as the (force out)/(force in) or in this case,

$$M.A. = \frac{F_4}{F_2}$$

or

$$\frac{F_4}{F_2} = \frac{\omega_2}{\omega_4}\frac{R_2}{R_4} \simeq 1.1\frac{R_2}{R_4}$$

Given a unit radius of the input force on the operator handle, a $28:1$ gear reduction in the worm gear, a 30% efficiency in the worm gear, and a length of 3.5 in. between the shoe (4,5) (which locates the resistance load on the window) and the instant center (1,4), the mechanical advantage of the system is (at 60° of the operator arm).

$$M.A. = \frac{F_{out}}{F_{in}} = \frac{(1,4 - 2,4)}{(1,2 - 2,4)}(\text{efficiency}) \times (\text{gear ratio}) \times \frac{\text{input radius}}{(1,4 - 4,5)}$$

$$= (1.1)(0.30)(28)\frac{1}{3.5} \qquad\qquad (8.102)$$

$$= 2.64$$

The mechanical advantage analysis was also performed at the fully open position (see Fig. 1.4a). Here the angular velocity ratio is reduced to

$$\frac{\omega_2}{\omega_4} = \frac{3.5}{10} = 0.35$$

In the open position, $(1,4 - 4,5)$ is about the same length as before and the mechanical advantage from Eq. (8.102) is

$$M.A. = (0.35)(0.30)(28)\frac{1}{3.5}$$

$$= 0.84$$

Methods of increasing the mechanical advantage of the current operator.   According to Eq. (8.102), there are several adjustments that could lead to an increased mechanical advantage of the current operator linkage:

1. Increase the efficiency of the operator.
2. Increase the gear ratio.
3. Reduce the distance $(4,6 - 4,5)$, which equals $(1,4 - 4,5)$ in the open position.
4. Increase the ratio $(1,4 - 2,4)/(1,2 - 2,4)$. Notice that these instant centers change their relationship with any change in linkage geometry.
5. Also, the operator will be more effective (i.e., F_{out} will cause more motion) if the frictional load at the shoe is decreased by reducing the coefficient of friction between the shoe (link 5) and the sill (link 1).

Adjustments 1, 2, and 5 involve higher cost, and 2 will also require more turns of the operator handle to open the sash.

Transmission-angle analysis.   Another measure of the mobility of the operator linkage is the transmission angle. For optimal mobility, the net force action on the output link (link 5 in this analysis) of a mechanism should be in the same direction as the output link velocity at the point of action of the resisting force. For our purposes, the transmission angle was defined in Chap. 3 as the smallest angle between the direction of the perpendicular to the shoe velocity (V_B) and the direction of the net static force F_{45} (since $|F_{45}| = |F_{54}|$) acting on the sash (link 4). A free-body diagram of link 5 is drawn (see Fig. 8.85). Forces F_{34}, F_{64}, and F_{54} must intersect at a single point. Although not shown, the frictional forces at the shoe could be included as well as the smaller frictional effects at A and C. The figure shows that the transmission angle in the open position is about 27°, which is marginal. As the linkage closes the sash, the transmission angle improves.

Figure 8.85 gives an indication of how the transmission angle can be improved in the present design. These improvements are

1. Move the pivot of the crank (center 1,6) out from the sill and/or over to the right. (Center 1,6 cannot be moved beyond the edge of the sill, however.) Moving the pivot A_0 to the right will have minor effect on the transmission angle and will introduce more cost due to the increased length of link 6.

2. Lower the connection between the crank (link 6) and the window (link 4). This will decrease the distance $(4,6 - 4,5)$, which will improve the mechanical advantage.

3. Decrease the length of the operator arm (link 2). This, however, is directly counter to what is advantageous for better mechanical advantage.

Thus the best change that would improve both the mechanical advantage and the transmission angle at the open position would be to follow suggestion 2. Unfortunately, a decrease in length $(4,6 - 4,5)$ directly decreases washability (the distance the base of the window slides out from its closed position), when the window is fully open

$$\text{washability} = (4,6 - 1,6) \cos \alpha \qquad (8.103)$$

where $\alpha = \sphericalangle AA_0B$.

The conclusion of both the mechanical advantage and the transmission-angle analyses is that even with some changes as suggested above, either the cost will increase or other desirable performance characteristics must be sacrificed. Thus this type of linkage has limited performance possibilities and other types of casement operators are to be investigated.

Type Synthesis

Kinematic synthesis of mechanisms can be separated into two steps: *type* and *dimensional* synthesis. The first helps determine the best linkage types, while the latter produces the significant dimensions of the mechanism that will best perform the desired task.

The simplest linkage chain, the four-bar, is the logical initial choice for a type of mechanism to suggest as a casement window operator. The sash could be connected to the coupler link of the four-bar linkage, as shown in Fig. 8.86a, for example. The drawbacks of this design are (1) the interference of the links with weather stripping in the open position, and (2) the inability of the sash to open to the 90° position without traveling through a toggle position if either of the links pinned to ground was designated as the input link. Here, a torque would be required on the sash to return it to the closed position (not possible without adding more linkage members). If the 90° opening requirement were relaxed (as in an awning window application), the four-bar design would be more acceptable. In both cases, however, the four-bar's ability to carry the weight of the sash is questionable since the entire sash moves away from the sill during its motion. In fact, the larger the requirement for washability the longer the input link and the higher the probability for sash sag.

The slider-crank linkage is the next logical choice for a casement operator linkage since the slider could move along the sill and thereby support the weight of the window (see Fig. 8.86b). Unfortunately, a torque on the sash will still be required to close it from a 90° position. Link 2 could not be used effectively as an input link since the transmission angle is 0° in the open position. Link 4 is also undesirable as an input link since the transmission angle is 0° in the closed position.

Since the four-link chain looks to have limited acceptability to the casement window application, mechanisms with more links must be pursued. Since the degree of freedom of a casement mechanism must be 1, a type synthesis effort can be simplified. For example, unless gear- or cam-type connections between links are used (those that allow two degrees of freedom between connecting members), the five-bar and seven-bar chains will not be appropriate. This is because these chains will result in two degrees of freedom if only pin or slider joints are used. Thus the six-bar chain is the next logical chain to investigate. As introduced in Chap. 1 (Figs. 1.9 to 1.13), there are only five types of six-bar linkages with a single degree of freedom—variations of the Stephenson and the Watt chains. The Stephenson six-bar has nonadjacent ternary links; in the Watt chain, the ternary links are pinned together. Since the window would be connected to a floating link, there are only 11 possible six-bar combinations with only revolute joints. If, however, sliders are allowed to replace one or more links, many more possible combinations result. For example, Fig. 8.87 shows some of the possible Stephenson III linkages that can be used for a casement operator. One can now appreciate the need for a systematic type synthesis.

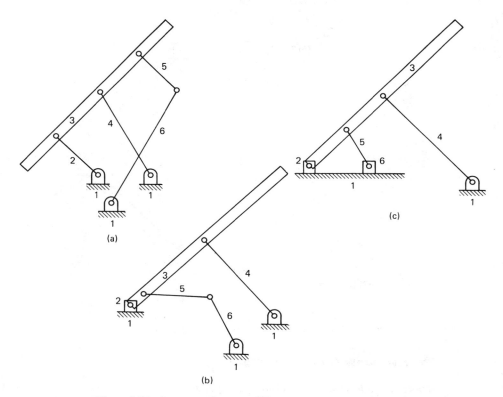

Figure 8.87   Some Stephenson III casement window operator types.

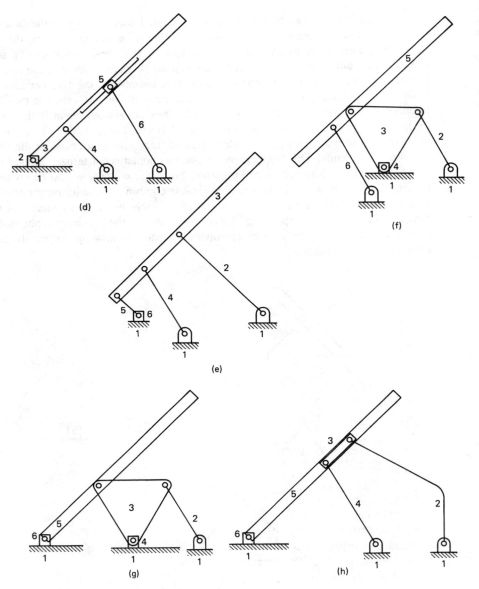

Figure 8.87 (cont.)

Other Window Operator Designs

Several popular casement and awning operators are six-bar chains. The "current" operator (Fig. 1.4) is a Stephenson III type (Fig. 8.87, linkage *d*). A metal awning linkage must move the window out from the frame before rotating the sash. Most do not allow

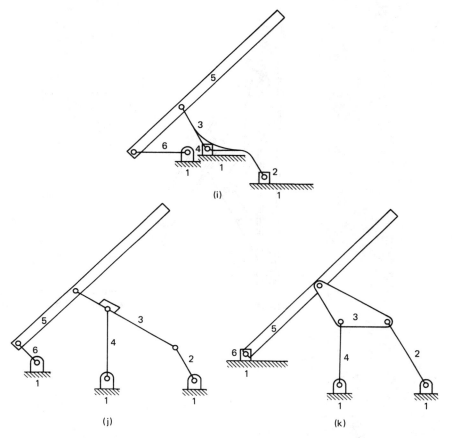

(i)

(j)

(k)

Figure 8.87 (cont.)

90° of rotation. The most popular awning linkage, the Anderberg (U.S. patent 2,784,459), shown in Figs. 8.88 and 8.89, is a Stephenson I type six-bar (see Fig. 1.11). There are several interesting variations of the basic Anderberg design. Figure 8.90 shows one (U.S. patent 3,345,777) where two pin joints on the slider are combined presumably for patent purposes (although kinematically, both of these link-ages are the same structural type). Figure 8.91 shows another proposed variation, where another link is added (dashed) for the purpose of reducing sag of the sash. From Eq. (8.98) we see that this design has zero degrees of freedom. The linkage will, however, have mobility (although extremely dependent on manufacturing accuracy) due to the special geometry of the new link and the clearances in the linkage joints.

Figure 8.92 shows a geared version (U.S. patent 3,838,537) which has the objective of helping "pull in" the awning window. This "overconstrained" linkage also maintains mobility due to geometry. The linkage is designed to have nearly constant angular veloc-ity ratio between links 3 and 5 so that the gears connecting these two links do not bind.

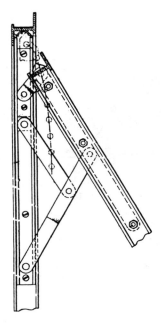

Figure 8.88 Awning linkage (A. W. Anderberg, U.S. Patent 2,784,459, 1957.)

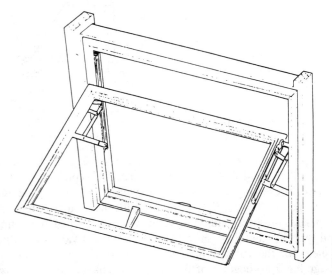

Figure 8.89 Awning linkage shown in open position.

Figure 8.93 shows the "gearless torque lock" awning window mechanism (U.S. patent 2,761, 674). This mechanism is a Watt II type six-bar. The Pella-type casement operator (U.S. patent 3,438,151) is shown in Fig. 8.94. Until the engagement of the pin and slot (discussed later), this linkage is a Stephenson III type six-bar.

Equations (8.98) to (8.100) will again be useful shortly, but first some functional aspects of the casement window application will be enumerated.

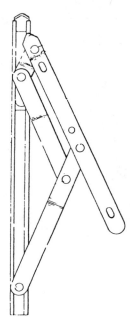

Figure 8.90 Awning linkage. (A. W. Anderberg, U.S. Patent 3,345,777, 1967.)

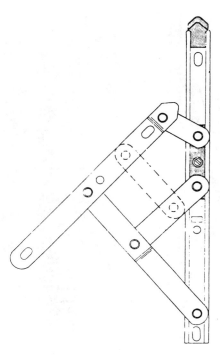

Figure 8.91 Awning linkage of Fig. 8.88 with dashed line added (Cotswald Catalog).

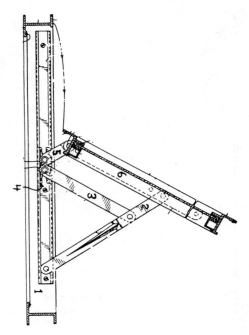

Figure 8.92 Awning linkage. (H. L. Stavenau and W. C. Bates, U.S. Patent 3,838,537, 1973.)

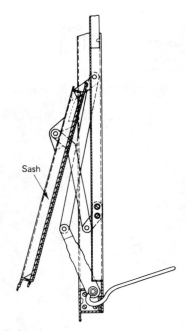

Figure 8.93 Awning linkage. (*N. C. Walberg et al., U.S. patent 2,761,674, 1956.*)

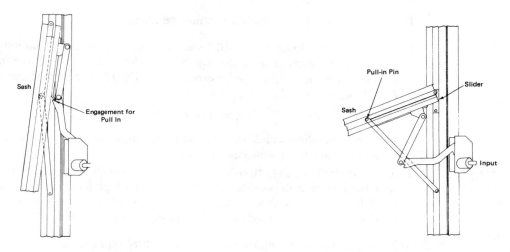

Figure 8.94   Pella linkage. (*R. Rivers and M. Minter, U.S. patent 3,438,151, 1969.*)

Functional Phase of Type Synthesis

The previous discussion clearly demonstrates that a four-link chain will not accomplish all the objectives of a new casement window operator. To help narrow the field of possible mechanisms with, say, five to eight links, we must investigate the function of an optimal mechanism. Based on observation of the four-bar chains and the current operator:

1. A multiloop linkage is preferred such that one loop guides the window through 90° while another loop acts as the driver or the operator. The concept followed here is that the guiding loop will probably run into transmission angle and/or mechanical advantage problems if one of its members is used as the input.

2. The sash should rest on a slider (which slides on the sill) so that the weight of the sash is supported in a simple manner as the window opens.

3. A slider-crank linkage (with the sash connected to its coupler) is the best guiding linkage, due to simplicity. Since the top of the window also must be guided, all one would need would be a duplicate slider-crank on top (without the driving loop) to keep the top of the sash coordinated with the bottom.

Based on these three simplifying decisions, the possible solutions have significantly decreased. For example, the five-bar chain (with one two-degree-of-freedom joint) is disregarded (since only a single loop is possible) in favor of a six-link chain. Also, several of the six-bar chain possibilities are set aside since they do not have a slider-crank loop with the window attached to the coupler link of that loop (e.g., mechanisms $a, c, e \rightarrow k$ of Fig. 8.87 would be discarded in the Stephenson III six-bar options shown).

Two other important observations can now be made:

1. Any mechanism loop(s) that will be added to the slider-crank chain must have total joint freedoms of three (for single loop addition), six (for two loops), or nine (for three loops), as can be derived from Eq. (8.100). For example, with $F = 1$, $\lambda = 3$, and $L_{IND} = 2$,

$$\Sigma f_i = 1 + 6 = 7$$

and since the slider-crank has four joint freedoms, the additional mechanism loop must have three total joint freedoms.

2. There are three ways of effectively driving the window through the 90° of motion:
   a. Push at or near the outside end of the sash (for high mechanical advantage).
   b. Pull at or near the shoe since the resistance to sash movements is at the shoe.
   c. Combination of methods a and b, causing a torque on the sash.

In the following sections, driving mechanisms of all three types are discussed.

New Casement Linkage Operators

Push-type operators. The current casement mechanism (Figs. 1.4 and 8.87d) has a push-type operator. The slider-crank is driven by a dyad (two links), a link pinned to a slider, in this case with $\Sigma f_i = 3$. The advantage of the current mechanism is good "pull-in" characteristics. Mechanisms b and c of Fig. 8.87 are two other possible single-loop dyad driver loops ($\Sigma f_i = 3$). Neither of these offers great improvement. Linkage b has much better characteristics at the open position but poor "pull-in." Linkage c offers poor mechanical advantage in all positions.

Several two-loop drivers with $\Sigma f_i = 6$, which had the possibility of extending the actuator arm, were investigated. Although the mechanical advantage of some of these extending-actuator-arm linkages was promising, due to their complexity (eight links) other problems (higher cost and multiple transmission angles) make them unacceptable.

Based on the desirability to have a less complex extending-arm actuator, mechanisms with higher-pair contact (gears) were investigated. The concept was that a cycloidal curve could possibly match the movement of the end of the window. This led to writing the equation of point P at the end of the window (Fig. 8.95):

$$\mathbf{Z}_p = Z(e^{i\phi_j} + \rho e^{i\alpha_j}), \quad \text{where } \alpha_j' = 180° - |\alpha_j| \tag{8.104}$$

$$\alpha_j = \sin^{-1}\frac{Z \sin \phi_j}{Z'} \le \frac{\pi}{2} \tag{8.105}$$

and

$$\rho = \frac{|\mathbf{W}|}{|\mathbf{Z}|} \tag{8.106}$$

An investigation of possible cycloidal driving mechanisms with $\Sigma f_i = 6$ or 9 (two or three loops) yielded a new hypocycloidal mechanism in Figs. 8.96 and 8.97. (U.S.

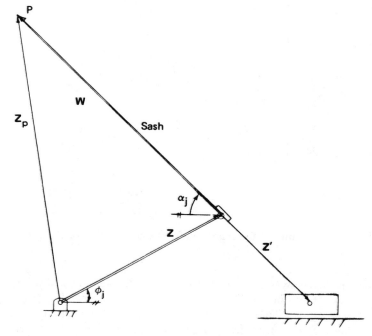

Figure 8.95   Slider-crank guiding linkage notation.

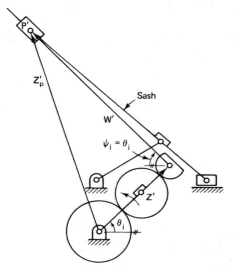

Figure 8.96   Hypocycloidal crank driving linkage notation. (*A. Erdman and J. Peterson, U.S. patent 4,266,371.*)

patent 4266371). The equation for the position vector of point P' at the end of this operator is

$$\mathbf{Z}'_p = \mathbf{Z}'(e^{i\theta_j} + \rho e^{i\psi_j}) \tag{8.107}$$

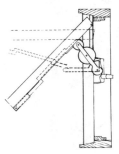

Figure 8.97 Hypocycloidal mechanism (slider-crank not shown).

where

$$\rho = \frac{|\mathbf{W}'|}{|\mathbf{Z}'|} \tag{8.108}$$

and since the radius of the outer planet is one-half that of the sun,

$$|\psi_j| = |\theta_j| \tag{8.109}$$

Equation (8.107) was set equal to Eq. (8.104) for a number of positions and the best solution built for testing purposes. The maximum difference between $|\mathbf{Z}_p'|$ and $|\mathbf{Z}_p|$ is only $\frac{1}{2}$ in. for a 24-inch window, yielding maximum mechanical advantage through the driving linkage. Figure 8.98 shows the level of input torque (experimentally determined) of this operator compared with the current operator. The potential improvement* is promising, especially near the fully open position. A force analysis performed on the driving linkage yields high forces between gear teeth and at the bearings. These high force demands, as well as the complexity of the hypocycloidal design, detract from the promising results shown in Fig. 8.98.

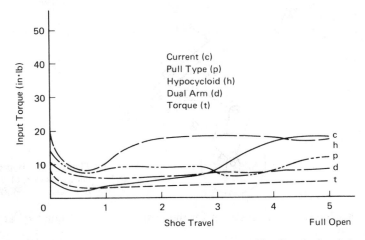

Figure 8.98 Required input torque for current operator (Fig. 2) and other designs.

* The hypocyclic model was binding during the test due to misalignment of gears.

Pull-type operators.   One of the most efficient methods of opening a window is driving near the point of frictional load—the shoe. The simplest way to pull the shoe is a dyad (see Fig. 8.87, linkage *b*). Unfortunately, the "pull-in" characteristics of this design are poor.

To achieve the desired window control near the closed position, a mechanism was proposed that operates in two modes: First, an over-constrained system ($F = 0$) was synthesized where a point on the crank follows a track on the sash for better pull-in and window control through the first 30° of motion (see the dashed position of Fig. 8.99). Beyond 30°, the mechanism has a single degree of freedom.

The dimensional synthesis of this new mechanism* (U.S. patent 4,253,276) was based on kinematics and kinetics. The relative sizes of the crank and connecting link were adjusted to allow good transmission angles throughout the operation and a toggle at closing. The cam slot was programmed to match the kinematic constraints of the rest of the mechanism. Experimental results for required input torque for the pull-type operator are shown in Fig. 8.98.

Push-pull-type operators.   The concept that inspired the evolution of the push-pull-type operator was to simultaneously push the window and pull near the slider. The problem with this objective is that the slider-crank linkage constrains the relationship between the slider velocity and the sash angular velocity. It is extremely difficult to match these kinematic relationships with a single-degree-of-freedom driving mechanism.

Thus a drive linkage had to be synthesized that would either *match* these constraints (i.e., $F = 0$) or incorporate strategic linkage geometry to capitalize on the hinge timing ($F = 1$). The former suggestion could be accomplished by using noncircular gears to achieve the desired shoe position and window angular position coordination. Unfortunately, the cost of the gears would probably be prohibitive and the mechanism would be too sensitive to inaccuracies, improper mounting, and wear. To convert this concept into one that employs gears of constant radius and has a mechanism with $F = 1$, from Eqs. (8.98) to (8.100), $l = 9$, $L_{\text{IND}} = 4$, $j = 12$ and $\Sigma f_i = 13$. A new casement operator linkage (shown in Fig. 8.100) called the "dual arm" operator (U.S.

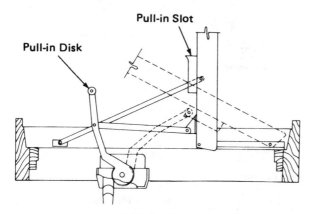

Pull-in Slot

Pull-in Disk

Figure 8.99 Pull-type operator. A disk enters a slot in the dashed position to aid pull-in. (*J. Peterson and E. Nelson, [123].*)

* Somewhat similar to the Pella concept of Fig. 8.94.

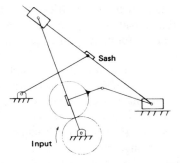

Figure 8.100 Dual-arm operator. (*Van Klompenburg, J. Peterson, E. Nelson, U.S. patent 4,241,541.*)

patent 4,241,541) satisfies these constraints. The optimal dimensions of this mechanism were derived based on kinematic and kinetic constraints. For example, an attempt was made to balance the forces applied on the window by both arms of the operator. Figure 8.98 shows the required input torque of this operator measured from an optimized prototype.

The push-pull concept is intuitively most attractive. Another attempt was made to design such an operator. Instead of having separate arms extending to the window, a concept was pursued to provide a torque directly on a window by fixing a gear segment to the sash. With a gear as part of the kinematic chain, the minimum number of links is

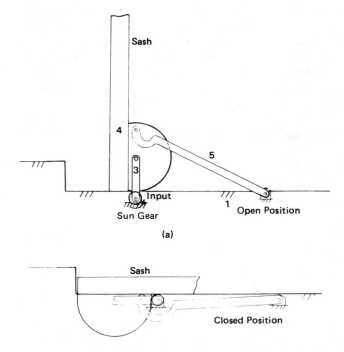

Figure 8.101 Torque operator shown in two positions.

$l = 5$ with $\Sigma f_i = 7$. The resulting concept is shown in Fig. 8.101. Notice that the sum of the diameters of the gears is approximately equal to the required opening. As shown, this design does not satisfy the requirement for supporting the weight of the window. One can add the slider-crank guiding linkage (not shown) by carefully designing the geared five-bar portion so that the inner edge of the sash follows a straight line. The dimensional synthesis was accomplished by noticing that links 3, 4, 5, and 1 make up a four-bar chain (similar to Fig. 8.86a). The objective then was to synthesize the four-bar such that the path of a point at the base of the window traveled a nearly straight line path. This can be approximated closely enough that the clearance in the slider will allow smooth operation of the entire mechanism. This new torque operator (patent pending) is shown in Fig. 8.101. Notice that $F = 1$ from Eq. (8.98) since $l = 7$ and $\Sigma f_i = 10$.

Figure 8.98 shows the experimentally obtained input torque requirements for the torque operator. These values are the lowest of the mechanisms tested. Unfortunately, the torque is too low. A nonperceived design constraint appeared after this mechanism was tested: a casement window mechanism should be self-locking. That is, when the input is set at a particular position, any reasonable wind load should not be able to close or further open the sash. This design overcame the mechanical advantage and transmission angle objectives so well that it is not self-locking (without addition of auxiliary components). The other drawback of this design (although not as critical) is the size of the planet gear and thus the size of the required sill cover.

Conclusions

A systematic type synthesis approach was applied to a casement window application which had many clearly defined design constraints. Separation of structure and function of mechanisms allowed a productive interchange of guiding rules that significantly narrowed the search for optimal solutions. The type synthesis yielded many casement window mechanism concepts, and the most feasible designs were investigated further in a dimensional synthesis step. Those that appeared to meet the constraints were designed and prototyped for evaluation. At least four new operator concepts were prototyped and exhibited favorable operating characteristics over operators currently on the market. At this point, marketing, manufacturing, and management input was utilized to select the best new mechanism. The "dual arm" concept was selected (and marketed since the mid-1980s) as the best alternative because:

1. It offered the best overall performance. (It pushed the window as well as pulled the window, to achieve a much lower operating torque from fully open to fully closed position.)

2. It could be manufactured at a reasonable cost.

3. It would require minimal changes for window manufacturers.

4. It would be compatible with the existing four-link slider-crank hinge.

5. Parts would be compatible with current manufacturing capabilities.

PROBLEMS*

8.1. (a) Determine the Chebyshev spacing for a function $y = 2x^2 - x$ for the range $0 \leq x \leq 2$ where four precision points are required.

   (b) Based on these precision points, find ϕ_2, ϕ_3, ϕ_4, and ψ_2, ψ_3, ψ_4 if $\Delta\phi = 45°$ and $\Delta\psi = 90°$.

8.2. It is desired to generate $y = e^x$, $0 \leq x \leq 4$, and specify three precision points. Using Chebyshev spacing, find:

   (a) x_1, x_2, x_3

   (b) ϕ_j, ψ_j ($j = 2, 3$) if $\Delta\phi = 80°$ and $\Delta\psi = 110°$

8.3. Find the Chebyshev spacing for three prescribed positions for the function $y = x^2 + 3x + 5$, $0 \leq x \leq 2$. Also find ϕ_j, ψ_j ($j = 2, 3$) if $\Delta\phi = \Delta\psi = 45°$.

8.4. (a) Determine the three-point Chebyshev spacing for the function $y = x^2$, $0 \leq x \leq 10$, and $\Delta\phi = \Delta\psi = 60°$.

   (b) Find ϕ_j, ψ_j for $j = 2, 3$.

   (c) If a four-bar linkage is to be designed to generate this function, determine the starting position of the linkage (if $\gamma_2 = 1°$ and $\gamma_3 = 30°$) by the complex-number method.

   (d) Draw the linkage in its three precision positions and determine if this is an acceptable linkage.

8.5. (a) Find the three-precision-point Chebyshev spacing for the function $y = x^{3/2}$, $0 \leq x \leq 100$ where $\Delta\phi = \Delta\psi = 60°$.

   (b) Find ϕ_j, ψ_j ($j = 2, 3$).

   (c) If a four-bar linkage is required for this task, solve for the resulting linkage using complex numbers if $\gamma_2 = 0.01°$ and $\gamma_3 = 12°$.

   (d) Draw the resulting four-bar in its three precision positions and determine if it is an acceptable linkage.

8.6. Design another pair of compound-lever snips from the suitable associated linkage of Fig. 8.31, which satisfies the objective set forth in the chapter and which is different from the design shown in Fig. 8.31.

8.7. Determine the associated linkage of the yoke-riveter configuration of Fig. P8.1.

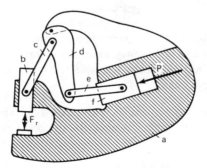

Figure P8.1 Yoke riveter.

8.8. Determine the associated linkage of the yoke-riveter configuration of Fig. P8.2a.

* Many of the examples in this chapter were generated from projects submitted by former students at the University of Minnesota. The creative ideas of these students are acknowledged.

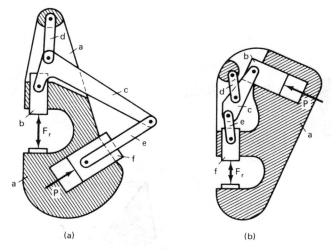

Figure P8.2 Two different types of yoke riveters.

    (a)   (b)

8.9. Determine the associated linkage of the yoke-riveter configuration of Fig. P8.2b.

8.10. Create other designs for the compound-lever snips that are different from those in the text or in Prob. 8.6.

8.11. Create other designs for the yoke-riveter that are different from those in the text.

8.12. Figure P8.3 shows one side of a container with a removable top in two required positions. Find the locations of acceptable ground pivots (A_0, B_0) of a four-bar linkage that will guide the top through these two positions without interfering with the side of the container.

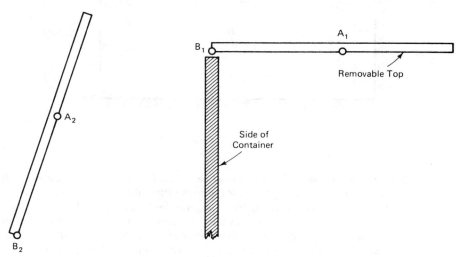

Figure P8.3

8.13. A student wishes to design a four-bar linkage that will store a bicycle above his bed. Two positions of the storage rack are shown in Fig. P8.4. Find acceptable ground and moving pivots for this design objective.

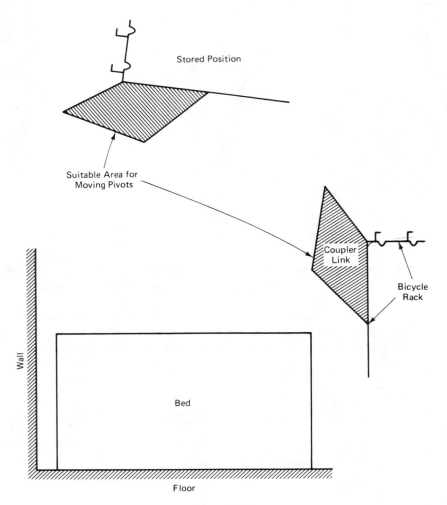

Figure P8.4

8.14. During maintenance a dust cup must be removed from a filter for dumping collected particles. Rather than bolting the dust cup to the filter, the dust cup is hinged at A_0 and acts as the output link of a four-bar function generator. Figure P8.5 shows the dust cup and the proposed coupler link in two positions. Determine an acceptable location of the fixed and moving pivots (B_0, B) such that B is located along the coupler AP, the ground pivot B_0 is within the filter dimensions, and the rotation of $B_0 B$ from position 1 to position 2 is 28° ccw.

8.15. Part of the design of an assembly line requires removing a box from one conveyor belt, rotating it by 90°, and placing it on another conveyor belt. Find an acceptable four-bar linkage to guide the box carrier through the two positions shown in Fig. P8.6.

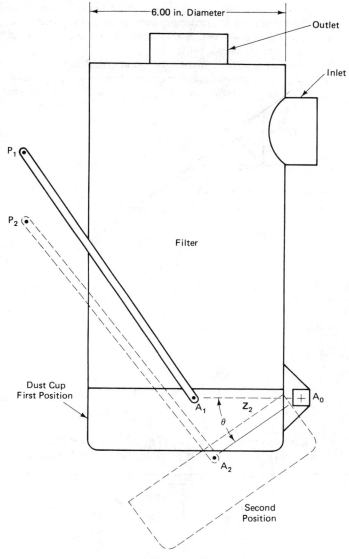

Figure P8.5

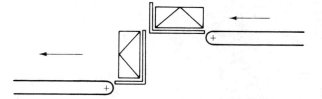

Figure P8.6

8.16. As part of an automation process, a four-bar linkage must be designed to remove boxes from one conveyer belt and deposit them on an upper conveyor belt as shown in Fig. P8.7 (three prescribed positions). Both ground and moving pivots must be located between the upper and lower conveyor belts.
   (a) Design an acceptable four-bar by the graphical method.
   (b) Design an acceptable four-bar by the complex-number method.
   (c) Design an acceptable four-bar by the ground pivot specification method.

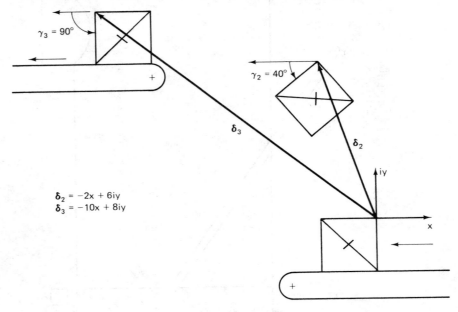

$$\delta_2 = -2x + 6iy$$
$$\delta_3 = -10x + 8iy$$

Figure P8.7

8.17. It is desired to synthesize a linkage to guide the movable shelf through the three positions shown in Fig. P8.8. The first position is level with the top of the cabinet for writing pur-

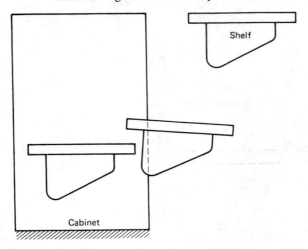

Figure P8.8

poses, and the third position is a stored position for the shelf. Ground pivots should fall within the cabinet while the linkage size should be minimized so as to take up the least amount of cabinet space. Find acceptable locations of ground and moving pivots by (a) the graphical method; (b) the complex-number method; (c) the ground-pivot specification method.

8.18. Design a compact linkage to be added to the farm vehicle of Fig. P8.9 so that the operator may maintain a vertical position as the tractor traverses the sloping terrain shown in Fig. P8.9.

(a) Use the graphical technique.

(b) Use the complex-number method.

(c) Use the ground-pivot specification method.

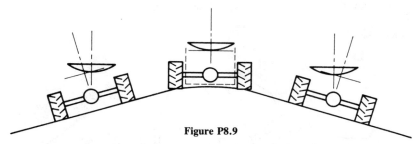

Figure P8.9

8.19. An avid foosball player wishes to design a ball-return linkage to be attached to the foosball table. Figure P8.10 shows three required positions for the coupler so that the ball may be

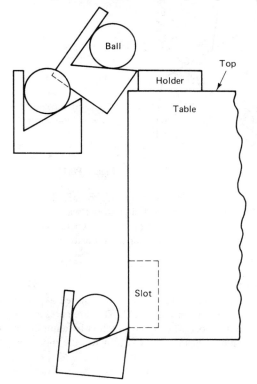

Figure P8.10

guided from the slot on the side of the table to the holder on the top edge. Ground pivots should fall within the table and the linkage should be compact.

(a) Use the graphical method.

(b) Use the complex-number method.

(c) Use the ground-pivot specification method.

8.20. Figure P8.11 shows three prescribed positions for the wing landing gear of a remote-controlled model aircraft. Design a four-bar motion generator for this task such that the moving pivots are within the wheel supporting member and the ground pivots are within the airframe.

(a) Use the graphical method.

(b) Use the complex-number method.

(c) Use the ground-pivot specification method.

Airframe

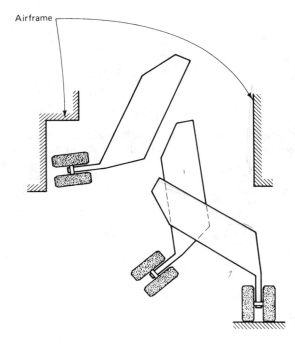

Figure P8.11

8.21. Rather than having to back a boat trailer into the water to unload a boat, a four-bar mechanism is sought to transfer the boat from the trailer to the water (see Fig. P8.12). Moving pivots should be connected to the cradle, and the fixed pivots should be close to the trailer platform. Design a four-bar motion generator for three positions.

(a) Use the graphical method.

(b) Use the complex-number method.

(c) Use the ground-pivot specification method.

8.22. A four-bar path generator (with prescribed timing) is required as part of an arm-actuated propulsion system for the wheelchair in Fig. P8.13. The three prescribed path points shown have been determined to be the most efficient arm motion by a number of individuals. This movement of the coupler path point (C_1, C_2, C_3) provides the input, while the output is a

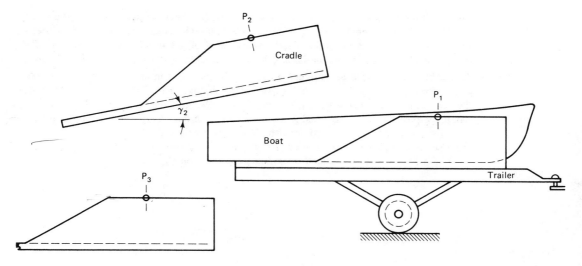

Figure P8.12

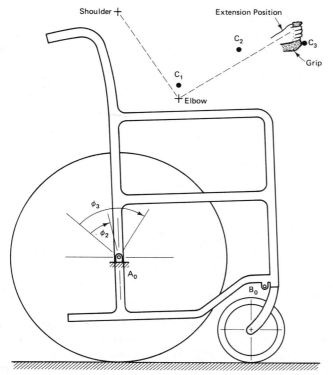

Figure P8.13

rotation of the large wheel with a ground pivot at A_0. (A clutch located at A_0 will slip when the grip is returned from C_3 to C_1 along the same path.) The other ground pivot B_0 is specified as well as the rotations of the wheel driving link A_0A ($\phi_2 = 38°$ cw, $\phi_3 = 80°$ cw). By the graphical method, find the initial position of an acceptable four-bar linkage for this task.

8.23. A four-bar linkage must be designed to accomplish one task in an automatic sewing machine (see Fig. P8.14). As input link (A_0A) rotates through $\phi_2 = 25°$ ccw, $\phi_3 = 135°$ ccw, the coupler point C must travel C_1, C_2, and C_3 to catch the thread loop.

  (a) If the positions of A_0, B_0, and A are prescribed (see the figure), find the location of B by the graphical method and draw the linkage in its three design positions.

  (b) Use the complex-number method to synthesize a new path generator (the same C_1, C_2, C_3, ϕ_2, and ϕ_3 are prescribed) with better transmission angles.

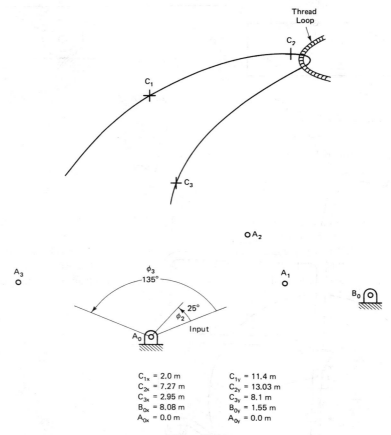

C_{1x} = 2.0 m	C_{1y} = 11.4 m
C_{2x} = 7.27 m	C_{2y} = 13.03 m
C_{3x} = 2.95 m	C_{3y} = 8.1 m
B_{0x} = 8.08 m	B_{0y} = 1.55 m
A_{0x} = 0.0 m	A_{0y} = 0.0 m

Figure P8.14

8.24. A handicapped individual is unable to turn the pages of a book but is able to depress a foot pedal. A four-bar linkage path generator (with prescribed timing) is the main component of a mechanism that will turn a page when actuated by the foot pedal. As the coupler point C of the four-bar moves from C_1 to C_3, one page flips over (see Fig. P8.15) and then C returns along the same path for the next cycle.

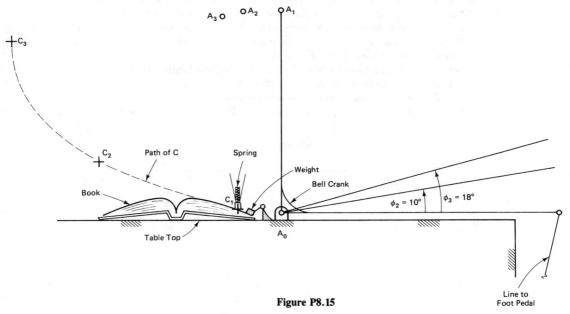

Figure P8.15

(a) With C_1, C_2, C_3, A_0, B_0, A_1, A_2, and A_3 prescribed, find the rest of the linkage by the graphical method.

(b) With only ϕ_2, ϕ_3 and C_1, C_2, C_3 prescribed, find acceptable alternative four-bars for this task by the complex number method.

8.25. A crank-rocker path generating four-bar is required to advance film in a camera as shown in Fig. P8.16.

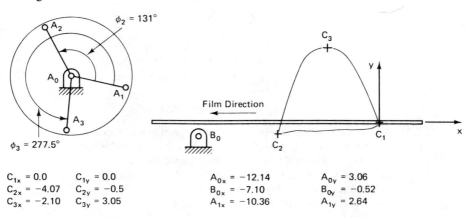

$C_{1x} = 0.0$	$C_{1y} = 0.0$		$A_{0x} = -12.14$	$A_{0y} = 3.06$
$C_{2x} = -4.07$	$C_{2y} = -0.5$		$B_{0x} = -7.10$	$B_{0y} = -0.52$
$C_{3x} = -2.10$	$C_{3y} = 3.05$		$A_{1x} = -10.36$	$A_{1y} = 2.64$

Figure P8.16

(a) Using the graphical method, find the four-bar linkage if A_0, A, B_0, C_1, C_2, C_3, ϕ_2, and ϕ_3 are given.

(b) Using the complex-number method, find other acceptable four-bar linkages given C_1, C_2, C_3, ϕ_2, and ϕ_3.

8.26. Figure P8.17a shows a butterfly valve in a tube that has a liquid flowing through it. A four-bar function generator is to be designed so that movement of the input link ($A_0 A$) in equal increments will produce equal incremental changes of the flow through the butterfly valve (the output). Figure P8.17b shows the angles required for this objective (ϕ_2, ϕ_3, ψ_2, and ψ_3) as well as the location of B_0, B, and A_0.

(a) Using the graphical method of Fig. 8.51, find the location of point A.

(b) Use the overlay technique to find point A.

(c) Use the loop-closure method to find point A.

(d) Use Freudenstein's equation to solve for this linkage ($\phi_1 = 128.5°$).

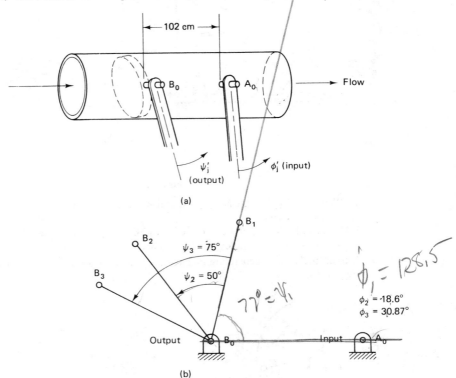

(a)

$\phi_1 = 128.5$

$\psi_1 = 77°$

$\phi_2 = -18.6°$

$\phi_3 = 30.87°$

(b)

Figure P8.17

8.27. A four-bar function generator is to be designed to guide an attic stairway from its stored position down to its deployed position where half of the ladder can slide down to meet the floor below. Figure P8.18 shows the proposed location of A_0, B_0, B, and angles ϕ_2, ϕ_3, ψ_2, and ψ_3 as well as the space constraints (roof joists). Find an acceptable four-bar linkage by:

(a) The graphical method of Fig. 8.51.

(b) The overlay method.

(c) The loop-closure method (only A_0, B_0, and the angles are prescribed in this case).

(d) Freudenstein's equation (with $\phi_1 = 119°$).

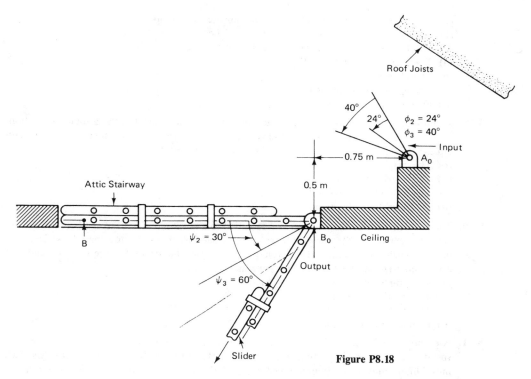

Figure P8.18

8.28. Figure P8.19 shows a conceptual drawing of a leg-driven recreational vehicle. Series of four-bar linkages transmit the leg movements of two occupants to the rear wheels. Offset-

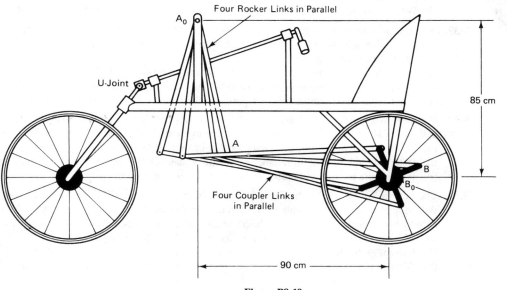

Figure P8.19

ting the function generators in phase should ensure smooth operation. If the input link $(A_0 A)$ rotates by $\phi_2 = 15°$ cw, $\phi_3 = 30°$ cw while the output $(B_0 B)$ rotates by $\psi_2 = 90°$ ccw, $\psi_3 = 180°$ ccw, find an acceptable four-bar function generator by:

(a) The graphical method of Fig. 8.51.

(b) The overlay method.

(c) The loop-closure method.

(d) Freudenstein's equation.

8.29. A four-bar linkage is to be designed to operate an artificial hand in the gripping operation. Figure P8.20 shows the angles that have been derived based on mechanical advantage principles. Design an acceptable four-bar linkage for this task by:

(a) The graphical method of Fig. 8.51.

(b) The overlay method.

(c) The loop-closure method.

(d) Freudenstein's equation.

8.30. The problem of a binding accelerator cable led to a proposed direct linkage between the accelerator pedal and the carburetor using a six-bar linkage — two four-bars in series. Figure P8.21 shows the required link rotations as well as the location of the three ground pivots. Synthesize the six-bar function generator by:

(a) The graphical method of Fig. 8.51.

(b) The overlay method.

(c) The loop-closure method.

(d) Freudenstein's equation.

8.31. A proposed tachometer (Fig. P8.22), which uses a rotating governor principle as an engine speed indicator, requires a function-generator linkage to convert the movement of the rack into a linear movement around a tachometer dial. The geared five-bar linkage is chosen to be synthesized for this task. Known quantities are ψ_j, ϕ_j, T_2/T_1, and Z_5.

(a) Write the loop-closure equation for this linkage in its first and jth position.

(b) Determine the maximum number of positions that this linkage can be synthesized for.

(c) What is the maximum number of positions for which a linear solution is obtainable?

8.32. A geared six-bar function generator (Fig. P8.23) is synthesized for the maximum number of positions allowable by linear solution techniques. Vector Z_6 as well as T_2/T_1, T_4/T_5, and $\phi_j = f(\psi_j)$ are known quantities.

(a) Write the loop-closure equation for this linkage in its first and jth positions.

(b) Determine the maximum number of positions that this linkage can be synthesized for.

(c) What is the maximum number of positions for which a linear solution is obtainable?

8.33. We wish to synthesize the linkage in Fig. P8.24 (Z_1 and Z_2) to guide a disk (A) in a slot defined by R_j ($j = 1, 2, \ldots, n$) such that the input angle (ϕ_j) and positions R_j are given.

(a) Write the standard-form equations for this linkage in its first and jth positions.

(b) Determine the maximum number of positions that this linkage can be synthesized for.

(c) What is the maximum number of positions for which a linear solution is obtainable?

8.34. The geared linkage shown in Fig. P8.25 is to be used as a function generator where ϕ_j is the independent (input) variable (a rotation of arm Z_1) and S_j is the dependent (output) variable (a linear displacement of the slider). ϕ_j and S_j are prescribed as well as T_2/T_1 and Z_5.

(a) Write the loop-closure equation for this linkage in its first and jth positions.

(b) Determine the maximum number of positions that this linkage can be synthesized for.

(c) What is the maximum number of positions for which a linear solution is obtainable?

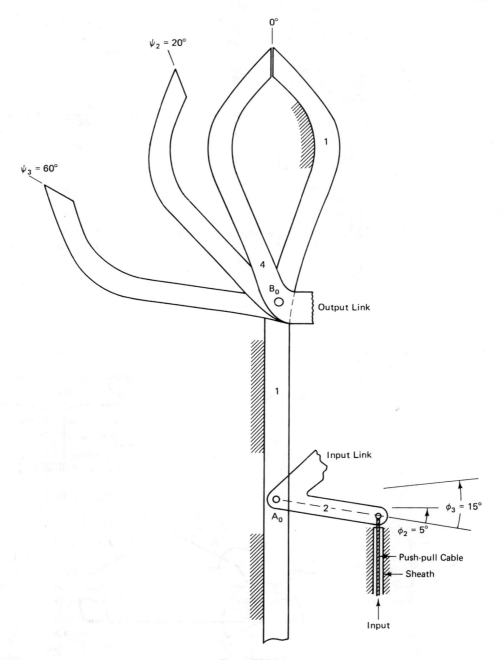

Figure P8.20

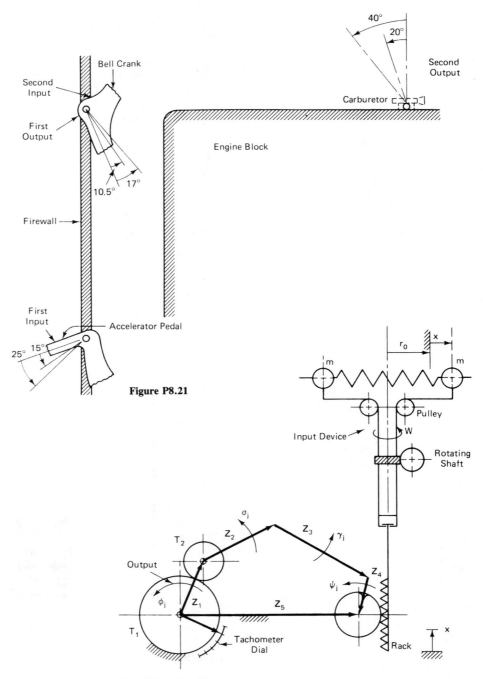

Figure P8.21

Figure P8.22

T_1 and T_2 — No. of Teeth on Gears

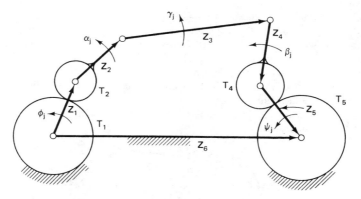

Figure P8.23

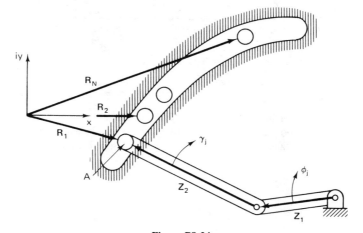

Figure P8.24

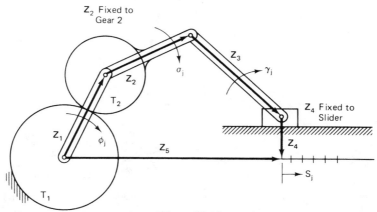

Figure P8.25

8.35. We wish to synthesize a function generator of Fig. P8.26 such that the rotations ϕ_j of the input crank and the displacement S_j of the output slider are prescribed (i.e., ϕ_j and S_j are known for $j = 1, 2, \ldots, n$). *Notice* that $\mathbf{Z}_4$ is fixed to the slider at an unknown constant angle α. (Also, the *initial* position of the slider $\mathbf{Z}_1$ is given as $1.0 + 0.0i$.)

  (a) Write the loop-closure equations (by two methods) for this linkage.
  (b) Determine the maximum number of positions that this linkage can by synthesized for (both methods).
  (c) What is the maximum number of positions (both methods) for which a linear solution is obtainable?

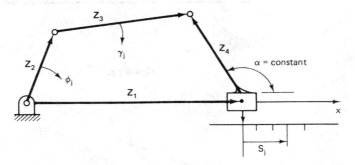

Figure P8.26

8.36. The six-bar linkage of Fig. P8.27 is to be synthesized for both path generation of point P and function generation $[\theta_j = f(\phi_j)]$. Write the standard-form equations for this so that the entire linkage may be synthesized by the standard form.

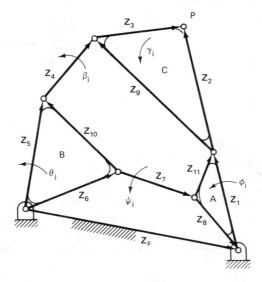

Figure P8.27

8.37. To comply with safety standards, the seven-bar motion generator linkage of Fig. P8.28 is being designed with two degrees of freedom so that the operator must use both hands to actuate the device. One hand-crank input is on link 3 (which rotates by specified angles β_j)

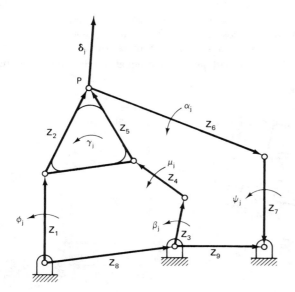

Figure P8.28

and the other hand input is on link 7 (which rotates by specified angles ψ_j). These two simultaneous rotations will cause point P to move along its path (specified by δ_j) while link 2 rotates by specified angles γ_j. Thus δ_j, β_j, ψ_j, and γ_j are prescribed.

Write the equations that describe this synthesis task in the standard form. Be sure to cover all independent loops. (One or more vectors may have to be chosen arbitrarily in order to utilize the standard form for all loops.)

8.38. The eight-bar linkage of Fig. P8.29 is to be synthesized such that δ_j, α_{2j}, α_{4j}, and α_{7j} are prescribed.

   (a) Write all the standard-form equations for this mechanism (making appropriate assumptions to assure the standard form).

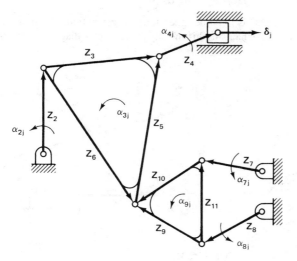

Figure P8.29

(b) Can we prescribe δ_j, α_{3j}, α_{4j}, and α_{7j} instead? Why or why not?

(c) Can we prescribe δ_j, α_{2j}, α_{3j}, and α_{9j} instead? Why or why not?

(d) Can we prescribe δ_j, α_{3j}, α_{4j}, and α_{9j} instead? Why or why not?

8.39. Write the displacement, velocity, and acceleration equations describing the linkage in Fig. P8.30.

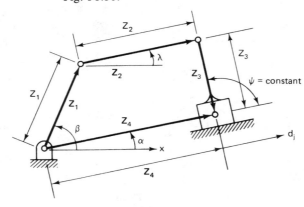

Figure P8.30

8.40. Synthesize a four-bar (see notation of Fig. 8.79) linkage such that the following are specified:

$$\mathbf{Z}_1 = -1.0 + 0.0i$$

$$\omega_2 = 1 \text{ rad/sec}, \qquad \omega_3 = -2 \text{ rad/sec}, \qquad \omega_4 = 3 \text{ rad/sec}$$

$$\alpha_2 = 3 \text{ rad/sec}^2, \qquad \alpha_3 = 1 \text{ rad/sec}^2, \qquad \alpha_4 = 2 \text{ rad/sec}^2$$

8.41. Design a four-bar linkage that will satisfy the following precision conditions:

$$\omega_2 = 8 \text{ rad/sec}, \qquad \alpha_2 = 0$$

$$\omega_3 = 1 \text{ rad/sec}, \qquad \alpha_3 = 20 \text{ rad/sec}^2$$

$$\omega_4 = 6 \text{ rad/sec}, \qquad \alpha_4 = 0$$

8.42. A rack and gear mechanism [31] is to be designed for path generation with prescribed timing. Figure P8.31 shows the mechanism plus the suggested vector diagram. Prescribed are

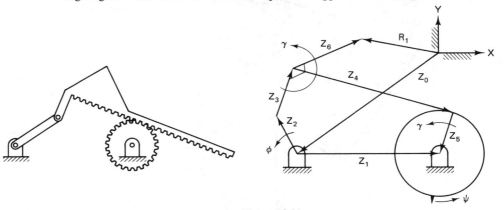

Figure P8.31

ϕ_j and $\mathbf{R}_j$. This mechanism has industrial applications in packaging as well as toys and leisure equipment.

(a) Write the synthesis equations for this mechanism in its first and jth position.

(b) For how many positions can this mechanism be designed?

(c) What are the maximum number of positions of synthesis yielding a linear solution method?

(d) If the task were motion generation of the rack, answer (a), (b), and (c) above.

Answers To Selected Problems

Chapter 1

1.28. **(b)** $F = 1$

1.36. Fig. P1.48; $F = 2$
Fig. P1.53; $F = 3$
Fig. P1.57; $F = 1$

Chapter 3

3.5. **(a)** Fig. P3.7
$\gamma_B = 40°$; $\delta_B = 50°$
$\gamma_C = 35°$; $\delta_C = 55°$

3.15. $\dfrac{\omega_4}{\omega_2} = 0.109$

3.23. $\mathbf{V}_F = 1.4$ cm/sec $\measuredangle\, 270°$
$\omega_7 = 0.103$ rad/sec *cw*

3.25. $\mathbf{V}_{DB} = 14.83$ in/sec $\measuredangle\, 180°$

3.30. $\omega_4 = 10.7$ rad/sec *cw*

3.45. **(a)** $V_C/\omega_{\text{in}} = 0.83$ in
(b) $F_{\text{out}}/F_{\text{in}} = 0.82$

3.47. **(b)** $\omega_4/\omega_2 = 0.36$
(c) $\omega_5/\omega_2 = -0.53$
(d) $V_{C6} = i\omega_2(\overrightarrow{1,2 - 2,6}) = 1.93 + \text{O}i$

3.55. **(a)** $\omega_4/\omega_6 = 1.72$
(b) M.A. = 1.05

3.61. $F_{\text{out}}/T_{\text{in}} = 8.8$ in^{-1}

3.75. **(a)** Four-bar Function Generator
(b) open position   M.A. = 10.3
closed position M.A. = 11.0

Chapter 4

4.6. $\mathbf{A}_C = 2250$ in/sec^2 $\measuredangle\, -117°$

4.12. $\mathbf{A}_D = 0.74$ in/$\measuredangle\, -120°$

4.23. $\alpha_2 = 8.7$ rad/sec^2 *cw*
$\mathbf{A}_B = 93.5$ cm/sec^2 $\measuredangle\, 338°$

4.29. $\mathbf{A}_{P_4} = 44{,}000$ in/sec^2 $\measuredangle\, 270°$

4.32. $\mathbf{A}_{P_4} = 145$ cm/sec^2 $\measuredangle\, 90°$

4.43. $\mathbf{A}_P = 341.5$ in/sec^2 $\measuredangle\, -110°$

Chapter 5

5.3. $T_2 = -2345$ lbf·in

5.7. $\mathbf{F}_{12} = 190$ lbf $\measuredangle\, -11.9°$
$\mathbf{F}_{23} = \mathbf{F}_{12}$
$\mathbf{F}_{34} = 110$ lbf $\measuredangle\, 40.6°$
$\mathbf{F}_{14} = 90$ lbf $\measuredangle\, 34.5°$
$T_{12} = 5544$ lbf·mm *cw*

5.14. (a) $\alpha_3 = 2.48$ rad/sec^2 ccw

              $\alpha_4 = 2.6$ rad/sec^2 ccw

      (b) $F_{02} = 0.232$ lbf

           $F_{03} = 0.97$ lbf

           $F_{04} = 0.28$ lbf

      (c) $\varepsilon_2 = 1.126$ ft

           $\varepsilon_3 = .166$ ft

           $\varepsilon_4 = .131$ ft

      (e) $F_{23} = 0.79$ lbf

           $F_{34} = 0.81$ lbf

      (f) $T_2 = 0.32$ lbf·ft ccw

5.15. $\mathbf{F}_{02} = 93.3$ N $\measuredangle -22.9°$

      $\mathbf{F}_A = 54.8$ N $\measuredangle -175°$

5.19. (a) $\mathbf{F}_A = 549$ lbf $\measuredangle -154.1°$

        $\mathbf{F}_B = 686$ lbf $\measuredangle -171°$

      (b) $T_S = 1,297$ lbf·in ccw

Chapter 6

6.2. $0° < \phi < 60°$, $\ddot{y} = 317 \, (10^{-6})$ in/deg^2

        $60° < \phi < 90°$, $\ddot{y} = 0$

        $90° < \phi < 180°$, $\ddot{y} = -212 \, (10^{-6})$

        $180° < \phi < 240°$, $\ddot{y} = -444 \, (10^{-6})$

        $240° < \phi < 330°$, $\ddot{y} = 296 \, (10^{-6})$

6.11. Let $\beta_1 = \beta_2$

      Use simple Harmonic motion.

6.27. $S = L - 10Lx^3 + 15Lx^4 - 6Lx^5$

      $\dot{S} = -30Lx^2 + 40Lx^3 - 30Lx^4$

      $\ddot{S} = -60Lx + 120Lx^2 - 120Lx^3$

Chapter 7

7.3. $\omega_E = 148.15$ rpm ccw

      $\omega_g = -37.04$ rpm cw

      $\omega_K = -296.3$ rpm cw

      $\omega_I = -222.25$ rpm cw

      (all viewed from right)

7.8. (a) $\omega_2/\omega_3 = 1 + N_1/N_2$

      (c) $\omega_2/\omega_3 = 1 - N_1/N_2$

      (f) $\omega_4/\omega_5 = 1 + (N_1 N_3)/(N_2 N_4)$

7.13. $\omega_6 = 38.56$ rpm cw

7.14. $\omega_2 = 0.52$ rpm ccw

7.18. $\omega_E = 0.104$ rpm cw

      (viewed from right)

Chapter 8

8.1. **(a)** $x_1 = 0.08$; $x_2 = 0.62$; $x_3 = 1.38$; $x_4 = 1.92$

     (b) $\phi_2 = 12.2°$ $\psi_2 = 1.3°$

              $\phi_3 = 29.3°$ $\psi_3 = 35.5°$

              $\phi_4 = 41.4°$ $\psi_4 = 80.8°$

8.16. Right side $\mathbf{Z} = 10.44 - 1.4i$

      $\mathbf{W} = -7.44 + 3.9i$

      Left side $\mathbf{Z} = 11.1 + 2.5i$

      $\mathbf{W} = -3.06 - 8i$

8.20. With $\boldsymbol{\delta}_2 = -0.2 + 3.3i$, $\boldsymbol{\delta}_3 = -1.02 + 7.4i$, left side $\mathbf{W} = 7.4 - 4.8i$

      $\mathbf{Z} = 0.77 - 1.5i$

      Right side $\mathbf{W} = 0.14 - 8.8i$

      $\mathbf{Z} = -4.4 + 4i$

8.25. Choose $\gamma_2 = 3.12°$, $\gamma_3 = 45.23°$

      $\psi_2 = 58.09°$, $\psi_3 = 52.82°$ then

      Right side $\mathbf{Z} = 8.27 - 0.75i$; $\mathbf{W} = 1.83 + 1.41i$

      Left side $\mathbf{Z} = 3.27 - 3.97i$; $\mathbf{W} = 1.52 + 4.2i$

8.32. **(b)** Nine

     (c) Five

References[*]

* References addressing contemporary design theory and methodology are found in the appendix of Chapter 2.

1. Alt, H., *Werkstattstech*, 26 (1932), pp. 61–64.
2. ———, "Gebriebetechnik," *VDI Tagungsheft,* Vol. 1 (1953).
3. Anton, C., C. Lentsch, A. Guggisberg, and A. G. Erdman, "Synthesis of a Letter Folding Mechanism," in "Mechanism Case Studies VII, Part 1" (compiled by R. Berkof), ASME Paper No. 84-DET-144, 1984.
4. Azimov, Morris, *Introduction to Design.* Englewood Cliffs, New Jersey: Prentice-Hall, Inc., 1962.
5. Bagci, C., "Force and Torque Analyses of Plane Mechanisms by Matrix Displacement-Direct Element Method," *International Symposium on Linkages and Computer Design Methods (Bucharest, Romania, June 1973),* pp. 77–97.
6. Barris, W. C., S. Kota, D. Riley, and A. G. Erdman, "Mechanism Synthesis Using the Workstation Environment," *Computer Graphics and Applications* (March 1988), pp. 39–50.
7. Beggs, J. S., *Mechanism.* New York: McGraw-Hill Book Company, 1955.
8. Benedict, C. E., and D. Tesar, "Analysis of a Mechanical System Using Kinematic Influence Coefficients," *Proceedings of the First Applied Mechanism Conference (Stillwater, Oklahoma, July 1969),* Paper No. 37. Stillwater, Oklahoma: Oklahoma State University, 1969.
9. Berkof, R. S., and D. Tesar, "Optimal Torque Balancing for a Complex Stamping and Indexing Machine," ASME Paper No. 70-Mech-82, 1970.
10. ———, A. G. Erdman, D. Hewitt, T. Bjorklund, D. Harvey, et al, "Mechanism Case Studies IV," ASME Paper No. 78-DET-19, 1978.
11. Bessler, W., D. Kohli, and G. N. Sandor, "Design of a Four-Link $P_1V_1A_1J_1$ Path Generator," NSF-ASME Technology Transfer Monograph, October 1975.
12. Bickford, J. H., *Mechanisms for Intermittent Motion.* New York: Industrial Press, Inc., 1972.
13. Bjorklund, T., and A. G. Erdman, "Case Study: The Thing," *Proceedings of the Fifth Applied Mechanisms Conference (Oklahoma City, Oklahoma, November 1977).* Stillwater, Oklahoma: Oklahoma State University, 1977.
14. Bloch, A., *VDI Bericht,* Vol. 29 (1958), p. 158.
15. Bonhom, W. E., "Calculating the Response of a Four-Bar Linkage," ASME Paper No. 70-Mech-69, 1970.
16. Buhl, Harold R., *Creative Engineering Design.* Ames, Iowa: The Iowa State University Press, 1960.
17. Burstall A. F., *A History of Mechanical Engineering.* Cambridge, Mass.: M.I.T. Press, 1965.
18. Carson, W., A. G. Erdman, and P. Starr, "An Overview of the Complete Mechanism Design Process," in *Monograph on Mechanism Design.* New York: McGraw-Hill Book Company, 1977; NSF Report No. GK-36624, Paper No. 1.
19. ———, and J. M. Trummel, "Time Response of Lower Pair Spatial Mechanisms Subjected to General Forces," ASME Paper No. 68-Mech-57, 1968.
20. Chace, M. A., "Using DRAM and ADAMS Programs to Simulate Machinery, Vehicles," *Agricultural Engineering,* (November 1978), pp. 16–18.
21. ———, and J. C. Angell, "Interactive Simulation of Machinery with Friction and Impact Using DRAM," SAE Paper No. 770050, 1977.
22. ———, and J. C. Angell, "User's Guide to DRAM (Dynamic Response of Articulated Machinery), Design Engineering Computer Aids Lab," University of Michigan, 1972.
23. Chase, T., A. G. Erdman, and D. Riley, "Synthesis of Six Bar Linkages Using an Interactive Package," *Proceedings of the 1981 OSU Applied Mechanisms Conference* (Kansas City, Missouri, December 1981), Paper No. LI, 5 pgs. Stillwater, Oklahoma: Oklahoma State University, 1981.
24. ———, "Burmester Theory for Four Precision Positions: An Extended Discourse with Application to the Dimensional Synthesis of Arbitrary Planar Linkages." Ph.D. diss., University of Minnesota, (1984).

25.   ———, A. G. Erdman, and D. Riley, "Improved Centerpoint Curve Generation Techniques for Four-Precision Position Synthesis Using the Complex Number Approach," *Journal of Mechanisms, Transmissions, and Automation in Design, ASME Transactions,* 107 (September 1985), pp. 370–76.

26.   ———, ———, ———, "Triad Synthesis for Up to Five Design Positions with Application to the Design of Arbitrary Planar Mechanisms," *ASME Journal of Mechanisms, Transmissions, and Automation in Design,* 109, No. 4 (December 1987), pp. 426–34.

27.   ———, ———, and G. Marier, "Computer Aided Design of an Improved Variable-Sheave V-Belt Drive," *Proceedings of the 10th OSU Applied Mechanisms Conference,* (December 1987).

28.   Chebyshev, P. L., "Theorie des mechanismes connus sous le nome de parallellograms (1953)," in *Oeuvres de P. L. Tchebychef,* Vol. 1. St. Petersburg: Markoff et Sonin, 1899; *Modern Mathematical Classics: Analysis,* S730. Ed. Richard Bellom. New York: Dover Publications, Inc., 1961.

29.   Chen, F. Y., "Gripping Mechanisms for Industrial Robots — An Overview," *Mechanism and Machine Theory,* 17, No. 5, (1982), pp. 299–311.

30.   Chironis, N. P., *Mechanisms, Linkages, and Mechanical Controls.* New York: McGraw-Hill Book Company, 1965.

31.   Claudio, M., and S. Kramer, "Kinematic Synthesis and Analysis of the Rack-and-Gear Mechanism for Four-Point Path Generation with Prescribed Input Timing," ASME Paper No. 85-DET-94, 1985.

32.   Crossley, F. R. E., "The Permutations of Kinematic Chains of 8 Members or Less from the Graph-Theoretic Viewpoint," in *Developments in Theoretical and Applied Mechanics,* Vol. 2, pp. 467–86. New York: Pergamon Press, 1965.

33.   ———, "A Systematic Approach to Creativity," *Machine Design,* (March 1980), pp. 150–53.

34.   ———, "Defining the Job: First Step in a Successful Design," *Machine Design,* (May 1980), pp. 128–31.

35.   Coy, J. J., "Geared Power Transmission Technology," in *Advanced Power Transmission Technology,* pp. 45–78. NASA CP 2210, (June 9-12), 1981.

36.   Dahlquist, G., and A. Bjorck, *Numerical Methods.* Englewood Cliffs, New Jersey: Prentice-Hall, Inc., 1974.

37.   Dhande, S. G., and G. N. Sandor, "Analytical Design of Cam-Type Angular-Motion Compensators," ASME Paper No. 76-DET-22, 1976.

38.   ———, and G. N. Sandor, "A Unified Approach to the Study of the Conjugate Gear Tooth Action," *Sixth Symposium on Mechanisms and Gear Abstracts,* Miskolc, Hungary (September 1978), p. 26.

38a.   Dimarogonas, A. D. and G. N. Sandor, "A General Method for Analysis of Mechanical Systems," *Proceedings,* Third World Congress for the Theory of Machines and Mechanisms, Dubrovnik, Yugoslavia, Sept. 13–19, 1971, pp. 121–32.

39.   Dixon, M. W., and C. O. Huey, Jr., "Fundamentals of Kinematic Synthesis," *1973 Textile Engineering Conference* (Charlotte, North Carolina, April 1973), pp. 1–31.

40.   Erdman, A. G., J. K. Mayfield, F. K. Dorman, J. Bowen, and W. J. Dahlof, "Design of a Wrench for Use in Spinal Surgery," *ASME 1979 Advances in Biomedical Engineering* (December 1979).

41.   ———, "Dynamic Synthesis of a Variable Speed Drive," *Proceedings of the Third Applied Mechanisms Conference (November 1973),* Paper No. 38. Stillwater, Oklahoma: Oklahoma State University, 1973.

42. ———, "A Guide to Mechanism Dynamics," *Proceedings of the Third Applied Mechanisms Conference (November 1973),* Paper No. 6. Stillwater, Oklahoma: Oklahoma State University, 1973.

43. ———, "Three and Four Precision Point Kinematic Synthesis of Planar Linkages," *Mechanism and Machine Theory,* 16, pp. 227–45.

44. ———, "Three Position Synthesis by Complex Numbers," in *Monograph on Mechanical Design.* New York: McGraw-Hill Book Company, 1977; NSF Report No. GK36624, Paper No. 49.

45. ———, and W. L. Carson, "Teaching Unit on Complex Numbers as Applied to Linkage Modeling," in *Monograph on Mechanical Design.* New York: McGraw-Hill Book Company, 1977; NSF Report No. GK36624, Paper No. 12.

46. ———, and J. E. Gustafson, "LINCAGES: *Li*nkage *IN*teractive *C*omputer *A*nalysis and *G*raphically *E*nhanced *S*ynthesis Package," ASME Paper No. 77-DTC-5, 1977.

47. ———, and Dana Lonn, "Synthesis of Planar Six-Bar Linkages for Five Precision Conditions by Complex Numbers," in *Monograph on Mechanism Design.* New York: McGraw-Hill Book Company, 1977; NSF Report No. GK36624, Paper No. 59.

48. ———, and Dana Lonn, "A Unified Synthesis of Planar Six-Bar Mechanisms Using Burmester Theory," *Proceedings of the Fourth World Congress on the Theory of Machines and Mechanisms* (Newcastle-upon-Tyne, England, September 1975).

49. ———, and A. Midha, "Man-Made Mechanism Models Multiply Mental Motivation," *Proceedings of the 3rd Applied Mechanisms Conference* (Oklahoma State University, November 1973), Paper No. 3. Stillwater, Oklahoma: Oklahoma State University, 1973.

50. ———, and D. Riley, "Computer-Aided Linkage Design Using the LINCAGES Package," ASME Paper No. 81-DET-121, 1981.

51. ———, and D. Riley, "Computer Graphics and Computer-Aided Design in Mechanical Engineering at the University of Minnesota," *Computers in Education,* 5 (Summer 1981) pp. 229–43.

52. ———, and G. N. Sandor, "Kinematic Synthesis of a Geared Five-Bar Function Generator," *Journal of Engineering for Industry,* 93, B, no. 1, pp. 157–64, Feb. 1971.

52a. ———, ———, "Teaching Unit on Complex Numbers as Applied to Linkage Modeling," NSF-ASME Technology Transfer Monograph, *Linkage Design Handbook,* July, 1975.

53. ———, and P. Starr, "Towards Technology Transfer: Kinematic and Dynamic Analysis of Linkages for the Design Engineer," *Proceedings of the First ASME Design Technology Conference in Machine Design (October 1974),* pp. 335–46. New York: American Society of Mechanical Engineers, 1974.

54. ———, E. Nelson, J. Peterson, and J. Bowen, "Type and Dimensional Synthesis of Casement Window Mechanisms," ASME Paper No. 80-DET-78 (1980) *Mechanical Engineering,* (December 1981), pp. 46–55.

55. ———, G. N. Sandor, and R. G. Oakberg, "A General Method for Kineto-Elastodynamic Analysis and Synthesis of Mechanisms," *Journal of Engineering for Industry,* 94, no. 4 (February 1971), pp. 11–16.

56. ———, R. K. Westby, G. R. Fichtinger, and F. R. Tepper, "Mechanisms Case Studies II: A New Derailleur Mechanism," ASME Paper No. 74-DET-56, 1974; *Mechanical Engineering,* 97 (July 1975), pp. 36–37.

57. ———, and D. R. Riley, *Linkage Design Using the LINCAGES^c Package.* SAE Paper No. 830801, (1983).

58. ———, M. Harber, and D. Riley, "Enhanced Computer Graphic Kinematic Synthesis for Three Prescribed Positions," *Proceedings of CAD/CAM, Robotics and Automation International Conference* (Tucson, Arizona, February 13–15, 1985).

59. ———, "Computer-Aided Design of Mechanisms: 1984 and Beyond," *Mechanism and Machine Theory Journal,* 20, No. 4, (Spring 1985), pp. 245–250.

60. ———, and D. Riley, "Computer Aided Design and Manufacturing," *Mechanical Design and Systems Handbook,* 2nd ed., ed. H. Rothbart, pp. 16.1–16.44. New York: McGraw-Hill, 1985.

61. ———, ———, "New Directions for Mechanism Kinematics and Dynamics," *Computers in Mechanical Engineering,* 3, No. 6 (May 1985), pp. 10–20.

62. ———, T. Chase, "New Software Synthesizes Complex Mechanisms," *Machine Design Magazine,* (August 22, 1985), pp. 107–113.

63. Esterline, A., D. Riley, and A. G. Erdman, "Design Theory and AI Implementations of Design Methodology," *Proceedings of 1989 NSF Engineering Design Research Conference,* (June 11–14, 1989), pp. 205–220. Amherst: University of Massachusetts.

64. Faik, S., and A. G. Erdman, "Sensitivity Distributions in the Three Position Synthesis Design Plane," To be published in *ASME Journal of Mechanical Design,* 1989.

65. Fiacco, A. V., and G. P. McCormick, *Nonlinear Programming — Sequential Unconstrained Minimization Technique.* New York: John Wiley & Sons, Inc., 1968.

66. Fichtinger, G., R. Westby, and A. Erdman, "Combination of Design Disciplines Offers Series of Novel Mechanisms," in *Product Engineering,* 45, no. 12 (December 1974), pp. 35–36.

67. Fletcher, R., and M. J. D. Powell, "A Rapidly Convergent Descent Method for Minimization," *Computer Journal,* 6, no. 2 (1963), pp. 163–68.

68. Fox, R. L., *Optimization Methods for Engineering Design,* 2nd ed. New York: Addison-Wesley Publishing Company, 1973.

69. ———, and K. C. Gupta, "Optimization Technology as Applied to Mechanism Design," *Journal of Engineering for Industry,* 95 (1973), pp. 657–63.

70. ———, and K. D. Willmert, "Optimum Design of Curve-Generating Linkage with Inequality Constraints," *Journal of Engineering for Industry,* 89, no. 1 (February 1967), pp. 144–52.

71. Freudenstein, F., "An Analytical Approach to the Design of Four-Link Mechanisms," *Transactions of the ASME,* 76 (1954), pp. 483–92.

72. ———, "Structural Error Analysis in Plane Kinematic Synthesis," *Journal of Engineering for Industry,* 81, no. 1 (January 1959), pp. 15–22.

73. ———, "Harmonic Analysis of Crank-and-Rocker Mechanisms with Applications," *Journal of Applied Mechanics,* 81E (1959), pp. 673–75.

74. ———, "On the Maximum and Minimum Velocities and Accelerations in Four-Link Mechanisms," *Transactions of the ASME,* 78 (1956), pp. 779–87.

75. ———, *Proceedings: International Conference on Mechanisms,* Yale 1961 (F. R. Erskine Crossley, Ed.), pp. 44–54, The Shoestring Press.

76. ———, and E. R. Maki, "The Creation of Mechanisms According to Kinematic Structure and Function," General Motors Research Publications, GMR-3073, September 1979; *International Journal for the Science of Architecture and Design,* (1980).

77. ———, and G. N. Sandor, "Kinematics of Mechanisms," in *Mechanical Design and System Handbook,* 2nd ed., New York: McGraw-Hill, 1985, pp. 4.1–4.80.

78. ———, and G. N. Sandor, "Synthesis of Path-Generating Mechanisms by Means of a Programmed Digital Computer," *Journal of Engineering for Industry,* 81B, no. 2 (May 1959), pp. 159–68.

79. Fuller, D., "Motivation is More than Gimmicks," *Machine Design* (November 1976), pp. 116–20.

80. Garrett, R. E., and A. S. Hall, Jr., "Effect of Tolerance and Clearance in Linkage Design," *Journal of Engineering for Industry,* 91B (1969), pp. 198–202.

81. Gupta, K. C., "A General Theory for Synthesizing Crank-Type Four Bar Function Generators with Transmission Angle Control," *Journal of Applied Mechanics,* 45, no. 2 (June 1968).

82. Hagen, D., A. G. Erdman, D. Harvey, and J. Tacheny, "Rapid Algorithms for Kinematic and Dynamic Analysis of Planar Rigid Linkages with Revolute Joints," ASME Paper No. 78-DET-64, 1978.

83. Hain, K., *Applied Kinematics,* 2nd ed. New York: McGraw-Hill, 1967.

84. Hall, A. S., *Kinematics and Linkage Design.* West Lafayette, Indiana: Balt Publishers, 1961.

85. Halter, J. M., "Force Synthesis to Produce a Desired Time Response of Mechanisms," Ph.D. Dissertation, University of Missouri-Columbia, May 1975.

86. Hartenberg, R. S., and J. Denavit, *Kinematic Synthesis of Linkages.* New York: McGraw-Hill Book Company, 1964.

87. Hinkle, R. T., *Design of Machines.* Englewood Cliffs, New Jersey: Prentice-Hall, Inc., 1957.

88. Hinkle, Rolland T., *Kinematics of Machines,* 2nd ed. Englewood Cliffs, New Jersey: Prentice-Hall, Inc., 1960.

89. Hrones, J. A., and G. L. Nelson, *Analysis of the Four-Bar Linkage.* New York: The Technology Press of MIT and John Wiley and Sons, Inc., 1951.

90. Johnson, R. C., "Impact Forces in Mechanisms," *Machine Design,* 30 (1958), pp. 138–46.

91. ———, and K. Towfigh, "Application of Number Synthesis to Practical Problems in Creative Design," ASME Paper No. 65-WA/MD-9, 1965.

92. ———, and K. Towfigh, "Application of Number Synthesis to Practical Problems in Creative Design," in *Mechanical Design Synthesis With Optimization Applications.* Ed. R. C. Johnson. New York: Van Nostrand Reinhold Company, 1971.

93. Kaufman, R. E., "Mechanism Design by Computer," *Machine Design,* (October 1978), pp. 94–100.

94. Kemler, E. N., and R. J. Howe, *Machine Design,* 23 (1951).

95. Kloomok, M., and R. V. Muffley, "Plate Cam Design — with Emphasis on Dynamic Effects," *Product Engineering,* (February 1955).

96. Kohli, D., A. E. Thompson, and G. N. Sandor, "Design of Four-Bar Linkages with Specified Motion Characteristics of All Moving Links," in *Monograph on Mechanism Design.* New York: McGraw-Hill Book Company, 1977, NSF Report no. GK36624, Paper No. 50.

97. Kota, S., D. Riley, and A. G. Erdman, "Development of Knowledge Base for Designing Linkage-Type Dwell Mechanisms; Part 1: Theory," *ASME Journal of Mechanisms, Transmissions, and Automation in Design,* 109, No. 3 (September 1987), pp. 308–15.

98. ———, ———, ———, "Development of Knowledge Base for Designing Linkage-Type Dwell Mechanisms; Part 2: Applications," *ASME Journal of Mechanisms, Transmissions, and Automation in Design,* 109, No. 3 (September 1987), pp. 316–21.

99. ———, and R. B. Gudapati, "Automatic Selection of Four-Bar Linkage Designs for Path Generation Task." *Proceedings of the ASME Design Automation Conference,* Montreal, Sept. 1989, pp. 395–402.

99a. Kounas, P. S., A. G. Erdman, and G. N. Sandor, "Kinematic Synthesis of Smoothly Stopping Mechanisms with Fast Moving Output Members," *Proceedings,* Third Applied Mechanisms Conference, Oklahoma State University, Stillwater, Oklahoma, Nov. 5–7, 1983.

100. Kramer, S. N., and G. N. Sandor, "Finite Kinematic Synthesis of a Cycloidal-Crank Mechanism for Function Generation," *Journal of Engineering for Industry,* 92, no. 3 (August 1970), pp. 531–36.

101. ———, and G. N. Sandor, "Kinematic Synthesis of Watt's Mechanism," *ASME Paper No. 70-Mech-50, 1970.*

102. ———, and G. N. Sandor, "Selective Precision Synthesis—A General Method of Optimization for Planar Mechanisms," *Journal of Engineering for Industry,* 97, no. 2 (May 1975), pp. 689–701.

103. Lee, T. W., "On the Kinematics and Dynamic Synthesis of a Variable-Speed Drive," ASME Paper No. 77-DET-124, 1977.

104. Lindholm, J. C., "A Survey of the Graphical Techniques in Designing for Specific Input-Output Relationships of a Four-Bar Mechanism," *Proceedings of the First Applied Mechanisms Conference (Tulsa, Oklahoma, July 1969).* Stillwater, Oklahoma: Oklahoma State University, 1969.

105. ———, "Design for Path Generation—Point Position Reduction," Linkage Design Monographs, Paper No. 38; NSF Final Report GK-36624.

106. Liou, F. W., and A. G. Erdman, "Analysis of a High-Speed Flexible Four-Bar Linkage, Part I: Formulation and Solution," *Advanced Topics in Vibrations,* ASME Pub. DE-Vol. 8, September 1987, pp. 128–138. And in *Journal of Vibration, Acoustics, Stress, and Reliability in Design,* 111, No. 1 (January 1989), pp. 35–41.

107. ———, ———, "Analysis of a High-Speed Flexible Four-Bar Linkage, Part II: Analytical and Experimental Results on the Apollo," *Advanced Topics in Vibrations,* ASME Pub. DE-Vol. 8, September 1987, pp. 139–146. And in *Journal of Vibration, Acoustics, Stress, and Reliability in Design,* 111, No. 1, (January 1989), pp. 42–47.

108. Loerch, R. J., A. G. Erdman, and G. N. Sandor, "On the Existence of Circle-Point and Center-Point Circles for Three-Precision-Point Dyad Synthesis," *Journal of Mechanical Design* (October 1979), pp. 554–62.

109. ———, A. G. Erdman, G. N. Sandor, and A. Midha, "Synthesis of Four-Bar Linkages With Specified Ground Pivots," *Proceedings of the 4th Applied Mechanisms Conference (Chicago, November 1975),* pp. 10.1–10.6. Stillwater, Oklahoma: Oklahoma State University, 1975.

110. Mabie, H. H., and C. F. Reinholtz, *Kinematics and Dynamics of Machinery.* New York: John Wiley and Sons, Inc., 1978, 4th ed.

111. Matthew, G., and D. Tesar, "Synthesis of Spring Parameters to Satisfy Specified Energy Levels in Planar Mechanisms," *Journal of Engineering for Industry* (May 1977), pp. 341–46.

112. McGovern, J. F., and G. N. Sandor, "Kinematic Synthesis of Adjustable Mechanisms, Part I: Function Generation; Part II: Path Generation," *Journal of Engineering for Industry,* 95, no. 2 (May 1973), pp. 417–29.

113. Meyer zur Capellen, W., "Kinematics—A Survey in Retrospect and Prospect," *Journal of Mechanisms,* 1 (1966), pp. 211–28.

114. Mittelstadt, W., D. Riley, and A. G. Erdman, "Integrated CAD of Mechanisms," *Mechanism and Machine Theory Journal,* 20, No. 4 (Spring 1985), pp. 303–12.

115. Modrey, J., "Analysis of Complex Kinematic Chains with Influence Coefficients," *Journal of Applied Mechanics,* 81E (1959), pp. 184–88.

116. Olson, D., A. G. Erdman, and D. Riley, "A New Graph Theory Representation for the Topological Analysis of Planetary Gear Trains," *Proceedings of the Seventh World Congress on the Theory of Machines and Mechanisms,* Vol. 3, (Sevilla, Spain, September 1987), pp. 1421–1426.

117. ———, ———, ———, "Formulation of Dimensional Synthesis Procedures for Complex Planar Linkages," *ASME Journal of Mechanisms, Transmissions, and Automation in Design,* 109, No. 3 (September 1987), pp. 322–28.

118. ———, ———, ———, "Topological Analysis of Single-Degree-of-Freedom Planetary Gear Trains," *Trends and Developments in Mechanisms, Machines, and Robotics — 1988,* DE-Vol. 15–1, pp. 125–131, being considered for ASME journal publication.

119. Orrell, M. G., "Runaway (Verge) Escapement Study," ASME Paper 84-DET-48, 1984.

120. Osman, M. O. M., and R. V. Dukkipati, "Kinematic Analysis of Planar Four-Link Mechanisms Using Complex Number Algebra," *Linkage Design Monographs;* NSF Final Report GK-36624, Paper No. 16.

121. Paul, B., "Analytical Dynamics of Mechanisms — A Computer Oriented Overview," *Mechanisms and Machine Theory,* 10, no. 6 (1975), pp. 481–507.

122. Paynter, Henry M., *Analysis and Design of Engineering Systems.* Cambridge, Massachusetts: The M.I.T. Press, 1960–61.

123. Peterson, J. A., and E. W. Nelson, "Operator for a Casement-Type Window," U.S. Patent No. 4,253,276, March 3, 1981.

124. Peterson, R., L. Logan, A. G. Erdman, and D. Riley, "LINCAGES-4: Computer-Aided Mechanism Synthesis and Analysis — Three Precision Point Program," *Proceedings of the 1988 Spring Design Engineering Conference* (Chicago, Illinois, March 7, 1988).

125. ———, ———, ———, ———, "Three Precision Point Synthesis of a Four-Bar Linkage: An Example Using the LINCAGES-4 Program," *Proceedings of ASME, Computers in Engineering 1988,* ASME, Vol. 2, pp. 91–96.

126. Pouliot, H. N., W. R. Delameter, and C. W. Robinson, "A Variable-Displacement Spark Engine," SAE Paper No. 770114, 1977.

127. Pryor, R. F., and G. N. Sandor, "On the Classification and Enumeration of Six-Link and Eight-Link Cam-Modulated Linkages," *Proceedings of the Fifth World Congress of IFToMM* (Montreal, July 1979), Paper No. USA-66.

128. Radcliffe, C. W., "Kinematics in Biomechanics Research," *Proceedings of the National Science Foundation Workshop on New Directions for Kinematics Research* (Stanford University, August 1976), pp. 174–98.

129. Rao, A. V. M., and G. N. Sandor, "Extension of Freudenstein's Equation to Geared Linkages," *Journal of Engineering for Industry,* 93, no. 1 (February 1971), pp. 201–10.

130. ———, A. G. Erdman, G. N. Sandor, et al, "Synthesis of Multi-Loop, Dual Purpose Planar Mechanisms Utilizing Burmester Theory," *Proceedings of the 2nd OSU Applied Mechanisms Conference (Stillwater, Oklahoma, October 1971),* pp. 7.1–7.23. Stillwater, Oklahoma: Oklahoma State University, 1971.

130a. ———, and G. N. Sandor, "Closed-Form Synthesis of Four-Bar Path Generators by Linear Superposition," *Proceedings,* Third World Congress for the Theory of Machines and Mechanisms, Dubrovnik, Yugoslavia, Sept. 13–19, 1971, pp. 383–94.

130b. ———, and G. N. Sandor, "Closed-Form Synthesis of Four-Bar Function Generators by Linear Superposition," *Proceedings,* Third World Congress for the Theory of Machines and Mechanisms, Dubrovnik, Yugoslavia, Sept. 13–19, 1971, pp. 395–405.

131. Raudsepp, E., "The Nitty-Gritty of Creativity," *Machine Design,* (April 1979), pp. 886–89.

132. Reinholtz, C. F., S. G. Dhande, and G. N. Sandor, "Kinematic Analysis of Planar Higher-Pair Mechanisms," *Mechanism and Machine Theory,* 13, no. 6, pp. 619–29.

133. Reuleaux, F., *Kinematics of Machinery: Outline of a Theory of Machines.* New York: Dover, 1963.

134. Rose, P. S., and G. N. Sandor, "Direct Analytical Synthesis of Four-Bar Function Generators With Optimal Structural Error," *Journal of Engineering for Industry,* 95, no. 2 (May 1973), pp. 563–71.

135. Rothbart, H. A., *Cams — Design, Dynamics, and Accuracy.* New York: John Wiley and Sons, 1956.

136. Rubel, A. J., and R. E. Kaufman, "KINSYN III: A New Human-Engineered System for Interactive Computer-Aided Design of Planar Linkages," *Journal of Engineering for Industry,* 99, no. 2 (May 1977).

137. Sandor, G. N., "Engineering and Humanistic Creativity — Is There a Difference?," *Mechanical Engineering Education Update,* Annual Conference of the ASEE, LSU, June 1979.

138. ———, "A General Complex-Number Method for Plane Kinematic Synthesis with Applications," Doctoral Dissertation, Columbia University in the City of New York, *University Microfilms,* Ann Arbor, Michigan, 305 pp., Library of Congress Card No. Mic 59–2596, 1959.

139. ———, "On Computer-Aided Graphical Kinematic Synthesis," Technical Seminar Series, Princeton University, 1962.

140. ———, "On the Loop Equations in Kinematics," *Transactions of the Seventh Conference on Mechanisms,* pp. 49–56. West Lafayette, Ind.: Purdue University, 1962.

141. ———, "The Seven Dangers of Designer Overspecialization and How to Avoid It by Design Education," *Mechanical Engineering* (October 1974), pp. 23–28.

142. ———, "The Seven Stages of Engineering Design," *Mechanical Engineering,* 86, no. 4 (April 1964), pp. 21–5.

143. ———, and Dan Perju, "Contributions to the Kinematic Synthesis of Adjustable Mechanisms," *Transactions of the International Symposium on Linkages and Computer Design Methods* (Bucharest, Romania, June 1973), Vol. A-46, pp. 636–50.

144. ———, A. G. Erdman, and E. Raghavacharyulu, "Coriolis-Acceleration Analysis in Planar Mechanisms — A Complex-Number Approach," *Mechanism and Machine Theory,* 17, no. 6, pp. 405–14, 1982.

145. ———, A. V. Mohan Rao, and A. G. Erdman, "A General Complex-Number Method of Synthesis and Analysis of Mechanisms Containing Prismatic and Revolute Pairs," *Proceedings of the Third World Congress on the Theory of Machines and Mechanisms* (Dubrovnik, Yugoslavia, September 1971), Vol. D, pp. 237–49. Beograd: Yugoslavian Committee on the Theory of Machines and Mechanisms, 1972.

145a. ———, A. V. M. Rao, and S. N. Kramer, "Geared Six-Bar Design," *Proceedings,* 2nd OSU Applied Mechanisms Conference, Stillwater, Oklahoma, Oct. 7–9, 1971, pp. 25–1 to 25–13.

145b. ———, A. V. M. Rao, and J. C. Kopanias, "Closed-Form Synthesis of Planar Single-Loop Mechanisms for Coordination of Diagonally Opposite Angles," *Proceedings,* "Mechanisms 1972" Conference, Institution of Mechanical Engineers, London, Sept. 5–7, 1972.

145c. ———, and A. V. M. Rao, "Synthesis of Function-Generator Mechanisms with Scale Factors as Unknown Design Parameters," *Transactions,* International Symposium on Linkages and Computer Design Methods, Bucharest, Romania, June 7–13, 1973, Vol. A-44, pp. 602–23.

145d. ———, J. F. McGovern, and C. Z. Smith, "The Design of Four-Bar Path-Generating Linkages by Fifth-Order Path Approximation in the Vicinity of a Single Point," *Proceedings,* Mechanisms 73 Conference, University of Newcastle upon Tyne, Sept. 11, 1973, Institution of Mechanical Engineers, London, England, pp. 65–77.

145e. ———, R. Alizade, and I. G. Novruzbekov, "Optimization of Four-Bar Function-Generating Mechanisms Using Penalty Functions with Inequality and Equality Constraints," *Mechanism and Machine Theory,* Vol. 10, No. 4, August 1975, pp. 327–36.

145f. ———, D. R. Hassel, and P. F. Marino, "Modern Mechanisms Make Manless Martian Mission Mobile — 'Spin-Off' Spells Stair-Climbing Self Sufficiency for Earthbound Handicapped," *Proceedings,* The Ninth Aerospace Mechanisms Symposium, Kennedy Space Center, Florida, Oct. 17–18, 1974, pp. 18–11 to 18–17; NASA TM X-3274, pp. 263–74.

145g. ——, A. G. Erdman, E. Raghavacharyulu, and C. F. Reinholtz, "On the Equivalence of Higher- and Lower-Pair Planar Mechanisms," *Proceedings*, Sixth World Congress of IFToMM on the Theory of Machines and Mechanisms, Delhi, India, Dec. 15–20, 1983.

146. Savage, M., "A Tooth for Tooth — The Designing of Gears," *Perspectives in Computing,* IBM Vol. 4, no. 1, Spring 1984.

147. Seth, P. N., and J. J. Uicker, "IMP (Integrated Mechanisms Program), A Computer-Aided Design Analysis System for Mechanisms and Linkages," *Journal of Engineering for Industry,* 94, no. 2 (May 1972), pp. 454–64.

148. Shigley, J. E., *Kinematic Analysis of Mechanisms.* New York: McGraw-Hill Book Company, 1969.

149. Shoup, T. E., and J. M. Herrera, "Design of Double Boom Cranes for Optimum Load Capacity," *Proceedings of the Fifth Applied Mechanisms Conference,* Paper No. 30. Stillwater, Oklahoma: Oklahoma State University, 1977.

150. ——, and R. S. Sodhi, "Designing Polycentric Mechanisms By Curve Matching," *Proceedings of the Sixth Applied Mechanisms Conference,* Paper No. IX. Stillwater, Oklahoma: Oklahoma State University, 1979.

151. Showlater, G., R. Giese, and A. G. Erdman, "Synthesis of Skylight Mechanisms," *Proceedings of the 1981 OSU Applied Mechanisms Conference,* Paper No. XXXVII, 8 pgs. Stillwater, Oklahoma: Oklahoma State University, 1981.

152. Struble, K. R., J. E. Gustafson, and A. G. Erdman, "Case Study: Synthesis of a Four-Bar Linkage to Pick and Place Filters Using the LINCAGES Computer Package," *Proceedings of the Fifth Applied Mechanisms Conference.* Stillwater, Oklahoma: Oklahoma State University, 1977.

153. Tacheny, J., G. Vetter, and A. G. Erdman, "Computer-Aided Design of Awning Window Deployment Linkages," *Proceedings of the Seventh OSU Applied Mechanisms Conference,* Paper No. XXIV, 5 pgs. Stillwater, Oklahoma: Oklahoma State University, 1981.

154. ——, A. G. Erdman, and D. L. Hagen, "Experimental Determination of Mechanism Time Response," *Proceedings of the Fifth World Congress on the Theory of Machines and Mechanisms* (Montreal, July 1979), pp. 130–38.

155. Tao, D. C., *Applied Linkage Synthesis.* Reading, Mass.: Addison-Wesley Publishing Company, Inc., 1964.

156. ——, *Fundamentals of Applied Kinematics.* Reading, Mass.: Addison-Wesley Publishing Company, Inc., 1967.

157. Tesar, D., and J. W. Sparks, "Multiply Separated Position Synthesis, Part 1: Point Synthesis," ASME Paper No. 68-Mech-66, 1968.

158. Thompson, T., A. G. Erdman, and D. Riley, "Type Selection of Robot and Gripper Kinematic Topology Using Expert Systems," *International Journal of Robotics Research,* 5, No. 2 (Summer 1986), pp. 183–89. And *Kinematics of Robot Manipulators,* ed. J. M. McCarthy, Cambridge, Mass.: M.I.T. Press, 1987, pp. 202–8. Also in *Expert Systems in Engineering,* ed. D. T. Pham, IFS Publication Ltd., 1988.

159. Thoreson, J., and A. G. Erdman, "Designing Mechanisms for Production Equipment," *Machine Design Magazine* (October 6, 1988), pp. 113–17.

160. Titus, J., A. G. Erdman, and D. Riley, "The Role of Type Synthesis in the Design of Machines," *Proceedings of the 1989 NSF Engineering Design Research Conference* (June 11–14, 1989), Amherst, Mass.: University of Massachusetts, pp. 451–74.

161. Turner, J., and A. G. Erdman, "Design of a Mechanism Clock," *Proceedings of the Fourth Applied Mechanisms Conference,* (November 1975). Stillwater, Oklahoma: Oklahoma State University, 1975.

162. Tuttle, E. R., S. W. Peterson, and J. E. Titus, "Emuneration of Basic Kinematic Chains Using the Theory of Finite Groups," *Trends and Developments in Mechanisms, Machines, and Robotics—1988,* Vol. 1, 1988 ASME Design Technology Conference, Kissimmee, Fla., pp. 165–172.

163. ———, ———, ———, "Further Applications of Group Theory to the Emuneration and Structural Analysis of Basic Kinematic Chains," *Trends and Developments in Mechanisms, Machines, and Robotics—1988,* Vol. 1, 1988 ASME Design Technology Conference, Kissimmee, Fla., pp. 173–177.

164. Uicker, J. J., and A. Raicu, "A Method for the Identification and Recognition of Equivalence of Kinematic Chains," *Mechanism and Machine Theory,* Vol. 10, New York: Pergamon Press, 1975, pp. 375–83.

165. von Fange, Eugene, *Professional Creativity.* Englewood Cliffs, New Jersey: Prentice-Hall, Inc., 1959.

166. Waldron, K. J., "Improved Solutions of the Branch and Order Problems of Burmester Linkage Synthesis," *Journal of Mechanism and Machine Theory,* 13 (1978), pp. 199–207.

167. White, R. J., *Solution Rectification For a Stephenson III Motion Generator,* Master's thesis, University of Minnesota, 1987.

168. Wilde, D. J., "Jacobians in Constrained Nonlinear Optimization," *Operations Research,* 13, no. 5 (September 1965), pp. 848–56.

169. Wiley, J. C., B. E. Romiz, N. Orlandea, T. A. Berenyi, and D. W. Smith, "Automated Simulation and Display of Mechanisms and Vehicle Behavior," *Proceedings of the Fifth World Congress on the Theory of Machines and Mechanisms (1979),* pp. 680–83.

170. Williams, R. J., and S. Rupprecht, "Dynamic Force Analysis of Planar Mechanisms," *Proceedings of the Sixth OSU Applied Mechanisms Conference,* Paper No. XLIII, 9 pgs. Stillwater, Oklahoma: Oklahoma State University, 1979.

171. ———, "Mechanism Design and Analysis with Micros," *Microcad News* (Sept./Oct. 1988), pp. 68–71.

Index